P9-BZR-069

DISCRETE-EVENT SYSTEM SIMULATION

Prentice-Hall International Series in Industrial and Systems Engineering

W. J. Fabrycky and J. H. Mize, Editors

Amos and Sarchet *Management for Engineers*
Banks and Carson *Discrete-Event System Simulation*
Beightler, Phillips, and Wilde *Fundamentals of Optimization*, 2/E
Blanchard *Logistics Engineering and Management*, 2/E
Blanchard and Fabrycky *Systems Engineering and Analysis*
Brown *Systems Analysis and Design for Safety*
Bussey *The Economic Analysis of Industrial Projects*
Fabrycky, Ghare, and Torgersen *Industrial Operations Research*
Frances and White *Facility Layout and Location: An Analytical Approach*
Gottfried and Weisman *Introduction to Optimization Theory*
Hammer *Product Safety Management and Engineering*
Ignizio *Programming in Single and Multiobjective Systems*
Kirkpatrick *Introductory Statistics and Probability for Engineering, Science, and Technology*
Mize, White, and Brooks *Operations Planning and Control*
Mundel *Improving Productivity and Effectiveness*
Mundel *Motion and Time Study: Improving Productivity*, 5/E
Ostwald *Cost Estimating for Engineering and Management*
Phillips and Garcia-Diaz *Fundamentals of Network Analysis*
Sivazlian and Stanfel *Analysis of Systems in Operations Research*
Sivazlian and Stanfel *Optimization Techniques in Operations Research*
Smalley *Hospital Management Engineering*
Thuesen, Fabrycky, and Thuesen *Engineering Economy*, 5/E
Turner, Mize, and Case *Introduction to Industrial and Systems Engineering*
Whitehouse *Systems Analysis and Design Using Network Techniques*

DISCRETE-EVENT SYSTEM SIMULATION

Jerry Banks

John S. Carson, II

Georgia Institute of Technology

PRENTICE-HALL, INC., Englewood Cliffs, New Jersey 07632

Library of Congress Cataloging in Publication Data

Banks, Jerry.
Discrete-event system simulation.

(Prentice-Hall international series in industrial and systems engineering)
Includes bibliographies and index.
1. Simulation methods. I. Carson, John S.
II. Title. III. Series.
T57.62.B35 1984 658.4'03'52 83-3340
ISBN 0-13-215582-6

Editorial/production supervision: Barbara Bernstein
Interior design: Anne Simpson
Manufacturing buyer: Anthony Caruso

To
Susie and David
Jonna and Jennifer

Printed in the United States of America

10 9 8 7 6 5 4 3 2

ISBN 0-13-215582-6

Prentice-Hall International, Inc., *London*
Prentice-Hall of Australia Pty. Limited, *Sydney*
Editora Prentice-Hall do Brasil, Ltda., *Rio de Janeiro*
Prentice-Hall Canada Inc., *Toronto*
Prentice-Hall of India Private Limited, *New Delhi*
Prentice-Hall of Japan, Inc., *Tokyo*
Prentice-Hall of Southeast Asia Pte. Ltd., *Singapore*
Whitehall Books Limited, *Wellington, New Zealand*

CONTENTS

PREFACE

The objective of this text is to provide a basic treatment of discrete-event simulation, one of the most widely used operations research tools presently available. The simulation of systems from the perspective of the engineer, manager, economist, or scientist has become a common application of the digital computer. Discrete-event simulation permits the evaluation of operating performance prior to the implementation of a system; it permits the comparison of various operational alternatives without perturbing the real system; it permits time compression so that timely policy decisions can be made. Finally, it can be used by many people because of its readily comprehended structure and the availability of special-purpose computer simulation languages.

However, simulation can be easily misused, and simulation results taken with more confidence than is justified. The proper collection and analysis of data, the use of analytic techniques when they will suffice, the verification and validation of models, and the appropriate design of simulation experiments are all potential pitfall areas which are treated extensively in this text.

The material in this book should be readily understandable to the reader who has a basic familiarity with differential and integral calculus, probability theory, and elementary statistics. No theorems or proofs appear in the text. However, statistical or mathematical correctness has not been sacrificed. Most topics in probability and statistics are thoroughly reviewed before being applied in the simulation context.

We believe in learning by example. Hence, after new material is introduced it is always followed by one or more examples of the subject matter. The examples comprise an integral part of the text. The reader will also find numerous figures and tables designed to enhance and motivate the material. There are many problems at the end of each chapter to further the reader's understanding of the material. A solutions manual is available from the publisher.

Chapters 3, 11, and 12 contain over 75 problems, ranging from the elementary to the complex, which provide simulation exercises for a student learning to use a simulation language. The simulation exercises in Chapters 11 and 12 (and a few in Chapter 3) specifically ask the student to address the issues of the design of a simulation experiment. Some problems specify the experimental design—run length, number of replications, initial conditions, system configurations to compare and so on—while others are more open-ended.

As a basic treatment of the fundamental concepts and principles of discrete-event simulation, the text is not language dependent. We emphasize the correct application of simulation technique, the design of simulation experiments, and the analysis of outputs, leaving the teaching of special-purpose languages to others. However, the concept and implementation of the event-scheduling method of advancing time in a discrete-event simulation is covered thoroughly in Chapter 3. Several manual simulation examples are used to illustrate the concepts of system state, entities and attributes, events, and activities, and the operation of the future event list. In addition, the reader is introduced to one general-purpose and four special-purpose simulation languages: FORTRAN and GASP, SIMSCRIPT, GPSS, and SLAM.

The book is divided into four sections. Part One, consisting of three chapters, is an introduction to discrete-event simulation. Chapter 1 introduces the concepts of system, simulation, model, and related ideas. Chapter 2 contains numerous worked examples of the simulation of simple systems. The goal of this chapter is to motivate the novice simulator and to illustrate the kinds of information that can be gained from a simulation. After one or two lectures, much of the material in Chapter 2 can be read independently by the student. Chapter 3 describes the event-oriented and process-interaction points of view and gives numerous examples to illustrate the main ideas of the event scheduling approach to simulation. This chapter also briefly describes and illustrates the simulation languages mentioned previously.

Part Two, consisting of three chapters, describes mathematical and statistical models that are useful in simulation. Chapter 4 is a review of several statistical distributions and their properties. Chapter 5 discusses the nature, and measures of performance, of queueing systems, the estimation of the performance measures from simulated observations, and the ideas of transient behavior and steady state. Chapter 6 describes a number of inventory models with emphasis on their measures of performance and the conditions under which they may be solved analytically. Chapters 5 and 6 are included because these types of systems, queueing and inventory, are so often simulated. It is important to understand the basic concepts and parameters in the models of these systems in order to conduct effective simulations with the level of expertise required to practice in the real world. In some situations, simplified assumptions can be made temporarily so that a queueing or inventory model can be used to give rough initial estimates of a performance measure or other parameter (such as the minimum number of servers needed, or a nearly optimal inventory policy). These initial estimates can then be refined by a simulation analysis. Queueing and inventory models, broadly conceived, include models of manufacturing and production facilities, service facilities, warehousing and material handling systems, transportation and distribution systems, communication and computer systems, health-care facilities, and many other types of systems.

Part Three, consisting of two chapters, concerns random numbers. Chapter 7 describes the generation and testing of random numbers, the basis of discrete-event stochastic simulation. Chapter 8 covers random variate generation, which is the use of random numbers to generate samples from arbitrary statistical distributions.

Part Four, consisting of four chapters, deals with the statistical analysis of simulation input data and output data. Chapter 9 concerns the collection and analysis of input data. This chapter is of great importance since so much of the total resources in an actual simulation are dedicated to this activity. Chapter 10 concerns the verification and validation of simulation models, a topic of utmost importance to practitioners. Chapters 11 and 12 concern the design and output analysis of simulation experiments. These topics have too often been ignored by simulators. The purpose of these chapters is to convey the notion that running a simulation model and getting output is far from sufficient to obtain reliable results.

Discrete-Event System Simulation can serve as a textbook in the following types of courses:

1. A junior- or senior-level introductory simulation course in engineering, management, or computer science (Chapters 1 through 8 and parts of 9 through 11, if no companion language text is used; parts of Chapters 3 through 6 can be omitted when a companion language text is used).
2. A second course in simulation (Chapters 9 through 12, a companion language text, and an outside project, which would be the simulation and analysis of a system in operation, and the comparison of alternative system configurations).
3. An introduction to stochastic models as part of an operations research course (Chapters 1, 2, and 4 through 6).

The strengths of this text lie in several areas. The material in manuscript form has been used in the classroom numerous times prior to its publication; the text contains many examples, figures, and tables; a solutions manual is available; there is an emphasis on discrete simulation methodology and a heavy emphasis on the statistical analysis of input and output data.

We decided to take what we consider to be the most important topics for an introduction to simulation and treat them in depth with many examples. For example, in Chapters 11 and 12, we treat output analysis by the method of replication, a technique readily comprehended by a student who has completed a course in elementary statistics. Several topics in output analysis, including time-series methods, spectral analysis, and the regenerative method, are more appropriate for an advanced simulation or statistics text. As another example, the variance reduction technique of common random numbers has been emphasized because it is easily understood and useful in practice, and the required statistical analysis is still at an elementary level.

Thanks are given to the many people who helped make this book possible. We are grateful to our colleagues, Donovan Young and Jim Swain, who taught from the manuscript during its development and made numerous suggestions and contributions. Special thanks are due the hundreds of students from numerous sections of our basic and advanced undergraduate discrete simulation course who provided criticism and encouragement during the several years while this book was in development. Many students were instrumental in solving the examples and checking the authors' work. Included among those to be thanked are Lee Blanton, Charles Bourquin, Michael Bruce, Irvin Lee, and Harvey Rickles. The solutions manual was prepared with the help of Francisco Ramis, Greg Cirincione, and K. K. Kuong-Lau. We are very appreciative of the assistance given by J. P. Spoerer, who prepared the index. Thanks are also due the typists who painstakingly waded through numerous revisions of the manuscript. Our special thanks in this regard go to Joene Owen and Betty Plummer. The School of Industrial and Systems Engineering provided numerous support services which facilitated the completion of the project. Many thanks are due Michael E. Thomas, Director, who continually encouraged us.

We have thoroughly enjoyed the professional manner in which Prentice-Hall, Inc. brought this text to production. Our special thanks go to Barbara J. Bernstein, our production editor, who responded favorably to numerous telephone calls at odd hours, requests for changes after the fact, authors' anxiety attacks, and more.

Finally, the valuable suggestions of the reviewers are appreciated.

Jerry Banks
John Carson

part one

INTRODUCTION TO DISCRETE-EVENT SYSTEM SIMULATION

INTRODUCTION TO SIMULATION

A *simulation* is the imitation of the operation of a real-world process or system over time. Whether done by hand or on a computer, simulation involves the generation of an artificial history of a system, and the observation of that artificial history to draw inferences concerning the operating characteristics of the real system.

The behavior of a system as it evolves over time is studied by developing a simulation *model.* This model usually takes the form of a set of assumptions concerning the operation of the system. These assumptions are expressed in mathematical, logical, and symbolic relationships between the *entities*, or objects of interest, of the system. Once developed and validated, a model can be used to investigate a wide variety of "what if" questions about the real-world system. Potential changes to the system can first be simulated in order to predict their impact on system performance. Simulation can also be used to study systems in the design stage, before such systems are built. Thus, simulation modeling can be used both as an analysis tool for predicting the effect of changes to existing systems, and as a design tool to predict the performance of new systems under varying sets of circumstances.

In some instances, a model can be developed which is simple enough to be "solved" by mathematical methods. Such solutions may be found by the use of differential calculus, probability theory, algebraic methods, or other mathematical techniques. The solution usually consists of one or more numerical parameters which are called measures of performance of the system. However, many

1

real-world systems are so complex that models of these systems are virtually impossible to solve mathematically. In these instances, numerical, computer-based simulation can be used to imitate the behavior of the system over time. From the simulation, data are collected as if a real system were being observed. This simulation-generated data is used to estimate the measures of performance of the system.

This book provides an introductory treatment to the concepts and methods of one form of simulation modeling—discrete-event simulation modeling. The first chapter initially discusses when to use simulation, its advantages and disadvantages, and actual areas of application. Then the concepts of system and model are explored. Finally, an outline is given of the steps in building and using a simulation model of a system.

1.1. When Is Simulation The Appropriate Tool?

The availability of special-purpose simulation languages, massive computing capabilities at a decreasing cost per operation, and advances in simulation methodologies have made simulation one of the most widely used and accepted tools in operations research and systems analysis. Circumstances under which simulation is the appropriate tool to use have been discussed by many authors, including Naylor et al. [1966]. Simulation can be used for the following purposes:

1. Simulation enables the study of, and experimentation with, the internal interactions of a complex system, or of a subsystem within a complex system.
2. Informational, organizational and environmental changes can be simulated and the effect of these alterations on the model's behavior can be observed.
3. The knowledge gained in designing a simulation model may be of great value toward suggesting improvement in the system under investigation.
4. By changing simulation inputs and observing the resulting outputs, valuable insight may be obtained into which variables are most important and how variables interact.
5. Simulation can be used as a pedagogical device to reinforce analytic solution methodolgies.
6. Simulation can be used to experiment with new designs or policies prior to implementation, so as to prepare for what may happen.
7. Simulation can be used to verify analytic solutions.

1.2. Advantages and Disadvantages of Simulation

Although simulation is an appropriate tool of analysis in many instances, the systems analyst should consider its advantages and disadvantages before pursuing the methodology in a particular instance. The primary advantages of simulation, as discussed by Schmidt and Taylor [1970], and others, are:

1. Once a model is built, it can be used repeatedly to analyze proposed designs or policies.
2. Simulation methods can be used to help analyze a proposed system even though the input data are somewhat sketchy.
3. It is usually the case that simulation data are much less costly to obtain than similar data from the real system.
4. Simulation methods are usually easier to apply than analytic methods. Thus, there are many more potential users of simulation methods than of analytic methods.
5. Whereas analytic models usually require many simplifying assumptions to make them mathematically tractable, simulation models have no such restrictions. With analytic models, the analyst usually can compute only a limited number of system performance measures. With simulation models, the data generated can be used to estimate any conceivable performance measure.
6. In some instances, simulation is the only means of deriving a solution to a problem.

Schmidt and Taylor also list disadvantages to consider before using simulation:

1. Simulation models for digital computers may be costly, requiring large expenditures of time in their construction and validation.

2. Numerous runs of a simulation model are usually required and this can result in high computer costs.
3. Simulation is sometimes used when analytic techniques will suffice. This situation occurs as users become familiar with simulation methodology and forget about their mathematical training.

In defense of simulation, the first two disadvantages mentioned by Schmidt and Taylor (and by others, e.g., Adkins and Pooch [1977]) are ameliorated somewhat by the availability of special-purpose simulation languages and ever more powerful computers which perform more operations per dollar. Several of the special-purpose languages are discussed in Chapter 3.

1.3. Areas of Application

There have been numerous applications of simulation in a wide variety of contexts. Hillier and Lieberman [1980] list the following examples to indicate the broad versatility of the technique:[1]

1. Simulation of the operations at a large airport by an airlines company to test changes in company policies and practices (e.g., amounts of maintenance capacity, berthing facilities, spare aircraft, and so on)
2. Simulation of the passage of traffic across a junction with time-sequenced traffic lights to determine the best time sequences
3. Simulation of a maintenance operation to determine the optimal size of repair crews
4. Simulation of the flux of uncharged particles through a radiation shield to determine the intensity of the radiation that penetrates the shield
5. Simulation of steelmaking operations to evaluate changes in operating practices and the capacity and configuration of the facilities
6. Simulation of the U.S. economy to predict the effect of economic policy decisions
7. Simulation of large-scale military battles to evaluate defensive and offensive weapon systems
8. Simulation of large-scale distribution and inventory control systems to improve the design of these systems
9. Simulation of the overall operation of an entire business firm to evaluate broad changes in the policies and operation of the firm and also to provide a business game for training executives
10. Simulation of a telephone communications system to determine the capacity of the respective components that would be required to provide satisfactory service at the most economical level
11. Simulation of the operation of a developed river basin to determine the best

[1]Reproduced with permission of F. S. Hillier and Gerald J. Lieberman, *Introduction to Operations Research*, 3rd ed., Holden-Day, San Francisco, 1980.

configuration of dams, power plants, and irrigation works that would provide the desired level of flood control and water-resource development
12. Simulation of the operation of a production line to determine the amount of in-process storage space that should be provided

1.4. Systems and System Environment

To model a system, it is necessary to understand the concept of a system and the system boundary. A *system* is defined as a group of objects that are joined together in some regular interaction or interdependence toward the accomplishment of some purpose. An example is a production system manufacturing automobiles. The machines, component parts, and workers operate jointly along an assembly line to produce a high-quality vehicle.

A system is often affected by changes occurring outside the system. Such changes are said to occur in the *system environment* [Gordon, 1978]. In modeling systems, it is necessary to decide on the *boundary* between the system and its environment. This decision may depend on the purpose of the study.

> In the case of the factory system, for example, the factors controlling the arrival of orders may be considered to be outside the influence of the factory and therefore part of the environment. However, if the effect of supply on demand is to be considered, there will be a relationship between factory output and arrival of orders, and this relationship must be considered an activity of the system. Similarly, in the case of a bank system, there may be a limit on the maximum interest rate that can be paid. For the study of a single bank, this would be regarded as a constraint imposed by the environment. In a study of the effects of monetary laws on the banking industry, however, the setting of the limit would be an activity of the system. [Gordon, 1978]

1.5. Components of a System

In order to understand and analyze a system, a number of terms are defined. An *entity* is an object of interest in the system. An *attribute* is a property of an entity. An *activity* represents a time period of specified length. If a bank is being studied, customers might be one of the entities, the balance in their checking accounts might be an attribute, and making deposits might be an activity.

The collection of entities that compose a system for one study might only be a subset of the overall system for another study [Law and Kelton, 1982]. For example, if the bank mentioned above is being studied to determine the number of tellers needed to provide for paying and receiving, the system can be defined as that portion of the bank consisting of the regular tellers and the customers waiting in line. If the purpose of the study is expanded to determine the number of special tellers needed (to prepare cashier's checks, to sell traveler's checks, etc.), the definition of the system must be expanded.

The *state* of a system is defined to be that collection of variables necessary to describe the system at any time, relative to the objectives of the study. In the study of a bank, possible state variables are the number of busy tellers, the number of customers waiting in line or being served, and the arrival time of the next customer. An *event* is defined as an instantaneous occurrence that may change the state of the system. The term *endogenous* is used to describe activities and events occurring within a system, and the term *exogenous* is used to describe activities and events in the environment that affect the system. In the bank study, the arrival of a customer is an exogenous event, and the completion of service of a customer is an endogenous event.

Table 1.1 lists examples of entities, attributes, activities, events, and state variables for several systems. Only a partial listing of the system components is shown. A complete list cannot be developed unless the purpose of the study is known. Depending on the purpose, various aspects of the system will be of interest, and then the listing of components can be completed.

1.6. Discrete and Continuous Systems

Systems can be categorized as discrete or continuous. "Few systems in practice are wholly discrete or continuous, but since one type of change predominates for most systems, it will usually be possible to classify a system as being either discrete or continuous" [Law and Kelton, 1982]. A *discrete system* is one in which the state variable(s) change only at a discrete set of points in time. The bank is an example of a discrete system since the state variable, the number of customers in the bank, changes only when a customer arrives or when the service provided a customer is completed. Figure 1.1 shows how the number of customers changes only at discrete points in time.

A *continuous system* is one in which the state variable(s) change continuously over time. An example is the head of water behind a dam. During and for some time after a rain storm, water flows into the lake behind the dam. Water is drawn from the dam for flood control and to make electricity. Evaporation also decreases the water level. Figure 1.2 shows how the state variable, head of water behind the dam, changes for this continuous system.

1.7. Model of a System

Sometimes it is of interest to study a system to understand the relationships between its components or to predict how the system will operate under a new policy. To study the system, it is sometimes possible to experiment with the sys-

Table 1.1. EXAMPLES OF SYSTEMS AND COMPONENTS

System	*Entities*	*Attributes*	*Activities*	*Events*	*State Variables*
Banking	Customers	Checking account balance	Making deposits	Arrival; departure	Number of busy tellers; number of customers waiting
Rapid rail	Riders	Origination; destination	Traveling	Arrival at station; arrival at destination	Number of riders waiting at each station; number of riders in transit
Production	Machines	Speed; capacity; breakdown rate	Welding; stamping	Breakdown	Status of machines (busy, idle, or down)
Communications	Messages	Length; destination	Transmitting	Arrival at destination	Number waiting to be transmitted
Inventory	Warehouse	Capacity	Withdrawing	Demand	Levels of inventory; backlogged demands

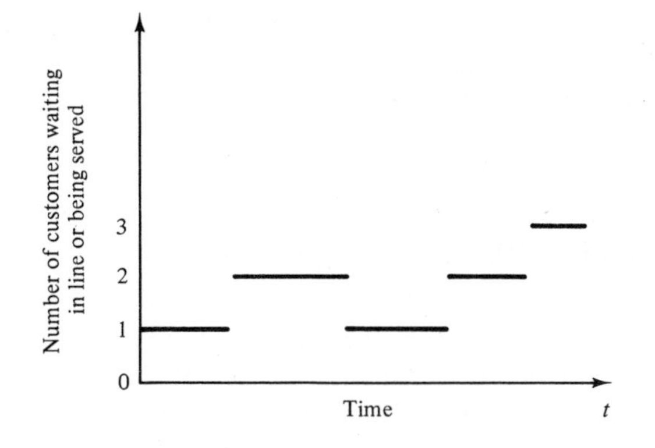

Figure 1.1. Discrete system state variable.

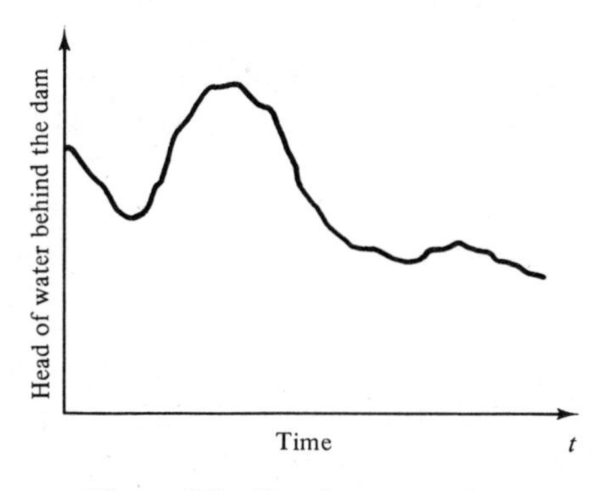

Figure 1.2. Continuous system state variable.

tem itself. However, this is not always possible. A new system may not yet exist; it may be only in hypothetical form or at the design stage. Even if the system exists, it may be impractical to experiment with it. For example, it may not be wise or possible to double the unemployment rate to determine the effect of employment on inflation. In the case of a bank, reducing the numbers of tellers to study the effect on the length of waiting lines may infuriate the customers so greatly that they move their accounts to a competitor. Consequently, studies of systems are often accomplished with a model of a system.

A *model* is defined as a representation of a system for the purpose of studying the system. For most studies, it is not necessary to consider all the details of a system; thus, a model is not only a substitute for a system, it is also a simplification of the system [Mihram and Mihram, 1974]. On the other hand, the model should be sufficiently detailed to permit valid conclusions to be drawn about the real system. Different models of the same system may be required as the purpose of investigation changes.

Just as the components of a system were entities, attributes, and activities, models are represented similarly. However, the model contains only those components that are relevant to the study. The components of a model are discussed more extensively in Chapter 3.

1.8. Types of Models

Models can be classified as being mathematical or physical. A mathematical model uses symbolic notation and mathematical equations to represent a system. A simulation model is a particular type of mathematical model of a system.

Simulation models may be further classified as being static or dynamic, deterministic or stochastic, and discrete or continuous. A *static* simulation model, sometimes called a Monte Carlo simulation, represents a system at a particular point in time. *Dynamic* simulation models represent systems as they change over time. The simulation of a bank from 9:00 A.M. to 4:00 P.M. is an example of a dynamic simulation.

Simulation models that contain no random variables are classified as *deterministic*. Deterministic models have a known set of inputs which will result in a unique set of outputs. Deterministic arrivals would occur at a dentist's office if all patients arrived at the scheduled appointment time. A *stochastic* simulation model has one or more random variables as inputs. Random inputs lead to random outputs. Since the outputs are random, they can be considered only as estimates of the true characteristics of a model. The simulation of a bank would usually involve random interarrival times and random service times. Thus, in a stochastic simulation, the output measures—the average number of people waiting, the average waiting time of a customer—must be treated as statistical estimates of the true characteristics of the system.

Discrete and continuous systems were defined in Section 1.6. Discrete and continuous models are defined in an analogous manner. However, a discrete simulation model is not always used to model a discrete system, nor is a continuous simulation model always used to model a continuous system. In addition, simulation models may be mixed, both discrete and continuous. The choice of whether to use a discrete or continuous (or both discrete and continuous) simulation model is a function of the characteristics of the system and the objective of the study. Thus, a communication channel could be modeled discretely if the characteristics and movement of each message were deemed important. Conversely, if the flow of messages in aggregate over the channel were of importance, modeling the system using continuous simulation could be more appropriate. The models considered in this text are discrete, dynamic, and stochastic.

1.9. Discrete-Event System Simulation

This is a textbook about discrete-event system simulation. Discrete-event systems simulation is the modeling of systems in which the state variable changes only at a discrete set of points in time. The simulation models are analyzed by numerical methods rather than by analytical methods. *Analytical* methods employ the deductive reasoning of mathematics to "solve" the model. For example, differential calculus can be used to determine the minimum-cost policy for some inventory models. *Numerical* methods employ computational procedures to "solve" mathematical models. In the case of simulation models, which employ numerical methods, models are "run" rather than solved; that is, an artificial history of the system is generated based on the model assumptions, and observations are collected to be analyzed and to estimate the true system performance measures. Since real-world simulation models are rather large, and since the amount of data stored and manipulated is so vast, the runs are usually conducted with the aid of a computer. However, much insight can be obtained by simulating small models manually. In summary, this textbook is about discrete-event system simulation in which the models of interest are analyzed numerically, usually with the aid of a computer.

1.10. Steps in a Simulation Study

Figure 1.3 shows a set of steps to guide a model builder in a thorough and sound simulation study. Similar figures and discussion of steps can be found in other sources [Shannon, 1975; Gordon, 1978; Law and Kelton, 1982]. The number beside each symbol in Figure 1.3 refers to the more detailed discussion in the text. The steps in a simulation study are as follows:

1. *Problem formulation.* Every study should begin with a statement of the problem. If the statement is provided by the policymakers, or those that have the problem, the analyst must ensure that the problem being described is clearly understood. If a problem statement is being developed by the analyst, it is important that the policymakers understand and agree with the formulation. Although not shown in Figure 1.3, there are occasions where the problem must be reformulated as the study progresses. In many instances, policymakers and analysts are aware that there is a problem long before the nature of the problem is known!

2. *Setting of objectives and overall project plan.* The objectives indicate the questions to be answered by simulation. At this point a determination should be made concerning whether simulation is the appropriate methodology for the problem as formulated and objectives as stated. Assuming that it is decided that simulation is appropriate, the overall project plan should include a state-

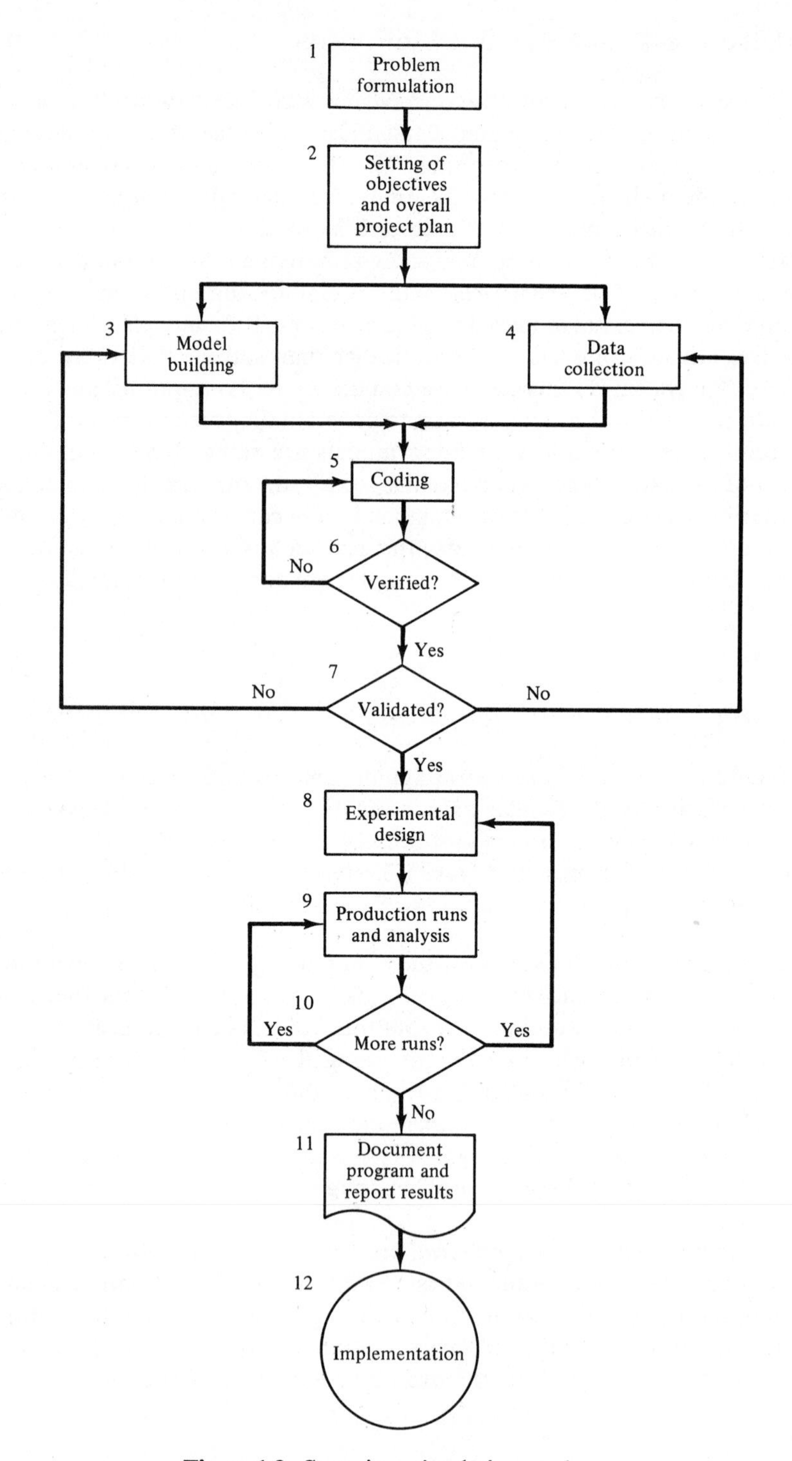

Figure 1.3. Steps in a simulation study.

ment of the alternative systems to be considered, and a method for evaluating the effectiveness of these alternatives. It should also include the plans for the study in terms of the number of people involved, the cost of the study, and the number of days required to accomplish each phase of the work with the anticipated results at the end of each stage.

3. *Model building.* The construction of a model of a system is probably as much art as science. Shannon [1975] provides a lengthy discussion of this step. "Although it is not possible to provide a set of instructions that will lead to building successful and appropriate models in every instance, there are some general guidelines that can be followed" [Morris, 1967]. The art of modeling is enhanced by an ability to abstract the essential features of a problem, to select and modify basic assumptions that characterize the system, and then to enrich and elaborate the model until a useful approximation results. Thus, it is best to start with a simple model and build toward greater complexity. However, the model complexity need not exceed that required to accomplish the purposes for which the model is intended. Violation of this principle will only add to model building and computer expenses. It is not necessary to have a one-to-one mapping between the model and the real system. Only the essence of the real system is needed.

It is advisable to involve the model user in the model construction. Involving the model user will both enhance the quality of the resulting model and increase the confidence of the model user in the application of the model. (Chapters 2 and 3 describe a number of simulation models. Chapters 5 and 6 describe queueing and inventory models that can be solved analytically. However, only experience with real systems—versus textbook problems—can "teach" the art of model building.)

4. *Data collection.* There is a constant interplay between the construction of the model and the collection of the needed input data [Shannon, 1975]. As the complexity of the model changes, the required data elements may also change. Also, since data collection takes such a large portion of the total time required to perform a simulation, it is necessary to begin it as early as possible, usually together with the early stages of model building.

The objectives of the study dictate, in a large way, the kind of data to be collected. In the study of a bank, for example, if the desire is to learn about the length of waiting lines as the number of tellers change, the types of data needed would be the distributions of interarrival times (at different times of the day), the service-time distributions for the tellers, and historic distributions on the lengths of waiting lines under varying conditions. These last data will be used to validate the simulation model. (Chapter 9 discusses data collection and data analysis; Chapter 4 discusses statistical distributions which occur frequently in simulation modeling.)

5. *Coding.* Since most real-world systems result in models that require a great deal of information storage and computation, the model must be programmed for a digital computer. The modeler must decide whether to program

the model in a general-purpose language such as FORTRAN or a special-purpose simulation language such as GPSS, SIMSCRIPT, or SLAM. A general-purpose language requires a much longer development time, but usually executes on the computer much faster than the special-purpose languages. Overall, however, special-purpose languages so speed up the coding (and verification) step that more and more model builders are using them.

6. *Verified?* Verification pertains to the computer program prepared for the simulation model. Is the computer program performing properly? With complex models, it is difficult, if not impossible, to code a model successfully in its entirety without a good deal of debugging. If the input parameters and logical structure of the model are correctly represented in the code, verification has been completed. For the most part, common sense is used in completing this step. (Chapter 10 discusses verification of simulation models.)

7. *Validated?* Validation is the determination that a model is an accurate representation of the real system. Validation is usually achieved through the calibration of the model, an iterative process of comparing the model to actual system behavior and using the discrepancies between the two, and the insights gained, to improve the model. This process is repeated until model accuracy is judged acceptable. In the example of a bank mentioned above, data were collected concerning the length of waiting lines under current conditions. Does the simulation model replicate this system measure? This is one means of validation. (Chapter 10 discusses the validation of simulation models.)

8. *Experimental design.* The alternatives that are to be simulated must be determined. Often, the decision concerning which alternatives to simulate may be a function of runs that have been completed and analyzed. For each system design that is simulated, decisions need to be made concerning the length of the initialization period, the length of simulation runs, and the number of replications to be made of each run. (Chapters 11 and 12 discuss issues associated with the experimental design.)

9. *Production runs and analysis.* Production runs, and their subsequent analysis, are used to estimate measures of performance for the system designs that are being simulated. (Chapters 11 and 12 discuss the analysis of simulation experiments.)

10. *More runs?* Based on the analysis of runs that have been completed, the analyst determines if additional runs are needed and what design those additional experiments should follow.

11. *Document program and report results.* Documentation is necessary for numerous reasons. If the program is going to be used again by the same or different analysts, it may be necessary to understand how the program operates. This will enable confidence in the program so that model users and policymakers can make decisions based on the analysis. Also, if the program is to be modified by the same or a different analyst, this can be greatly facilitated by adequate documentation. One experience with an inadequately documented program is usually enough to convince an analyst of the necessity of this important step.

Another reason for documenting a model is so that model users can change parameters of the model at will in an effort to determine the relationships between input parameters and output measures of performance, or to determine the input parameters that "optimize" some output measure of performance.

The result of all the analysis should be reported clearly and concisely. This will enable the model users (now, the decision makers) to review the final formulation, the alternative systems that were addressed, the criterion by which the alternatives were compared, the results of the experiments, and the recommended solution to the problem. Furthermore, if decisions have to be justified at a higher level, the report should provide a vehicle of certification for the model user/ decision maker and add to the credibility of the model and the model building process.

12. *Implementation.* The success of the implementation phase depends on how well the previous 11 steps have been performed. It is also contingent upon how thoroughly the analyst has involved the ultimate model user during the entire simulation process. If the model user has been thoroughly involved during the model-building process and if the model user understands the nature of the model and its outputs, the likelihood of a vigorous implementation is enhanced [Pritsker and Pedgen, 1979]. Conversely, if the model and its underlying assumptions have not been properly communicated, implementation will probably suffer, regardless of the simulation model's validity.

The simulation model-building process shown in Figure 1.3 can be broken down into four phases. The first phase, consisting of steps 1 (Problem Formulation) and 2 (Setting of Objective and Overall Design), is a period of discovery or orientation. The initial statement of the problem is usually quite "fuzzy," the initial objectives will usually have to be reset, and the original project plan will usually have to be fine-tuned. These recalibrations and clarifications may occur in this phase, or perhaps after or during another phase (i.e., the analyst may have to restart the process).

The second phase is related to model building and data collection, and includes steps 3 (Model Building), 4 (Data Collection), 5 (Coding), 6 (Verification), and 7 (Validation). A continuing interplay is required among the steps. Exclusion of the model user during this phase can have dire implications at the point of implementation.

The third phase concerns running the model. It involves steps 8 (Experimental Design), 9 (Production Runs and Analysis), and 10 (Additional Runs). This phase must have a thoroughly conceived plan for experimenting with the simulation model. A discrete-event stochastic simulation is in fact a statistical experiment. The output variables are estimates that contain random error, and therefore a proper statistical analysis is required. Such a philosophy is in contrast to the analyst who makes a single run and draws an inference from that single data point.

The fourth and last phase, implementation, involves steps 11 (Document

Program and Report Results) and 12 (Implementation). Successful implementation depends on continual involvement of the model user and the successful completion of every step in the process. Perhaps the most crucial point in the entire process is step 7 (Validation), because an invalid model is going to lead to erroneous results, which if implemented could be dangerous, costly, or both.

REFERENCES

ADKINS, GERALD, AND UDO W. POOCH [1977], "Computer Simulation: A Tutorial," *Computer*, Vol. 10, No. 4, pp. 12–17.

GORDON, GEOFFREY [1978], *System Simulation*, 2nd ed., Prentice-Hall, Englewood Cliffs, N.J.

HILLIER, FREDERICK S., AND GERALD J. LIEBERMAN [1980], *Introduction to Operations Research*, 3rd ed., Holden-Day, San Francisco.

LAW, AVERILL M., AND W. DAVID KELTON [1982], *Simulation Modeling and Analysis*, McGraw-Hill, New York.

MIHRAM, DANIELLE, AND G. ARTHUR MIHRAM [1974], "Human Knowledge, The Role of Models, Metaphors and Analogy," *International Journal of General Systems*, Vol. 1, No. 1, pp. 41–60.

MORRIS, W. T. [1967], "On the Art of Modeling," *Management Science*, Vol. 13, No. 12.

NAYLOR, T. H., J. L. BALINTFY, D. S. BURDICK, AND K. CHU [1966], *Computer Simulation Techniques*, Wiley, New York.

PRITSKER, A. ALAN B., AND CLAUDE D. PEDGEN [1979], *Introduction to Simulation and SLAM*, Halsted Press, New York.

SCHMIDT, J. W., AND R. E. TAYLOR [1970], *Simulation and Analysis of Industrial Systems*, Irwin, Homewood, Ill.

SHANNON, ROBERT E. [1975], *Systems Simulation: The Art and Science*, Prentice-Hall, Englewood Cliffs, N.J.

EXERCISES

1. Name several entities, attributes, activities, events, and state variables for the following systems:
(a) A small appliance repair shop
(b) A cafeteria
(c) A grocery store
(d) A laundromat
(e) A fast-food restaurant
(f) A hospital emergency room
(g) A taxicab company with 10 taxis
(h) An automobile assembly line

2. Consider the simulation process shown in Figure 1.3.
   (a) Reduce the steps by at least two by combining similar activities. Give your rationale.
   (b) Increase the steps by at least two by separating current steps or enlarging on existing steps. Give your rationale.

3. A simulation of a major traffic intersection is to be conducted with the objective of improving the current traffic flow. Provide three iterations, in increasing order of complexity, of steps 1 and 2 in the simulation process of Figure 1.3.

4. In what ways and at what steps might a personal computer be used to support the simulation process of Figure 1.3?

5. Consider the list of problem areas shown in Section 1.3. Relate the problem areas to the courses of study in your curriculum.

6. A simulation is to be conducted of cooking a spaghetti dinner to determine what time a person should start in order to have the meal on the table by 7:00 P.M. Read a recipe for preparing a spaghetti dinner (or ask a friend or relative, etc., for the recipe). As best you can, trace what you understand to be needed in the data collection phase of the simulation process of Figure 1.3, in order to perform a simulation in which the model includes each step in the recipe. What are the events, activities, and state variables in this system?

7. What are the events and activities associated with the operation of your checkbook?

SIMULATION EXAMPLES

The objective of this chapter is the presentation of several examples of simulations that can be performed directly—without the aid of a computer. These examples provide an insight into the methodology of discrete system simulation and into some of the accompanying analysis that must be performed. By placing simulation examples at this early stage of the text, the reader will appreciate the many fine points that are offered in later chapters. The simulations in this chapter are accomplished by following three steps:

1. Determine the characteristics of each of the inputs to the simulation. Quite often, these may be modeled as probability distributions, either continuous or discrete.
2. Construct a simulation table. Each simulation table is different, for each is developed for the purpose at hand. An example of a simulation table is shown in Table 2.1. In this example there are p inputs, $x_{ij}, j = 1, 2, \ldots, p$, and one response, y_i, for each of repetitions $i = 1, 2, \ldots, n$.
3. For each repetition i, generate a value for each of the p inputs, and evaluate the function, calculating a value of the response y_i. This step is accomplished by sampling values from the distributions determined in step 1.

Simulation is a powerful tool which can be used to analyze many complex problems. However, before simulation is chosen as the solution method, every

2

effort should be made to solve the problem mathematically, perhaps with mathematical models developed for queueing problems, perhaps with inventory theory, and so on. Simulation modeling can be time consuming, and if a closed-form solution exists, it can be much less expensive than simulation.

In this chapter, numerous simulation examples are given. The first two problem areas involve queueing and inventory models. Simulation has been very useful in solving real problems in these two areas. In order that the reader will be able to understand the context in which these problems exist, some prior explanation is given. Later, in Chapters 5 and 6, queueing models and inventory systems are described in greater detail.

Table 2.1. SIMULATION TABLE

	Inputs						
Repetitions	x_{i1}	x_{i2}	$\cdots$	x_{ij}	$\cdots$	x_{ip}	*Response,* y_i
1							
2							
3							
.							
.							
.							
n							

Other interesting examples are given in this chapter. The first is a reliability problem—another area in which simulation has been useful. Also, there is an example that introduces the concept of random normal numbers. Finally, there is an example of the determination of lead-time demand.

2.1. Simulation of Queueing Systems

A queueing system is described by its calling population, the nature of the arrivals and services, the system capacity, and the queueing discipline. These attributes of a queueing system are described in detail in Chapter 5. A simple queueing system is portrayed in Figure 2.1.

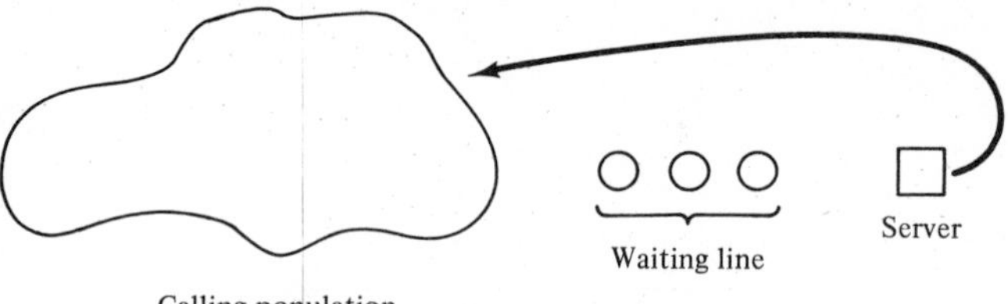

Figure 2.1. Queueing system.

In this system the calling population is infinite; that is, if a unit leaves the calling population and joins the waiting line or enters service, there is no change in the arrival rate of other units that may need service. Also, in this system, arrivals for service occur one at a time in a random fashion and once they join the waiting line, they are eventually served. In addition, service times are of some random length according to a probability distribution which does not change over time. Also, the system capacity is unlimited. (The system includes the unit in service plus those waiting in line.) Finally, callers are served in the order of their arrival (often called FIFO, for first in, first out) by a single server, or channel.

Arrivals and services are described by the distributions of the time between arrivals and service times. The overall effective arrival rate must be less than the maximum service rate, or the waiting line will grow without bound. When queues grow without bound, they are termed "explosive" or unstable. An exceptional situation would be arrival rates that are greater than service rates for short periods of time. However, such a situation is more complex than that described in this chapter.

Prior to introducing several simulations of queueing systems, it is necessary to understand the concepts of state of the system, events, and clock time. The state of the system is the number of units in the system and the status of the server, busy or idle. An event is a set of circumstances that cause an instantaneous change in the state of the system. In a single-channel queueing system there are only two possible events that can affect the state of the system. The two events are the entry of a unit into the system (the arrival event) or the completion of service on a unit (the departure event). The queueing system includes the server, the unit being serviced (if one is being serviced), and units in the queue (if any are waiting).

If service has just been completed, the simulation proceeds in the manner shown in the flow diagram of Figure 2.2. Note that the server has only two possible conditions in Figure 2.2: it is either busy or idle.

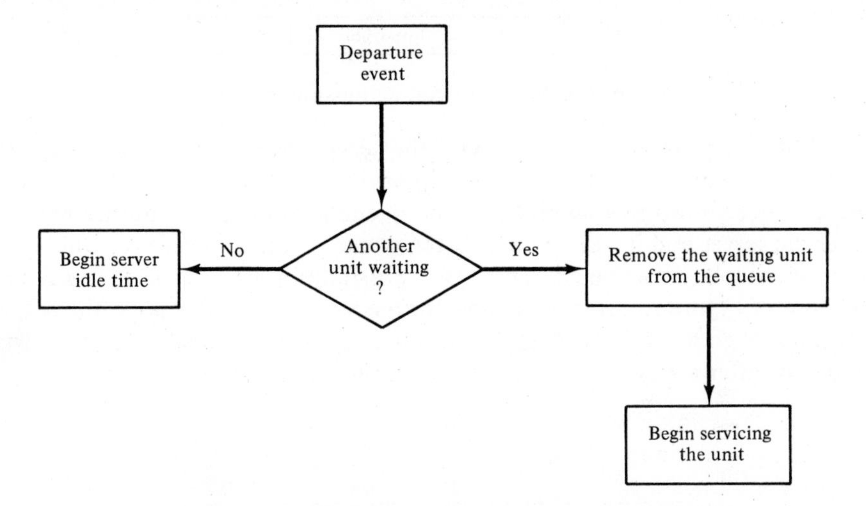

Figure 2.2. Service just completed flow diagram.

The second event occurs when a unit enters the system. The flow diagram in such a case is shown in Figure 2.3. The unit may find the server either idle or busy; therefore, either the unit enters the server, or it enters the queue for

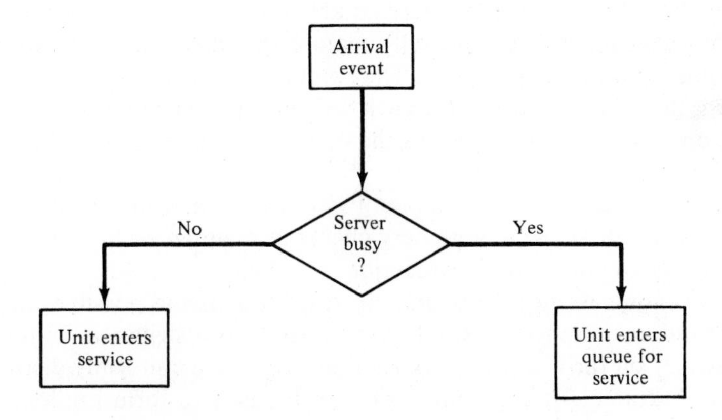

Figure 2.3. Unit entering system flow diagram.

the server. The unit follows the course of action shown in Figure 2.4. If the server is busy, the unit enters the queue. If the server is idle and the queue is empty, the unit enters the server. It is impossible for the server to be idle and the queue to be not empty.

		Queue status	
		Not empty	Empty
Server status	Busy	Enter queue	Enter queue
	Idle	Impossible	Enter service

Figure 2.4. Potential unit actions upon arrival.

After the completion of a service the server may become idle, or remain busy with the next unit. The relationship of these two outcomes to the status of the queue is shown in Figure 2.5. If the queue is not empty, another unit will enter the server and it will be busy. If the queue is empty, the server will be idle after a service is completed. These two possibilities are shown as the shaded portions of Figure 2.5. It is impossible for the server to become busy if the queue is empty when a service is completed. Similarly, it is impossible for the server to be idle after a service is completed when the queue is not empty.

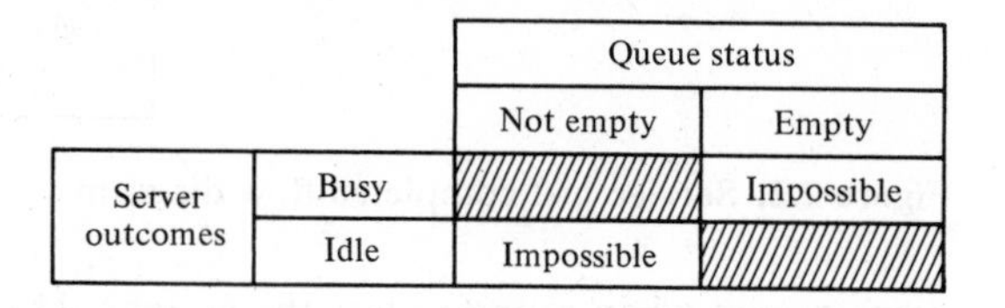

Figure 2.5. Server outcomes after service completion.

Now, how can the events described above occur in simulated time? Simulations of queueing systems generally require the maintenance of an event list for determining what happens next. The event list indicates the times at which the different types of events occur for each unit in the queueing system. The times are kept on a "clock," which marks the occurrences of events in time. In simulation, the events usually occur at random. The randomness imitates real life to portray uncertainty. For example, it is not known with certainty when the next customer will arrive at a grocery checkout counter, or how long the bank teller will take to complete a transaction.

The randomness needed to imitate real life is made possible through the use of "random numbers." Random numbers are distributed uniformly and independently on the interval (0, 1). Random digits are uniformly distributed on the set {0, 1, 2, . . . , 9}. Random digits can be used to form random numbers by selecting the proper number of digits for each random number and placing a decimal point to the left of the value selected. The proper number of digits is dictated by the accuracy of the data being used for input purposes. If the input distribution has values with two decimal places, two digits are taken from a random digits table (such as Table A.1) and the decimal point is placed to the left to form a random number.

Random numbers can also be generated. When numbers are generated using a set procedure they are often referred to as pseudo-random numbers. Since the method is known, it is always possible to know what the sequence of numbers will be prior to the simulation. Various methods for generating random numbers are discussed in Chapter 8.

In a single-channel queueing system interarrival times and service times are determined (generated) from the distributions of these random variables. The examples that follow show how such times are generated. For simplicity, assume that the times between arrivals were generated by rolling a die five times and recording the up face. Table 2.2 contains a set of five interarrival

Table 2.2. INTERARRIVAL AND CLOCK TIMES

Customer	*Interarrival Time*	*Arrival Time on Clock*
1	—	0
2	2	2
3	4	6
4	1	7
5	2	9
6	6	15

times generated in this manner. These five interarrival times are used to compute the arrival times of six customers at the queueing system.

The first customer is assumed to arrive at clock time 0. This starts the clock in operation. The second customer arrives two time units later, at a clock time of 2. The third customer arrives four time units later, at a clock time of 6; and so on.

The second time of interest is the service time. Table 2.3 contains service times generated at random from a distribution of service times. The only possible

Table 2.3. SERVICE TIMES

Customer	*Service Time*
1	2
2	1
3	3
4	2
5	1
6	4

service times are one, two, three, and four time units. Assuming that all four values are equally likely to occur, these values could have been generated by placing the numbers one through four on chips and drawing the chips from a hat with replacement, being sure to record the numbers selected. Now, the inter-arrival times and service times must be meshed to simulate the single-channel queueing system. As shown in Table 2.4, the first customer arrives at clock time zero, and immediately begins service, which requires 2 minutes. Service is completed at clock time 2. The second customer arrives at clock time 2 and is finished at clock time 3. Note that the fourth customer arrived at clock time 7, but service could not begin until clock time 9. This occurred because customer 3 did not finish service until clock time 9.

Table 2.4. SIMULATION TABLE EMPHASIZING CLOCK TIMES

Customer Number	*Arrival Time (Clock)*	*Time Service Begins (Clock)*	*Service Time (Duration)*	*Time Service Ends (Clock)*
1	0	0	2	2
2	2	2	1	3
3	6	6	3	9
4	7	9	2	11
5	9	11	1	12
6	15	15	4	19

Table 2.4 was designed specifically for a single-channel queue which serves customers on a first in, first out (FIFO) basis. It keeps track of the clock time at which each event occurs. The second column of Table 2.4 records the clock time of each arrival event, while the last column records the clock time of each departure event. The occurrence of the two types of events in chronological order is shown in Table 2.5 and Figure 2.6.

Table 2.5. CHRONOLOGICAL ORDERING OF EVENTS

Event Type	*Customer Number*	*Clock Time*
Arrival	1	0
Departure	1	2
Arrival	2	2
Departure	2	3
Arrival	3	6
Arrival	4	7
Departure	3	9
Arrival	5	9
Departure	4	11
Departure	5	12
Arrival	6	15
Departure	6	19

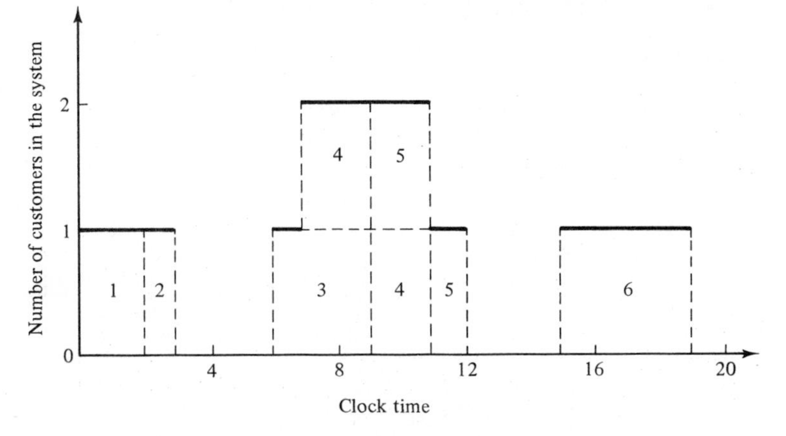

Figure 2.6. Number of customers in the system.

It should be noted that Table 2.5 is ordered by clock time, in which case the events may or may not be ordered by customer number. The chronological ordering of events is the basis of the approach to discrete-event simulation described in Chapter 3.

Figure 2.6 depicts the number of customers in the system at the various clock times. It is a visual image of the event listing of Table 2.5. Customer 1 is in the system from clock time 0 to clock time 2. Customer 2 arrives at clock time 2 and departs at clock time 3. No customers are in the system from time 3 to time 6. During some time periods two customers are in the system, such as at clock time 8, when both customers 3 and 4 are in the system. Also, there are times at which events occur simultaneously, such as at clock time 9, when customer 5 arrives and customer 3 departs.

Example 2.1 follows the logic described above while keeping track of a number of attributes of the system. Example 2.2 is concerned with a two-channel queueing system. The flow diagrams for a multichannel queueing system are slightly different from those for a single-channel system. The development and interpretation of these flow diagrams is left as an exercise for the reader.

Example 2.1 Single-channel queue

A small grocery store has only one checkout counter. Customers arrive at this checkout counter at random from 1 to 8 minutes apart. Each possible value of inter-arrival time has the same probability of occurrence, as shown in Table 2.6. The service times vary from 1 to 6 minutes with the probabilities shown in Table 2.7. The last two columns of Tables 2.6 and 2.7 will be discussed later. The problem is to analyze the system by simulating the arrival and service of 20 customers.

In actuality, 20 customers is too small a run length to use to draw any final conclusions. The accuracy of the results is enhanced by increasing the sample size, as discussed in Chapter 11. However, the purpose of the exercise is to demonstrate how hand simulations are conducted, not to recommend changes in the grocery store. A second issue that will be raised now and discussed thoroughly in Chapter 11 is that

Table 2.6. DISTRIBUTION OF TIME BETWEEN ARRIVALS

Time between Arrivals (Minutes)	*Probability*	*Cumulative Probability*	*Random Digit Assignment*
1	0.125	0.125	001–125
2	0.125	0.250	126–250
3	0.125	0.375	251–375
4	0.125	0.500	376–500
5	0.125	0.625	501–625
6	0.125	0.750	626–750
7	0.125	0.875	751–875
8	0.125	1.000	876–000

Table 2.7. SERVICE-TIME DISTRIBUTION

Service Time (Minutes)	*Probability*	*Cumulative Probability*	*Random Digit Assignment*
1	0.10	0.10	01–10
2	0.20	0.30	11–30
3	0.30	0.60	31–60
4	0.25	0.85	61–85
5	0.10	0.95	86–95
6	0.05	1.00	96–00

of initial conditions. A simulation of a grocery store that starts with an empty system is not realistic unless the intention is to model the system from startup or to model until steady-state operation is reached. However, to facilitate the teaching purpose of this example, starting conditions and concerns are overlooked.

A set of uniformly distributed random numbers is needed to generate the arrivals at the checkout counter. Random numbers have the following properties:

1. The set of random numbefs is uniformly distributed between 0 and 1.
2. Successive random numbers are independent.

As mentioned previously, uniformly distributed random numbers can be generated by many methods, some of which are discussed in Chapter 7. In addition, tables of random digits have been developed. These tables have been used to generate random numbers in this example.

Random digits are converted to random numbers by placing the decimal at the appropriate point. In this case, three-place random numbers will suffice. It is only necessary to list 19 random numbers to generate times between arrivals. Why only 19 numbers? The first arrival is assumed to occur at time zero, so only 19 more arrivals need to be generated to end up with 20 customers.

The last two columns of Tables 2.6 and 2.7 are used to generate random arrivals and random service times. The next-to-last column in each table contains the cumulative probability for the distribution. The last column contains the random digit assignment. Consider Table 2.6 only. The first random digit assignment is 001–125. There are 1000 three-digit values possible (001 through 000). The probability of a time between arrival of 1 minute is 0.125, and 125 of the 1000 random digit values are assigned to such an occurrence. Three digits are needed since the probability distribution is described with three-digit accuracy; for example, the probability that the time between arrivals is 4 minutes is 0.125. Times between arrival for 19 customers are generated by listing 19 three-digit values from Table A.1 and comparing them to the random digit assignment of Table 2.6.

It is good practice to start at a random position in the random digit table and proceed in a systematic direction, never re-using the same stream of digits in a given problem. If the same pattern is used repeatedly, bias could result because the same event pattern would be generated. The time-between-arrival determination is shown in Table 2.8. Note that the first random digits are 913. To obtain the corresponding time between

Table 2.8. TIME-BETWEEN-ARRIVAL DETERMINATION

Customer	*Random Digits*	*Time between Arrivals (Minutes)*	*Customer*	*Random Digits*	*Time between Arrivals (Minutes)*
1	—	—	11	109	1
2	913	8	12	093	1
3	727	6	13	607	5
4	015	1	14	738	6
5	948	8	15	359	3
6	309	3	16	888	8
7	922	8	17	106	1
8	753	7	18	212	2
9	235	2	19	493	4
10	302	3	20	535	5

arrivals, enter the fourth column of Table 2.6 and read 8 minutes from the first column of the table.

Service times for all 20 customers are shown in Table 2.9. These service times were generated based on the methodology described above, together with the aid of

Table 2.9. SERVICE TIMES GENERATED

Customer	*Random Digits*	*Service Time (Minutes)*	*Customer*	*Random Digits*	*Service Time (Minutes)*
1	84	4	11	32	3
2	10	1	12	94	5
3	74	4	13	79	4
4	53	3	14	05	1
5	17	2	15	79	5
6	79	4	16	84	4
7	91	5	17	52	3
8	67	4	18	55	3
9	89	5	19	30	2
10	38	3	20	50	3

Table 2.7. The first customer's service time is 4 minutes because the random digits 84 fall in the bracket 61–85.

The essence of a manual simulation is the simulation table. These tables are designed for the situation at hand and further structured to answer the questions posed. The simulation table for this problem is shown in Table 2.10 and is an extension of the type of table already seen in Table 2.4. The first customer is assumed to arrive at time 0. Service begins immediately and finishes at time 4. The customer was in the system for 4 minutes. The second customer arrives at time 8. Thus, the server (checkout person) was idle for 4 minutes. Skipping down to the fourth customer, it is seen that this customer arrived at time 15 but could not be served until time 18. This customer had to wait in the queue for 3 minutes. This process continues for all 20 customers. Totals are formed as shown for service times, time customers spend in the system, idle time of the server, and time the customers wait in the queue.

Some of the findings from this short-term simulation are as follows:

1. The average waiting time for a customer is 2.8 minutes.
   This is determined in the following manner:

$$\text{Average waiting time (minutes)} = \frac{\text{total time customers wait in queue (minutes)}}{\text{total numbers of customers}}$$

$$= \frac{56}{20} = 2.8 \text{ minutes}$$

2. The probability that a customer has to wait in the queue is 0.65.
   This is determined in the following manner:

$$\text{Probability (wait)} = \frac{\text{numbers of customers who wait}}{\text{total number of customers}}$$

$$= \frac{13}{20} = 0.65$$

3. The proportion of idle time of the server is 0.21. This is determined in the following manner:

$$\text{Probability of idle server} = \frac{\text{total idle time of server (minutes)}}{\text{total run time of simulation (minutes)}}$$

$$= \frac{18}{86} = 0.21$$

   The probability of the server being busy is the complement of 0.21, or 0.79.

4. The average service time is 3.4 minutes. This is determined in the following manner:

$$\text{Average service time} = \frac{\text{total service time (minutes)}}{\text{total number of customers}}$$

$$= \frac{68}{20} = 3.4 \text{ minutes}$$

   This result can be compared with the expected service time by finding the mean of the service-time distribution using the equation

$$E(S) = \sum_{s=0}^{\infty} s p(s)$$

Table 2.10. SIMULATION TABLE FOR QUEUEING PROBLEM

Customer	*Time Since Last Arrival (Minutes)*	*Arrival Time*	*Service Time (Minutes)*	*Time Service Begins*	*Time Customer Waits in Queue (Minutes)*	*Time Service Ends*	*Time Customer Spends in System (Minutes)*	*Idle Time of Server (Minutes)*
1	—	0	4	0	0	4	4	0
2	8	8	1	8	0	9	1	4
3	6	14	4	14	0	18	4	5
4	1	15	3	18	3	21	6	0
5	8	23	2	23	0	25	2	2
6	3	26	4	26	0	30	4	1
7	8	34	5	34	0	39	5	4
8	7	41	4	41	0	45	4	2
9	2	43	5	45	2	50	7	0
10	3	46	3	50	4	53	7	0
11	1	47	3	53	6	56	9	0
12	1	48	5	56	8	61	13	0
13	5	53	4	61	8	65	12	0
14	6	59	1	65	6	66	7	0
15	3	62	5	66	4	71	9	0
16	8	70	4	71	1	75	5	0
17	1	71	3	75	4	78	7	0
18	2	73	3	78	5	81	8	0
19	4	77	2	81	4	83	6	0
20	5	82	3	83	1	86	4	0
			68		56		124	18

Applying the expected value equation to the distribution in Table 2.7 gives

Expected service time =

$$1(0.10) + 2(0.20) + 3(0.30) + 4(0.25) + 5(0.10) + 6(0.05) = 3.2 \text{ minutes}$$

The expected service time is slightly lower than the average service time in the simulation. The longer the simulation, the closer the average will be to $E(S)$.

5. The average time between arrivals is 4.3 minutes. This is determined in the following manner:

$$\text{Average time between arrivals (minutes)} = \frac{\text{Sum of all times between arrivals (minutes)}}{\text{number of arrivals} - 1}$$

$$= \frac{82}{19} = 4.3 \text{ minutes}$$

One is subtracted from the denominator because the first arrival is assumed to occur at time 0. This result can be compared to the expected time between arrivals by finding the mean of the discrete uniform distribution whose endpoints are $a = 1$ and $b = 8$. The mean is given by

$$E(A) = \frac{a+b}{2} = \frac{1+8}{2} = 4.5 \text{ minutes}$$

The expected time between arrivals is slightly higher than the average. However, in a longer simulation the average value of the time between arrivals should approach the theoretical mean, $E(A)$.

6. The average waiting time of those who wait is 4.3 minutes. This is determined in the following manner:

$$\text{Average waiting time of those who wait (minutes)} = \frac{\text{total time customers wait in queue (minutes)}}{\text{total numbers of customers who wait in queue}}$$

$$= \frac{56}{13} = 4.3 \text{ minutes}$$

7. The average time a customer spends in the system is 6.2 minutes. This can be determined in two ways. First, the computation can be achieved by the following relationship:

$$\text{Average time customer spends in the system (minutes)} = \frac{\text{total time customers spend in the system (minutes)}}{\text{total number of customers}}$$

$$= \frac{124}{20} = 6.2 \text{ minutes}$$

The second way of computing this same result is to realize that the following relationship must hold:

$$\text{Average time customer spends in the system (minutes)} = \text{average time customer spends waiting in the queue (minutes)} + \text{average time customer spends in service (minutes)}$$

Findings 1 and 4 of the list above provide the data for the right-hand side of this equation to give the following:

Average time customer spends in the system (minutes) $= 2.8 + 3.4 = 6.2$ minutes

A decision maker would be interested in results of this type, but a longer simulation would increase the accuracy of the findings. However, some subjective inferences can be drawn at this point. Most customers have to wait; however, the average waiting time is not excessive. The server does not have an undue amount of idle time. Objective statements about the results would depend on balancing the cost of waiting with the cost of additional servers. (Simulations requiring variations of the arrival and service distribution are presented as exercises for the reader.)

EXAMPLE 2.2 THE ABLE-BAKER CARHOP PROBLEM

The purpose of this example is to indicate the simulation procedure when there is more than one channel. Consider a drive-in restaurant where carhops take orders and bring food to the car. Cars arrive in the manner shown in Table 2.11. There are two carhops—Able and Baker. Able is better able to do the job, and works somewhat faster than Baker. The distribution of service times is shown in Tables 2.12 and 2.13.

Table 2.11. INTERARRIVAL DISTRIBUTION OF CARS

Time between Arrivals (Minutes)	*Probability*	*Cumulative Probability*	*Random Digit Assignment*
1	0.25	0.25	01–25
2	0.40	0.65	26–65
3	0.20	0.85	66–85
4	0.15	1.00	86–00

Table 2.12. SERVICE DISTRIBUTION OF ABLE

Service Time (Minutes)	*Probability*	*Cumulative Probability*	*Random Digit Assignment*
2	0.30	0.30	01–30
3	0.28	0.58	31–58
4	0.25	0.83	59–83
5	0.17	1.00	84–00

Table 2.13. SERVICE DISTRIBUTION OF BAKER

Service Time (Minutes)	*Probability*	*Cumulative Probability*	*Random Digit Assignment*
3	0.35	0.35	01–35
4	0.25	0.60	36–60
5	0.20	0.80	61–80
6	0.20	1.00	81–00

Table 2.14. SIMULATION TABLE FOR CARHOP EXAMPLE

					Able			Baker			
Customer No.	*Random Digits for Arrival*	*Time between Arrivals*	*Clock Time of Arrival*	*Random Digits for Service*	*Time Service Begins*	*Service Time*	*Time Service Ends*	*Time Service Begins*	*Service Time*	*Time Service Ends*	*Time in Queue*
1	—	—	0	95	0	5	5				0
2	26	2	2	21				2	3	5	0
3	98	4	6	51	6	3	9				0
4	90	4	10	92	10	5	15				0
5	26	2	12	89				12	6	18	0
6	42	2	14	38	15	3	18				1
7	74	3	17	13	18	2	20				1
8	80	3	20	61	20	4	24				0
9	68	3	23	50				23	4	27	0
10	22	1	24	49	24	3	27				0
11	48	2	26	39	27	3	30				1
12	34	2	28	53				28	4	32	0
13	45	2	30	88	30	5	35				0
14	24	1	31	01				32	3	35	1
15	34	2	33	81	35	4	39				2
16	63	2	35	53				35	4	39	0
17	38	2	37	81	39	4	43				2
18	80	3	40	64				40	5	45	0
19	42	2	42	01	43	2	45				1
20	56	2	44	67	45	4	49				1
21	89	4	48	01				48	3	51	0
22	18	1	49	47	49	3	52				0
23	51	2	51	75				51	5	56	0
24	71	3	54	57	54	3	57				0
25	16	1	55	87				56	6	62	1
26	92	4	59	47	59	3	62				0
						56			43		11

The simulation proceeds in a manner similar to that of Example 2.1, except that it is more complex. A simplifying rule is that Able gets the customer if both carhops are idle; perhaps Able has seniority. (The solution would be different if the decision were made at random concerning who gets a car when both are idle.)

The problem is to find out how well the current arrangement is working. To estimate the system measures, a simulation of 1 hour of operation will be used. A longer simulation would be much more reliable, but for purposes of illustration, a 1-hour period has been selected.

The simulation proceeds much in the manner of Example 2.1. There are more events with this simulation example because there is a second server. The possible events include the following: customers arrive, customers begin service from Able, customers complete service from Able, customers begin service from Baker, customers complete service from Baker. The simulation table is shown in Table 2.14.

The analysis of Table 2.14 results in the following findings:

1. Over the 62-minute period Able is busy 90% of the time.
2. Baker was busy only 69% of the time. The seniority rule keeps Baker less busy.
3. Nine of 26 or about 35% of the arrivals had to wait. The average waiting time for all customers was only about 0.42 minute, or 25 seconds, which is very small.
4. Those 9 who did have to wait only waited an average of 1.22 minutes, which is quite low.
5. In summary, this system seems well balanced. One server cannot handle all the diners, and three servers would probably be too many. Adding an additional server would surely reduce the waiting time to nearly zero. However, the cost of waiting would have to be quite high to justify an additional worker.

2.2. Simulation of Inventory Systems

An important class of simulation problems involves inventory systems. A simple inventory system is shown in Figure 2.7. This inventory system has a periodic review of length N at which time the inventory level is checked. An order is made which will bring the inventory up to the level M. At the end of the first review period, an order quantity, Q_1, is placed. In this inventory system the lead time (i.e., the length of time between the placement and receipt of an order) is zero. Since demands are not usually known with certainty, the order quantities are probabilistic. Demand is shown as being uniform over the time period in Figure 2.7. In actuality, demands are not usually uniform and do fluctuate over time. One possibility is that demands all occur at the beginning of the cycle. Another realistic possibility is that the lead time is other than zero and probabilistic.

Notice that in the second cycle, the amount in inventory drops below zero, indicating a shortage. In Figure 2.7, these units are backordered; when the order arrives, the demand for the backordered items is satisfied first. To avoid shortages, a buffer, or safety, stock would need to be carried.

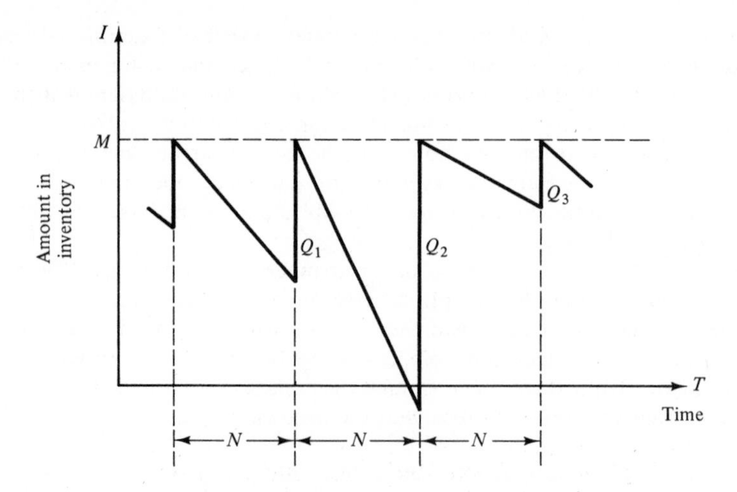

Figure 2.7. Probabilistic order-level inventory system.

Carrying stock in inventory has an associated cost attributed to the interest paid on the funds borrowed to buy the items (this also could be considered as the loss from not having the funds available for other investment purposes). Other costs can be placed in the carrying or holding cost column: renting of storage space, hiring guards, and so on. An alternative to carrying so much inventory is to make more frequent reviews, and consequently, more frequent purchases or replenishments. This has an associated cost, the ordering cost. Also, there is a cost in being short. Customers may get angry, with a subsequent loss of good will. Larger inventories decrease the possibilities of shortages. These costs must be traded off in order to minimize the total cost of an inventory system.

The total cost (or total profit) of an inventory system is the measure of performance. This can be affected by the policy alternatives. For example, in Figure 2.7, the decision maker can control the maximum inventory level, M, and the length of the cycle, N. What effect does changing N have on the various costs?

In an (M, N) inventory system, the events that may occur are the demand for items in the inventory, the review of the inventory position, and the receipt of an order at the end of each review period. When the lead time is zero, as in Figure 2.7, the last two events occur simultaneously.

Much has been written about inventory systems. In Chapter 6, a synopsis of some of the more popular inventory models is given. However, the goal of this chapter is to illustrate, by example, how the simulation of an inventory system proceeds.

EXAMPLE 2.3 THE NEWSPAPER SELLER'S PROBLEM

(In this example, only a single time period of finite length is relevant and only a single procurement is made. Inventory remaining at the end of the single time period is sold for scrap or discarded. A wide variety of real-world problems are of this form, including the stocking of spare parts, perishable items, style goods, and special seasonal items [Hadley and Whitin, 1963]).

A classical inventory problem concerns the purchase and sale of newspapers. The paper seller buys the papers for 13 cents each and sells them for 20 cents each. Newspapers not sold at the end of the day are sold as scrap for 2 cents each. Newspapers can be purchased in bundles of 10. Thus, the paper seller can buy 50, 60, and so on. There are three types of newsdays, "good," "fair," and "poor," with probabilities of 0.35, 0.45, and 0.20, respectively. The distribution of papers demanded on each of these days is given in Table 2.15. The problem is to determine the optimal number of papers the newspaper seller should purchase. This will be accomplished by simulating demands for 20 days and recording profits from sales each day. This problem is easily solved by inventory theory in Chapter 6.

Table 2.15. DISTRIBUTION OF NEWSPAPERS DEMANDED

	Demand Probability Distribution		
Demand	*Good*	*Fair*	*Poor*
40	0.03	0.10	0.44
50	0.05	0.18	0.22
60	0.15	0.40	0.16
70	0.20	0.20	0.12
80	0.35	0.08	0.06
90	0.15	0.04	0.00
100	0.07	0.00	0.00

The profits are given by the following relationship:

$$\text{Profit} = \begin{matrix}\text{revenue}\\ \text{from sales}\end{matrix} - \begin{matrix}\text{cost of}\\ \text{newspapers}\end{matrix} - \begin{matrix}\text{lost profit from}\\ \text{excess demand}\end{matrix} + \begin{matrix}\text{salvage from sale}\\ \text{of scrap papers}\end{matrix}$$

From the problem statement, the revenue from sales is 20 cents for each paper sold. The cost of newspapers is 13 cents for each paper purchased. The lost profit from excess demand is 7 cents for each paper demanded that could not be provided. Such a shortage cost is somewhat controversial, but makes the problem much more interesting. The salvage value of scrap papers is 2 cents each.

Tables 2.16 and 2.17 provide the random digit assignments for the types of newsdays and the demands for those newsdays. To solve this problem by simulation requires setting a policy of buying a certain number of papers each day, then simulating the demands for papers over the 20-day time period to determine the total profit. The

Table 2.16. RANDOM DIGIT ASSIGNMENT FOR TYPE OF NEWSDAY

Type of Newsday	*Probability*	*Cumulative Probability*	*Random Digit Assignment*
Good	0.35	0.35	01–35
Fair	0.45	0.80	36–80
Poor	0.20	1.00	81–00

Table 2.17. RANDOM DIGIT ASSIGNMENTS FOR NEWSPAPERS DEMANDED

	Cumulative Distribution			*Random Digit Assignment*		
Demand	*Good*	*Fair*	*Poor*	*Good*	*Fair*	*Poor*
40	0.03	0.10	0.44	01–03	01–10	01–44
50	0.08	0.28	0.66	04–08	11–28	45–66
60	0.23	0.68	0.82	09–23	29–68	67–82
70	0.43	0.88	0.94	24–43	69–88	83–94
80	0.78	0.96	1.00	44–78	89–96	95–00
90	0.93	1.00	1.00	79–93	97–00	
100	1.00	1.00	1.00	94–00		

policy (number of newspapers purchased) is moved to other values until the daily profit decreases at the level before and the level after that value. The value in between is the optimal number of papers the newspaper seller should purchase.

The simulation table for purchasing 70 newspapers is shown in Table 2.18. On day 1 the demand is for 60 newspapers. The revenue from the sale of 60 newspapers is $12.00. Ten newspapers are left over at the end of the day. The salvage value at 2 cents each is 20 cents. The profit for the first day is determined as follows:

$$\text{Profit} = \$12.00 - \$9.10 - 0 + \$.20 = \$3.10$$

On the fifth day the demand is greater than the supply. The revenue from sales is $14.00, since only 70 papers are available under this policy. An additional 20 papers could have been sold. Thus, a lost profit of $1.40 ($20 \times 7$ cents) is assessed. The daily profit is determined as follows:

$$\text{Profit} = \$14.00 - \$9.10 - \$1.40 + 0 = \$3.50$$

The profit for the 20-day period is the sum of the daily profits, $72.60. It can also be computed from the totals for the 20 days of the simulation as follows:

$$\text{Total profit} = \$258.00 - \$182.00 - \$5.60 + \$2.20 = \$72.60$$

The determination of the optimal number of newspapers to purchase is left as an exercise for the reader.

EXAMPLE 2.4 SIMULATION OF AN (M, N) INVENTORY SYSTEM

This example follows the pattern of the probabilistic order-level inventory system shown in Figure 2.7. Suppose that the maximum inventory level, M, is 11 units and the review period, N, is 5 days. The problem is to estimate, by simulation, the average ending units in inventory and the number of days when a shortage condition existed. The distribution of the number of units demanded per day is shown in Table 2.19. In this example, lead time is a random variable, as shown in Table 2.20. Assume that orders are placed at the close of business and are received for inventory at the beginning of business as determined by the lead time.

To make an estimate of the mean units in ending inventory, many cycles would have to be simulated. For purposes of this example, only five cycles will be shown. The student is asked to continue the example as an exercise at the end of the chapter.

Table 2.18. SIMULATION TABLE FOR PURCHASE OF 70 NEWSPAPERS

Day	*Random Digits for Type of Newsday*	*Type of Newsday*	*Random Digits for Demand*	*Demand*	*Revenue from Sales*	*Lost Profit from Excess Demand*	*Salvage from Sale of Scrap*	*Daily Profit*
1	94	Poor	80	60	$ 12.00	—	$0.20	$ 3.10
2	77	Fair	20	50	10.00	—	0.40	1.30
3	49	Fair	15	50	10.00	—	0.40	1.30
4	45	Fair	88	70	14.00	—	—	4.90
5	43	Fair	98	90	14.00	$1.40	—	3.50
6	32	Good	65	80	14.00	0.70	—	4.20
7	49	Fair	86	70	14.00	—	—	4.90
8	00	Poor	73	60	12.00	—	0.20	3.10
9	16	Good	24	70	14.00	—	—	4.90
10	24	Good	60	80	14.00	0.70	—	4.20
11	31	Good	60	80	14.00	0.70	—	4.20
12	14	Good	29	70	14.00	—	—	4.90
13	41	Fair	18	50	10.00	—	0.40	1.30
14	61	Fair	90	80	14.00	0.70	—	4.20
15	85	Poor	93	70	14.00	—	—	4.90
16	08	Good	73	80	14.00	0.70	—	4.20
17	15	Good	21	60	12.00	—	0.20	3.10
18	97	Poor	45	50	10.00	—	0.40	1.30
19	52	Fair	76	70	14.00	—	—	4.90
20	78	Fair	96	80	14.00	0.70	—	4.20
					$258.00	$5.60	$2.20	$72.60

Table 2.19. RANDOM DIGIT ASSIGNMENTS FOR DAILY DEMAND

Demand	*Probability*	*Cumulative Probability*	*Random Digit Assignment*
0	0.10	0.10	01–10
1	0.25	0.35	11–35
2	0.35	0.70	36–70
3	0.21	0.91	71–91
4	0.09	1.00	92–00

Table 2.20. RANDOM DIGIT ASSIGNMENTS FOR LEAD TIME

Lead Time (Days)	*Probability*	*Cumulative Probability*	*Random Digit Assignment*
1	0.6	0.6	1–6
2	0.3	0.9	7–9
3	0.1	1.0	0

The random digit assignments for daily demand and lead time are shown in the rightmost columns of Tables 2.19 and 2.20. The resulting simulation table is shown in Table 2.21. The simulation has been started with the inventory level at 3 units and an order of 8 units scheduled to arrive in 2 days' time.

Following the simulation table for several selected days indicates how the process operates. The order for 8 units is available on the morning of the third day of the first cycle, raising the inventory level from 1 unit to 9 units. Demands during the remainder of the first cycle reduced the ending inventory level to 2 units on the fifth day. Thus, an order for 9 units was placed. The lead time for this order was 1 day. The order of 9 units was added to inventory on the morning of day 2 of cycle 2.

Notice that the beginning inventory on the second day of the third cycle was zero. An order for 2 units on that day led to a shortage condition. The units were backordered on that day and the next day also. On the morning of day 4 of cycle 3 there was a beginning inventory of 9 units. The 4 units that were backordered and the 1 unit demanded that day reduced the ending inventory to 4 units.

Based on five cycles of simulation, the average ending inventory is approximately 3.5 ($87 \div 25$) units. On 2 of 25 days a shortage condition existed.

2.3. Other Examples of Simulation

This section includes examples of the simulation of a reliability problem, a bombing mission, and the generation of the lead-time demand distribution given the distributions of demand and lead time.

Example 2.5 A Reliability Problem

A large milling machine has three different bearings that fail in service. The cumulative distribution function of the life of each bearing is identical as shown in Table 2.22. When a bearing fails, the mill stops, a repairperson is called, and a new

Table 2.21. SIMULATION TABLE FOR (M, N) INVENTORY SYSTEM

Cycle	*Day*	*Beginning Inventory*	*Random Digits for Demand*	*Demand*	*Ending Inventory*	*Shortage Quantity*	*Order Quantity*	*Random Digits for Lead Time*	*Days Until Order Arrives*
1	1	3	24	1	2	0	—	—	1
	2	2	35	1	1	0	—	—	0
	3	9	65	2	7	0	—	—	—
	4	7	81	3	4	0	—	—	—
	5	4	54	2	2	0	9	5	1
2	1	2	03	0	2	0	—	—	0
	2	11	87	3	8	0	—	—	—
	3	8	27	1	7	0	—	—	—
	4	7	73	3	4	0	—	—	—
	5	4	70	2	2	0	9	0	3
3	1	2	47	2	0	0	—	—	2
	2	0	45	2	0	2	—	—	1
	3	0	48	2	0	4	—	—	0
	4	9	17	1	4	0	—	—	—
	5	4	09	0	4	0	7	3	1
4	1	4	42	2	2	0	—	—	0
	2	9	87	3	6	0	—	—	—
	3	6	26	1	5	0	—	—	—
	4	5	36	2	3	0	—	—	—
	5	3	40	2	1	0	10	4	1
5	1	1	07	0	1	0	—	—	0
	2	11	63	2	9	0	—	—	—
	3	9	19	1	8	0	—	—	—
	4	8	88	3	5	0	—	—	—
	5	5	94	4	1	0	10	8	2
					87				

Table 2.22. BEARING-LIFE DISTRIBUTION

Bearing Life (Hours)	*Probability*	*Cumulative Probability*	*Random Digit Assignment*
1000	0.10	0.10	01–10
1100	0.13	0.23	11–23
1200	0.25	0.48	24–48
1300	0.13	0.61	49–61
1400	0.09	0.70	62–70
1500	0.12	0.82	71–82
1600	0.02	0.84	83–84
1700	0.06	0.90	85–90
1800	0.05	0.95	91–95
1900	0.05	1.00	96–00

bearing is installed. The delay time of the repairperson's arriving at the milling machine is also a random variable, with the distribution given in Table 2.23. Downtime for the mill is estimated at $5 per minute. The direct on-site cost of the repairperson is $12 per hour. It takes 20 minutes to change one bearing, 30 minutes to change two bearings, and 40 minutes to change three bearings. The bearings cost $16 each. A proposal has been made to replace all three bearings whenever a bearing fails. Management needs an evaluation of this proposal.

Table 2.23. DELAY-TIME DISTRIBUTION

Delay Time (Minutes)	*Probability*	*Cumulative Probability*	*Random Digit Assignment*
5	0.6	0.6	1–6
10	0.3	0.9	7–9
15	0.1	1.0	0

Table 2.24 represents a simulation of 20,000 hours of operation under the current method of operation. Note that there are instances where more than one bearing fails at the same time. This is unlikely to occur in practice, and is due to using a rather coarse grid of 100 hours. It will be assumed in this example that the times are never exactly the same, and thus no more than one bearing is changed at any breakdown. Sixteen bearing changes were made for bearings 1 and 2, but only 14 bearing changes were required for bearing 3. The cost of the current system is estimated as follows:

Cost of bearings = 46 bearings × \$16/bearing = \$736
Cost of delay time = (110 + 125 + 95) minutes × \$5/minute = \$1650
Cost of downtime during repair =
46 bearings × 20 minutes/bearing × \$5/minute = \$4600
Cost of repairpersons = 46 bearings × 20 minutes/bearing × \$12/60 minutes
= \$184
Total cost = \$736 + \$1650 + \$4600 + \$184 = \$7170

Table 2.24. BEARING REPLACEMENT USING CURRENT METHOD

	Bearing 1					*Bearing 2*					*Bearing 3*				
	RD[a]	*Life (Hours)*	*Accumulated Life (Hours)*	*RD*	*Delay (Minutes)*	*RD*	*Life (Hours)*	*Accumulated Life (Hours)*	*RD*	*Delay (Minutes)*	*RD*	*Life (Hours)*	*Accumulated Life (Hours)*	*RD*	*Delay (Minutes)*
1	67	1,400	1,400	2	5	70	1,500	1,500	0	15	76	1,500	1,500	0	15
2	08	1,000	2,400	3	5	43	1,200	2,700	7	10	65	1,400	2,900	2	5
3	49	1,300	3,700	1	5	86	1,700	4,400	3	5	61	1,400	4,300	7	10
4	84	1,600	5,300	7	10	93	1,800	6,200	1	5	96	1,900	6,200	1	5
5	44	1,200	6,500	8	10	81	1,600	7,800	2	5	65	1,400	7,600	3	5
6	30	1,200	7,700	1	5	44	1,200	9,000	8	10	56	1,300	8,900	3	5
7	10	1,000	8,700	2	5	19	1,100	10,100	1	5	11	1,100	10,000	6	5
8	63	1,400	10,100	8	10	51	1,300	11,400	1	5	86	1,700	11,700	3	5
9	02	1,000	11,100	3	5	45	1,300	12,700	7	10	57	1,300	13,000	1	5
10	02	1,000	12,100	8	10	12	1,100	13,800	8	5	49	1,300	14,300	4	5
11	77	1,500	13,600	7	10	48	1,300	15,100	0	15	36	1,200	15,500	8	10
12	59	1,300	14,900	5	5	09	1,000	16,100	8	10	44	1,200	16,700	2	5
13	23	1,100	16,000	5	5	44	1,200	17,300	1	5	94	1,800	18,500	1	5
14	53	1,300	17,300	9	10	46	1,200	18,500	2	5	78	1,500	20,000	7	10
15	85	1,700	19,000	6	5	40	1,200	19,700	8	10					
16	75	1,500	20,500	4	5	52	1,300	21,000	5	5					
					110					125					95

[a]RD, random digits.

Table 2.25 is a simulation using the proposed method. Notice that, as long as possible, the same lifetimes appear for all three bearings. It is assumed that the bearings are in order on a shelf and they are taken sequentially and placed on the mill. (The effect of using different random numbers versus common random numbers is discussed in Chapter 12.) The random digits that lead to the lives of the additional bearings are shown above the slashed line beginning with the 15th replacement of bearing 3.

Table 2.25. BEARING REPLACEMENT USING METHOD PROPOSED

	Bearing 1 Life (Hours)	*Bearing 2 Life (Hours)*	*Bearing 3 Life (Hours)*	*First Failure (Hours)*	*Accumulated Life (Hours)*	*RD*	*Delay (Minutes)*
1	1,400	1,500	1,500	1,400	1,400	3	5
2	1,000	1,200	1,400	1,000	2,400	7	10
3	1,300	1,700	1,400	1,300	3,700	5	5
4	1,600	1,800	1,900	1,600	5,300	1	5
5	1,200	1,600	1,400	1,200	6,500	4	5
6	1,200	1,200	1,300	1,200	7,700	3	5
7	1,000	1,100	1,100	1,000	8,700	7	10
8	1,400	1,300	1,700	1,300	10,000	8	10
9	1,000	1,300	1,300	1,000	11,000	8	10
10	1,000	1,100	1,300	1,000	12,000	3	5
11	1,500	1,300	1,200	1,200	13,200	2	5
12	1,300	1,000	1,200	1,000	14,200	4	5
13	1,100	1,200	1,800	1,100	15,300	1	5
14	1,300	1,200	1,500	1,200	16,500	6	5
15	1,700	1,200	63/1,400	1,200	17,700	2	5
16	1,500	1,300	21/1,100	1,100	18,800	7	10
17	85/1,700	53/1,300	23/1,100	1,100	19,900	0	15
18	05/1,000	29/1,200	51/1,300	1,000	20,900	5	5
							125

When the new policy is used, some 18 sets of bearings were required. In the two simulations, repairperson delays were not duplicated but were generated independently. The total cost of the new policy is computed as follows:

Cost of bearings = 54 bearings × \$16/bearing = \$864
Cost of delay time = 125 minutes × \$5/minute = \$625
Cost of downtime during repairs = 18 sets × 40 minutes/set × \$5/minute = \$3600
Cost of repairpersons = 18 sets × 40 minutes/set × \$12/60 minutes = \$144
Total cost = \$864 + \$625 + \$3600 + \$144 = \$5233

The new policy generates a savings of \$1937 over a 20,000-hour simulation. If the machine runs continuously, the simulated time is about $2\frac{1}{4}$ years. Thus, the savings are about \$860 per year.

EXAMPLE 2.6 RANDOM NORMAL NUMBERS

A classic simulation problem is that of a squadron of bombers attempting to destroy an ammunition depot which is shaped as shown in Figure 2.8. If a bomb lands anywhere on the depot, a hit is scored. Otherwise, the bomb is a miss. The aircraft is flying in the horizontal direction. Ten bombers are in each squadron. The aiming point is the dot located in the heart of the ammunition dump. The point of impact is assumed to be normally distributed around the aiming point with a standard deviation of 600 meters in the horizontal direction and 300 meters in the vertical direction. The problem is to simulate the operation and make statements about the number of bombs on target.

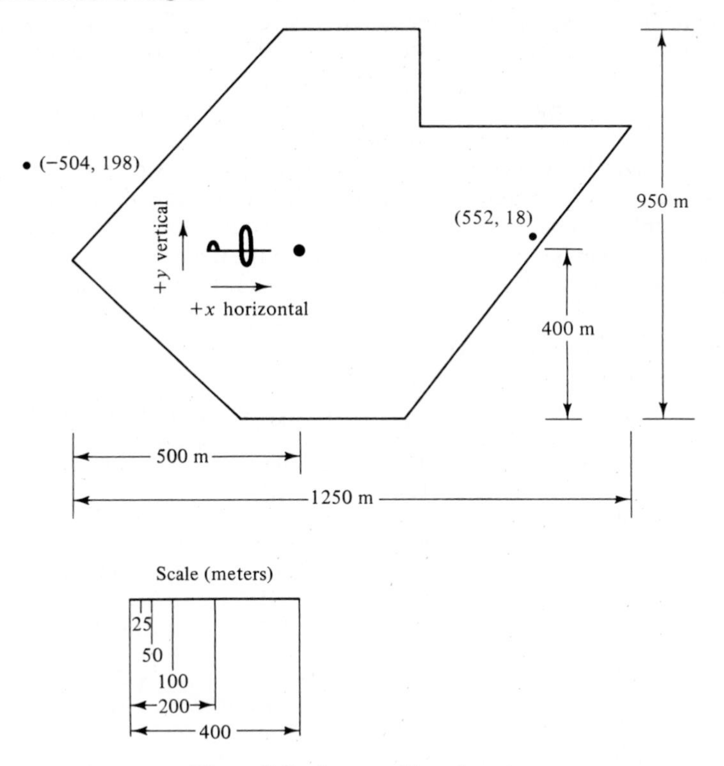

Figure 2.8. Ammunition depot.

Recall that the standardized normal variate, Z, is distributed as

$$Z = \frac{X - \mu}{\sigma}$$

where X is a normal random variable, μ is the true mean of the distribution of X, and σ is the standard deviation of X. Then,

$$X = Z\sigma + \mu$$

In this example the aiming point can be considered as (0, 0); that is, the μ value in the horizontal direction is 0, and similarly for the μ value in the vertical direction. Then,

$$X = Z\sigma_X$$
$$Y = Z\sigma_Y$$

where (X, Y) are the simulated coordinates of the bomb after it has fallen. Now, $\sigma_X = 600$ and $\sigma_Y = 300$. Then,

$$X = 600Z_i$$
$$Y = 300Z_j$$

The i and j subscripts have been added to indicate that the values of Z should be different. What are these Z values and where can they be found? The values of Z are random normal numbers. These can be generated from uniformly distributed random numbers as discussed in Chapter 8. Alternatively, tables of random normal numbers have been generated. A small sample of random normal numbers is given in Table A.2.

To understand what happens in these bombing missions, a simulation of perhaps 10 or 20 runs might be conducted. However, space limitations do not permit such an extensive simulation. An example of one run will indicate how the simulation is performed. The table of random normal numbers is used in the same way as the table of random numbers: that is, start at a random place in the table and proceed in a systematic direction avoiding overlap. Table 2.26 shows the results of a simulated run. The

Table 2.26. SIMULATED BOMBING RUN

Bomber	*RNN_x*	*x Coordinate (600 RNN_x)*	*RNN_y*	*y Coordinate (300 RNN_y)*	*Result*[a]
1	−0.84	−504	0.66	198	Miss
2	1.03	618	−0.13	−39	Miss
3	0.92	552	0.06	18	Hit
4	−1.82	−1,092	−1.40	−420	Miss
5	−0.16	−96	0.23	69	Hit
6	−1.78	−1,068	1.33	399	Miss
7	2.04	1,224	0.69	207	Miss
8	1.08	648	−1.10	−330	Miss
9	−1.50	−900	−0.72	−216	Miss
10	−0.42	−252	−0.60	−180	Hit

[a]Total: 3 hits, 7 misses.

mnemonic RNN_x stands for "random normal number to compute the x coordinate" and corresponds to Z_i above. The first random normal number used was −0.84, generating an x coordinate $600(-0.84) = -504$. The random normal number to generate the y coordinate was 0.66, resulting in a y coordinate of 198. Taken together (−504, 198) is a miss, for it is off the target. The resulting point and that of the third bomber are plotted on Figure 2.8. The 10 bombers had 3 hits and 7 misses. Many more runs are needed to assess the potential for destroying the dump. Additional runs appear as an exercise for the reader. This is an example of a Monte Carlo, or static, simulation, since time is not an element of the solution.

EXAMPLE 2.7 LEAD-TIME DEMAND

Lead-time demand may occur in an inventory system when the lead time is other than instantaneous. The lead time is the time from placement of an order until the order is received. In a realistic situation, lead time is a random variable. During the lead time, demands also occur at random. Lead-time demand is thus a random variable defined as the sum of the demands over the lead time, or $\sum_{i=0}^{T} D_i$, where i is the time period of the lead time, $i = 0, 1, 2, \ldots$; D_i is the demand during the ith time period; and T is the lead time. The distribution of lead-time demand is determined by simulating many cycles of lead time and building a histogram based on the results.

A firm sells bulk rolls of newsprint. The daily demand is given by the following probability distribution:

Daily Demand (Rolls)	3	4	5	6
Probability	0.20	0.35	0.30	0.15

The lead time is the number of days from placing an order until the firm receives the order from the supplier. In this instance, lead time is a random variable given by the following distribution:

Lead Time (Days)	1	2	3
Probability	0.36	0.42	0.22

Table 2.27 shows the random digit assignment for demand and Table 2.28 does the same for lead time. The incomplete simulation table is shown in Table 2.29. The random

Table 2.27. RANDOM DIGIT ASSIGNMENT FOR DEMAND

Daily Demand	*Probability*	*Cumulative Probability*	*Random Digit Assignment*
3	0.20	0.20	01–20
4	0.35	0.55	21–55
5	0.30	0.85	56–85
6	0.15	1.00	86–00

Table 2.28. RANDOM DIGIT ASSIGNMENT FOR LEAD TIME

Lead Time (Days)	*Probability*	*Cumulative Probability*	*Random Digit Assignment*
1	0.36	0.36	01–36
2	0.42	0.78	37–78
3	0.22	1.00	79–00

Table 2.29. SIMULATION TABLE FOR LEAD-TIME DEMAND

Cycle	*Random Digits for Lead Time*	*Lead Time (Days)*	*Random Digits for Demand*	*Demand*	*Lead-Time Demand*
1	57	2	87	6	
			34	4	10
2	33	1	82	5	5
3	93	3	28	4	
			19	3	
			63	5	12
4	55	2	91	6	
			26	4	10
⋮	⋮	⋮	⋮	⋮	⋮

digits for the first cycle were 57. This generates a lead time of 2 days. Thus, two pairs of random digits must be generated for the daily demand. The first of these pairs is 87, which leads to a demand of 6. This is followed by a demand of 4. The lead-time demand for the first cycle is 10. After many cycles are simulated, a histogram is formulated. The histogram might appear as shown in Figure 2.9. This example illustrates how simulation can be used to study an unknown distribution by generating a random sample from the distribution.

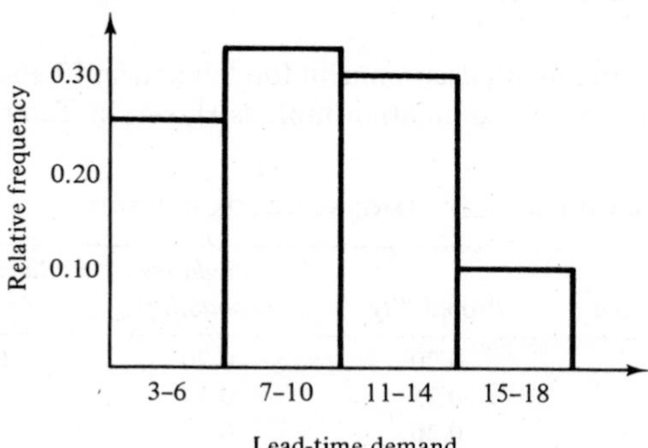

Figure 2.9. Histogram for lead-time demand.

2.4. Summary

The purpose of this chapter is to introduce simulation concepts via examples, to illustrate general areas of application, and to motivate the remaining chapters. By placing the examples early in the text, the reader's appreciation of the concepts, methods, and analytic techniques in subsequent chapters will be enhanced.

Ad hoc simulation tables were used in completing each example. Events in the tables were generated using uniformly distributed random numbers and, in one case, random normal numbers. The examples illustrate the need for determining the characteristics of the input data, generating random variables from the input models, and analyzing the resulting response. These subjects are treated in more detail in the remaining chapters of the text. Thus, this chapter serves to motivate the need for the remaining 10 chapters.

Examples are drawn principally from queueing and inventory systems, because a large number of simulations concern problems in these areas. Additional examples are given in the areas of reliability, static simulation, and the generation of a random sample from an unknown distribution.

REFERENCE

Hadley G., and T. M. Whitin [1963], *Analysis of Inventory Systems*, Prentice-Hall, Englewood Cliffs, N.J.

EXERCISES

Use different random digits from Table A.1 for all problems referring to examples in the text.

1. In Example 2.1, let the arrival distribution be uniformly distributed between 1 and 10 minutes. Develop the simulation table and the analysis for 20 customers. What is the effect of changing the arrival time distribution?

2. In Example 2.1, let the service distribution be changed to the following:

Service time (Minutes)	1	2	3	4	5	6
Probability	0.05	0.10	0.20	0.30	0.25	0.10

Develop the simulation table and the analysis for 20 customers. What is the effect of changing the service-time distribution?

3. Perform the simulation in Example 2.1 for 20 more customers (customers 21 through 40). Compare the results of Example 2.1 to your results.

4. In Example 2.1, determine the time-weighted-average number of customers in the system and the time-weighted-average number of customers waiting. (*Hint:* Use Figure 2.6.)

5. In Example 2.2, change the arrival distribution of cars to the following:

Time between Arrivals (Minutes)	0	1	2	3	4
Probability	0.10	0.20	0.35	0.20	0.15

Develop the simulation and subsequent analysis for a period of 1 hour. What is the effect of changing the arrival distribution?

6. Again, with respect to Example 2.2, Able has a kneecap injury and cannot move as fast. Consequently, two things happen. Able's service distribution changes and Baker gets first shot at the customer if both carhops are idle. Able's new service distribution is as follows:

Service Time (Minutes)	3	4	5	6
Probability	0.30	0.30	0.25	0.15

(a) Develop a simulation and subsequent analysis for 30 service completions. What is the effect of Able's injury and the new rule?

(b) What would be the effect of adding a new employee that works at Baker's speed? The new employee would have all leftover work after Baker and Able.

7. Modify Example 2.2 so that Able has a probability of 0.45 of getting a customer in case both are idle. What is the effect of such a change?

8. Determine the optimal number of newspapers to be purchased daily in Example 2.3. Should the same random digits be used for each level of newspapers purchased by the seller? Why or why not? What effect does using new random numbers have?

9. A baker is trying to determine how many dozens of bagels to bake each day. The probability distribution of the number of bagel customers is as follows:

Number of Customers/Day	8	10	12	14
Probability	0.35	0.30	0.25	0.10

Customers order 1, 2, 3, or 4 dozen bagels according to the following probability distribution.

Number of Dozen Ordered/Customer	1	2	3	4
Probability	0.4	0.3	0.2	0.1

Bagels sell for $2.25 per dozen. They cost $1.58 per dozen to make. All bagels not sold at the end of the day are sold at half-price to a local grocery store. Based on 5 days of simulation, how many dozen (to the nearest 10 dozen) bagels should be baked each day?

10. Demand for widgets follows the probability distribution shown:

Daily Demand	0	1	2	3	4
Probability	0.33	0.25	0.20	0.12	0.10

Stock is examined every 7 days (the plant is in operation every day) and if the stock level has reached 6 units, or less, an order for 10 widgets is placed. The lead time (days until delivery) is probabilistic and follows the following distribution:

Lead Time (Days)	1	2	3
Probability	0.3	0.5	0.2

When the simulation begins, it is the beginning of the week, 12 widgets are on hand, and no orders have been backordered. (Backordering is allowed.) Simulate 6 weeks of operation of this system. Analyze the system. Perform additional simulations to determine the effect on shortages if increases or decreases occur in (1) the review period, (2) the reorder quantity, and (3) the reorder point.

11. A plumbing supply firm is interested in the distribution of lead-time demand of industrial sinks. The frequency function for daily demand is known and occurs as shown:

Daily Demand	0	1	2	3	4
Probability	0.18	0.39	0.29	0.09	0.05

The distribution of lead time has been reconstructed from past records as follows:

Lead Time (Days)	0	1	2	3	4	5
Probability	0.135	0.223	0.288	0.213	0.118	0.023

Develop the distribution of lead-time demand based on 20 cycles of lead time. Prepare a histogram of the distribution using intervals 0–2, 3–5, ... Then, prepare a histogram using intervals 0–1, 2–3, 3–4, ... Does changing the width of the interval width have a pronounced effect on the form of the histogram of the lead-time demand distribution?

12. Develop and interpret flow diagrams analogous to Figures 2.2 and 2.3 for a queueing system with i channels.

13. Rework the simulation of the proposed method of Example 2.5 using new random digits. What is the effect on the total cost of generating a new set of events? When is it satisfactory to use the same random digits (and events) for competing proposals?

14. Smalltown Taxi operates one vehicle during the 9:00 A.M. to 5:00 P.M. period. Currently, consideration is being given to the addition of a second vehicle to the fleet. The demand for taxis follows the distribution shown:

Time between Calls (Minutes)	15	20	25	30	35
Probability	0.14	0.22	0.43	0.17	0.04

The distribution of time to complete a service is as follows:

Service Time (Minutes)	5	15	25	35	45
Probability	0.12	0.35	0.43	0.06	0.04

Simulate five individual days of operation of the current system and the system with an additional taxicab. Compare the two systems with respect to the waiting times of the customers and any other measures that might shed light on the situation.

15. Continue Example 2.6 for nine more runs and estimate how well the simulated bombers will do when they make their raid on the ammunition depot.

16. The random variables X, Y, and Z are distributed as follows:

$$X \sim N(\mu = 100, \sigma^2 = 100)$$
$$Y \sim N(\mu = 300, \sigma^2 = 225)$$
$$Z \sim N(\mu = 40, \sigma^2 = 64)$$

Simulate 50 values of the random variable

$$W = \frac{X + Y}{Z}$$

Prepare a histogram of the resulting values using class intervals of width equal to 3.

17. Lead time for a stock item is normally distributed with a mean of 7 days and a standard deviation of 2 days. Daily demand is distributed as follows:

Daily Demand	0	1	2	3	4
Probability	0.367	0.368	0.184	0.062	0.019

Determine the lead time demand for 20 order cycles.

18. Consider Example 2.4.
(a) Extend the example for 15 more cycles and draw conclusions.
(b) Rework the example for 10 cycles with $M = 10$.
(c) Rework the example for 10 cycles with $N = 6$.

19. Estimate, by simulation, the average number of lost sales per week for an inventory system that functions as follows:
(a) Whenever the inventory level falls to or below 10 units, an order is placed. Only one order can be outstanding at a time.
(b) The size of each order is equal to $20 - I$, where I is the inventory level when the order is placed.
(c) If a demand occurs during a period when the inventory level is zero, the sale is lost.
(d) Daily demand is normally distributed with a mean of 5 units and a standard deviation of 1.5 units. (Round off demands to the closest integer during the simulation.)
(e) Lead time is distributed uniformly between zero and 5 days, integers only.
(f) The simulation will start with 18 units in inventory.
(g) For simplicity, assume that all demands occur at 12 noon and that all orders are placed at the same time. Assume further that orders are received at 5:00 P.M., or after the demand which occurred on that day.
(h) Let the simulation run for 5 weeks.

20. An elevator in a manufacturing plant carries exactly 400 kilograms of material.

There are three kinds of material, and these are in boxes, that arrive for a ride on the elevator. These materials and their distributions of time between arrivals are as follows:

Material	*Weight (Kilograms)*	*Interarrival Time (Minutes)*
A	200	5 ± 2 (uniform)
B	100	6 (constant)
C	50	$P(2) = 0.33$
		$P(3) = 0.67$

It takes the elevator 1 minute to go up to the second floor, 2 minutes to unload, and 1 minute to return to the first floor. The elevator does not leave the first floor unless it has a full load. Simulate 1 hour of operation of the system. What is the average transit time for a box of material A (time from its arrival until it is unloaded)? What is the average waiting time for a box of material B? How many boxes of material C made the trip in 1 hour?

21. The random variables X and Y are distributed as follows:

$$X \sim 10 \pm 10 \text{ (uniform)}$$
$$Y \sim 10 \pm 8 \text{ (uniform)}$$

(a) Simulate 200 values of the random variable

$$Z = XY$$

Prepare a histogram of the resulting values. What is the range of Z and what is its average value?

(b) Same as (a) except

$$Z = X/Y$$

22. Perform one of the foregoing simulation exercises in FORTRAN, GPSS, SIMSCRIPT, GASP, SLAM, or any other computer language.

DISCRETE-EVENT SIMULATION: GENERAL PRINCIPLES AND COMPUTER SIMULATION LANGUAGES

This chapter develops a common framework for the modeling of complex systems, and introduces some of the major discrete-event simulation languages currently in use. The modeling approach is called discrete-event simulation. As briefly discussed in Chapter 1, a system is modeled in terms of its state at each point in time, and various events whose occurrence causes a change in state. Discrete-event modeling is appropriate for those systems for which significant changes in system state occur at discrete points in time.

Every discrete-event simulation language has its own world view, or way of looking at the system being modeled. The languages described in this chapter can be classified as taking the event-scheduling approach or the process-interaction approach to discrete-event modeling. The event-scheduling approach requires that the analyst concentrate on the events and how they affect system state. The process-interaction approach allows the analyst to concentrate on a single entity (such as a customer) and the sequence of events and activities it undergoes as it "passes through" the system. When using a general-purpose language, such as FORTRAN, ALGOL, BASIC, or Pascal, a simulator would most likely adopt the event-scheduling approach. Languages such as GASP facilitate the use of the event-scheduling approach, while GPSS provides one implementation of the process-interaction approach. Some more modern languages, such as SIMSCRIPT and SLAM, allow the simulator to use either approach or a mixture of the two, whichever is more appropriate for the problem at hand.

3

Section 3.1 discusses the general principles of the event-scheduling and process-interaction approaches, and gives several examples by means of hand simulations. Section 3.2 gives examples of modeling a simple system using FORTRAN, SIMSCRIPT, GPSS, and SLAM, as well as briefly discussing GASP.

3.1. Concepts in Discrete-Event Simulation

The concept of a system and a model of a system were discussed briefly in Chapter 1. This chapter deals exclusively with dynamic, stochastic systems (i.e., involving time and containing random elements) which change in a discrete manner. This section expands on these concepts and develops a framework for the development of a discrete-event model of a system. The major concepts are briefly defined and then illustrated by examples:

System — A collection of entities (e.g., people and machines) that interact together over time to accomplish one or more goals

Model — An abstract representation of a system, usually containing logical and/or mathematical relationships which describe a system in terms of state, entities and their attributes, sets, events, activities, and delays

System state	A collection of variables that contain all the information necessary to describe the system at any time
Entity	Any object or component in the system which requires explicit representation in the model (e.g., a server, a customer, a machine)
Attributes	The properties of a given entity (e.g., the priority of a waiting customer, the routing of a job through a job shop)
Set	A collection of (permanently or temporarily) associated entities, ordered in some logical fashion (such as all customers currently in a waiting line, ordered by first come, first served, or by priority)
Event	An instantaneous occurrence that changes the state of a system (such as an arrival of a new customer)
Activity	A duration of time of specified length (e.g., a service time or interarrival time), which length is known when it begins (although it may be defined in terms of a statistical distribution)
Delay	A duration of time of unspecified length, which length is not known until it ends (e.g., a customer's delay in a last in, first out waiting line, which when it begins, depends on future arrivals)

Sets are sometimes called lists, queues, or chains. An activity can be deterministic (e.g., a service time that always takes 5 minutes), or probabilistic (e.g., 5 ± 3 minutes, uniformly distributed), or, in general, any type of mathematical function. However it is characterized, the duration of an activity is computable in the model at the instant it begins. By contrast, a delay typically ends when some logical condition becomes true; this logical condition is usually a result of the interaction of many events. A customer's time spent in a waiting line is a typical example of delay. A delay is sometimes called a conditional wait, in contrast to an activity, which is called an unconditional wait. Note that the ending of an activity is an event, often termed a primary event. The beginning and ending of a delay is called a conditional event. (The beginning of an activity may be a primary or a conditional event.) The term "event" in this text refers to a primary event.

The systems considered here are dynamic, that is, changing over time. Therefore, system state, entity attributes and the number of active entities, the contents of sets, and the activities and delays currently in progress are all functions of time and are constantly changing over time. Time itself is represented by a variable called CLOCK.

EXAMPLE 3.1 (ABLE AND BAKER, REVISITED)

Consider the Able–Baker carhop system of Example 2.2. A discrete-event model has the following components:

System state	$L_Q(t)$, the number of cars waiting to be served at time t
	$L_A(t)$, 0 or 1 to indicate Able being idle or busy at time t
	$L_B(t)$, 0 or 1 to indicate Baker being idle or busy at time t
Entities	Neither the customers (i.e., cars) nor the servers need to be explicitly represented, except in terms of the state variables, unless certain customer averages are desired (compare Examples 3.2 and 3.3)
Events	Arrival event
	Service completion by Able
	Service completion by Baker
Activities	Interarrival time, defined in Table 2.11
	Service time by Able, defined in Table 2.12
	Service time by Baker, defined in Table 2.13
Delay	The wait in queue until Able or Baker becomes free

The definition of the model components provides a static description of the model. In addition, a description of the dynamic relationships and interactions between the components is also needed. Some questions that need answers include:

1. How does each event affect system state, entity attributes, and set contents?
2. How are activities defined (i.e., deterministic, probabilistic, or some other mathematical equation)? What event marks the beginning or end of each activity? Can the activity begin regardless of system state, or is its beginning conditioned on the system being in a certain state? (For example, a machining "activity" cannot begin unless the machine is idle, not broken, and not in maintenance.)
3. Which events trigger the beginning (and end) of each type of delay? Under what conditions does a delay begin, or end?
4. What is the system state at time 0? What events should be generated at time 0 to "prime" the model—that is, to get the simulation started?

A discrete-event simulation is the modeling over time of a system all of whose state changes occur at discrete points in time—those points when an event occurs. A discrete-event simulation (hereafter called a simulation) proceeds by producing a sequence of system snapshots (or system images) which represent the evolution of the system through time. A given snapshot at a given time (CLOCK $= t$) includes not only the system state at time t, but also a list (called the *future event list*) of all activities currently in progress and when each such activity will end, the status of all entities and current membership of all sets, plus the current values of cumulative statistics and counters that will be used to calculate summary statistics at the end of the simulation. A prototype system snapshot is shown in Figure 3.1. (Not all models will contain every element exhibited in Figure 3.1. Further illustrations are provided in the examples in this chapter.)

The mechanism for advancing simulation time and guaranteeing that all events occur in correct chronological order is based on the future event list. This list is a special set which contains all events that have been scheduled to occur at a future time. Scheduling a future event means that at the instant an activity begins, its duration is computed (perhaps it is generated in "random" fashion) and the end-activity event, together with its event time, is placed on the future event list. In the real world, most

Clock	System state	Entities and attributes	Set 1	Set 2	...	Future event list, FEL	Cumulative statistics and counters
t	$(x, y, z, \ldots)$					$(3, t_1)$ – Type 3 event to occur at time t_1 $(1, t_2)$ – Type 1 event to occur at time t_2	
	⋮					⋮ ⋮	

Figure 3.1. Prototype system snapshot at simulation time t.

future events are not scheduled but merely happen—such as random breakdowns or random arrivals. In the model, such random events are represented by the end of some activity, which in turn is represented by a statistical distribution.

At any given time t, the future event list (FEL) contains all previously scheduled future events and their associated event times (called $t_1, t_2, \ldots$ in Figure 3.1). The FEL is ordered by event time, meaning that the events are arranged chronologically; that is, the event times satisfy

$$t < t_1 \leq t_2 \leq t_3 \leq \cdots \leq t_n$$

Time t is the value of CLOCK, the current value of simulation time. The event associated with time t_1 is called the imminent event; that is, it is the next event that will occur. After the system snapshot at simulation time CLOCK $= t$ is complete, the CLOCK is advanced to simulation time CLOCK $= t_1$, and the imminent event is removed from the FEL and executed. Execution of the imminent event means that a new system snapshot for time t_1 is created based on the old snapshot at time t and the nature of the imminent event. At time t_1, new future events may or may not be generated, but if any are, they are scheduled by putting them in their proper position on the FEL. After the new system snapshot for time t_1 is completed, the clock is advanced to the time of the new imminent event and that event is executed. This process repeats until the simulation is over. The sequence of actions which a simulator (or simulation language) must perform to advance the clock and build a new system snapshot is called the *event-scheduling/time-advance algorithm*, whose steps are listed in Figure 3.2 (and explained below).

The length and contents of the FEL are constantly changing as the simulation progresses, and thus its efficient management in a computerized simulation will have a major impact on the efficiency of the computer program representing the model. The management of a list is called list processing. The major list processing operations performed on a FEL are removal of the imminent event, addition of a new event to the list, and occasionally removal of some event (called cancellation of an event). As the imminent event is usually at the top of the list, its removal is as efficient as possible. Addition of a new event (and cancellation of an old event) requires a search of the list. The efficiency of this search depends on the logical organization of the list and on how the search is conducted. In addition to the FEL, all the sets in a model are

Old system snapshot at time t

CLOCK	System state	. . .	Future event list	. . .
t	(5, 1, 6)		$(3, t_1)$ – Type 3 event to occur at time t_1 $(1, t_2)$ – Type 1 event to occur at time t_2 $(1, t_3)$ – Type 1 event to occur at time t_3 ⋮ ⋮ ⋮ $(2, t_n)$ – Type 2 event to occur at time t_n	

Event-scheduling/time-advance algorithm

Step 1. Remove imminent event (event 3, time t_1) from FEL.
Step 2. Advance CLOCK to imminent event time (i.e., advance CLOCK from t to t_1).
Step 3. Execute imminent event: update system state, change entity attributes, and set membership as needed.
Step 4. Generate future events (if necessary) and place on FEL in correct position. (Example: Event 4 to occur at time t^*, where $t_2 < t^* < t_3$.)
Step 5. Update cumulative statistics and counters.

New system snapshot at time t_1

CLOCK	System state	. . .	Future event list	. . .
t_1	(5, 1, 5)		$(1, t_2)$ – Type 1 event to occur at time t_2 $(4, t^*)$ – Type 4 event to occur at time t^* $(1, t_3)$ – Type 1 event to occur at time t_3 ⋮ ⋮ ⋮ $(2, t_n)$ – Type 2 event to occur at time t_n	

Figure 3.2. Advancing simulation time and updating system image.

maintained in some logical order, and the operations of addition and removal of entities from the set also require efficient list-processing techniques. A brief introduction to list processing in simulation is given by Law and Kelton [1982, Chap. 2].

The removal and addition of events from the FEL is illustrated in Figure 3.2. Event 3 with event time t_1 represents, say, a service completion event at server 3. Since it is the imminent event at time t, it is removed from the FEL in step 1 (Figure 3.2) of the event-scheduling/time-advance algorithm. When event 4 (say, an arrival event) with event time t^* is generated at step 4, one possible way to determine its correct position on the FEL is to conduct a top-down search:

If $t^* < t_2$, place event 4 at the top of the FEL.
If $t_2 \leq t^* < t_3$, place event 4 second on the list.
If $t_3 \leq t^* < t_4$, place event 4 third on the list.

.
.
.

If $t_n \leq t^*$, place event 4 last on the list.

(In Figure 3.2, it was assumed that t^* was between t_2 and t_3.) Another way is to conduct a bottom-up search. The least efficient way to maintain the FEL is to leave it as an unordered list (additions placed arbitrarily at the top or bottom), which would require at step 1 of Figure 3.2 a complete search of the list for the imminent event before each clock advance. (The imminent event is the event on the FEL with the lowest event time.)

The system snapshot at time 0 is defined by the initial conditions and the generation of the so-called exogenous events. The specified initial conditions define the system state at time 0. For example, in Figure 3.2, if $t = 0$, then the state (5, 1, 6) might represent the initial number of customers at three different points in the system. An exogenous event is a happening "outside the system" which impinges on the system. An important example is an arrival to a queueing system. At time 0, the first arrival event is generated, and scheduled on the FEL. The interarrival time is an example of an activity. When the clock eventually is advanced to the time of this first arrival, a second arrival event is generated. First, an interarrival time is generated, say a^*; it is added to the current time, say CLOCK $= t$; the resulting (future) event time, $t + a^* = t^*$, is used to position the new arrival event on the FEL. This method of generating an external arrival stream is called bootstrapping, and provides one example of how future events are generated in step 4 of the event-scheduling/time-advance algorithm. Bootstrapping is illustrated in Figure 3.3. The first three interarrival times generated are 3.7, 0.4, and 3.3 time units. The beginning and end of an interarrival interval are examples of primary events.

A second example of how future events are generated (step 4 of Figure 3.2) is provided by a service completion event in a queueing simulation. When one customer completes service, say at current time CLOCK $= t$, if the next customer is present, then

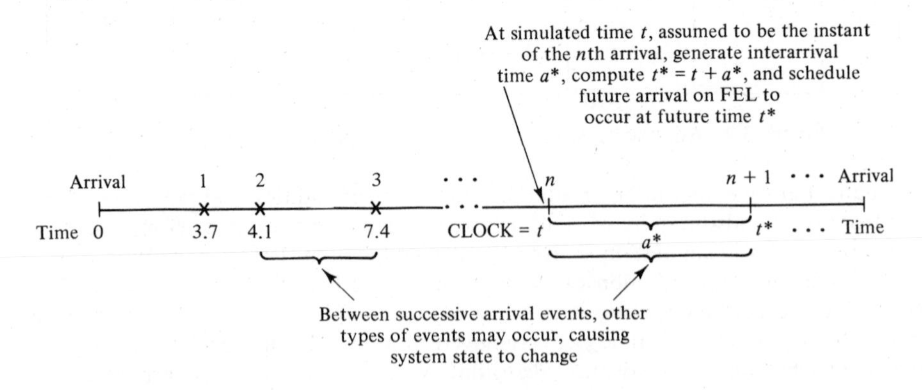

Figure 3.3. Generation of an external arrival stream by bootstrapping.

a new service time, say s^*, will be generated for the next customer. The next service completion event will be scheduled to occur at future time $t^* = t + s^*$, by placing onto the FEL a new service completion event with event time t^*. In addition, a service completion event will be generated and scheduled at the time of an arrival event provided that, upon arrival, there is at least one idle server in the server group. A service time is an example of an activity. Beginning service is a conditional event, because its occurrence is triggered only on the condition that a customer is present and a server is free. Service completion is an example of a primary event. Note that a conditional event, such as beginning service, is triggered by a primary event occurring and certain conditions prevailing in the system.

A third important example is the alternate generation of runtimes and downtimes for a machine subject to breakdowns. At time 0, the first runtime will be generated and an end-of-runtime event scheduled. Whenever an end-of-runtime event occurs, a downtime will be generated and an end-of-downtime event scheduled on the FEL. When the CLOCK is eventually advanced to the time of this end-of-downtime event, a runtime is generated and an end-of-runtime event scheduled on the FEL. In this way, runtimes and downtimes continually alternate throughout the simulation. A runtime and a downtime are examples of activities, and end of runtime and end of downtime are primary events.

Every simulation must have a stopping event, here called E, which defines how long the simulation will run. There are generally two ways to stop a simulation:

1. At time 0, schedule a stop simulation event at a future time T_E. Thus, before simulating it is known that the simulation will run over the time interval $[0, T_E]$. Example: Simulate a job shop for $T_E = 40$ hours.
2. Run length T_E is determined by the simulation itself. Generally, T_E is the time of occurrence of some specified event E. Examples: T_E is the time of the 100th service completion at a certain service center. T_E is the time of breakdown of a complex system. T_E is the time of disengagement or total kill (whichever occurs first) in a combat simulation.

In case 2, T_E is not known ahead of time. Indeed, it may be one of the statistics of primary interest to be produced by the simulation.

A systematic approach to simulation which concentrates on events and their effects on system state is called an *event-scheduling* approach to discrete-event simulation. This approach will be illustrated by the manual simulations of Section 3.1.1 and the FORTRAN and SIMSCRIPT simulations of Section 3.2. A different outlook is provided by the *process-interaction* approach. A process is a time-ordered collection of events, activities, and delays which are somehow related, say to some entity. An example of a "customer process" is shown in Figure 3.4. Many processes are usually simultaneously active in a model, and the interaction among processes may be quite complex. Figure 3.4 also illustrates the interaction between two successive customer processes in a first come, first served single-server queue. Languages based on the process-interaction approach include GPSS and SLAM; also, SIMSCRIPT II.5 (Release 8) optionally allows the use of the process-interaction approach. Illustrations of these process-based languages are given in Section 3.2.

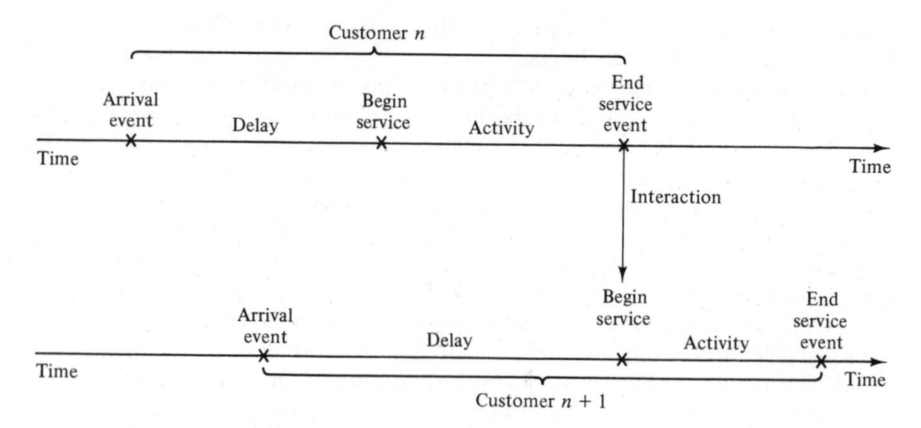

Figure 3.4. Two interacting customer processes in a single-server queue.

3.1.1. Manual Simulation Using Event Scheduling

In conducting an event-scheduling simulation, a simulation table is used to record the successive system snapshots as time advances.

Example 3.2 (Single-Channel Queue)

Reconsider the grocery store with one checkout counter which was simulated in Example 2.1 by an ad hoc method. The system consists of those customers in the waiting line plus the one (if any) checking out. The model has the following components:

System state	(LQ(t), LS(t)), where LQ(t) is the number of customers in the waiting line, and LS(t) is the number being served (0 or 1), at time t
Entities	The server and customers are not explicitly modeled, except in terms of the state variables above
Events	Arrival (A)
	Departure (D)
	Stopping event (E), scheduled to occur at time 60
Activities	Interarrival time, defined in Table 2.6
	Service time, defined in Table 2.7
Delay	Customer time spent in waiting line

The events on the FEL are written as (event type, event time). In this model, the FEL will always contain either two or three events. The effect of the arrival and departure events was first shown in Figures 2.2 and 2.3, and is shown in more detail in Figures 3.5 and 3.6.

The simulation table for the checkout counter is given in Table 3.1. The reader should cover all system snapshots except one, starting with the first, and attempt to construct the next snapshot from the previous one and the event logic in Figures 3.5 and 3.6. The interarrival times and service times will be identical to those used in Table 2.10, namely:

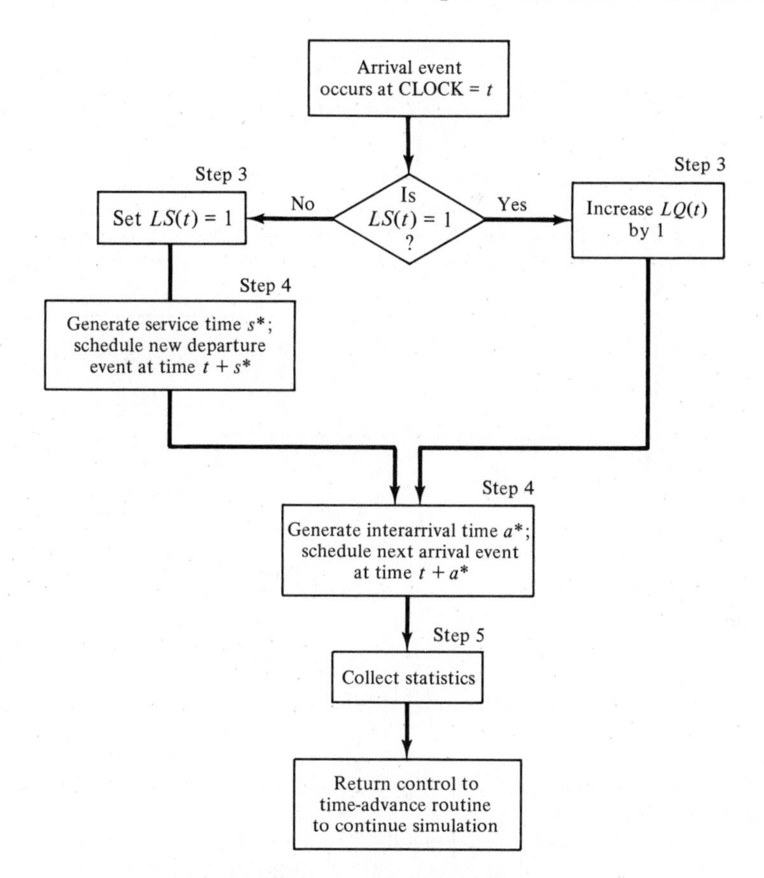

Figure 3.5. Execution of the arrival event.

Interarrival Times	8	6	1	8	3	8	...
Service Times	4	1	4	3	2	4	...

Initial conditions are that the first customer arrives at time 0 and begins service. This is reflected in Table 3.1 by the system snapshot at time zero (CLOCK = 0), with LQ(0) = 0, LS(0) = 1, and both a departure event and arrival event on the FEL. Also, the simulation is scheduled to stop at time 60. Only two statistics, server utilization and maximum queue length, will be collected. Server utilization is defined by total server busy time (B) divided by total time (T_E). Total busy time, B, and maximum queue length, MQ, will be cumulated as the simulation progresses. A column headed "comments" is included to aid the reader. (a^* and s^* are the generated interarrival and service times, respectively.)

As soon as the system snapshot at time CLOCK = 0 is complete, the simulation begins. At time 0, the imminent event is (D, 4). The CLOCK is advanced to time 4, and (D, 4) is removed from the FEL. Since LS(t) = 1 for $0 \leq t \leq 4$ (i.e., the server was busy for 4 minutes), the cumulative busy time is increased from B = 0 to B = 4. By the event logic in Figure 3.6, set LS(4) = 0 (the server becomes idle). The FEL is

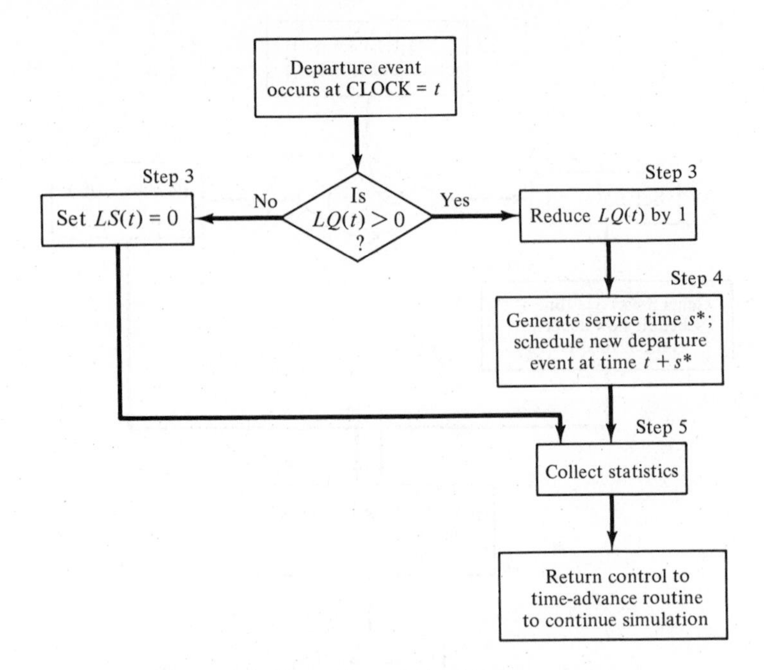

Figure 3.6. Execution of the departure event.

left with only two future events, (A, 8) and (E, 60). The simulation CLOCK is next advanced to time 8 and an arrival event executed. The interpretation of the remainder of Table 3.1 is left to the reader.

The simulation in Table 3.1 covers the time interval [0, 21]. At simulated time 21, the system is empty, but the next arrival will occur at future time 23. The server was busy for 12 of the 21 time units simulated, and the maximum queue length was one. This simulation is, of course, too short to draw any reliable conclusions. Exercise 1 asks the reader to continue the simulation and to compare the results to those in Example 2.1. Note that the simulation table gives the system state at all times, not just the listed times. For example, from time 15 to time 18, there is one customer in service and one in the waiting line.

When an event scheduling algorithm is computerized, only one snapshot (the current one or partially updated one) is kept in computer memory. With the idea of implementing event scheduling in FORTRAN or some other general-purpose language, the following rule should be followed. A new snapshot can be derived only from the previous snapshot, newly generated random variables, and the event logic (Figures 3.5 and 3.6). Past snapshots should be ignored when advancing the clock. The current snapshot should contain all information necessary to continue the simulation.

Example 3.3 (The Checkout Counter Simulation, Continued)

Suppose that in the simulation of the checkout counter in Example 3.2 the simulator desires to estimate mean response time and mean proportion of customers who spend 4 or more minutes in the system. A response time is the length of time a customer

Table 3.1. SIMULATION TABLE FOR CHECKOUT COUNTER (EXAMPLE 3.2)

	System State				*Cumulative Statistics*	
Clock	LQ(t)	LS(t)	*Future Event List*	*Comment*	B	MQ
0	0	1	(D, 4) (A, 8) (E, 60)	First A occurs (a^* = 8) Schedule next A (s^* = 4) Schedule first D	0	0
4	0	0	(A, 8) (E, 60)	First D occurs: (D, 4)	4	0
8	0	1	(D, 9) (A, 14) (E, 60)	Second A occurs: (A, 8) (a^* = 6) Schedule next A (s^* = 1) Schedule next D	4	0
9	0	0	(A, 14) (E, 60)	Second D occurs: (D, 9)	5	0
14	0	1	(A, 15) (D, 18) (E, 60)	Third A occurs: (A, 14) (s^* = 4) Schedule next D	5	0
15	1	1	(D, 18) (A, 23) (E, 60)	Fourth A occurs: (A, 15) (Customer delayed)	6	1
18	0	1	(D, 21) (A, 23) (E, 60)	Third D occurs: (D, 18) (s^* = 3) Schedule next D	9	1
21	0	0	(A, 23) (E, 60)	Fourth D occurs: (D, 21)	12	1

spends in the system. In order to estimate these customer averages, it is necessary to expand the model in Example 3.2 to explicitly represent the individual customers. In addition, to be able to compute an individual customer's response time when that customer departs, it will be necessary to know that customer's arrival time. Therefore, a customer entity with arrival time as an attribute will be added to the list of model components in Example 3.2. These customer entities will be stored in a set to be called "CHECKOUT LINE"; they will be called C1, C2, C3, ... Finally, the notation for events on the FEL will be expanded to indicate which customer is affected. For example, (D, 4, C1) means that customer C1 will depart at time 4. The additional model components are listed below:

Entities	(Ci, *t*), representing customer Ci who arrived at time *t*
Events	(A, *t*, Ci), the arrival of customer Ci at time *t*
	(D, *t*, Cj), the departure of customer Cj at time *t*
Set	"CHECKOUT LINE," the set of all customers currently at the checkout counter (being served or waiting to be served), ordered by time of arrival

Three new cumulative statistics will be collected: S, the sum of customer response times for all customers who have departed by the current time; F, the total number of customers who spend 4 or more minutes at the checkout counter; and N_D, the total number of departures up to the current simulation time. These three cumulative statistics will be updated whenever the departure event occurs; the logic for collecting these statistics would be incorporated into step 5 of the departure event in Figure 3.6.

The simulation table for Example 3.3 is shown in Table 3.2. The same data for interarrival and service times will be used again, so that Table 3.2 essentially repeats Table 3.1, except that the new components are included (and the comment column has been deleted). These new components are needed for the computation of S, F, and N_D. For example, at time 4 a departure event occurs for customer C1. The customer entity C1 is removed from the set "CHECKOUT LINE"; the attribute "time of arrival" is noted to be 0, so the response time for this customer was 4 minutes. Hence, S is incremented by 4 minutes, and F and N_D are incremented by one customer. Similarly, at time 21 when the departure event (D, 21, C4) is being executed, the response time for customer C4 is computed by

$$\begin{aligned}\text{Response time} &= \text{CLOCK TIME} - \text{attribute "time of arrival"} \\ &= 21 - 15 \\ &= 6 \text{ minutes}\end{aligned}$$

Then S is incremented by 6 minutes, and F and N_D by one customer.

For a simulation run length of 21 minutes, the average response time was $S/N_D = 15/4 = 3.75$ minutes, and the observed proportion of customers who spent 4 or more minutes in the system was $F/N_D = 0.75$. Again, this simulation was far too short to regard these estimates with any degree of accuracy. The purpose of Example 3.3, however, was to illustrate the notion that in many simulation models the information desired from the simulation (such as the statistics S/N_D and F/N_D) determine to some extent the structure of the model.

Table 3.2. SIMULATION TABLE FOR EXAMPLE 3.3

	System State		*Set*	*Future Event*	*Cumulative Statistics*		
Clock	LQ(t)	LS(t)	"*CHECKOUT LINE*"	*List*	S	N_D	F
0	0	1	(C1, 0)	(D, 4, C1) (A, 8, C2) (E, 60)	0	0	0
4	0	0		(A, 8, C2) (E, 60)	4	1	1
8	0	1	(C2, 8)	(D, 9, C2) (A, 14, C3) (E, 60)	4	1	1
9	0	0		(A, 14, C3) (E, 60)	5	2	1
14	0	1	(C3, 14)	(A, 15, C4) (D, 18, C3) (E, 60)	5	2	1
15	1	1	(C3, 14) (C4, 15)	(D, 18, C3) (A, 23, C5) (E, 60)	5	2	1
18	0	1	(C4, 15)	(D, 21, C4) (A, 23, C5) (E, 60)	9	3	2
21	0	0		(A, 23, C5) (E, 60)	15	4	3

EXAMPLE 3.4 (THE DUMP TRUCK PROBLEM)

Six dump trucks are used to haul coal from the entrance of a small mine to the railroad. Figure 3.7 provides a schematic of the dump truck operation. Each truck is loaded by one of two loaders. After loading, a truck immediately moves to the scale to be weighed as soon as possible. Both the loaders and the scale have a first come,

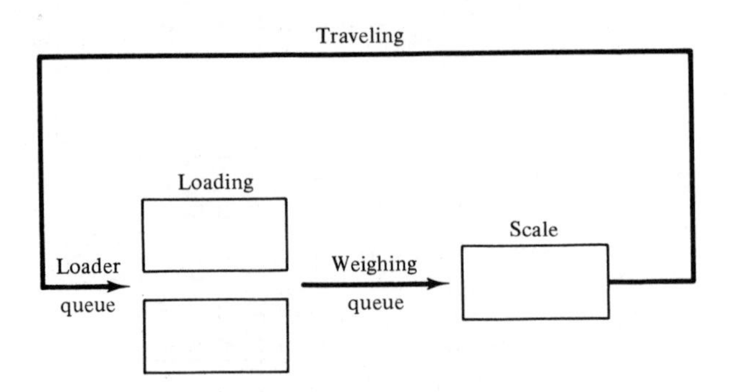

Figure 3.7. Dump truck problem.

first served waiting line (or queue) for trucks. Travel time from a loader to the scale is considered negligible. After being weighed, a truck begins a travel time (during which time the truck unloads), and then afterward returns to the loader queue. The distributions of loading time, weighing time, and travel time are given in Tables 3.3, 3.4, and 3.5, respectively, together with the random digit assignment for generating these variables using random digits from Table A.1. The purpose of the simulation is to estimate the loader and scale utilizations (percentage of time busy).

Table 3.3. DISTRIBUTION OF LOADING TIME FOR THE DUMP TRUCKS

Loading Time	*Probability*	*Cumulative Probability*	*Random Digit Assignment*
5	0.30	0.30	1–3
10	0.50	0.80	4–8
15	0.20	1.00	9–0

Table 3.4. DISTRIBUTION OF WEIGHING TIME FOR THE DUMP TRUCKS

Weighing Time	*Probability*	*Cumulative Probability*	*Random Digit Assignment*
12	0.70	0.70	1–7
16	0.30	1.00	8–0

Table 3.5. DISTRIBUTION OF TRAVEL TIME FOR THE DUMP TRUCKS

Travel Time	*Probability*	*Cumulative Probability*	*Random Digit Assignment*
40	0.40	0.40	1–4
60	0.30	0.70	5–7
80	0.20	0.90	8–9
100	0.10	1.00	0

The model has the following components:

System state [LQ(t), L(t), WQ(t), W(t)], where
LQ(t) = number of trucks in loader queue
L(t) = number of trucks (0, 1, or 2) being loaded
WQ(t) = number of trucks in weigh queue
W(t) = number of trucks (0 or 1) being weighed, all at simulation time t

Events (ALQ, t, DTi), dump truck i arrives at loader queue (ALQ) at time t
(EL, t, DTi), dump truck i ends loading (EL) at time t
(EW, t, DTi), dump truck i ends weighing (EW) at time t

Entities The six dump trucks (DT1, . . . , DT6)

Sets Loader queue, all trucks waiting to begin loading, ordered on a first come, first served basis
Weigh queue, all trucks waiting to be weighed, ordered on a first come, first served basis

Activities Loading time, weighing time, and travel time

Delays Delay at loader queue, and delay at scale

The simulation table is given in Table 3.6. It has been assumed that five of the trucks are at the loaders and one is at the scale at time 0. The activity times are taken from the following list as needed:

Loading Time	10	5	5	10	15	10	10
Weighing Time	12	12	12	16	12	16	
Travel Time	60	100	40	40	80		

When an end-loading (EL) event occurs, say for truck j at time t, other events may be triggered. If the scale is idle [W(t^-) = 0], truck j begins weighing and an end-weighing event (EW) is scheduled on the FEL; otherwise, truck j joins the weigh queue. If at this time there is another truck waiting for a loader, it will be removed from the loader queue and will begin loading by the scheduling of an end-loading event (EL) on the FEL. This logic for the occurrence of the end-loading event, as well as the appropriate logic for the other two events, should be incorporated into an event diagram as in

Table 3.6. SIMULATION TABLE FOR DUMP TRUCK OPERATION (EXAMPLE 3.4)

Clock	System State				Sets		Future Event	Cumulative Statistics	
t	$LQ(t)$	$L(t)$	$WQ(t)$	$W(t)$	Loader Queue	Weigh Queue	List	B_L	B_S
0	3	2	0	1	DT4 DT5 DT6		(EL, 5, DT3) (EL, 10, DT2) (EW, 12, DT1)	0	0
5	2	2	1	1	DT5 DT6	DT3	(EL, 10, DT2) (EL, 5 + 5, DT4) (EW, 12, DT1)	10	5
10	1	2	2	1	DT6	DT3 DT2	(EL, 10, DT4) (EW, 12, DT1) (EL, 10 + 10, DT5)	20	10
10	0	2	3	1		DT3 DT2 DT4	(EW, 12, DT1) (EL, 20, DT5) (EL, 10 + 15, DT6)	20	10
12	0	2	2	1		DT2 DT4	(EL, 20, DT5) (EW, 12 + 12, DT3) (EL, 25, DT6) (ALQ, 12 + 60, DT1)	24	12
20	0	1	3	1		DT2 DT4 DT5	(EW, 24, DT3) (EL, 25, DT6) (ALQ, 72, DT1)	40	20
24	0	1	2	1		DT4 DT5	(EL, 25, DT6) (EW, 24 + 12, DT2) (ALQ, 72, DT1) (ALQ, 24 + 100, DT3)	44	24

25	0	0	3	1	DT4	(EW, 36, DT2)	45	25
					DT5	(ALQ, 72, DT1)		
					DT6	(ALQ, 124, DT3)		
36	0	0	2	1	DT5	(EW, 36 + 16, DT4)	45	36
					DT6	(ALQ, 72, DT1)		
						(ALQ, 36 + 40, DT2)		
						(ALQ, 124, DT3)		
52	0	0	1	1	DT6	(EW, 52 + 12, DT5)	45	52
						(ALQ, 72, DT1)		
						(ALQ, 76, DT2)		
						(ALQ, 52 + 40, DT4)		
						(ALQ, 124, DT3)		
64	0	0	0	1		(ALQ, 72, DT1)	45	64
						(ALQ, 76, DT2)		
						(EW, 64 + 16, DT6)		
						(ALQ, 92, DT4)		
						(ALQ, 124, DT3)		
						(ALQ, 64 + 80, DT5)		
72	0	1	0	1		(ALQ, 76, DT2)	45	72
						(EW, 80, DT6)		
						(EL, 72 + 10, DT1)		
						(ALQ, 92, DT4)		
						(ALQ, 124, DT3)		
						(ALQ, 144, DT5)		
76	0	2	0	1		(EW, 80, DT6)	49	76
						(EL, 82, DT1)		
						(EL, 76 + 10, DT2)		
						(ALQ, 92, DT4)		
						(ALQ, 124, DT3)		
						(ALQ, 144, DT5)		

Figures 3.5 and 3.6 of Example 3.2. The construction of these event logic diagrams is left as an exercise for the reader (Exercise 2).

As an aid to the reader, in Table 3.6 whenever a new event is scheduled, its event time is written as "t + (activity time)." For example, at time 0 the imminent event is an EL event with event time 5. The clock is advanced to time $t = 5$, dump truck 3 joins the weigh queue (since the scale is occupied), and truck 4 begins to load. Thus, an EL event is scheduled for truck 4 at future time 10, computed by (present time) + (loading time) $= 5 + 5 = 10$.

In order to estimate the loader and scale utilizations, two cumulative statistics are maintained:

$$B_L = \text{total busy time of both loaders from time 0 to time } t$$
$$B_S = \text{total busy time of the scale from time 0 to time } t$$

Since both loaders are busy from time 0 to time 20, $B_L = 40$ at time $t = 20$. But from time 20 to time 24, only one loader is busy; thus, B_L increases by only 4 minutes over the time interval [20, 24]. Similarly, from time 25 to time 36, both loaders are idle ($L(25) = 0$), so B_L does not change. For the relatively short simulation in Table 3.6, the utilizations are estimated as follows:

$$\text{Average loader utilization} = \frac{49/2}{76} = 0.32$$

$$\text{Average scale utilization} = \frac{76}{76} = 1.00$$

These estimates cannot be regarded as accurate estimates of the long-run "steady-state" utilizations of the loader and scale; a considerably longer simulation would be needed to reduce the effect of the assumed conditions at time 0 (five of the six trucks at the loaders) and to realize accurate estimates. On the other hand, if the simulator were interested in the so-called transient behavior of the system over a short period of time (say 1 or 2 hours), given the specified initial conditions, then the results in Table 3.6 can be considered representative (or constituting one sample) of that transient behavior. Additional samples can be obtained by conducting additional simulations, each one having the same initial conditions but using a different stream of random digits to generate the activity times.

Table 3.6, the simulation table for the dump truck operation, could have been simplified somewhat by not explicitly modeling the dump trucks as entities. That is, the events could be written as (EL, t), and so on, and the state variables used to keep track merely of the number of trucks in each part of the system, not which trucks were involved. With this representation, the same utilization statistics could be collected. On the other hand, if mean "system" response time, or proportion of trucks spending more than 30 minutes in the "system," were being estimated, where "system" refers to the loader queue and loaders and the weigh queue and scale, then dump truck entities (DTi), together with an attribute equal to arrival time at the loader queue, would be needed. Whenever a truck left the scale, that truck's response time could be computed as current simulation time (t) minus the arrival-time attribute. This new response time would be used to update the cumulative statistics: S = total response time of all trucks which have been through the "system" and F = number of truck response times which have been greater than 30 minutes. This example again illustrates the notion that to some extent the complexity of the model depends on the performance measures being estimated.

3.2. Programming Languages for Discrete-Event Systems Simulation

Computer simulation languages greatly facilitate the development and execution of simulations of complex real-world systems. Each language generally possesses an orientation to real-world situations, or "world view," which may be classified as event-oriented or process-oriented, as discussed in Section 3.1. When using one of these languages, the resulting model will be event or process oriented, or will possibly have a combination of the two orientations. Four languages—FORTRAN, SIMSCRIPT, GPSS, and SLAM—will be discussed in some detail in the following subsections; GASP will be briefly described.

FORTRAN is a scientific programming language and was not specifically designed for use in simulation. When using FORTRAN, the analyst will probably adopt the event-scheduling orientation. At the other extreme, GPSS is a highly structured special-purpose simulation language which is transaction oriented, a special case of the more general process orientation. GPSS was designed for relatively easy simulation of queueing systems, such as job shops. FORTRAN and GPSS are widely used for discrete-event simulation modeling.

SIMSCRIPT and SLAM are high-level simulation programming languages which have constructs specifically designed to facilitate model building. Both SIMSCRIPT II.5 and SLAM provide the analyst with a choice of orientations, or a model may be built using a mixture of the two orientations. Unlike FORTRAN, these two languages provide management of the future event list and other sets, built-in random variate generators, and built-in statistics gathering routines. Unlike GPSS, complex computations are easily accomplished in either language. SLAM and an extended version of SIMSCRIPT (called C-SIMSCRIPT) provide the additional capability to conduct continuous simulations, that is, to model systems having continuously changing state variables. SLAM is FORTRAN-based and contains a GASP-like language as a subset. GASP, described in Section 3.2.2, is a set of FORTRAN subroutines for facilitating event-oriented simulations written in FORTRAN. In Section 3.2.5, the process-oriented component of SLAM is briefly described. SIMSCRIPT, on the other hand, contains as a subset a complete scientific programming language comparable to FORTRAN, PL/1, or ALGOL. Section 3.2.3 gives a complete example of the use of SIMSCRIPT from the event scheduling point of view. All four languages will be used to simulate a modified version of Example 2.1.

EXAMPLE 3.5 (THE CHECKOUT COUNTER: A TYPICAL SINGLE-SERVER QUEUE)

The system, a grocery checkout counter, is modeled as a single server queue. The simulation will run until 1000 customers have been served. In addition, assume that the interarrival times of customers are exponentially distributed with a mean of 4.5 minutes, and that the service times are (approximately) normally distributed with a mean of 3.2 minutes and a standard deviation of 0.6 minute. (The approximation is that service times are always positive.) When the cashier is busy, a queue forms with no customers turned away. This example was manually simulated in Examples 3.2 and 3.3 using the event scheduling point of view. The model contains two events, the arrival and departure events. Figures 3.5 and 3.6 provide the event logic. The next four subsections illustrate the simulation of this single-server queue in FORTRAN, SIMSCRIPT, GPSS, and SLAM. Although this example is much simpler than models that arise in the study of complex systems, its simulation contains the essential components of the simulation of more complex systems.

3.2.1. Simulation in FORTRAN

FORTRAN is a widely known and widely available programming language, and has been used extensively in simulation. It does not, however, provide any facilities directly aimed at aiding the simulator. The simulator is forced to program all details of the event-scheduling/time-advance algorithm, the statistics gathering capability, the generation of samples from specified probability distributions, and the report generator. (However, several scientific subroutine libraries, such as IMSL, contain numerous random variate generators.) For large models, the use of FORTRAN can become quite cumbersome; additionally, it can result in models which are difficult to debug and run slowly unless the programmer takes a carefully organized approach and is knowledgeable in efficient list-processing techniques. For small models, simulation with FORTRAN (or any other general-purpose language) can be used as a learning tool to reinforce the concepts of the event-scheduling/time-advance algorithm. For the most part, the special-purpose simulation languages hide the details of event scheduling.

Any discrete-event simulation model written in FORTRAN contains the components discussed in Section 3.1: system state, entities and attributes, sets, events, activities and delays, plus the components listed below. To facilitate development and debugging, it is best to organize the FORTRAN model in a modular fashion using subroutines. The following components are common to almost all models written in FORTRAN:

CLOCK	A variable defining simulated time
Initialization subroutine	A routine used to define the system state at time 0
Time-advance subroutine	A routine that searches the future event list (FEL) for the next event (called the imminent event and denoted by IMEVT) and advances the clock to the time of occurrence of the imminent event
Scheduling subroutine	A routine that places generated future events on the FEL (not used in Example 3.6)
Event subroutines	For each event type, a subroutine to update system state (and cumulative statistics) when that event occurs

Random variate generators	Routines to generate samples of desired probability distributions
Main program	Provides overall control of the event-scheduling algorithm
Report generator	A routine that computes summary statistics from cumulative statistics and prints a report at the end of the simulation

The overall structure of a FORTRAN simulation program is shown in Figure 3.8. This flowchart is an expansion of the event-scheduling/time-advance algorithm outlined in Figure 3.2. (The steps mentioned in Figure 3.8 refer to the five steps in Figure 3.2.)

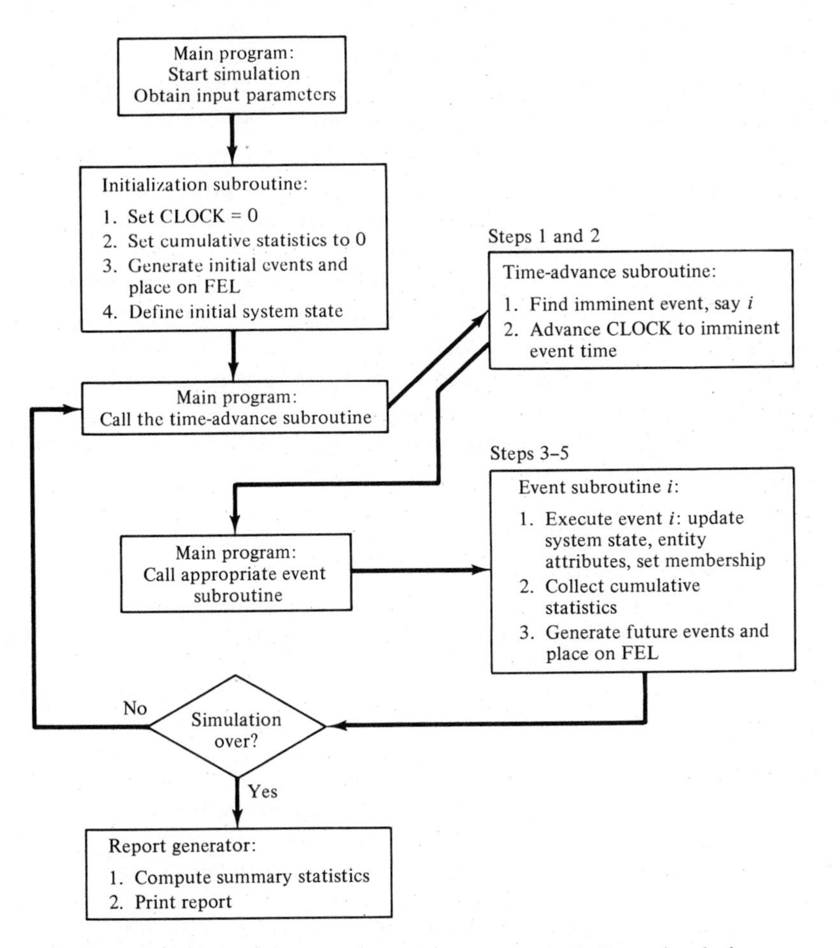

Figure 3.8. Overall structure of an event scheduling simulation program.

The simulation begins by setting the simulation CLOCK to zero, initializing cumulative statistics to zero, generating any initial events (there will always be at least one) and placing them on the FEL, and defining the system state at time 0. The simulation program then cycles between the time-advance subroutine and the appropriate event subroutines until the simulation is over. The time-advance subroutine searches the FEL to find the imminent event, which, say, is an event of type *i*. The simulation CLOCK is then advanced to the time of the imminent event. (Recall that during this duration of time between the occurrence of two successive events, the system state and entity attributes do not change in value. Indeed, this is the definition of discrete-event simulation: The system state changes only when an event occurs.) Next, event subroutine *i* is called to execute the imminent event, update cumulative statistics, and generate future events (to be placed on the FEL). Executing the imminent event means that the system state, entity attributes, and set membership are changed to reflect the fact that event *i* has occurred. Note that all actions in an event subroutine take place at one instant of simulated time. The value of the variable CLOCK does not change in an event routine. If the simulation is not over, control passes again to the time-advance subroutine, then the appropriate event subroutine, and so on. When the simulation is over, control passes to the report generator, which computes the desired summary statistics from the collected cumulative statistics and prints a report.

The efficiency of a simulation model in terms of computer runtime is determined to a large extent by the techniques used to manipulate the FEL and other sets. As discussed earlier in Section 3.1, removal of the imminent event and addition of a new event are the two main operations performed on the FEL. In the following example, which has only two event types, the FEL is handled by a method which is relatively simple but which would be extremely inefficient in a model with many events.

EXAMPLE 3.6 (SINGLE-SERVER QUEUE SIMULATION IN FORTRAN)

A grocery checkout counter, defined in detail in Example 3.5, is now simulated using FORTRAN. A version of this example was simulated manually in Examples 3.2 and 3.3, where the system state, entities and attributes, sets, events, activities, and delays were analyzed and defined.

The two events, the arrival event and the departure event, are called events of type 1 and type 2, respectively. The type of the imminent event is designated by the variable IMEVT. The subroutines for this model and the flow of control are shown in Figure 3.9, which is an adaptation of Figure 3.8 for this particular problem. Table 3.7 lists the FORTRAN variables used for system state, entity attributes and sets, activity durations, and cumulative and summary statistics; the FORTRAN functions used to generate samples from the exponential and normal distributions; and all the other subroutines needed.

The main program, shown in Figure 3.10, controls the overall flow of the event-scheduling/time-advance algorithm. Global variables, whose values are known to all subroutines, are listed in two COMMON blocks called SIM and TIMEKP. The main

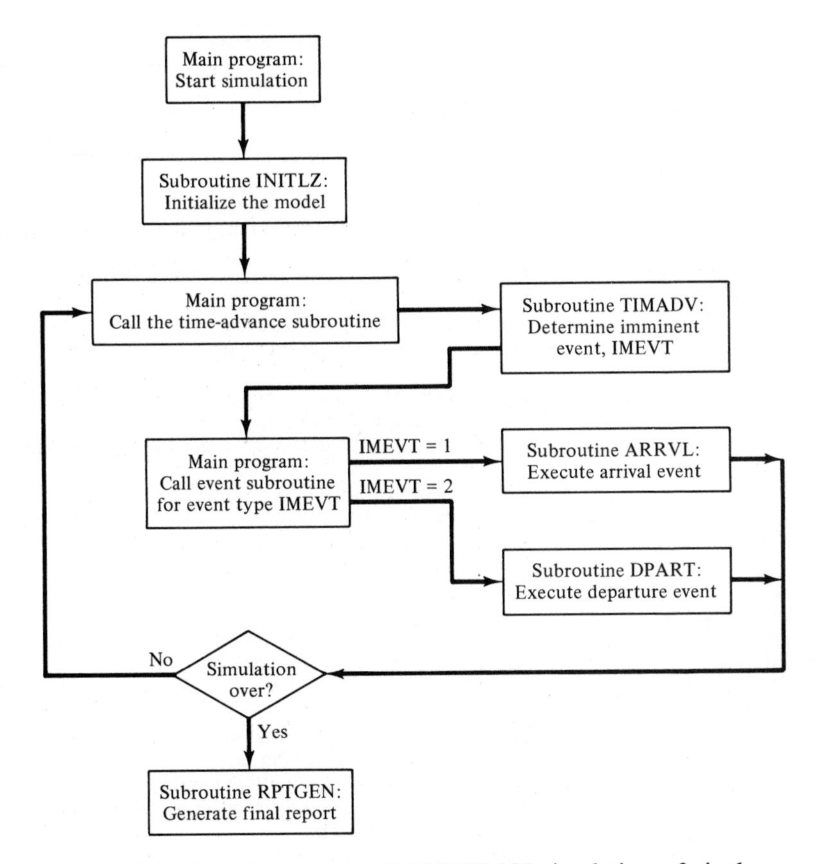

Figure 3.9. Overall structure of FORTRAN simulation of single-server queue.

program follows the logic of Figure 3.9. First, the input parameters—MIAT, MSVT, SIGMA, and NCUST—and the number of events, NUMEVS, are specified. Then subroutine INITLZ is called to initialize the model. Next subroutine TIMADV is called to determine the imminent event and to advance the clock to the time of occurrence of that event. Then the appropriate event subroutine is called. (If IMEVT = 1, the GO TO passes control to CALL ARRVL; if IMEVT = 2, control is passed to CALL DPART.) After the appropriate event is executed, a check is made to see if the simulation is over. If it is not over, the program cycles from time advance to event execution repeatedly until the stopping criterion is satisfied. When the simulation is finally over, a call is made to subroutine RPTGEN to generate the final report.

A listing for subroutine INITLZ is given in Figure 3.11. The simulation clock, system state, and other variables are initialized. Note that the time of the first arrival is generated randomly by FUNCTION EXPON and stored in FEL(1). Thus, it is assumed that the system is empty at simulated time CLOCK = 0. Since the system is empty (at CLOCK = 0), no departure can be scheduled. Thus, the time of occurrence of the next departure, FEL(2), is set equal to "infinity" (i.e., to a very large value, here

Table 3.7. DEFINITIONS OF VARIABLES, FUNCTIONS, AND SUBROUTINES IN THE FORTRAN MODEL OF THE SINGLE-SERVER QUEUE

Variables	*Description*
System state	
LQT	Number of customers in the waiting line at current simulated time
LST	Number being served (0 or 1) at current simulated time
Entity attributes and sets	
CHKOUT(I)	Arrival time of (I − 1)st customer in the checkout line (so CHKOUT(2) is arrival time of first customer in line not being served)
CHKOUT(1)	Arrival time of customer currently being served
Future event list	
FEL(I)	Time of occurrence of next event of type I, I = 1, 2
IMEVT	Type of the imminent event (1 or 2)
Activity durations	
IAT	The interarrival time between the previous customer's arrival and the next arrival
SVT	The service time of the last customer to begin service
Input parameters	
MIAT	Mean interarrival time (4.5 minutes)
MSVT	Mean service time (3.2 minutes)
SIGMA	Standard deviation of service time (0.6 minute)
NCUST	The stopping criteria—number of customers to be served (1000)
Simulation variables	
CLOCK	The current value of simulated time
NUMEVS	Number of event types (NUMEVS = 2)
Variables used to gather statistics	
TLE	Time of occurrence of the last event (used to update B)
B	Total busy time of server (so far)
MQ	Maximum length of waiting line (so far)
S	Sum of customer response times for all customers who have departed (so far)
ND	Number of departures (so far)
F	Number of customers who spent 4 or more minutes at the checkout counter (so far)
Summary statistics	
RHO = B/CLOCK	Proportion of time server is busy (here the value of CLOCK is the final value of simulated time)
MQ	Maximum length of the waiting line
AVGR = S/ND	Average response time
PC4 = F/ND	Proportion of customers who spent 4 or more minutes at the checkout counter

Functions	*Description*
EXPON(FMEAN)	Function to generate samples from an exponential distribution with mean FMEAN
NORML(XMU,SIGMA)	Function to generate samples from a normal distribution with mean XMU and standard deviation SIGMA

Subroutines	*Description*
INITLZ	Initialization subroutine
TIMADV	Time-advance routine
ARRVL	Event subroutine that executes the arrival event
DPART	Event subroutine that executes the departure event
RPTGEN	Report generator

```
      PROGRAM FTNSIM(OUTPUT,TAPE6=OUTPUT)
C
C
C   MAIN PROGRAM
C
C     1-INITIALIZES MODEL
C     2-CALLS TIME-ADVANCE AND EVENT ROUTINES
C     3-CALLS REPORT GENERATOR TO TERMINATE SIMULATION NORMALLY
C
      REAL MIAT,MSVT
      COMMON /SIM/ MIAT,MSVT,SIGMA,NCUST,LQT,LST,TLE,
     1            CHKOUT(100),B,MQ,S,F,ND
      COMMON /TIMEKP/ CLOCK,IMEVT,NUMEVS,FEL(2)
        NUMEVS=2
C
C ASSIGN VALUES TO INPUT PARAMETERS
C (THESE VALUES COULD BE STORED IN A FILE
C  AND READ INTO THE PROGRAM, FOR CONVENIENCE.)
C
      MIAT = 4.5
      MSVT = 3.2
      SIGMA = .6
      NCUST = 1000
C
C CALL INITIALIZATION ROUTINE
C
      CALL INITLZ
C
C CALL TIME-ADVANCE ROUTINE TO DETERMINE IMMINENT
C EVENT AND ADVANCE CLOCK TO THE IMMINENT
C EVENT TIME.
C
  30    CALL TIMADV
C
C THE VARIABLE "IMEVT" INDICATES THE IMMINENT
C EVENT.    IMEVT = 1 FOR AN ARRIVAL,
C           IMEVT = 2 FOR A DEPARTURE.
C
      GO TO (40,50),IMEVT
C
C CALL THE APPROPRIATE EVENT SUBROUTINE
C
  40    CALL ARRVL
C
      GO TO 30
C
  50    CALL DPART
C
C
C CHECK TO SEE IF SIMULATION IS OVER.
C IF NOT, RETURN TO TIME-ADVANCE ROUTINE
C
        IF(ND .LT. NCUST)GO TO 30
C
C WHEN SIMULATION IS OVER, CALL REPORT GENERATOR
C
        CALL RPTGEN
      STOP
      END
```

Figure 3.10. FORTRAN main program for single-server queue simulation.

1.0E+30 = 10^{30}). Therefore, the first event to occur will be an arrival event. On the other hand, if it were assumed that one customer was present at simulated time CLOCK = 0, and that this customer was just beginning a service time, the following changes would have to be made:

```
C
C
C INITIALIZATION ROUTINE
C
      SUBROUTINE INITLZ
      REAL MIAT,MSVT
      COMMON /SIM/ MIAT,MSVT,SIGMA,NCUST,LQT,LST,TLE,
     1             CHKOUT(100),B,MQ,S,F,ND
      COMMON /TIMEKP/ CLOCK,IMEVT,NUMEVS,FEL(2)
C
C INITIALIZE SIMULATION
C   1-SET SIMULATION CLOCK TO ZERO
C   2-ASSUME SYSTEM IS EMPTY AND IDLE
C     AT TIME ZERO
C   3-INITIALIZE CUMULATIVE STATISTICS TO 0.
C
        CLOCK=0.0
        IMEVT = 0
        LQT = 0
        LST = 0
        TLE = 0
        B = 0
        MQ = 0
        S = 0
        F = 0
        ND = 0
C
C GENERATE TIME OF FIRST ARRIVAL, IAT, AND
C SCHEDULE FIRST ARRIVAL IN FEL(1).
C SET FEL(2) TO "INFINITY" TO INDICATE THAT
C A DEPARTURE IS NOT POSSIBLE WHILE THE
C SYSTEM IS EMPTY.
C
        FEL(1)=CLOCK + EXPON(MIAT)
        FEL(2)=1.0E+30
      RETURN
      END
```

Figure 3.11. FORTRAN initialization routine for single-server queue simulation.

LST = 1
CHKOUT(1) = CLOCK
FEL(2) = CLOCK + NORML(MSVT,SIGMA)

Other initialization assumptions could be handled in a similar fashion.

A listing for the time-advance routine, SUBROUTINE TIMADV, is given in Figure 3.12. It uses a "brute-force" approach for the processing of the FEL: That is, the array FEL, namely FEL(1), FEL(2), . . . , FEL(NUMEVS), is searched for its minimum value, which occurs, say, at index IMEVT. Then event IMEVT is the imminent event and it will occur at time FEL(IMEVT). Simulation time, called CLOCK, is advanced to the imminent event time and control is returned to the main program. Note that if FEL(I) is equal to "infinity" for all event types I, a message is printed that the future event list is empty and the simulation cannot continue. (Why?) Either a mistake in programming logic has occurred, or the programmer canceled all future events on purpose. In any case, the report generator is called. (If a mistake has been made, the model output may help in locating it.)

Figure 3.13 gives a listing of event subroutine ARRVL, which executes the arrival event each time it occurs. The basic logic of the arrival event for a single-server queue was given earlier in Figure 3.5. First, a determination is made of the server's status (i.e., busy or idle), indicated by variable LST = 1 or 0, respectively. If the server was

```
C
C
C TIME ADVANCE ROUTINE - FINDS NEXT EVENT ON FUTURE EVENT
C                        LIST AND ADVANCES CLOCK
C
C
      SUBROUTINE TIMADV
      REAL MIAT,MSVT
      COMMON /SIM/ MIAT,MSVT,SIGMA,NCUST,LQT,LST,TLE,
     1             CHKOUT(100),B,MQ,S,F,ND
      COMMON /TIMEKP/ CLOCK,IMEVT,NUMEVS,FEL(2)
        FMIN=1.E+29
        IMEVT=0
C
C       SEARCH THE FUTURE EVENT LIST FOR THE NEXT EVENT
C
      DO 30 I=1,NUMEVS
        IF(FEL(I).GE.FMIN)GO TO 30
        FMIN=FEL(I)
        IMEVT=I
   30 CONTINUE
      IF(IMEVT.GT.0)GO TO 50
C
C       ERROR CONDITION - FUTURE EVENT LIST EMPTY
C
        WRITE(6,40)
   40   FORMAT(1X,"****FUTURE EVENT LIST EMPTY****",
     1        1X,"**SIMULATION CANNOT CONTINUE**")
        CALL RPTGEN
        STOP
C
C ADVANCE SIMULATION CLOCK
C
C NEXT EVENT IS TYPE "IMEVT", WHICH WILL
C OCCUR AT TIME FEL(IMEVT)
C
   50 CLOCK = FEL(IMEVT)
      RETURN
      END
```

Figure 3.12. FORTRAN time-advance routine for single-server queue simulation.

idle, the cumulative statistics B, MQ, S, ND, and F do not need updating. The server's status is changed to busy (LST = 1), and the arrival time is recorded in CHKOUT(1). Note that CHKOUT represents a set that stores the customer attribute "arrival time," which is used later in subroutine DPART to compute a customer's response time. Since a service is just beginning, a service time (SVT) is generated by FUNCTION NORML, and a departure is scheduled by placing the departure time into FEL(2). Note that FEL(2) is set equal to the current simulated time (CLOCK) plus the duration of the service time (SVT) just beginning. Control is passed to statement 100 (Figure 3.13), at which point an interarrival time (IAT) is generated using FUNCTION EXPON, and the next arrival is scheduled by computing the next arrival time (CLOCK + IAT) and storing it on the future event list at FEL(1). Control is then returned to the main program. On the other hand, if the server was busy (LST = 1) when the arrival occurred, control is passed to statement 20 (Figure 3.13). The number in queue (LQT) is increased by one, and the customer's arrival-time attribute is recorded at the "back" of the set CHKOUT. (The array CHKOUT is dimensioned to 100. Therefore, as long as the number of customers (LQT + LST) in the system is 100 or less, no problems will occur. Note that a check is made and control is passed to statement 200 if I = LQT + LST is greater than 100. An error message is printed and the report generator

```
C
C
C ARRIVAL EVENT ROUTINE
C
C
      SUBROUTINE ARRVL
      REAL MIAT,MSVT,IAT
      COMMON /SIM/ MIAT,MSVT,SIGMA,NCUST,LQT,LST,TLE,
     1             CHKOUT(100),B,MQ,S,F,ND
      COMMON /TIMEKP/ CLOCK,IMEVT,NUMEVS,FEL(2)
C
C
C DETERMINE IF SERVER IS BUSY
C
      IF(LST .EQ. 1) GO TO 20
C
C SERVER IS IDLE.  UPDATE SYSTEM STATE AND
C RECORD ARRIVAL TIME OF NEW CUSTOMER
C
        LST = 1
        CHKOUT(1) = CLOCK
C
C GENERATE A SERVICE TIME FOR THE NEW ARRIVAL
C AND SCHEDULE THE DEPARTURE FOR THIS ARRIVAL
C
        SVT = NORML(MSVT,SIGMA)
        FEL(2) = CLOCK + SVT
C
C UPDATE CUMULATIVE STATISTIC, MQ (AND ALSO TLE)
C
        TLE = CLOCK
        IF(LQT .GT. MQ) MQ = LQT
      GO TO 100
C
C
C SERVER IS BUSY.  UPDATE SYSTEM STATE AND
C RECORD ARRIVAL TIME OF NEW CUSTOMER
C
  20    LQT = LQT + 1
        I = LQT + LST
        IF(I .GT. 100) GO TO 200
        CHKOUT(I) = CLOCK
C
C UPDATE CUMULATIVE STATISTICS, B AND MQ.
C (S, ND, AND F ARE NOT UPDATED WHEN AN
C  ARRIVAL OCCURS.)
C
        B = B + (CLOCK - TLE)
        TLE = CLOCK
        IF(LQT .GT. MQ) MQ = LQT
C
C GENERATE AN INTERARRIVAL TIME AND
C SCHEDULE THE NEXT ARRIVAL EVENT
C
 100  IAT = EXPON(MIAT)
      FEL(1) = CLOCK + IAT
      RETURN
C
C
C ERROR CONDITION HAS OCCURRED.  ARRAY CHKOUT
C HAS OVERFLOWED.  INCREASE DIMENSION.
C
 200  WRITE(6,205)
 205  FORMAT(1X,"****OVERFLOW IN ARRAY CHKOUT. INCREASE ",
     1          "DIMENSION.****",
     2      /,1X,"****SIMULATION CANNOT CONTINUE****")
C
      CALL RPTGEN
      STOP
      END
```

Figure 3.13. FORTRAN arrival event routine for single-server queue simulation.

called, after which the simulation stops. It could be that the simulation is correct but the system is much more congested than anticipated, in which case the dimension of CHKOUT should be increased, say to 200. Also, the value in the IF statement should be increased from 100 to 200. It could also be that the simulation contains an error in logic or in data specifications.) Next the cumulative statistics B and MQ are updated. The total busy time, B, is updated by

$$\text{“new B”} = \text{“old B”} + \text{“busy time since last event”}$$

or in FORTRAN,

$$B = B + (CLOCK - TLE)$$

Recall that TLE is the time of occurrence of the previous event. Since it is known that the server was busy over the time interval (TLE, CLOCK), it follows that total busy time should be increased by the amount (CLOCK − TLE). After B is updated, TLE and MQ are updated. As before, the next arrival is generated and scheduled on the FEL, after which control passes to the main program.

Subroutine DPART, which executes the departure event, is listed in Figure 3.14. A flowchart for the logic of the departure event was given in Figure 3.6. First, the cumulative statistics, B, S, ND, and F are updated. (Note that the maximum queue length, MQ, cannot change in value when a departure occurs.) The response time, RT, of the departing customer is computed by

$$\begin{aligned} RT &= \text{(current time)} - \text{(time of arrival of customer now departing)} \\ &= CLOCK - CHKOUT(1) \end{aligned}$$

Next, cumulative response time, S, and number of departures, ND, are incremented. The number of customers who experience a response time of 4 or more minutes, namely F, is incremented if RT $\geq$ 4.0. The waiting line is then checked (LQT $\geq$ 1, or not) to see if there is a customer waiting to begin service. If not (i.e., LQT = 0), the server's status is set to zero (LST = 0), and the next departure event is set to "infinity" (i.e., 1.0E+30) to guarantee that an arrival occurs next; then control is returned to the main program. If there is a customer in line (i.e., LQT $\geq$ 1), all customers are "moved up one space"; that is, the arrival-time attributes are moved forward in the array CHKOUT. When this "moving up" is finished, CHKOUT(1) again contains the arrival time of the (new) customer in service; CHKOUT(2) the arrival time of the first customer in line (behind the one being served); and so on. Then system state is updated by reducing number in line, LQT, by one. Finally, the service time, SVT, of the customer just beginning service is generated, and a departure event for this customer is scheduled by setting FEL(2) equal to

$$\begin{aligned} FEL(2) &= \text{(current time)} + \text{(duration of service time)} \\ &= CLOCK + SVT \end{aligned}$$

Control is then returned to the main program.

The report generator, subroutine RPTGEN, is listed in Figure 3.15. The summary statistics, RHO, AVGR, and PC4, are computed by the formulas in Table 3.7. Then the input parameters, MIAT, MSVT, SIGMA, and NCUST, are printed, followed by the summary statistics. It is a good idea to print the input parameters at the end of the simulation in order to verify that their values are correct, and that these values have not been inadvertently changed.

```
C
C
C DEPARTURE EVENT ROUTINE
C
C
      SUBROUTINE DPART
      REAL MIAT,MSVT
      COMMON /SIM/ MIAT,MSVT,SIGMA,NCUST,LQT,LST,TLE,
     1             CHKOUT(100),B,MQ,S,F,ND
      COMMON /TIMEKP/ CLOCK,IMEVT,NUMEVS,FEL(2)
C
C
C UPDATE CUMULATIVE STATISTICS, B, S, ND, AND F
C (LQT IS DECREASING, SO MQ DOES NOT CHANGE NOW)
C
      B = B + (CLOCK - TLE)
      TLE = CLOCK
      RT = CLOCK - CHKOUT(1)
      S = S + RT
      ND = ND + 1
      IF(RT .GE. 4.0) F = F + 1
C
C CHECK CONDITION OF WAITING LINE
C
      IF(LQT .GE. 1) GO TO 20
C
C NO CUSTOMERS IN LINE, SO SERVER BECOMES IDLE
C AND NEXT DEPARTURE TIME IS SET EQUAL TO "INFINITY"
C
        LST = 0
        FEL(2) = 1.E+30
      RETURN
C
C AT LEAST ONE CUSTOMER IS IN LINE, SO MOVE EACH
C CUSTOMER IN LINE FORWARD ONE SPACE
C
   20 DO 30 I = 1,LQT
        I1 = I + 1
        CHKOUT(I) = CHKOUT(I1)
   30 CONTINUE
C
C UPDATE SYSTEM STATE
C
      LQT = LQT - 1
C
C GENERATE NEW SERVICE TIME FOR CUSTOMER BEGINNING
C SERVICE, AND SCHEDULE THE NEXT DEPARTURE EVENT
C
      SVT = NORML(MSVT,SIGMA)
      FEL(2) = CLOCK + SVT
      RETURN
      END
```

Figure 3.14. FORTRAN departure event routine for single-server queue simulation.

Figure 3.16 provides a listing of function EXPON, and Figure 3.17 a listing of function NORML. Both of these functions first call function RANF, which is a library routine (available on some computers) which generates samples uniformly distributed on the interval (0, 1). Such routines, called random number generators, are discussed in Chapter 7. The techniques for generating exponentially and normally distributed random variates, discussed in Chapter 8, are based on first generating a $U(0, 1)$ random number, R. For further explanation, the reader is referred to Chapters 7 and 8.

The output from the grocery checkout counter simulation is shown in Figure 3.18. It should be emphasized that the output statistics are statistical estimates that contain random error. The values shown are influenced by the particular random numbers that

```
C
C
C REPORT GENERATOR
C
C
      SUBROUTINE RPTGEN
      REAL MIAT,MSVT
      COMMON /SIM/ MIAT,MSVT,SIGMA,NCUST,LQT,LST,TLE,
     1            CHKOUT(100),B,MQ,S,F,ND
      COMMON /TIMEKP/ CLOCK,IMEVT,NUMEVS,FEL(2)
C
C COMPUTE SUMMARY STATISTICS
C
      RHO = B/CLOCK
      AVGR = S/ND
      PC4 = F/ND
C
C WRITE OUTPUT
C
      WRITE(6,10)
   10 FORMAT(6X,"SINGLE SERVER QUEUE SIMULATION - GROCERY",
     1       " STORE CHECKOUT COUNTER")
C
      WRITE(6,20) MIAT,MSVT,SIGMA,NCUST
   20 FORMAT(//,17X,"MEAN INTERARRIVAL TIME          ",F10.2,
     1       /,17X,"MEAN SERVICE TIME               ",F10.2,
     2       /,17X,"STANDARD DEVIATION OF SERVICE TIMES",F5.2,
     3       /,17X,"NUMBER OF CUSTOMERS SERVED         ",I6)
C
      WRITE(6,30)RHO,MQ,AVGR,PC4,CLOCK,ND
   30 FORMAT(///,17X,"SERVER UTILIZATION          ",F8.2,
     1       /,17X,"MAXIMUM LINE LENGTH         ",I8
     2       /,17X,"AVERAGE RESPONSE TIME       ",F8.2," MINUTES"
     3       /,17X,"PROPORTION WHO SPEND FOUR",
     4       /,19X,"MINUTES OR MORE IN SYSTEM",F6.2,
     5       /,17X,"SIMULATION RUNLENGTH        ",F8.2," MINUTES"
     6       /,17X,"NUMBER OF DEPARTURES        ",I8)
      RETURN
      END
```

Figure 3.15. FORTRAN report generator for single-server queue simulation.

happened to have been used, by the initial conditions at time 0, and by the run length (in this case, 1000 departures). Methods for estimating the standard error of such estimates are discussed in Chapter 11.

In some simulations it is desired to stop the simulation after a fixed length of time, say TE = 12 hours = 720 minutes. In this case, an additional event, the stopping event (say, type 3), is defined and scheduled to occur by setting FEL(3) = TE. When the stopping event does occur, the cumulative statistics will be updated and the report generator called. The main program and subroutine INITLZ will require minor changes. In particular, NUMEVS must be set to 3 and the computed GO TO statement in the main program must be changed. Exercise 4 asks the reader to make these changes. Exercise 5 considers the additional change that any customer at the checkout counter at simulated time CLOCK = TE should be allowed to exit the store, but no new arrivals are allowed after time TE.

3.2.2. Simulation in GASP

GASP IV is a collection of FORTRAN subroutines designed to facilitate an event-scheduling simulation written in FORTRAN. It consists of over 30 subroutines and functions which provide numerous needed facilities, including

```
C
C
C EXPONENTIAL RANDOM VARIATE GENERATOR
C
C
      FUNCTION EXPON(FMEAN)
C
C GENERATE A U(0,1) RANDOM NUMBER
      R=RANF(DUMMY)
C
C GENERATE AN EXPONENTIAL RANDOM VARIABLE WITH MEAN FMEAN
C (SEE EQUATION (8-2B).)
C
      EXPON = -FMEAN*ALOG(R)
      RETURN
      END
```

Figure 3.16. Exponential random variate generator for single-server queue simulation.

```
C
C NORMAL RANDOM VARIATE GENERATOR
C
      FUNCTION NORML(MEAN,SIGMA)
      REAL MEAN,SIGMA
      DATA K/0/,PI/3.14159/
C
C CHECK TO SEE WHICH N(0,1) RANDOM VARIABLE TO USE
C
      IF(K.EQ.1)GO TO 10
C
C GENERATE TWO UNIFORM(0,1) RANDOM NUMBERS
C
        RONE=RANF(DUMMY)
        RTWO=RANF(DUMMY)
C
C GENERATE TWO NORMAL(0,1) RANDOM VARIABLES
C (SEE EQUATION (8-26).)
C
        ZONE=SQRT(-2*ALOG(RONE)) * COS(2*PI*RTWO)
        ZTWO=SQRT(-2*ALOG(RONE)) * SIN(2*PI*RTWO)
C
C COMPUTE NORMAL RANDOM VARIABLE WITH PARAMETERS
C (MEAN, SIGMA) FOR MEAN AND STANDARD DEVIATION
C
        NORML = ZONE*SIGMA + MEAN
C
        K = 1
      RETURN
C
C COMPUTE NORMAL RANDOM VARIABLE, N(MEAN, SIGMA**2)
C
  10    NORML = ZTWO*SIGMA + MEAN
        K = 0
C
      RETURN
      END
```

Figure 3.17. Normal random variate generator for single-server queue simulation.

a time-advance routine (called GASP); routines to manage the future event list (i.e., to add new events to the future event list); routines to add and remove entities from sets; routines to collect statistics; random variate generation routines; and a standard report generator. The programmer must provide a main program, an initialization routine, event routines, and, if desired, a report

```
SINGLE SERVER QUEUE SIMULATION - GROCERY STORE CHECKOUT COUNTER

          MEAN INTERARRIVAL TIME                     4.50
          MEAN SERVICE TIME                          3.20
          STANDARD DEVIATION OF SERVICE TIMES         .60
          NUMBER OF CUSTOMERS SERVED                 1000


          SERVER UTILIZATION             .60
          MAXIMUM LINE LENGTH              5
          AVERAGE RESPONSE TIME         4.59 MINUTES
          PROPORTION WHO SPEND FOUR
            MINUTES OR MORE IN SYSTEM    .48
          SIMULATION RUNLENGTH       4460.68 MINUTES
          NUMBER OF DEPARTURES          1000
```

Figure 3.18. Output from the FORTRAN single-server queue simulation.

generator, plus a subroutine called EVNTS. The main program must have a statement "CALL GASP" to start the simulation. Subroutine GASP determines the imminent event and calls a user-written subroutine EVNTS with an index (such as IMEVT in the FORTRAN model, Section 3.2.1, but called NEXT in GASP IV). This index (NEXT) indicates which event is the imminent event, to be called by EVNTS.

GASP IV is available for most computers that have a FORTRAN compiler. A complete description is given by Pritsker [1974]. A brief description with one example is given by Law and Kelton [1982].

3.2.3. Simulation in SIMSCRIPT

SIMSCRIPT II.5 is a high-level programming language with facilities specifically designed to aid in the development of a discrete-event simulation model. As a simulation language, it allows the event-scheduling or the process-interaction point of view. As a scientific language, it is at least as powerful as FORTRAN, PL/1, ALGOL, or Pascal. Many programming tasks can be done more efficiently (i.e., less programmer's time) in SIMSCRIPT than in FORTRAN. A SIMSCRIPT program can be written in English-like statements in free format style; such a program is almost self-documenting, and can be easily explained to a nonprogrammer. In contrast to FORTRAN, SIMSCRIPT provides automatic maintenance of the future event list and the time-advance/event-scheduling algorithm; automatic maintenance of sets, including the operations of adding and removing entities from sets; automatic collection of requested statistics; and numerous random variate generators for a wide variety of probability distributions.

SIMSCRIPT was initially developed by the RAND corporation in the 1960s, and was originally FORTRAN-based. SIMSCRIPT II.5, the latest version of the language, is owned by CACI, Inc. (Los Angeles and Washington, D.C.). It is available for most large computer systems (IBM 360/370, CDC/6000-7000,

VAX, Honeywell H/600-6000, and Univac/1100) and can be leased or purchased through CACI. It has the capability to allow a process-interaction approach, as well as the event-scheduling approach to be described here, and an extension is available to allow continuous simulations [Delfosse, 1976]. CACI has several booklets available which give a general description of the language plus simple examples [Russell and Annino, 1979; Russell, 1976]. The following more complete references are also available: Kiviat et al. [1973], Russell [1983], and the SIMSCRIPT II.5 Reference Handbook by CACI [1976]. In addition, CACI periodically gives seminars on SIMSCRIPT II.5.

The world view taken by SIMSCRIPT is based on *entities*, *attributes*, and *sets*. Entities are classified as being either permanent or temporary. *Permanent entities* represent objects in a system that will remain in the system for the duration of the simulation. Examples include a fixed number of servers in a queueing model, a fixed number of ships in a shipping model, or the fixed number of dump trucks in Example 3.4. *Temporary entities* represent objects, such as customers in a queueing model, which "arrive" to the system, remain for awhile, and then "leave" the system. The number of temporary entities active in the model may vary considerably over the course of the simulation. Entities may have attributes, and associated entities may belong to a set (just as in the manual simulations, Examples 3.3 and 3.4). A SIMSCRIPT program consists of a preamble, a main program, event routines, and ordinary subroutines. As mentioned, the time-advance routine, random variate generation routines, and statistics gathering routines are provided automatically. The preamble gives a static description of the system by defining all entities, their attributes, and the sets to which they may belong. It also defines the global variables (used to partially define system state) and lists the statistics to be collected on certain variables. A large collection of variables are automatically maintained. For example, TIME.V represents the simulation clock. If QUEUE is the name of a set, N.QUEUE is the number of entities in the set. (Note that names in SIMSCRIPT can be any length and can contain embedded periods.) The main program reads (or assigns) the value of input parameters, initializes the system state, and generates the first events. The event routines are automatically called by the time-advance subroutine, which in turn is activated by the statement "START SIMULATION" in the main program. Ordinary subroutines may be called from any event routine or the main program.

Example 3.7 (Single-Server Queue Simulation in SIMSCRIPT)

A SIMSCRIPT model of the grocery store checkout counter (Example 3.5) will now be described. In the FORTRAN solution, the first arrival time was a random time chosen from the exponential interarrival distribution; in the SIMSCRIPT model, it is assumed that the first arrival occurs at time zero. Otherwise, identical assumptions are made in the two models.

The preamble is listed in Figure 3.19. The statements in the preamble will now be explained. The mode of variables may be integer or real; the "NORMALLY, MODE IS . . ." statement sets a background mode for all undeclared variables. Although names in SIMSCRIPT may be of any length, the first five or six characters must be unique (depending on the type of name and the computer being used); the "DEFINE word to MEAN expression" statement instructs the compiler to substitute the "expression" for every occurrence of "word" in the program. In this example, line 3 of the preamble guarantees that the name "DEPARTING.CUSTOMER" will be compiled as "ATRB1" and thus will not be confused with the event "DEPARTURE." Next, the two events are defined: ARRIVAL and DEPARTURE. In SIMSCRIPT, an event notice is a record that a certain type of event is going to occur at some (usually future) time. Event notices are filed on the future event list, just as entities may be filed in a set.

```
PREAMBLE
    NORMALLY, MODE IS INTEGER
    DEFINE DEPARTING.CUSTOMER TO MEAN ATRB1
    DEFINE ARRIVAL.TIME TO MEAN ATRB2

    EVENT NOTICES INCLUDE ARRIVAL
         EVERY DEPARTURE HAS A DEPARTING.CUSTOMER
    TEMPORARY ENTITIES
         EVERY CUSTOMER HAS AN ARRIVAL.TIME
              AND MAY BELONG TO THE QUEUE
    THE SYSTEM OWNS THE QUEUE
    DEFINE QUEUE AS FIFO SET
    DEFINE REPORT.GENERATOR AS A ROUTINE

    DEFINE MIAT, MSVT, SIGMA, ARRIVAL.TIME, AND RESPONSE.TIME
         AS REAL VARIABLES
    DEFINE SERVER, NCUST, IS.RT.4 AND NUMBER.OF.DEPARTURES
         AS INTEGER VARIABLES
    DEFINE IDLE TO MEAN 0
    DEFINE BUSY TO MEAN 1

    ACCUMULATE RHO AS THE AVERAGE OF SERVER
    TALLY MAX.Q.LENGTH AS THE MAXIMUM OF N.QUEUE
    TALLY AVG.RT AS THE AVERAGE OF RESPONSE.TIME
    TALLY PROB.RT.GE.4 AS THE AVERAGE OF IS.RT.4
END
```

Figure 3.19. SIMSCRIPT preamble for the single-server queue model.

Additionally, event notices may have attributes. The "ARRIVAL" event has no attributes, but the "DEPARTURE" event has the attribute "DEPARTING.CUSTOMER." For each class of events, there is an event routine, to be discussed later. One type of temporary entity, a "CUSTOMER," is defined with one attribute, "ARRIVAL.TIME." A "CUSTOMER" entity may belong to the "QUEUE," which is a first in, first out set "owned" by the system (which essentially means that there is only one set named "QUEUE"). (In this model, the set called "QUEUE" contains waiting customers but not the one being served. This definition was a matter of programmer's choice.) "REPORT.GENERATOR" is an ordinary subroutine to be called by the programmer (it will not be called by the time-advance/event-scheduling routine). Next, certain global variables are explicitly defined as being in real or integer mode. The two "DEFINE . . . TO MEAN . . ." statements mean that all occurrences of the word IDLE (BUSY) will be replaced with the number 0(1) before the program is compiled. Such symbolic substitutions improve the readability of the program. The "ACCUMULATE" and "TALLY" statements request certain statistics to be collected. "SERVER" is a system state variable to which the programmer assigns values of 1 (BUSY) or 0 (IDLE) to represent the status of the server. With the "ACCUMULATE" statement, SIMSCRIPT will automatically maintain a time-weighted cumulative sum (called "B" in the FORTRAN model) of the variable "SERVER." At the end of the simulation, this cumulative sum divided by TIME.V will yield "RHO." The average "RHO," automatically produced by the "ACCUMULATE" statement, is an example of a time-weighted average (discussed further in Section 5.4). RHO will equal the proportion of time that the server is busy. The first "TALLY" will yield MAX.Q.LENGTH as the maximum number of entities on the set "QUEUE," since "N.set-name" is automatically maintained by SIMSCRIPT for each set. When a DEPARTURE event occurs, the programmer will compute the "RESPONSE.TIME" of the "DEPARTING. CUSTOMER" and SIMSCRIPT will automatically add its value to a cumulative statistic, which will be used at the end of the simulation to compute the average response time, "AVG.RT." Whenever a response time is 4 or more minutes in length, the variable "IS.RT.4" is set to 1; otherwise, its value is 0; thus, PROB.RT.GE.4 will be the proportion of departures whose response time is greater than or equal to 4 minutes.

The main program is shown as Figure 3.20. The SCHEDULE statement is used to place event notices on the future event list. The first ARRIVAL event is SCHEDULED to occur at time zero. The input parameters, MIAT, MSVT, SIGMA, and NCUST (having the same meaning as in the FORTRAN model), are assigned values;

```
MAIN
    SCHEDULE AN ARRIVAL NOW

    LET MIAT = 4.5  ''MINUTES, THE MEAN INTERARRIVAL TIME
    LET MSVT = 3.2  ''MINUTES, THE MEAN SERVICE TIME
    LET SIGMA = 0.6  ''MINUTE, THE STANDARD DEVIATION OF SERVICE TIME
    LET NCUST = 1000  ''CUSTOMERS TO BE SERVED (THE STOPPING CRITERIA)


    LET SERVER = IDLE  ''SO THAT THE FIRST ARRIVAL WILL FIND THE SERVER IDLE
    LET NUMBER.OF.DEPARTURES = 0

    START SIMULATION

END
```

Figure 3.20. SIMSCRIPT main program for the single-server queue model.

alternatively, they could have been read from a file using SIMSCRIPT's free field or formatted input. The system state variable "SERVER" is set to "IDLE" (0), but it will be set to "BUSY" immediately, by the ARRIVAL event routine. The START SIMULATION statement begins the execution of the event-scheduling/time-advance algorithm. Note that if it were desired to have the first arrival occur at a random time, the SCHEDULE statement could be replaced by

SCHEDULE AN ARRIVAL IN EXPONENTIAL.F(MIAT,1) MINUTES

The exponential random variate generator (as well as many others) is built in to SIMSCRIPT. (MIAT is the mean; the second argument may be any integer from 1 to 10, to indicate which of 10 random number streams is desired.)

The event routine for the ARRIVAL event is shown in Figure 3.21. The "CREATE . . ." statement creates a copy, or instance, of the named temporary entity. The attribute "ARRIVAL.TIME" of this newly created CUSTOMER is set equal to the present simulated time. SIMSCRIPT contains a structured IF statement of the form

IF condition is true
    then do these statements
OTHERWISE (or ELSE)
    do these statements
ALWAYS (or REGARDLESS)

which considerably improves the readability of a model. If the SERVER is IDLE (i.e., if the variable SERVER equals 0), the SERVER is made BUSY and a departure of the newly arriving CUSTOMER is scheduled to occur at the end of a generated service time (assumed to be normally distributed). What this scheduling means is that a DEPARTURE event notice is placed on the future event list; this event notice contains extra information (called a pointer or index) which points to the particular CUSTOMER who will be departing. Each temporary entity CREATEd will have a pointer associated with it to distinguish it from other temporary entities of the same type. The values of these pointers must be stored, else they are lost. The pointer to the CUSTOMER just beginning service is stored in the DEPARTURE event notice. The pointers to any other CUSTOMER entities, namely, to those CUSTOMERs waiting to be served, are stored in the set called "QUEUE" by the statement "FILE CUSTOMER IN QUEUE," whenever an ARRIVAL event occurs with the SERVER in the BUSY state. Under all circumstances, when one arrival occurs, the next arrival is SCHEDULEd by generating an interarrival time. Note that the normal random variate generator, namely NORMAL.F, requires three arguments: the mean (MSVT), the standard deviation (SIGMA), and the random number stream desired (an integer from 1 to 10).

```
EVENT ARRIVAL
    CREATE A CUSTOMER
         LET ARRIVAL.TIME = TIME.V

    IF SERVER IS EQUAL TO IDLE,
         LET SERVER = BUSY
         SCHEDULE A DEPARTURE GIVEN CUSTOMER IN
              NORMAL.F(MSVT, SIGMA, 1) MINUTES

    ELSE
         FILE CUSTOMER IN QUEUE

    REGARDLESS
         SCHEDULE AN ARRIVAL IN EXPONENTIAL.F(MIAT, 1) MINUTES

    RETURN
    END
```

Figure 3.21. SIMSCRIPT arrival event routine for the single-server queue model.

The DEPARTURE event routine is shown in Figure 3.22. When this event occurs, the attribute DEPARTING.CUSTOMER will contain the pointer previously stored in the event notice (by the SCHEDULE statement in the ARRIVAL routine or the SCHEDULE statement in this routine). This pointer indicates which CUSTOMER

```
EVENT DEPARTURE GIVEN DEPARTING.CUSTOMER
    LET RESPONSE.TIME = TIME.V - ARRIVAL.TIME(DEPARTING.CUSTOMER)
    DESTROY A CUSTOMER CALLED DEPARTING.CUSTOMER

    IF RESPONSE.TIME*HOURS.V*MINUTES.V IS NOT LESS THAN 4,
           LET IS.RT.4 = 1
    ELSE
           LET IS.RT.4 = 0
    REGARDLESS

    ADD 1 TO NUMBER.OF.DEPARTURES

    IF NUMBER.OF.DEPARTURES IS GE NCUST,
           CALL REPORT.GENERATOR
    ALWAYS

    IF QUEUE IS EMPTY,
         LET SERVER = IDLE

    OTHERWISE
         REMOVE FIRST CUSTOMER FROM QUEUE
         SCHEDULE A DEPARTURE GIVEN CUSTOMER IN
              NORMAL.F(MSVT, SIGMA, 1) MINUTES

    REGARDLESS

RETURN
END
```

Figure 3.22. SIMSCRIPT departure event routine for the single-server queue model.

entity is associated with this DEPARTURE event, and is used to retrieve the ARRIVAL. TIME of the DEPARTING.CUSTOMER, so that this customer's RESPONSE.TIME can be computed. Next the CUSTOMER entity called DEPARTING.CUSTOMER is DESTROYED, meaning that the portion of computer memory previously allocated for storing the entity and its attributes is now available to SIMSCRIPT to be used for other purposes. (All temporary entities should be destroyed when they leave the system; otherwise, the computer memory could fill with information which would be of no use to the simulation. If memory completely filled, the simulation could not continue.) Next, the indicator variable IS.RT.4 is set to 0 or 1, depending on whether RESPONSE. TIME is less than 4 minutes, or greater than or equal to 4 minutes. The (default) time unit in SIMSCRIPT is days, so RESPONSE.TIME in days is converted to minutes (in the IF statement) by multiplying by 24 (HOURS.V) and by 60 (MINUTES.V). The NUMBER.OF.DEPARTURES is an output statistic collected by the programmer. (ADD 1 TO X is equivalent to LET X=X+1.) In the next IF statement, the NUMBER.OF.DEPARTURES is compared to NCUST = 1000 to see if the REPORT. GENERATOR should be called. (Note that ELSE is optional in the IF . . . ALWAYS construction.) The final IF statement executes the logic of the DEPARTURE event. If the set "QUEUE" is EMPTY, the SERVER is made IDLE and control is returned to the time-advance/event-scheduling routine. (EMPTY is a keyword.) If the "QUEUE" is not empty, there must be at least one CUSTOMER entity in the set, so the first one is REMOVEd. Recall that the set "QUEUE" was defined as a FIFO (first in, first out) set in the preamble. A DEPARTURE event for the removed CUSTOMER entity is then SCHEDULEd and control is returned to the time-advance routine.

Figure 3.23 provides a listing of the REPORT.GENERATOR routine. (This

```
ROUTINE REPORT.GENERATOR
PRINT 1 LINE THUS
          SINGLE SERVER QUEUE SIMULATION - GROCERY STORE CHECKOUT COUNTER
SKIP 2 OUTPUT LINES
PRINT 4 LINES WITH MIAT, MSVT, SIGMA, NCUST THUS
                   MEAN INTERARRIVAL TIME                        **.**
                   MEAN SERVICE TIME                             **.**
                   STANDARD DEVIATION OF SERVICE TIMES           **.**
                   NUMBER OF CUSTOMERS TO BE SERVED               ****
SKIP 3 OUTPUT LINES
PRINT 7 LINES WITH RHO, MAX.Q.LENGTH, AVG.RT*HOURS.V*MINUTES.V,
    PROB.RT.GE.4, TIME.V*HOURS.V*MINUTES.V, AND
    NUMBER.OF.DEPARTURES  THUS
                   SERVER UTILIZATION                      **.**
                   MAXIMUM LINE LENGTH                      ****
                   AVERAGE RESPONSE TIME                   **.** MINUTES
                   PROPORTION WHO SPEND FOUR
                     MINUTES OR MORE IN SYSTEM              *.**
                   SIMULATION RUNLENGTH                  ****.**
                   NUMBER OF DEPARTURES                     ****
STOP
END
```

Figure 3.23. SIMSCRIPT report generator for the single-server queue model.

routine is not an event routine.) This routine is called from the event routine DEPARTURE at the first instant that the stopping criterion is met. The general form of a PRINT statement is

PRINT n LINES WITH variable-names THUS

which must be followed by n lines exactly as the programmer wants them to look in the output. Note the "picture" formatting for variables: the value of MIAT will be placed precisely where **.** is located, and so on. (This "picture" formatting makes the coding of a report much easier than when using FORTRAN-like format statements.) The STOP statement at the end of the routine causes the simulation to stop.

The simulation output is exhibited in Figure 3.24. Note the differences between the SIMSCRIPT model's output in Figure 3.24 and the FORTRAN model's output in Figure 3.18. These differences are attributable to the fact that the two languages used different streams of random numbers. Again it is seen that a simulation run provides

```
SINGLE SERVER QUEUE SIMULATION - GROCERY STORE CHECKOUT COUNTER


          MEAN INTERARRIVAL TIME                        4.50
          MEAN SERVICE TIME                             3.20
          STANDARD DEVIATION OF SERVICE TIMES            .60
          NUMBER OF CUSTOMERS TO BE SERVED              1000



          SERVER UTILIZATION                     .67
          MAXIMUM LINE LENGTH                      8
          AVERAGE RESPONSE TIME                 6.51 MINUTES
          PROPORTION WHO SPEND FOUR
            MINUTES OR MORE IN SYSTEM            .63
          SIMULATION RUNLENGTH               4794.13
          NUMBER OF DEPARTURES                  1000
```

Figure 3.24. Output from SIMSCRIPT model of single-server queue.

an *estimate* of system performance, and this estimate may contain random error due to the random fluctuations inherent in the system. If the FORTRAN and SIMSCRIPT models could both be run for an "infinite" length of time (instead of 4460 and 4794 minutes, respectively), both models would produce identical estimates, and these estimates would contain no random error. In practice, two estimates of the same parameter produced by two runs using different random numbers will eventually become closer in value as simulation runlength increases.

If the simulator desired to simulate for a fixed length of time, say 80 hours, then a third event, called STOP.SIMULATION, could be defined in the preamble. In the main program, this event could be scheduled by the statement

SCHEDULE A STOP.SIMULATION IN 80 HOURS

The event routine STOP.SIMULATION would merely call the REPORT.GENERATOR routine. In addition, the IF statement that checks NUMBER.OF.DEPARTURES in the DEPARTURE event routine should be removed.

In summary, it should be apparent that SIMSCRIPT is a powerful language for use in discrete-event simulation. If thoroughly understood, the development and debugging of a SIMSCRIPT model should take considerably less time than a FORTRAN model. SIMSCRIPT has several aids to debugging which can further shorten model development time. In addition, the use of the process-interaction capability of SIMSCRIPT, when appropriate, can considerably reduce the number of statements in a model, and thus can speed model development. This capability is described briefly by Law and Kelton [1982] and Russell [1976] and in detail by Russell [1983].

3.2.4. Simulation in GPSS

GPSS is a highly structured, special-purpose simulation language using the process-interaction approach and oriented toward queueing simulations. A *block diagram* provides a description of the system. Temporary entities called *transactions* are created and may be pictured as flowing through the block diagram. Thus, GPSS can be used for any situation in which entities (e.g., customers) can be viewed as passing through a system (e.g., a network of queues). GPSS is not a procedural language, such as FORTRAN or SIMSCRIPT; it is a structured method for describing, in detail, certain types of systems. The GPSS processor then takes this description (i.e., the block diagram) and automatically performs a simulation. The time-advance/event-scheduling mechanism is totally hidden from view.

GPSS (General Purpose Simulation System) was originally developed by Geoffrey Gordon of IBM about 1960. It has evolved through several versions, the latest being called GPSS/360 and GPSS V. It is available not only for IBM/360 and 370 series computers, but in some form for most large computer systems. An introduction to GPSS can be found in Gordon [1978, 1975] or Schriber [1974]. In recent years, Professor Schriber has given a week-long GPSS seminar every summer at the University of Michigan. A new implementation of GPSS, called GPSS/H, has recently been developed by Henriksen [1979] of Wolverine Software, Falls Church, Virginia. Models written in GPSS/360 (or GPSS V, which contains GPSS/360 as a subset) reportedly will run with little or no modification under GPSS/H from 4 to 30 times faster (CPU time). Furthermore, GPSS/H has additional features which remove some of the limitations of GPSS/360. It is presently available for IBM machines only, but eventually will be made available for other computers.

Since the 1960s, GPSS has probably been the most widely used discrete-event simulation language. Two reasons for this popularity are the ease of learning the language (especially for nonprogrammers) and the relatively short time required for building a complex model. On the other hand, GPSS has several shortcomings, some of which may lead the unwary programmer to develop apparently valid but actually invalid models. Other shortcomings make the modeling of certain systems extremely cumbersome:

1. The simulation clock, called ACl, can only assume the discrete values 1, 2, 3, ... The final result of any calculations is always truncated (rounded down) before it is used as an event time. Therefore, a suitably small time unit must be chosen so that generated activity times and the clock will be sufficiently accurate.
2. Complex numerical calculations and/or logical actions can be extremely difficult, if not impossible, to execute in GPSS. Specifically, GPSS does not have the capability to directly compute logarithms, sines, exponentials, absolute values, maximum, and other commonly used mathematical functions. In theory, GPSS can approximate these functions by a piecewise linear function, or by a GPSS HELP block which allows calling a FORTRAN routine. In practice, these approximations have not been developed or are not extremely accurate, and the capability to call a FORTRAN routine is quite cumbersome and limited in usefulness.
3. GPSS has no built-in random variate generators. Without the availability of the mathematical functions mentioned in item 2, GPSS must resort to interpolation in a GPSS Function in order to generate random variates. This piecewise linear approximation is described in Section 8.1.6. Standard approximations have been developed for the exponential distribution with mean 1 and for the standard normal distribution [Gordon, 1975].
4. All eight random number streams produce the same sequence of random numbers unless the programmer explicitly defines distinct "seeds" for each stream. A "seed" is analogous to a starting point in a stream of random numbers such as those in Table A.1. The RMULT statement is used to specify the seeds for each stream. (If RMULT is not used, all eight seeds default to a common value of 37 in GPSS V.)

It should be noted that GPSS/H has eliminated most of these shortcomings.

The two main concepts in GPSS are *transactions* and *blocks*. A block can be represented by a pictorial symbol, or a single statement in the GPSS language.

There are over 40 standard blocks in GPSS. Each block represents a specific action or event that can occur in a typical system. These blocks are arranged into a block diagram which represents a "customer" process. Figure 3.4 showed one example of a "customer" process for a single-server queue. A GPSS description of such a process is shown in the block diagram of Figure 3.25, to be explained in Example 3.8 below. Transactions representing active, dynamic entities may be pictured as flowing through the block diagram. Every path that a transaction can take through the system must be represented in the block diagram, which can have branches. The limited resources of a system are represented by the predefined GPSS entities, facilities and storages. A facility is essentially a single server. A storage is a group of parallel servers. Statistics are automatically collected on the utilization of facilities and storages. In addition,

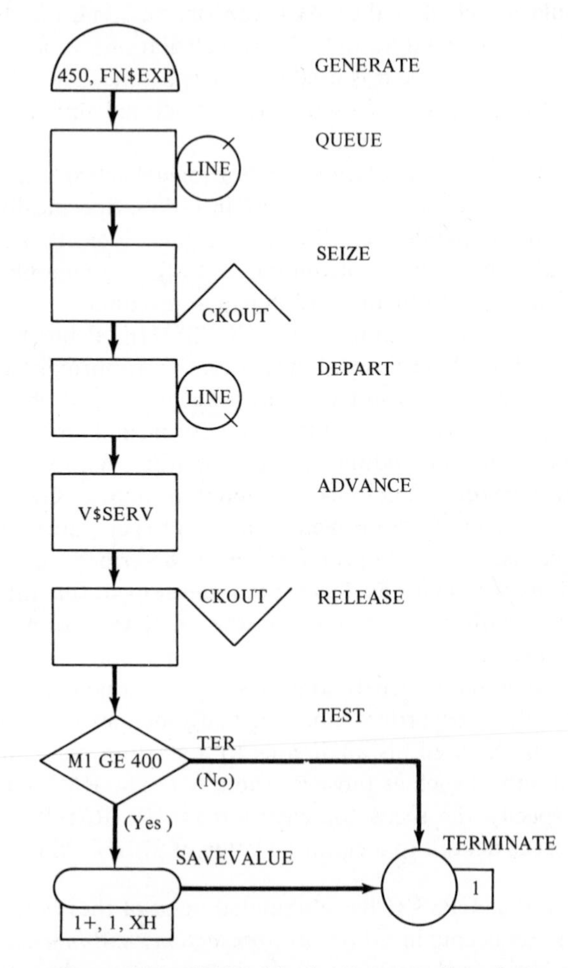

Figure 3.25. GPSS block diagram for single-server queue simulation.

queue and table entities are available to collect statistics on waiting lines or transit times.

EXAMPLE 3.8 (SINGLE-SERVER QUEUE SIMULATION IN GPSS)

Figure 3.25 exhibits the block diagram and Figure 3.26 the GPSS program for the grocery store checkout counter model described in Example 3.5. Note that the program (Figure 3.26) is a translation of the block diagram together with additional definition and control cards.

In Figure 3.25, the GENERATE block represents the arrival event, and also defines the interarrival times by its operands "450, FN$EXP." The time unit is 1/100 minute; so 450 time units = 4.5 minutes is the mean interarrival time. The second operand, FN$EXP, is a reference to FUNCTION EXP, which is used to generate approximately exponentially distributed random variates with mean one. (Multiplying 450 times FN$EXP changes the mean to 450.) The GENERATE block creates transactions with interarrival times defined by its operands, and upon arrival these transactions begin to pass through the block diagram.

The single server is represented by a facility called CKOUT. A facility is modeled by a pair of blocks, SEIZE and RELEASE, as follows:

SEIZE	CKOUT	Begin service as soon as possible
· · ·	(blocks to model service time)	Service activity
RELEASE	CKOUT	End service

In GPSS, many transactions can simultaneously be in the block diagram; each active transaction is always in a specific block; the action of a block takes place when a transaction enters a block; some blocks (such as QUEUE, DEPART, RELEASE, ADVANCE) always accept transactions, allowing them to enter the block and continue to the next block, if possible; other blocks sometimes refuse admittance to a transaction. The SEIZE block refuses admittance to a transaction whenever the facility is busy, or occupied, with another transaction. In such a case, transactions attempting to SEIZE the facility remain in the block immediately preceding the SEIZE block. The transactions will be admitted on a first come, first served basis. (The queue discipline can be changed by the use of priorities, or by so-called user chains, which are sets of transactions which the programmer can order in almost any fashion.)

Activities are represented by ADVANCE blocks. In Figure 3.25, the ADVANCE block represents the normally distributed service times. A service time is computed by the formula in the FVARIABLE statement, which makes a reference to FUNCTION NORM. This FUNCTION generates (approximately) standard normal random variates with mean zero and standard deviation 1. The FVARIABLE statement converts this standard normal random variate to one having mean 320 and standard deviation 60 time units.

Statistics are collected on the waiting line by the pair of blocks QUEUE . . . DEPART, which taken together define a GPSS queue entity. Note that a GPSS queue entity does *not* cause a waiting line to form; the SEIZE block causes the waiting line to form; the QUEUE . . . DEPART blocks merely measure various statistics on this waiting line. A GPSS queue entity can be used more generally to gather transit time and congestion statistics on any subsystem of the system under study. Upon entering

```
 BLOCK                                                                            CARD
NUMBER  *LOC   OPERATION  A,B,C,D,E,F,G,H,I,J        COMMENTS                       NUMBER

        *                                                                             1
               SIMULATE                                                               2
        * SIMULATION OF A SINGLE SERVER QUEUE                                         3
        *                                                                             4
1        EXP   FUNCTION   RN1,C24         EXPONENTIAL GENERATOR                       5
1       0,0/.1,.104/.2,.222/.3,.355/.4,.509/.5,.69                                    6
1       .6,.915/.7,1.2/.75,1.38/.8,1.6/.84,1.83/.88,2.12                              7
1       .9,2.3/.92,2.52/.94,2.81/.95,2.99/.96,3.2/.97,3.5                             8
1       .98,3.9/.99,4.6/.995,5.3/.998,6.2/.999,7/.9997,8                              9
        *                                                                            10
2        NORM  FUNCTION   RN1,C25         NORMAL GENERATOR                           11
2       0,-5/.00003,-4/.00135,-3/.00621,-2.5/.02275,-2/.06681,-1.5                   12
2       .11507,-1.2/.15866,-1/.21186,-.8/.27425,-.6/.34458,-.4/.42074,-.2            13
2       .5,0/.57926,.2/.65542,.4/.72575,.6/.78814,.8/.84134,1/.88493,1.2             14
2       .93319,1.5/.97725,2/.99379,2.5/.99865,3/.99997,4/1,5                         15
        *                                                                            16
1        SERV  FVARIABLE  320+60*FN$NORM            GENERATE SERVICE TIME            17
        *                                                                            18
     1         GENERATE   450,FN$EXP      CUSTOMERS ARRIVE AT RANDOM, ON THE         19
        *                                  AVERAGE EVERY 4.5 MINUTES                 20
        *                                 (TIME UNIT = 1/100 MINUTE)                 21
     2         QUEUE      LINE            CUSTOMER JOINS WAITING LINE                22
     3         SEIZE      CKOUT           BEGIN CHECKOUT AT CASH REGISTER            23
     4         DEPART     LINE            CUSTOMER STARTING SERVICE LEAVES QUEUE     24
     5         ADVANCE    V$SERV          CUSTOMER'S SERVICE TIME                    25
     6         RELEASE    CKOUT           CUSTOMER LEAVES CHECKOUT AREA              26
     7         TEST GE    M1,400,TER      IS RESPONSE TIME GE 4 MINUTES?             27
     8         SAVEVALUE  1+,1,XH         IF SO, ADD 1 TO COUNTER(XH1)               28
     9   TER   TERMINATE  1                                                          29
        *                                                                            30
               START      1000            SIMULATE FOR 1000 DEPARTURES               31
```

Figure 3.26. GPSS program for single-server queue simulation.

the QUEUE block in Figure 3.25, a transaction is said to be a member of the queue called "LINE." Upon entering the DEPART block, a transaction is no longer a member of the queue. Average time spent in the queue, plus the time-averaged number of transactions in the queue, are automatically collected statistics and are printed at the end of the simulation.

Figure 3.25 may now be briefly described as follows: A transaction arrives at the GENERATE block from time to time, joins the queue "LINE" immediately, SEIZEs the facility "CKOUT" as soon as possible, immediately DEPARTs the queue, spends some amount of simulated time in the ADVANCE block, after which the facility is RELEASEd, thus becoming available for the next transaction attempting to SEIZE it. After RELEASEing the facility, a transaction enters the TERMINATE block, which destroys it.

In Figure 3.26, the SIMULATE statement is a control card that instructs GPSS to conduct the simulation. (If not present, GPSS merely checks for syntax errors.) Next follows two GPSS Function definitions for generating approximate exponential and standard normal random variates. The FVARIABLE statement computes the desired normally distributed service time from a standard normal random variate. The block statements corresponding to the block diagram then follow. The last statement, START 1000, starts, and stops, the simulation. A counter with an initial value of 1000 is decremented by "A" each time that a transaction enters the TERMINATE block, where "A" is the operand on the TERMINATE block ($A = 1$ in Figure 3.26). When the counter reaches 0, the simulation stops and an output report (Figure 3.27) is automatically produced.

Note that all the desired statistics (as collected in the FORTRAN model) are included in the standard output report, with the exception of the proportion of customers whose response time was 4 minutes or longer. GPSS automatically maintains many system and transaction attributes, one of which, M1, represents a transaction's transit time in the model from the instant of generation. The following blocks have been added to the model to count the number of transactions whose response time (M1) was 4 minutes or longer.

```
        .
        .
        .
TEST GE       M1, 400, TER    Test if M1 is GE 400
SAVEVALUE     1+, 1, XH       If so, increment XH1 by 1.
        .
        .
        .
```

The final count will be stored (and printed) in the "savevalue" XH1.

The standard GPSS output report for the model with the addition above is shown in Figure 3.27. Note that:

Proportion of time server is busy	$= 0.688$
Maximum queue length	$= 11$
Average response time	$= 422.671 + 318.196 = 740.867$
Proportion of customers who spend 4 or more minutes at the checkout counter	$= \dfrac{648}{1000} = 0.648$

```
FINAL TERMINATION PRINTOUT

RELATIVE CLOCK      462628   ABSOLUTE CLOCK     462628 TERMINATIONS TO GO          0
BLOCK COUNTS
BLOCK     CURRENT  TOTAL  BLOCK     CURRENT  TOTAL  BLOCK     CURRENT  TOTAL  BLOCK     CURRENT  TOTAL  BLOCK     CURRENT  TOTAL
    1         0     1001
    2         1     1001
    3         0     1000
    4         0     1000
    5         0     1000
    6         0     1000
    7         0     1000
    8         0      648
TER           0     1000
                                         *****************************************
                                         *                                       *
                                         *              FACILITIES               *
                                         *                                       *
                                         *****************************************

                                       - AVERAGE  UTILIZATION  DURING -

FACILITY        NUMBER      AVERAGE      TOTAL     AVAIL.     UNAVAIL.     CURRENT     PERCENT        TRANSACTION NUMBER
                ENTRIES     TIME/TRAN    TIME      TIME       TIME         STATUS    AVAILABILITY     SEIZING  PREEMPTING
 CKOUT            1000        318.196    0.688                               A         100.000


                                         *****************************************
                                         *                                       *
                                         *                QUEUES                 *
                                         *                                       *
                                         *****************************************

QUEUE      MAXIMUM    AVERAGE     TOTAL      ZERO     PERCENT     AVERAGE     $AVERAGE     TABLE     CURRENT
           CONTENTS   CONTENTS    ENTRIES    ENTRIES   ZEROS     TIME/TRANS   TIME/TRANS   NUMBER    CONTENTS
LINE           11       0.915      1001       326      32.6       422.671      626.806                   1
 $AVERAGE TIME/TRANS = AVERAGE TIME/TRANS EXCLUDING ZERO ENTRIES

                                         *****************************************
                                         *                                       *
                                         *          HALFWORD SAVEVALUES          *
                                         *                                       *
                                         *****************************************

NUMBER .. CONTENTS    NUMBER .. CONTENTS    NUMBER .. CONTENTS    NUMBER .. CONTENTS    NUMBER .. CONTENTS    NUMBER .. CONTENTS:
    1         648
```

Figure 3.27. Standard GPSS output report for single-server queue simulation.

In summary, one striking aspect of GPSS, compared to the FORTRAN and SIMSCRIPT models, is the small number of statements required to model a single-server queue. In general, a process-interaction-based language, whether GPSS, or the process portion of SIMSCRIPT or SLAM, takes considerably fewer statements to model commonly occurring queueing phenomena. On the other hand, GPSS is less flexible and less powerful than SIMSCRIPT or SLAM. Some complex phenomena can be difficult and cumbersome to represent in the limited number of available GPSS blocks. Nevertheless, GPSS has been widely and successfully used in a large number of simulation projects.

3.2.5 Simulation in SLAM

SLAM is a high-level, FORTRAN-based simulation language which allows an event-scheduling or process-interaction orientation, or a combination of both approaches. The event-scheduling portion of SLAM is quite similar to GASP, which was briefly described in Section 3.2.2. The process-interaction portion of SLAM is similar in many respects to GPSS. This section briefly describes the process-interaction portion of SLAM. A good reference and textbook for SLAM is provided by Pritsker and Pegden [1979], the developers of SLAM. SLAM is marketed by Pritsker and Associates, Inc., West Lafayette, Indiana, who also give short courses in its use.

To use the process-interaction approach with SLAM, the simulator develops a network, consisting of nodes and branches, which represents the processes in the system pictorially. The objects flowing through the system are called *entities*. (Note that the SLAM definition of an entity is more restricted than the definition used in this text. A SLAM entity is analogous to a GPSS transaction. Such entities are dynamic; they flow through a process.) Recall that a process is the sequence of events and activities that confront an entity as it flows through the system. A complete SLAM network model of a system represents all possible paths that an entity can take as it passes through the system. To run the model, the simulator translates the network representation directly into computer statements which are input to the SLAM processor.

SLAM automatically handles the event-scheduling/time-advance algorithm, set (or file) operations such as the addition and deletion of entities, the collection of numerous statistics, and the generation of random samples. In SLAM, sets are called files. With its automatic file handling, SLAM can easily manage queues on a first in, first out or last in, first out basis, or entities can be ranked (and served) in order of some attribute such as priority. Unlike GPSS, SLAM has built-in random variate generators for a wide variety of statistical distributions.

A SLAM network consists of *branches* and *nodes*. A branch represents the passage of time; that is, it represents an activity. In addition, a branch may represent a limited number of servers. A branch is coded as an ACTIVITY statement. Nodes are used to represent the arrival event (CREATE node), delays or conditional waits (QUEUE node), the departure event (TERMINATE node), and other typical system actions.

EXAMPLE 3.9 (SINGLE-SERVER QUEUE SIMULATION IN SLAM)

A SLAM network model for the grocery store checkout counter is shown in Figure 3.28. Note that each node has an associated symbol plus operands. The translation of that network into SLAM statements is shown in Figure 3.29.

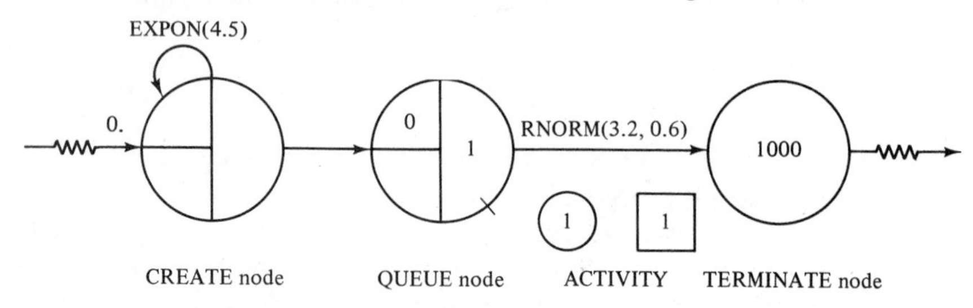

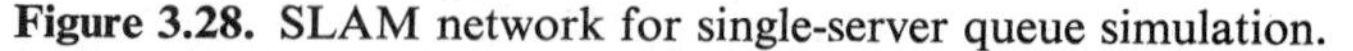

Figure 3.28. SLAM network for single-server queue simulation.

```
GEN,BANKS AND CARSON,SINGLE SERVER QUEUE EXAMPLE,5/31/1982,1;
LIMITS,1,0,30;  MODEL CAN USE 1 FILE,MAX NO. OF SIMULTANEOUS ENTRIES 30
NETWORK;        BEGINNING OF MODEL
      CREATE,EXPON(4.5);             CUSTOMERS ARRIVE AT CHECKOUT
      QUEUE(1);                      CUSTOMERS WAIT FOR SERVICE IN QUEUE FILE ONE (1)
      ACTIVITY(1)/1,RNORM(3.2,.6);   CHECKOUT SERVICE TIME IS N(3.2, 0.6)
      TERMINATE,1000;                SIMULATE UNTIL 1000 CUSTOMERS ARE CHECKED OUT
      ENDNETWORK;                    END OF MODEL
FIN;            END OF SIMULATION
```

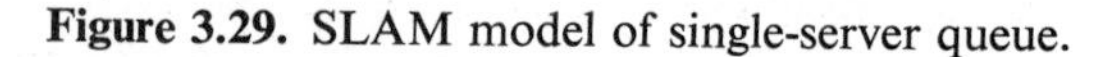

Figure 3.29. SLAM model of single-server queue.

Figure 3.28 shows that a single-server queue can be modeled with three nodes and one nonzero branch. The CREATE node represents the arrival event. An entity is CREATEd at specified intervals and begins to flow through the network. The symbol for the CREATE node, the corresponding SLAM statement, and selected operands are shown below:

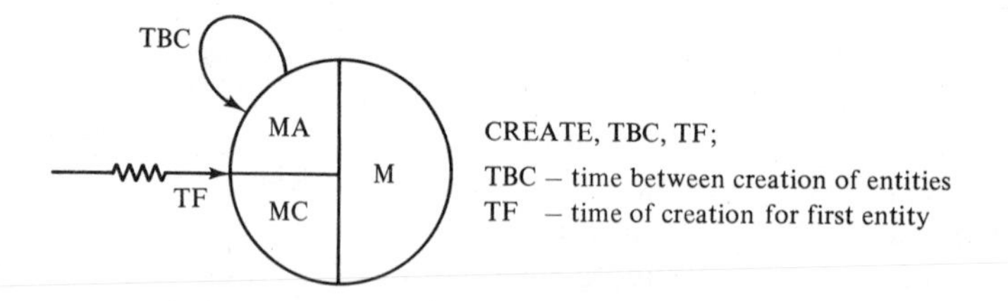

(The other operands, MA, MC, and M, are explained in Pritsker and Pegden [1979]. The discussion in this section is limited to the nodes and operands needed for Example 3.9.) In Figure 3.28, it is seen that TBC = EXPON(4.5), which means that the interarrival times are exponentially distributed with mean 4.5 time units. The operand TF is omitted. It defaults to a value of 0.; hence the first arrival occurs at simulated time zero. Alternatively, TF could be a constant, say TF = 100, to indicate that the first arrival occurs at time 100; or TF could have been probabilistic, say TF =EXPON (4.5).

In the network of Figure 3.28, the CREATE node is followed by a branch that requires no amount of simulated time. Such branches are called connectors and are placed in the pictorial network to connect two nodes. No statement is required for a connector.

The next node, a QUEUE node, represents a delay or conditional wait; that is, it represents a location where entities wait until service can begin. The symbol for the QUEUE node, the corresponding SLAM statement, and selected operands are as follows:

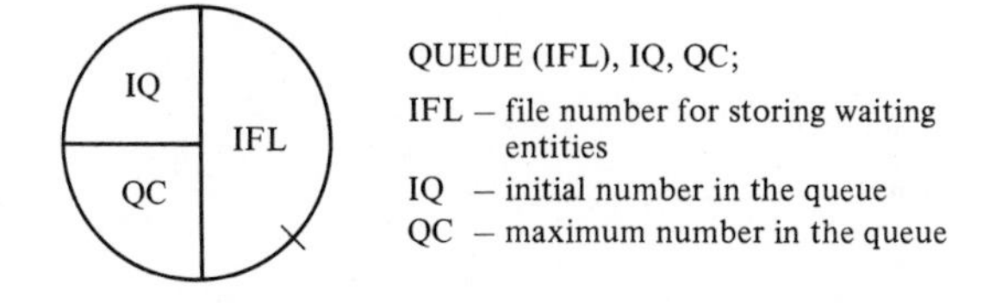

The operand, IFL, is the file number that identifies the file (or set) in which entities (and their attributes) are stored until service can begin. Entities are stored on the file only if all the servers are busy. Otherwise, an entity begins service immediately. The default queue discipline for files is first in, first out, but other, more complex disciplines can be used. The operand, IQ, defaults to 0, and QC defaults to "infinity." The use of IQ allows the queue to be nonempty at time 0, while the use of QC allows for finite waiting room. [If the QUEUE node is at its capacity, QC, when an entity arrives, the other operands (not listed above) determine the new arrival's disposition, which will be to balk or be blocked.] In Figure 3.28, IFL has the value 1, and IQ and QC are omitted. Thus, the waiting line called "file 1," is initially empty, and has unlimited capacity. Statistics are automatically collected on the average number of entities in the file (i.e., average length of the waiting line) and average waiting times of entities on the file.

In Figure 3.28, the QUEUE node is followed by an ACTIVITY branch, whose symbol, corresponding statement, and selected operands are as follows:

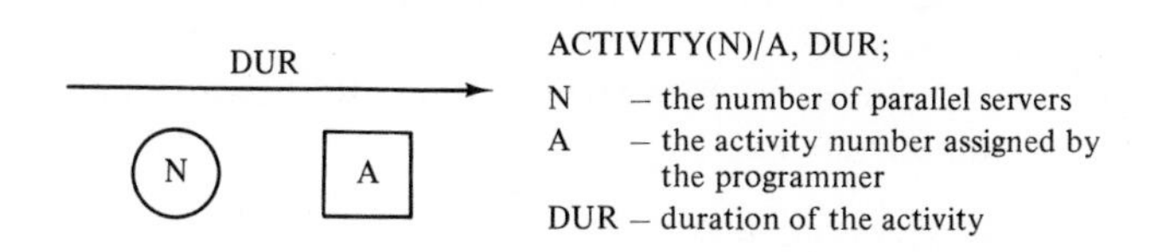

A branch represents an activity, that is, an explicitly defined duration of time such as a service time. The operand, N, specifies the number of identical parallel servers. The operand, A, is an activity number assigned by the programmer and used to uniquely identify a group of servers. Utilization statistics are automatically collected for each such group of servers. The operand, DUR, is used to define the duration of the activity. The duration, DUR, may be constant, a function of system attributes (e.g., a function of queue length), or random. In Figure 3.28, N = 1 indicates a single server, A = 1 indicates activity number 1, and DUR = RNORM(3.2, 0.6) indicates that service times are generated as samples from a normal distribution with mean 3.2 and standard deviation 0.6 time unit (minutes).

Each branch in a SLAM model must have a start node and an end node. Typically, a QUEUE node is the start node for a particular branch. The end node may be any other type of node. It is either the sequential node or some other node whose label is specified as an operand on the ACTIVITY statement. [For example, "ACTIVITY (1)/1, RNORM(3.2, 0.6),, NLBL" specifies that NLBL is the label of the end node. Any node can be labeled. Note that a missing operand is specified by successive commas with no operand between them.] In Figure 3.28, the end node for the branch is a TERMINATE node, which represents a departure event; that is, the entity leaves the system and the entity record is destroyed. The symbol and operand for this node are as follows:

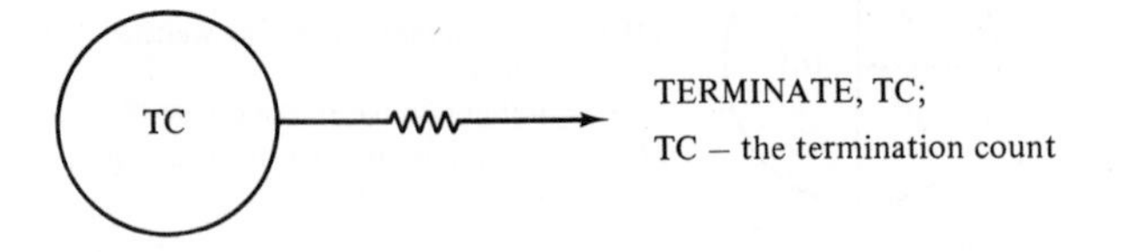

If the operand TC is a positive integer, the simulation will stop when TC entities have been TERMINATEd at this node. (When there are several TERMINATE nodes, the simulation ends when the first termination count, TC, is reached. If TC is left blank on a particular TERMINATE node, the node destroys arriving entities but will not be used to stop the simulation.) In Figure 3.28, the termination count TC = 1000, which means that the simulation will run until 1000 customers have been served.

The network model in Figure 3.28 is translated into SLAM statements as shown in Figure 3.29. Note that all SLAM statements end with a semicolon. Besides the network statements, there are additional control statements. The first three statements provide general information to set up the model. The first required statement, GEN . . . , provides identification and other general information. The second required statement, LIMITS, specifies the number of files, the maximum number of attributes per entity, and the maximum length of all files combined. The third statement, NETWORK, and the next to last statement, ENDNETWORK, must surround all SLAM network statements (nodes and branches). The FIN statement is always the last statement in the deck.

The output statistics from the SLAM simulation are shown in Figure 3.30. The following estimates were obtained:

Maximum length of waiting line = 8
Server utilization = 0.7694

The average response time and proportion of customers having response time greater than or equal to 4 minutes were not estimated. Additional SLAM statements could be added to measure these statistics.

In summary, it is seen that a SLAM model, like a GPSS model, requires many fewer statements than a FORTRAN model. SLAM is more versatile than GPSS, because if the network representation proves inadequate, SLAM has the capability to combine GASP-like modeling with the process-oriented network statements. Overall, SLAM offers many desirable features to the simulation analyst.

S L A M   S U M M A R Y   R E P O R T

SIMULATION PROJECT SINGLE SERVER QUEUE   BY BANKS AND CARSON

DATE 5/31/1982   RUN NUMBER   1 OF   1

CURRENT TIME   .4164E+04
STATISTICAL ARRAYS CLEARED AT TIME 0.

FILE STATISTICS

FILE NUMBER	ASSOCIATED NODE TYPE	AVERAGE LENGTH	STANDARD DEVIATION	MAXIMUM LENGTH	CURRENT LENGTH	AVERAGE WAITING TIME
1	QUEUE	1.1221	1.4672	8	0	4.6677
2		1.7694	.4212	3	2	2.6523

SERVICE ACTIVITY STATISTICS

ACTIVITY INDEX	START NODE LABEL/TYPE	SERVER CAPACITY	AVERAGE UTILIZATION	STANDARD DEVIATION	CURRENT UTILIZATION	AVERAGE BLOCKAGE	MAXIMUM IDLE TIME/SERVERS	MAXIMUM BUSY TIME/SERVERS	ENTITY COUNT
1	QUEUE	1	.7694	.4212	1	0.0000	21.4036	232.2203	1000

Figure 3.30. SLAM output report for single-server queue.

Table 3.8. COMPARISON OF LANGUAGES FOR DISCRETE-EVENT SIMULATION

	Language				
Criteria	*FORTRAN*	*GASP*	*SIMSCRIPT II.5*	*GPSS V*	*SLAM*
Ease of learning	Good	Good	Good	Excellent	Excellent
Ease of conceptualizing a problem	Poor	Fair	Good	Excellent[a]	Excellent[a]
Systems oriented toward	None[b]	All	All	Queueing	All
Modeling approach					
Event-scheduling	No[b]	Yes	Yes	No	Yes
Process-interaction	No[b]	No	Yes	Yes	Yes
Continuous	No[b]	Yes	Yes	No	Yes
Support					
Random sampling built in	No[c]	Yes	Yes	No[d]	Yes
Statistics-gathering capability	Poor	Excellent	Excellent	Good	Excellent
List-processing capability	Poor	Good	Excellent	Fair[d]	Good
Ease of getting standard report	Poor	Excellent	Fair	Excellent	Excellent
Ease of designing special report	Fair	Good	Excellent	Poor[d]	Good
Debugging aids	Fair	Good	Excellent	Fair[d]	Good
Computer runtime	Excellent[e]	Good	Good	Poor[d]	Good
Documentation for learning language and for reference	Very good	Very good	Fair	Very good	Very good
Self-documenting code	Poor	Good	Good	Excellent	Good
Cost	Low[f]	Low	High	Low (GPSS/H, high)	Medium

[a]For queueing models, the block diagram (network) conceptualization is excellent.
[b]FORTRAN is not oriented toward system simulation. The programmer develops any desired orientation and takes any desired modeling approach.
[c]Several scientific subroutine libraries (e.g., IMSL) have FORTRAN routines for random variate generation.
[d]GPSS/H is much improved over GPSS V in these respects.
[e]FORTRAN will be fast assuming that the model is programmed in the most efficient manner.
[f]Usually available at most computer installations.

3.3. Summary and Comparison of Simulation Languages

This chapter has presented a brief introduction to five languages—FORTRAN, GASP IV, SIMSCRIPT II.5, GPSS V, and SLAM—that could be used in a simulation project. There are many criteria that are relevant when deciding which language to use on a given project. Some of these criteria are presented in Table 3.8, which provides a comparison of the five languages discussed in this chapter. However, if the simulator already knows a language, and that language has the capability to model the given system, familiarity may become the overriding criteria. The learning of any language requires considerable time and effort.

In general, GPSS and the process-interaction portion of SLAM are easiest to learn, especially for nonprogrammers. The trade-off is that the process-interaction languages are generally less versatile and less flexible. In SLAM and GASP (which are FORTRAN-based), and in SIMSCRIPT, the simulator has the full power and flexibility of a complete programming language, which makes possible complex numerical calculations and the modeling of unusual and diverse situations. Shannon [1975] provides a detailed discussion of choosing a simulation language and the criteria for making that choice.

REFERENCES

CACI, Inc. [1976], *SIMSCRIPT II.5 Reference Handbook*, Los Angeles.

Delfosse, C. M. [1976], *Continuous Simulation and Combined Simulation in SIMSCRIPT II.5*, CACI, Inc., Arlington, Va.

Gordon, Geoffrey [1975], *The Application of GPSS V to Discrete System Simulation*, Prentice-Hall, Englewood Cliffs, N.J.

Gordon, Geoffrey [1978], *System Simulation*, 2nd ed., Prentice-Hall, Englewood Cliffs, N.J.

Henriksen, J. O. [1979], *The GPSS/H User's Manual*, Wolverine Software, Falls Church, Va.

Kiviat, P. J., R. Villanueva, and H. M. Markowitz [1973], *SIMSCRIPT II.5 Programming Language*, ed. by E. C. Russell, CACI, Inc., Los Angeles.

Law, A. M., and W. D. Kelton [1982], *Simulation Modeling and Analysis*, McGraw-Hill, New York.

Pritsker, A. A. B. [1974], *The GASP IV Simulation Language*, Wiley, New York.

Pritsker, A. A. B., and C. Dennis Pegden [1979], *Introduction to Simulation and SLAM*, Wiley, New York.

Russell, E. C. [1983], *Building Simulation Models with SIMSCRIPT II.5*, CACI, Inc., Los Angeles.

Russell, E. C. [1976], *Simulating with Processes and Resources in SIMSCRIPT II.5*, CACI, Inc., Los Angeles.

RUSSELL, E. C., AND J. S. ANNINO [1979], *A Quick Look at SIMSCRIPT II.5*, CACI, Inc., Los Angeles.

SCHRIBER, THOMAS J. [1974], *Simulation Using GPSS*, Wiley, New York.

SHANNON, R. E. [1975], *Systems Simulation: The Art and Science*, Prentice-Hall, Englewood Cliffs, N.J.

EXERCISES

Instructions to the student: For most exercises, the student should first construct a model by explicitly defining:

1. System state
2. System entities and their attributes
3. Sets and the entities that may be put into the sets
4. Events and activities
5. Variables needed to collect cumulative statistics

Second, the student should either (1) develop the event logic (as in Figures 3.5 and 3.6 for Example 3.2) in preparation for using the event-scheduling approach, or (2) develop the system processes (as in Figure 3.4) in preparation for using the process-interaction approach. Finally, unless otherwise stated in the problem, the student should code the model in a general-purpose language (such as FORTRAN) or a special-purpose simulation language (such as GPSS, SIMCRIPT, or SLAM), or some of the simpler problems may be simulated manually.

Most problems contain activities that are uniformly distributed over an interval $[a, b]$. When using a simulation language, assume that all values between a and b are possible; that is, the activity time is a *continuous* random variable. When conducting a manual simulation, assume that $a, a + 1, a + 2, \ldots, b$ are the only possible values; that is, the activity time is a *discrete* random variable. The discrete assumption will simplify the manual simulation.

The uniform distribution is denoted by $U(a, b)$, where a and b are the endpoints of the interval, or by $m \pm h$, where m is the mean and h is the "spread" of the distribution. These four parameters are related by the equations

$$m = \frac{a+b}{2} \qquad h = \frac{b-a}{2}$$

$$a = m - h \qquad b = m + h$$

Some of the uniform random variate generators available in simulation languages require specification of a and b; others require m and h.

Some problems have activities that are assumed to be normally distributed, which is denoted by $N(\mu, \sigma^2)$, where μ is the mean and σ^2 the variance. (Since activity times are nonnegative, the normal distribution is appropriate only if

$\mu \geq k\sigma$, where k is at least 4 and preferably 5 or larger. If a negative value is generated, it is discarded.) Other problems use the exponential distribution with some rate λ, or mean $1/\lambda$. Chapter 4 reviews these distribtuions; Chapter 8 covers the generation of random variables having these distributions. Most languages have a facility to easily generate samples from these distributions. For FORTRAN simulations, the student may use the functions given in Section 3.2.1 for generating samples from the normal and exponential distributions.

1. (a) Using the event-scheduling approach, continue the (manual) checkout counter simulation which was begun in Example 3.2, Table 3.1. Use the same inter-arrival and service times that were previously generated and used in Example 2.1. When the last interarrival time is used, continue the simulation (allowing no new arrivals) until the system is empty. Compare the results obtained here to those obtained in Example 2.1. The results should be identical.
(b) Do exercise 1(a) again, adding the model components necessary to estimate mean response time and proportion of customers who spend 4 or more minutes in the system. (*Hint:* See Example 3.3, Table 3.2.)
(c) Comment on the relative merits of manual versus computerized simulations.

2. Construct the event logic diagrams for the dump truck problem, Example 3.4.

3. In the dump truck problem of Example 3.4, it is desired to estimate mean response time and the proportion of response times which are greater than 30 minutes. A response time for a truck begins when that truck arrives at the loader queue, and ends when the truck finishes weighing. Add the model components and cumulative statistics needed to estimate these two measures of system performance. Simulate for 8 hours.

4. Make the necessary modifications to the FORTRAN model of the checkout counter (Example 3.6) so that the simulation will run for exactly 60 hours. [*Note:* The $U(0, 1)$ random number generator RANF(·) will have to be changed, depending on the computer being used.]

5. In addition to the change in Exercise 4, also assume that any customers still at the counter at time 60 hours will be served, but no arrivals after time 60 hours are allowed. Make the necessary changes to the FORTRAN code and run the model.

6. Implement the changes in Exercises 4 and 5 in any language (SIMSCRIPT, GASP, GPSS, or SLAM).

7. Redo Example 2.2 (the Able–Baker carhop problem) by a manual simulation using the event-scheduling approach.

8. Redo Example 2.4 [the (M, N) inventory system] by a manual simulation using the event-scheduling approach.

9. Redo Example 2.5 (the bearing replacement problem) by a manual simulation using the event-scheduling approach.

10. Ambulances are dispatched at a rate of one every 15 ± 10 minutes in a large metropolitan area. Fifteen percent of the calls are false alarms which require 12 ± 2 minutes to complete. All other calls can be one of two kinds. The first kind are classified as serious. They constitute 15% of the non-false alarm calls and take 25 ± 5 minutes to complete. The remaining calls take 20 ± 10 minutes to complete. Assume that there are a very large number of available ambulances, and that any number can be on call at any time. Simulate the system for 500 calls to be completed.

11. In Exercise 10, estimate the mean time that an ambulance takes to complete a call.

12. (a) In Exercise 10, suppose that there is only one ambulance available. Any calls that arrive while the ambulance is out must wait. Can one ambulance handle the work load?

(b) Simulate with x ambulances, where $x = 1, 2, 3$, or 4, and compare the alternatives on the basis of length of time a call must wait, percentage of calls that must wait, and percentage of time the ambulance is out on call.

13. A hunter is hunting migratory birds in Jesup. She must remain in her present position until she has successfully killed 20 birds. It takes 2 ± 1 seconds to fire the gun and 3 ± 1 seconds to reload. The hunter is using a double-barreled shotgun, fires at most twice at each bird and reloads after firing at each bird. Birds pass over at a rate of one every 10 ± 2 seconds and the hunter has a 75% success rate on each shot. How long does it take the hunter to kill 20 birds?

14. A superhighway connects one large metropolitan area to another. A vehicle leaves the first city every 20 ± 15 seconds. Twenty percent of the vehicles have 1 passenger, 30% of the vehicles have 2 passengers, 10% have 3 passengers, and 10% have 4 passengers. The remaining 30% of the vehicles are buses which carry 40 people. It takes 60 ± 10 minutes for a vehicle to travel between the two metropolitan areas. How long does it take for 5000 people to arrive in the second city?

15. People arrive at a meat counter at a rate of one every 25 ± 10 seconds. There are two sections: one for beef and one for pork. People want goods from them in the following proportion: beef only, 50%; pork only, 30%; beef and pork, 20%. It takes 45 ± 20 seconds for a butcher to serve one customer for one order. All customers place one order, except "beef and pork" customers place two orders. Assume that there are enough butchers available to handle all customers present at any time. Simulate until 200 customers are served.

16. In Exercise 15, what is the maximum number of butchers needed during the course of simulation? Would this number always be sufficient to guarantee that no customer ever had to wait?

17. In Exercise 15, simulate with x butchers, where $x = 1, 2, 3, 4$. When all butchers are busy, a line forms. For each value of x, estimate the mean number of busy butchers.

18. A one-chair unisex hair shop has arrivals at the rate of one every 20 ± 15 minutes. One-half of the arriving customers want a dry cut, 30% want a style, and 20% want a trim only. A dry cut takes 15 ± 5 minutes, a style cut takes 25 ± 10 minutes, and a trim takes 10 ± 3 minutes. Simulate 50 customers coming through the hair shop. Compare the given proportion of service requests of each type with the simulated outcome. Are the results reasonable? Base your answer on the binomial distribution.

19. An airport has two concourses. Concourse 1 passengers arrive at a rate of one every 15 ± 2 seconds. Concourse 2 passengers arrive at a rate of one every 10 ± 5 seconds. It takes 30 ± 5 seconds to walk down concourse 1 and 35 ± 10 seconds to walk down concourse 2. Both concourses empty into the main lobby, adjacent to the baggage claim. It takes 10 ± 3 seconds to reach the baggage claim area from the main lobby. Only 60% of the passengers go to the baggage claim area. Simulate the passage of 500 passengers through the airport system. How many of these passengers went through the baggage claim area? How close can one expect this simulation estimate to be to the expected number who visit the baggage claim area? [The expected number is 0.60(500) = 300.]

20. In a multiphasic screening clinic, patients arrive at a rate of one every 5 ± 2 minutes to enter the audiology section. Examination takes 3 ± 1 minutes. Eighty percent of the patients were passed on to the next test with no problems. Of the remaining 20%, one-half require simple procedures which take 2 ± 1 minutes and are then sent for reexamination with the same probability of failure. The other half are sent home with medication. Simulate the system to determine how long it takes to screen and pass 200 patients. (*Note:* Persons sent home with medication are not considered "passed.")

21. Consider a bank with four tellers. Tellers 3 and 4 deal only with business acounts, while Tellers 1 and 2 deal with general accounts. Clients arrive at the bank at a rate of one every 3 ± 1 minutes. Of the clients, 33% are business accounts. Clients randomly choose between the two tellers available for each type of account. (Assume that a customer chooses a line without regard to its length and does not change lines.) Business accounts take 15 ± 10 minutes to complete and general accounts take 6 ± 5 minutes to complete. Simulate the system for 500 transactions to be completed. What percentage of time is each type of teller busy? What is the average time that each type of customer spends in the bank?

22. Repeat Exercise 21 assuming that customers join the shortest line for the teller handling their type of account.

23. In Exercises 21 and 22, estimate the mean delay of business customers and of regular customers. (Delay is time spent in the waiting line, and is exclusive of service time.) Also estimate the mean length of the waiting line, and the mean proportion of customers who are delayed longer than 1 minute.

24. Three different machines are available for machining a special type of part for 1 hour of each day. The processing time data are as follows:

Machine	*Time to Machine One Part*
1	20 ± 4 seconds
2	10 ± 3 seconds
3	15 ± 5 seconds

Assume that parts arrive by conveyor at a rate of one every 15 ± 5 seconds for the first 3 hours of the day. Machine 1 is available for the first hour, machine 2 for the second hour, and machine 3 for the third hour of each day. How many parts are produced in a day? How large a storage area is needed for parts waiting for a machine? Do parts "pile up" at any particular time? Why?

25. People arrive at a self-service cafeteria at the rate of one every 30 ± 20 seconds. Forty percent go to the sandwich counter, where one worker makes a sandwich in 60 ± 30 seconds. The rest go to the main counter, where one server spoons the prepared meal onto a plate in 45 ± 30 seconds. All customers must pay a single cashier, which takes 25 ± 10 seconds. For all customers, eating takes 20 ± 10 minutes. After eating, 10% of the people go back for dessert, spending an additional 10 ± 2 minutes altogether in the cafeteria. Simulate until 100 people have left the cafeteria. How many people are left in the cafeteria, and what are they doing, at the time the simulation stops?

26. Thirty trucks carrying bits and pieces of a C-5N cargo plane leave Atlanta at the same time for the Port of Savannah. From past experience, it is known that it takes 6 ± 2 hours for a truck to make the trip. Forty percent of the drivers stop for coffee, which takes an additional 15 ± 5 minutes.

(a) Model the situation as follows: For each driver, there is a 40% chance of stopping for coffee. When will the last truck reach Savannah?

(b) Model it so that exactly 40% of the drivers stop for coffee. When will the last truck reach Savannah?

27. Customers arrive at the Last National Bank every 40 ± 35 seconds. Currently, the customers pick one of two tellers at random. A teller services a customer in 75 ± 25 seconds. Once a customer joins a line, the customer stays in that line until the transaction is complete. Some customers want the bank to change to the single-line method used by the Lotta Trust Bank. Which is the faster method for the customer? Simulate for 15 minutes before collecting any summary statistic, then simulate for a 2-hour period. Compare the two queueing disciplines on the basis of teller utilization (percentage of busy time), mean delay of customers, and proportion of customers who must wait (before service begins) more than 1 minute, and more than 3 minutes.

28. Loana Tool Company rents chain saws. Customers arrive to rent chain saws at the rate of one every 30 ± 30 minutes. Dave and Betty handle these customers. Dave can rent a chain saw in 14 ± 4 minutes. Betty takes 10 ± 5 minutes. Customers returning chain saws arrive at the same rate as those renting chain saws. Dave and Betty spend 2 minutes with a customer to check in the returned chain saw. Service is first come, first served. When no customers are present, or Betty alone is busy, Dave gets these returned saws ready for re-renting. For each saw, this maintenance and cleanup takes him 6 ± 4 minutes and 10 ± 6 minutes, respectively. Whenever Dave is idle, he begins the next maintenance or cleanup. Upon finishing a maintenance or cleanup, Dave begins serving customers if one or more is waiting. Betty is always available for serving customers. Simulate the operation of the system starting with an empty shop at 8:00 A.M., closing the doors at 6:00 P.M., and getting chain saws ready for re-renting until 7:00 P.M. From 6:00 until 7:00 P.M., both Dave and Betty do maintenance and cleanup. Estimate the mean delay of customers who are renting chain saws.

29. In Exercise 28, change the shop rule regarding maintenance and cleanup to get a chain saw ready for re-rental. Now Betty does all this work. Upon finishing a cleanup on a saw, she helps Dave if a line is present. (That is, Dave and Betty both serve new customers and check in returned saws until Dave alone is busy, or the shop is empty.) Then she returns to her maintenance and cleanup duties.

(a) Estimate the mean delay of customers who are renting chain saws. Compare the two shop rules on this basis.

(b) Estimate the proportion of customers who must wait more than 5 minutes. Compare the two shop rules on this basis.

(c) Discuss the pros and cons of the two criteria in parts (a) and (b) for comparing the two shop rules. Suggest other criteria.

30. U of LA (University of Lower Altoona) has one computer terminal. Students arrive at the terminal every 15 ± 10 minutes to use the terminal for 12 ± 6 minutes. If the terminal is busy, 60% will come back in 10 minutes to use the terminal. If the terminal is still busy, 50% (of the 60%) will return in 15 minutes. How many students fail to use the terminal compared to 500 that actually finish? Demand and service occur 24 hours a day.

31. A warehouse holds 1000 cubic meters of cartons. These cartons come in three sizes: little (1 cubic meter), medium (2 cubic meters), and large (3 cubic meters). The cartons arrive at the following rates: little, every 10 ± 10 minutes; medium, every 15 minutes; and large, every 8 ± 8 minutes. If no cartons are removed, how long will it take to fill an empty warehouse?

32. Go Ape! buys a Banana II computer to handle all of its data processing needs. Jobs arrive every 10 ± 10 minutes to be batch processed one at a time. Processing takes 7 ± 7 minutes. The monkeys that run their computer cause a system failure every 60 ± 60 minutes. The failure lasts for 8 ± 4 minutes. When a failure occurs, the job that was being run resumes processing where it was left off. Simulate the operation of this system for 24 hours. Estimate the mean system response time. (A system response time is the length of time from arrival until processing is completed.) Also estimate the mean delay for those jobs that are in service when a computer system failure occurs.

33. Able, Baker, and Charlie are three carhops at the Sonic Drive-In (service at the speed of sound!). Cars arrive every 5 ± 5 minutes. The carhops service customers at the rate of one every 10 ± 6 minutes. However, the customers prefer Able over Baker, and Baker over Charlie. If the carhop of choice is busy, the customers choose the first available carhop. Simulate the system for 200 service completions. Estimate Able's, Baker's, and Charlie's utilization (percentage of time busy).

34. Jiffy Car Wash is a five-stage operation that takes 2 ± 1 minutes for each stage. There is room for 6 cars to wait to begin the car wash. The car wash facility holds 5 cars which move through the system in order, one car not being able to move until the car ahead of it moves. Cars arrive every 2.5 ± 2 minutes for a wash. If the car cannot get into the system, it drives across the street to Speedy Car Wash. Estimate the balking rate per hour. That is, how many cars drive off per hour? Simulate for one 12-hour day.

35. Consider the three machines A, B, and C pictured below. Arrivals of parts and processing times are as indicated (times in minutes).

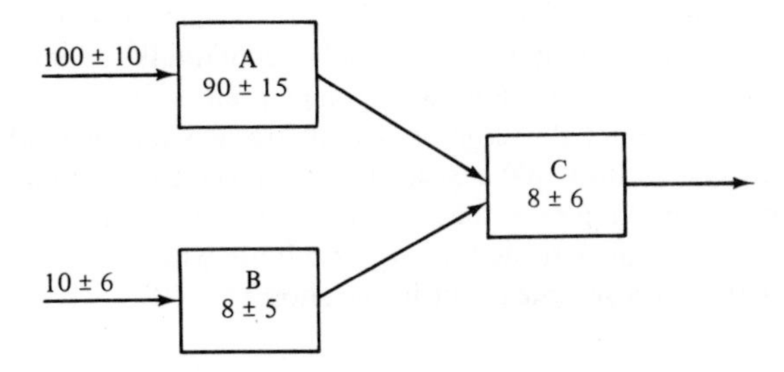

Machine A processes type A parts, machine B processes type B parts, and machine C processes both types of parts. All machines are subject to random breakdown: machine A every 400 $\pm$ 350 minutes with a downtime of 15 $\pm$ 14 minutes, machine B every 200 $\pm$ 150 minutes with a downtime of 10 $\pm$ 8 minutes, and machine C almost never, so its downtime is ignored. Parts from A are processed at C as soon as possible, ahead of any parts from B. When machine A breaks down, any part in it is sent to machine B and processed as soon as B becomes free, but processing begins over again, taking 100 $\pm$ 20 minutes. Again, parts from A are processed ahead of any parts waiting at B, but after any part currently being processed. When machine B breaks down, any part being processed resumes processing as soon as B becomes available. All machines handle one part at a time. Make two independent replications of the simulation of this system. Each replication will consist of an 8-hour initialization phase to load the system with parts, followed by a 40-hour steady-state run. (Independent replications means that each run uses a different stream of random numbers.) Management is interested in long-run production level [i.e., number of parts of each type (A or B) produced per 8-hour day], long-run utilization of each machine, and the existence of bottlenecks (long "lines" of waiting parts). Report the output date in a table as follows:

	Run 1	*Run 2*	*Average of Two Runs*
Utilization A			
Utilization B			
Etc.			

Include a brief statement summarizing the important results.

36. Workers come to a supply store at the rate of one every 10 $\pm$ 4 minutes. Their requisitions are handled by one of three clerks; a clerk takes 22 $\pm$ 10 minutes to handle a requisition. All requisitions are then passed to a single cashier, who spends 7 $\pm$ 6 minutes per requisition. Simulate the system for 50 hours.
(a) Estimate the utilization of each clerk, based on the 50-hour simulation.
(b) How many workers are completely served? How many do the three clerks serve? How many workers arrive? Are all three clerks ever busy at the same time? What is the average number of busy clerks?

37. People arrive at a barbershop at the rate of one every 4.5 minutes. If the shop is full (it can hold five people altogether), 30% of the potential customers leave and come back in 60 $\pm$ 20 minutes. The others leave and do not return. One barber gives a haircut in 8 $\pm$ 2 minutes, whereas the second talks a lot and it takes 12 $\pm$ 4 minutes. If both barbers are idle, a customer prefers the first barber. (Treat customers trying to reenter the shop as if they are new customers.)
Simulate this system until 300 customers have received a haircut.
(a) Estimate the balking rate, that is, the number turned away per minute.
(b) Estimate the number turned away per minute who do not try again.
(c) What is the average time spent in the shop?

(d) What is the average time spent getting a haircut (not including delay)?
(e) What is the average number of customers in the shop?

38. People arrive at a microscope exhibit at a rate of one every 8 ± 2 minutes. Only one person can see the exhibit at a time. It takes 5 ± 2 minutes to see the exhibit. A person can buy a "privilege" ticket for $1 which gives him or her priority in line over those who are too cheap to spend the buck. Some 50% of the viewers do this, but they make their decision to do so only if one or more people are in line when they arrive. The exhibit is open continuously from 10:00 A.M. to 4:00 P.M. Simulate the operation of the system for one complete day. How much money is generated from the sale of privilege tickets?

39. Messages are sent from the emergency room of Northside Hospital to central supply. Replies are sent back to the emergency room. Messages are created every 6 ± 3 minutes. They go through a pneumatic tube which takes 2 minutes to traverse. Only one message can be carried through the tube at a time. Some 70% of these messages require a reply. To prepare a reply takes 2 ± 1 minute. Messages from the emergency room have priority. Simulate until 100 replies have been received at the emergency room. Estimate the utilization of the pneumatic tube. How many messages (and replies) are waiting to be sent, on the average?

40. Two machines are available for drilling parts (A-type and B-type). A-type parts arrive at a rate of one every 10 ± 3 minutes and B-type parts arrive at a rate of one every 3 ± 2 minutes. For B-type parts, workers choose an idle machine, or if both drills, the Dewey and the Truman, are busy, they choose a machine at random and stay with their choice. A-type parts must be drilled as soon as possible; therefore, if a machine is available, preferably the Dewey, it is used; otherwise the part goes to the head of the line for the Dewey drill. All jobs take 4 ± 3 minutes to complete. Simulate the completion of 100 A-type parts. Estimate the mean number of A-type parts waiting to be drilled.

41. A telephone in a police precinct is used for both emergency calls and personal calls. Personal calls are on a first come, first served basis, and are made at a rate of one every 5 ± 1 minutes. Emergency calls have priority and can preempt other calls. They arrive at a rate of one every 15 ± 5 minutes. Emergency calls take 3 ± 1 minutes to complete while personal calls take 2 ± 2 minutes. Twenty percent of the people using the phone on a nonemergency basis wish to make another call as soon as possible, but they are given the lowest priority for their second call. Simulate until 200 calls of all types have been completed. Estimate the phone utilization.

42. A computer center substation has two terminals. Students arrive at a rate of one every 8 ± 2 minutes. They can be interrupted by professors, who arrive at a rate of one every 12 ± 2 minutes. There is one systems analyst who can interrupt anyone, but students are interrupted before professors. The systems analyst spends 6 ± 4 minutes on the terminal and then returns in 20 ± 5 minutes. Professors and students spend 4 ± 2 minutes on the terminal. If a person is interrupted, that person joins the head of the queue and resumes service as soon as possible. Simulate for 50 professor or analyst jobs. Estimate the interruption rate per hour, and the mean length of the waiting line of students.

43. Parts are machined on a drill press. They arrive at a rate of one every 5 ± 3 minutes and it takes 3 ± 2 minutes to machine them. Once every hour, a rush job arrives which takes 12 ± 3 minutes to complete. The rush job interrupts the present job, which will become the next normal job on the machine (it begins the machining process from the start). Simulate the machining of 10 rush jobs. Estimate the mean system response time for each type of part. (A response time is the total time that a part spends in the system.)

44. Messages are generated at a rate of one every 35 ± 10 seconds for transmission one at a time. Transmission takes 20 ± 5 seconds. At intervals of 6 ± 3 minutes, urgent messages lasting 10 ± 3 seconds take over the transmission line. Any message in progress must be reprocessed for 2 minutes before it can be resubmitted for transmission. When resubmitted, it goes to the head of the line. Simulate for 90 minutes. Estimate the percentage of time the line is busy with ordinary messages.

45. A worker packs boxes that arrive at a rate of one every 15 ± 3 minutes. It takes 8 ± 3 minutes to pack a box. Once every hour the worker is in:errupted to wrap specialty orders that take 16 ± 3 minutes to pack. The order which is preempted is continued in progress. Simulate for 40 hours. Estimate the mean proportion of time that the number of boxes waiting to be packed is more than five.

46. A patient arrives at the Emergency Room at Hello-Hospital about every 40 ± 19 minutes. They will be treated by either Doctor Slipup or Doctor Gutcut. Twenty percent of the patients are classified as NIA (need immediate attention) and the rest as CW (can wait). NIA patients are given the highest priority, 3, see a doctor as soon as possible for 40 ± 37 minutes, but then their priority is reduced to 2 and they wait until a doctor is free again, when they receive further treatment for 30 ± 25 minutes and are then discharged. CW patients initially receive a priority of 1 and are treated (when their turn comes) for 15 ± 14 minutes; their priority is then increased to 2, they wait again until a doctor is free and receive 10 ± 8 minutes of final treatment, and are then discharged. Simulate for 20 days of continuous operation, 24 hours per day. Precede this by a 2-day initialization period to load up the system with patients. Report conditions at times 0 days, 2 days, and 22 days. Does a 2-day initialization appear long enough to load the system to a level reasonably close to steady-state conditions?

(a) Measure the average and maximum queue length of NIA patients from arrival to first seeing a doctor. What percent do not have to wait at all? Also tabulate and plot the distribution of this initial waiting time for NIA patients. What percent wait less than 5 minutes before seeing a doctor?

(b) Tabulate and plot the distribution of total time in system for all patients. Estimate the 90% quantile. That is, 90% of the patients spend less than x amount of time in the system. Estimate x.

(c) Tabulate and plot the distribution of remaining time in system from after the first treatment to discharge, for all patients. Estimate the 90% quantile.

(*Note:* Most simulation languages provide the facility to automatically tabulate the distribution of any specified variable.)

47. People arrive at a newspaper stand with an interarrival time that is exponentially distributed with a mean of 0.5 minute. Fifty-five percent of the people buy just the morning paper, while 25% buy the morning paper and a *Wall Street Journal*. The remainder buy only the *Wall Street Journal*. One clerk handles the *Wall Street Journal* sales, while another clerk handles morning paper sales. A person buying both goes to the *Wall Street Journal* clerk. The time it takes to serve a customer is normally distributed with a mean of 40 seconds and a standard deviation of 4 seconds for all transactions. Collect statistics on queues for each type of transaction. Suggest ways for making the system more efficient. Simulate for 4 hours.

48. Bernie remodels houses and makes room additions. The time it takes to finish a job is normally distributed with a mean of 17 elapsed days and a standard deviation of 3 days. Homeowners sign contracts for jobs at exponentially distributed intervals having a mean of 20 days. Bernie has only one crew. Estimate the mean waiting time (from signing the contract until work begins) for those jobs where a positive wait occurs. Also estimate the percentage of time the crew is idle. Simulate until 100 jobs have been completed.

49. Parts arrive at a machine in random fashion with exponential interrival times having a mean of 60 seconds. All parts require 5 seconds to prepare and align for machining. There are three different types of parts, in the proportions shown below. The times to machine each type of part are normally distributed with mean and standard deviation as follows:

Part Type	*Percent*	*Mean (Seconds)*	σ *(Seconds)*
1	50	48	8
2	30	55	9
3	20	85	12

Find the distribution of total time to complete processing for all types of parts. What proportion of parts take more than 60 seconds for complete processing? How long do parts have to wait, on the average? Simulate for one 8-hour day.

50. Shopping times at a department store have been found to have the following distribution:

Shopping Time (Minutes)	*Number of Shoppers*
0–10	90
10–20	120
20–30	270
30–40	145
40–50	88
50–60	28

After shopping, the customers choose one of six checkout counters. Checkout times are normally distributed with a mean of 5.1 minutes and a standard deviation of 0.7 minute. Interarrival times are exponentially distributed with a mean of 1 minute. Gather statistics for each checkout counter (including queues). Tabulate the distribution of time to complete shopping, and the distribution of time to complete shopping and checkout procedures. What proportion of customers spend more than 45 minutes in the store? Simulate for one 16-hour day.

51. The interarrival time for parts needing processing is given as follows:

Interarrival Time (Seconds)	*Proportion*
10–20	0.20
20–30	0.30
30–40	0.50

There are three types of parts: A, B, and C. The proportion of each part, and the mean and standard deviation of the normally distributed processing times are as follows:

Part Type	*Proportion*	*Mean*	*Standard Deviation*
A	0.5	30 seconds	3 seconds
B	0.3	40 seconds	4 seconds
C	0.2	50 seconds	7 seconds

Each machine processes any type of part, one part at a time. Use simulation to compare one to two to three machines working in parallel. What criteria would be appropriate for such a comparision?

52. Orders are received for one of four types of parts. The interarrival time between orders is exponentially distributed with a mean of 10 minutes. The table that follows shows the proportion of the parts by type and the time to fill each type of order by the single clerk.

Part Type	*Percentage*	*Service Time (Minutes)*
A	40	N(6.1, 1.3)
B	30	N(9.1, 2.9)
C	20	N(11.8, 4.1)
D	10	N(15.1, 4.5)

Orders of types A and B are picked up immediately after they are filled, but orders of types C and D must wait 10 ± 5 minutes to be picked up. Tabulate the distribution of time to complete delivery for all orders combined. What proportion take less than 15 minutes? What proportion take less than 25 minutes? Simulate for an 8-hour initialization period, followed by a 40-hour run. Do not use any data collected in the 8-hour initialization period.

53. Three independent widget-producing machines all require the same type of vital part which needs frequent maintenance. To increase production it is decided to keep two spare parts on hand (for a total of $2 + 3 = 5$ parts). After 2 hours of use, the part is removed from the machine and taken to a single technician, who can do the required maintenance in 30 ± 20 minutes. After maintenance, the part is placed in the pool of spare parts, to be put into the first machine that requires it. The technician has other duties, namely, repairing other items which have a higher priority and which arrive every 60 ± 20 minutes requiring 15 ± 15 minutes to repair. Also, the technician takes a 15-minute break in each 2-hour time period. That is, the technician works 1 hour 45 minutes, takes off 15 minutes, works 1 hour 45 minutes, takes off 15 minutes, and so on.

(a) What are the model's initial conditions? That is, where are the parts at time 0 and what is their condition? Are these conditions typical of "steady state"?

(b) Make each replication of this experiment consist of an 8-hour initialization phase followed by a 40-hour data collection phase. Make four statistically independent replications of the experiment all in one computer run. (That is, make four runs with each using a different set of random numbers.)

(c) Estimate the mean number of busy machines and the proportion of time the technician is busy.

(d) Parts are estimated to cost the company $40 per part per 8-hour day (regardless of how much they are in use). The cost of the technician is $10 per hour. A working machine produces widgets worth $80 for each hour of production. Develop an expression to represent total cost per hour which can be attributed to widget production (i.e., not all of the technician's time is due to widget production). Evaluate this expression based on the results of the simulation.

54. The Wee Willy Widget Shop overhauls and repairs all types of widgets. The shop consists of five work stations and the flow of jobs through the shop is as depicted here:

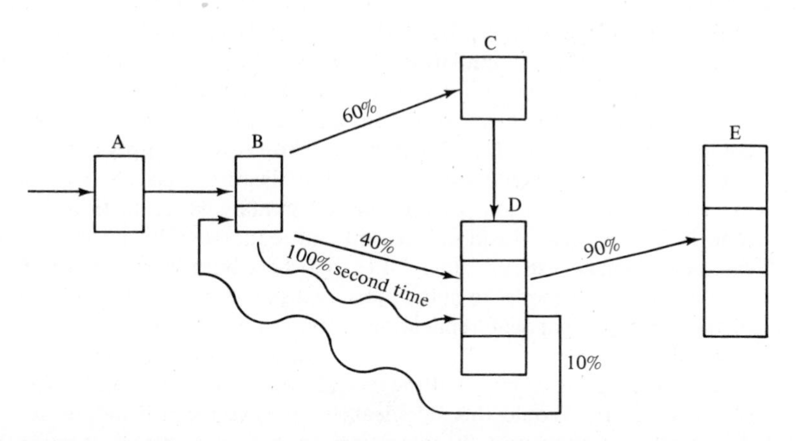

Regular jobs arrive at station A at the rate of one every 15 ± 13 minutes. Rush jobs arrive every 4 ± 3 hours and are given a higher priority except at station C, where they are put on a conveyor and sent through a cleaning and degreasing operation along with all other jobs. For jobs the first time through a station, processing and repair times are as follows:

Station	*Number Machines or Workers*	*Processing and/or Repair Times (Minutes)*	*Description*
A	1	12 ± 2	Receiving clerk
B	3	40 ± 20	Disassembly and parts replacement
C	1	20	Degreaser
D	4	50 ± 40	Reassembly and adjustments
E	3	40 ± 5	Packing and shipping

The times listed above hold for all jobs that follow one of the two sequences A → B → C → D → E or A → B → D → E. However, about 10% of the jobs coming out of station D are sent back to B for further work (which takes 30 ± 10 minutes) and then are sent to D and finally to E. The path of these jobs is as follows:

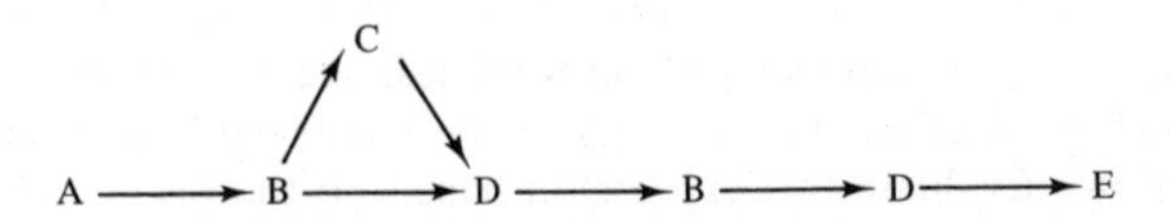

Every 2 hours, beginning 1 hour after opening, the degreasing station C shuts down for routine maintenance, which takes 10 ± 1 minute. However, this routine maintenance does not begin until the current widget, if any, has completed its processing.

(a) Make three independent replications of the simulation model, where one replication equals an 8-hour simulation run, preceded by a 2-hour initialization run. The three sets of output represent three typical days. The main performance measure of interest is mean response time per job, where a response time is the total time a job spends in the shop. The shop is never empty in the morning, but the model will be empty without the initialization phase. So run the model for a 2-hour initialization period and collect statistics from time 2 hours to time 10 hours. This "warm-up" period will reduce the downward bias in the estimate of mean response time. Note that the 2-hour warm-up is a device to load a simulation model to some more realistic level than empty. From each of the three independent replications, obtain an estimate of mean response time. Also obtain an overall estimate, the sample average of the three estimates.

(b) Management is considering putting one additional worker at the busiest station (A, B, D, or E). Would this significantly improve mean response time?

(c) As an alternative to part (b), management is considering replacing machine C with a faster one that processes a widget in only 14 minutes. Would this significantly improve mean response time?

55. A building materials firm loads trucks with two payloader tractors. The distribution of truck loading times has been determined to be exponential with a mean loading time of 6 minutes. The truck interarrival time is exponentially distributed with an arrival rate of 16 per hour. The waiting time of a truck and driver is estimated to cost $40 per hour. How much (if any) could the firm save (per 10 hour day) if an overhead hopper system that would fill any truck in a constant time of 2 minutes is installed? (Assume that the present tractors could and would adequately service the conveyors loading the hoppers.)

56. A milling machine department has 10 machines. The runtime until failure occurs on a machine is exponentially distributed with a mean of 20 hours. Repair times are uniformly distributed between 3 and 7 hours. Select an appropriate run length and appropriate initial conditions.

(a) How many repair persons are needed to ensure that the mean number of machines running is greater than eight?

(b) If there are two repair persons, estimate the expected number of machines that are either running or being served.

part two

MATHEMATICAL AND STATISTICAL MODELS

STATISTICAL MODELS IN SIMULATION

In modeling real-world phenomena there are few situations where the actions of the entities within the system under study can be completely predicted in advance. The world the model builder sees is probabilistic rather than deterministic. There are many causes of variation. The time it takes a repairperson to fix a broken machine is a function of the complexity of the breakdown, whether the repairperson brought the proper replacement parts and tools to the site, whether another repairperson asks for assistance during the course of the repair, whether the machine operator receives a lesson in preventive maintenance, and so on. To the model builder, these variations appear to occur by chance and cannot be predicted. However, some statistical model may well describe the time to make a repair.

An appropriate model can be developed by sampling the phenomenon of interest. Then, through educated guesses, the model builder would select a known distribution form, make an estimate of the parameter(s) of this distribution, and then test to see how good a fit has been obtained. Through continued efforts in the selection of an appropriate distribution form, a postulated model may be accepted. This multistep process is described in Chapter 9.

Section 4.1 contains a review of probability terminology and concepts. Some typical applications of statistical models, or distribution forms, are given in Section 4.2. Then, a number of selected discrete and continuous distributions are discussed in Sections 4.3 and 4.4. The selected distributions are those which describe a wide variety of probabilistic events and, further, appear in different

4

contexts in other chapters of this text. Additional discussion about the distribution forms appearing in this chapter, and about distribution forms mentioned but not described, is available from a number of sources [Hines and Montgomery, 1980; Ross, 1981; Feller, 1968; Mood and Graybill, 1963; Fishman, 1973]. Section 4.5 describes the Poisson process and its relationship to the exponential distribution. Section 4.6 discusses empirical distributions.

4.1. Review of Terminology and Concepts

1. *Discrete random variables.* Let X be a random variable. If the number of possible values of X is finite, or countably infinite, X is called a discrete random variable. The possible values of X may be listed as $x_1, x_2, \ldots$. In the finite case the list terminates. In the countably infinite case the list continues indefinitely.

EXAMPLE 4.1

The number of jobs arriving each week at a job shop is observed. The random variable of interest is X, where

$$X = \text{number of jobs arriving each week}$$

The possible values of X are given by the range space of X, which is denoted by R_X. Here $R_X = \{0, 1, 2, \ldots\}$.

Let X be a discrete random variable. With each possible outcome x_i in R_X, a number $p(x_i) = P(X = x_i)$ gives the probability that the random variable equals the value of x_i. The numbers $p(x_i)$, $i = 1, 2, \ldots$, must satisfy the following two conditions:

1. $p(x_i) \geq 0$ for all i.
2. $\sum_{i=1}^{\infty} p(x_i) = 1$.

The collection of pairs $(x_i, p(x_i))$, $i = 1, 2, \ldots$ is called the probability distribution of X, and $p(x_i)$ is called the probability mass function (pmf) of X.

EXAMPLE 4.2

Consider the experiment of tossing a single die. Define X as the number of spots on the up face of the die after a toss. Then $R_X = \{1, 2, 3, 4, 5, 6\}$. Assume the die is loaded so that the probability that a given face lands up is proportional to the number of spots showing. The discrete probability distribution for this random experiment is given by

x_i	1	2	3	4	5	6
$p(x_i)$	1/21	2/21	3/21	4/21	5/21	6/21

The earlier stated conditions are satisfied. That is, $p(x_i) \geq 0$ for $i = 1, \ldots, 6$ and $\sum_{i=1}^{\infty} p(x_i) = 1/21 + \cdots + 6/21 = 1$. The distribution is shown graphically in Figure 4.1.

2. *Continuous random variables.* If the range space R_X of the random variable X is an interval or a collection of intervals, X is called a continuous random variable. For a continuous random variable X, the probability that X lies in the interval $[a, b]$ is given by

$$P(a \leq X \leq b) = \int_a^b f(x)\, dx \tag{4.1}$$

The function $f(x)$ is called the probability density function (pdf) of the random variable X. The pdf satisfies the following conditions:

a. $f(x) \geq 0$ for all x in R_X.

b. $\int_{R_X} f(x)\, dx = 1$.

c. $f(x) = 0$ if x is not in R_X.

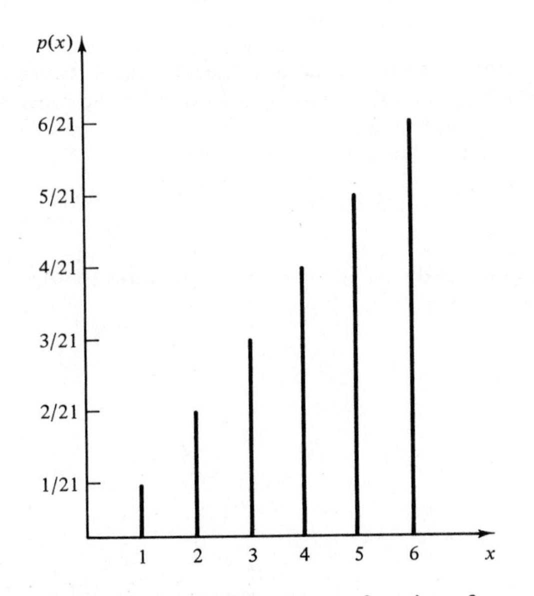

Figure 4.1. Probability mass function for loaded die example.

As a result of equation (4.1), for any specified value x_0, $P(X = x_0) = 0$ since

$$\int_{x_0}^{x_0} f(x)\, dx = 0$$

Since $P(X = x_0) = 0$, the following equations also hold:

$$P(a \leq X \leq b) = P(a < X \leq b) = P(a \leq X < b) = P(a < X < b) \tag{4.2}$$

The graphical interpretation of equation (4.1) is shown in Figure 4.2. The shaded area represents the probability that X lies in the interval $[a, b]$.

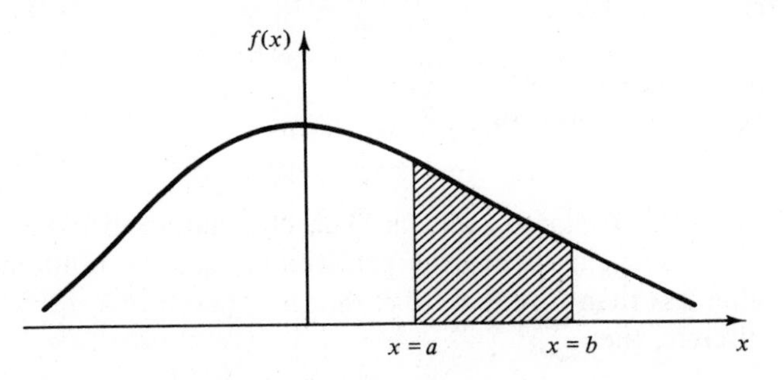

Figure 4.2. Graphical interpretation of $P(a < X < b)$.

EXAMPLE 4.3

The life of a cathode ray tube used to inspect cracks in aircraft wings is given by X, a continuous random variable assuming all values in the range $x \geq 0$. The pdf of the lifetime, in years, is as follows:

$$f(x) = \begin{cases} \frac{1}{2}e^{-x/2}, & x \geq 0 \\ 0, & \text{otherwise} \end{cases}$$

This pdf is shown graphically in Figure 4.3. The random variable X is said to have an exponential distribution with mean 2 years.

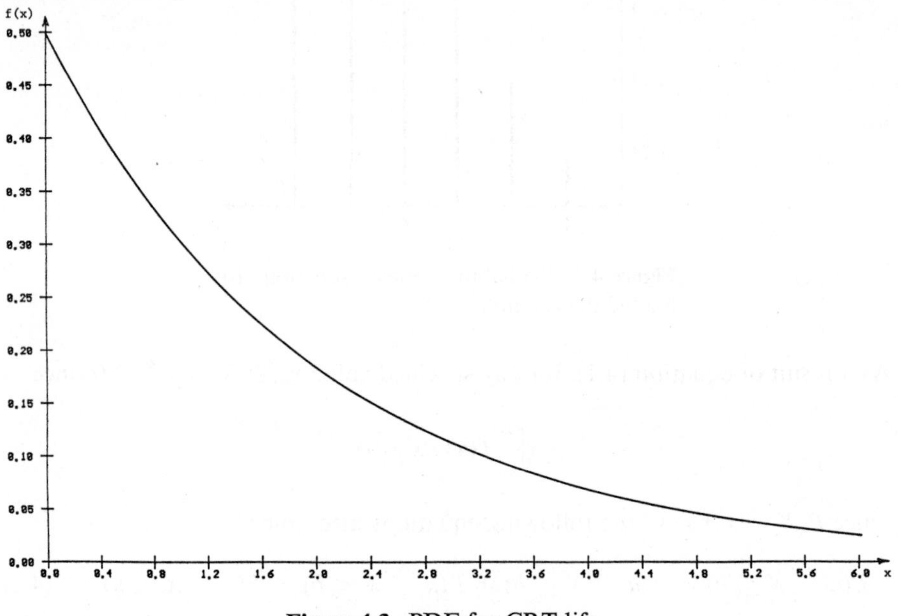

Figure 4.3. PDF for CRT life.

The probability that the life of the cathode ray tube is between 2 and 3 years is determined from

$$\begin{aligned} P(2 \leq X \leq 3) &= \tfrac{1}{2}\int_2^3 e^{-x/2}\,dx \\ &= -e^{-3/2} + e^{-1} = -0.223 + 0.368 = 0.145 \end{aligned}$$

3. *Cumulative distribution function.* The cumulative distribution function (cdf), denoted by $F(x)$, measures the probability that the random variable X assumes a value less than or equal to x, that is, $F(x) = P(X \leq x)$.

If X is discrete, then

$$F(x) = \sum_{\substack{\text{all} \\ x_i \leq x}} p(x_i) \tag{4.3a}$$

If X is continuous, then

$$F(x) = \int_{-\infty}^{x} f(t)\, dt \tag{4.3b}$$

Some properties of the cdf are listed here:

a. F is a nondecreasing function. If $a < b$, then $F(a) \leq F(b)$.
b. $\lim_{x\to\infty} F(x) = 1$.
c. $\lim_{x\to-\infty} F(x) = 0$.

All probability questions about X can be answered in terms of the cdf. For example,

$$P(a < X \leq b) = F(b) - F(a) \qquad \text{for all } a < b \tag{4.4}$$

For continuous distributions not only does equation (4.4) hold, but also the probabilities in equation (4.2) are equal to $F(b) - F(a)$.

EXAMPLE 4.4

The die-tossing experiment described in Example 4.2 has a cdf given as follows:

x	$(-\infty, 1)$	$[1, 2)$	$[2, 3)$	$[3, 4)$	$[4, 5)$	$[5, 6)$	$[6, \infty)$
$F(x)$	0	1/21	3/21	6/21	10/21	15/21	21/21

where $[a, b) = \{a \leq x < b\}$. The cdf for this example is shown graphically in Figure 4.4.

If X is a discrete random variable with possible values $x_1, x_2, \ldots$, where $x_1 < x_2 < \cdots$, the cdf is a step function. The value of the cdf is constant in the interval $[x_{i-1}, x_i)$ and then takes a step, or jump, of size $p(x_i)$ at x_i. Thus, in Example 4.4, $p(3) = 3/21$, which is the size of the step when $x = 3$.

EXAMPLE 4.5

The cdf for the cathode ray tube described in Example 4.3 is given by

$$F(x) = \tfrac{1}{2} \int_{0}^{x} e^{-t/2}\, dt = 1 - e^{-x/2}$$

The probability that the cathode ray tube will last for less than 2 years is given by

$$P(0 \leq X \leq 2) = F(2) - F(0) = F(2) = 1 - e^{-1} = 0.632$$

The probability that the life of the cathode ray tube is between 2 and 3 years is determined from

$$\begin{aligned} P(2 \leq X \leq 3) &= F(3) - F(2) = (1 - e^{-3/2}) - (1 - e^{-1}) \\ &= -e^{-3/2} + e^{-1} = -0.223 + 0.368 = 0.145 \end{aligned}$$

as found in Example 4.3.

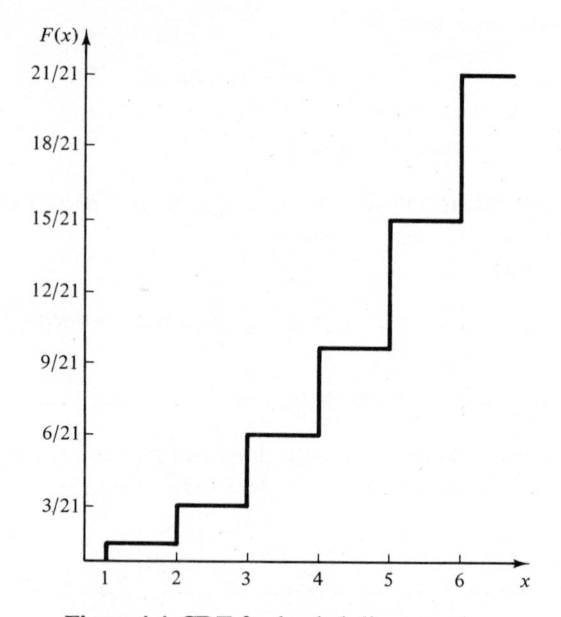

Figure 4.4 CDF for loaded-die example.

4. *Expectation.* An important concept in probability theory is that of the expectation of a random variable. If X is a random variable, the expected value of X, denoted by $E(X)$, for discrete and continuous variables is defined as follows:

$$E(X) = \sum_{\text{all } i} x_i p(x_i) \qquad \text{if } X \text{ is discrete} \tag{4.5a}$$

and

$$E(X) = \int_{-\infty}^{\infty} x f(x)\, dx \qquad \text{if } X \text{ is continuous} \tag{4.5b}$$

The expected value $E(X)$ of a random variable X is also referred to as the mean, μ, or the first moment of X. The quantity $E(X^n)$, $n \geq 1$, is called the nth moment of X, and is computed as follows:

$$E(X^n) = \sum_{\text{all } i} x_i^n p(x_i) \qquad \text{if } X \text{ is discrete} \tag{4.6a}$$

and

$$E(X^n) = \int_{-\infty}^{\infty} x^n f(x)\, dx \qquad \text{if } X \text{ is continuous} \tag{4.6b}$$

The variance of a random variable, X, denoted by $V(X)$ or σ^2, is defined by

$$V(X) = E[(X - E[X])^2]$$

A useful identity in computing $V(X)$ is given by

$$V(X) = E(X^2) - [E(X)]^2 \tag{4.7}$$

The mean $E(X)$ is a measure of the central tendency of a random variable. The variance of X measures the expected square of the random variable from its expected value. Thus, the variance, $V(X)$, is a measure of the spread or variation of the possible values of X around the mean $E(X)$. The standard deviation, σ, is defined to be the square root of the variance, σ^2. The mean, $E(X)$, and the standard deviation, $\sigma = \sqrt{V(X)}$, are expressed in the same units.

EXAMPLE 4.6

The mean and variance of the die-tossing experiment described in Example 4.2 are determined as follows:

$$E(X) = 1\left(\frac{1}{21}\right) + 2\left(\frac{2}{21}\right) + \cdots + 6\left(\frac{6}{21}\right) = \frac{91}{21} = 4.33$$

To compute $V(X)$ using equation (4.7), first compute $E(X^2)$ from equation (4.6a) as follows:

$$E(X^2) = 1^2\left(\frac{1}{21}\right) + 2^2\left(\frac{2}{21}\right) + \cdots + 6^2\left(\frac{6}{21}\right) = 21$$

Thus,

$$V(X) = 21 - \left(\frac{91}{21}\right)^2 = 21 - 18.78 = 2.22$$

and

$$\sigma = \sqrt{V(X)} = 1.49$$

EXAMPLE 4.7

The mean and variance of the life of the cathode ray tube described in Example 4.3 are determined as follows:

$$E(X) = \frac{1}{2}\int_0^\infty xe^{-x/2}\,dx = -xe^{-x/2}\Big|_0^\infty + \int_0^\infty e^{-x/2}\,dx$$
$$= 0 + \frac{1}{1/2}e^{-x/2}\Big|_0^\infty = 2 \text{ years}$$

To compute $V(X)$ using equation (4.7), first compute $E(X^2)$ from equation (4.6b) as follows:

$$E(X^2) = \frac{1}{2}\int_0^\infty x^2e^{-x/2}\,dx$$

Thus,

$$E(X^2) = -x^2e^{-x/2}\Big|_0^\infty + 2\int_0^\infty xe^{-x/2}\,dx = 8$$

Thus,

$$V(X) = 8 - 2^2 = 4 \text{ years}^2$$

and

$$\sigma = \sqrt{V(X)} = 2 \text{ years}$$

With a mean life of 2 years and a standard deviation of 2 years, most analysts would conclude that actual lifetimes, X, have a fairly large variability.

5. *The mode.* The mode is used in describing several statistical models which appear in this chapter. In the discrete case, the mode is the value of the random variable that occurs most frequently. In the continuous case, the mode is the maximum value of the pdf. The mode may not be unique; if the modal value occurs at two values of the random variable, the distribution is said to be bimodal.

4.2. Useful Statistical Models

Numerous situations arise in the conduct of a simulation where an investigator may choose to introduce probabilistic events. In Chapter 2, queueing, inventory, and reliability examples were given. In a queueing system, interarrival and service times are often probabilistic. In an inventory model the time between demands and the lead times (time between placing and receiving an order) may be probabilistic. In a reliability model, the time to failure may be probabilistic. In each of these instances, the simulator desires to generate random events, and to use a known statistical model if the underlying distribution can be found. In the following paragraphs, statistical models appropriate to these application areas will be discussed. Additionally, statistical models useful in the case of limited data are mentioned.

1. *Queueing systems.* In Chapter 2, examples of waiting-line problems were given. In both Chapters 2 and 3, these problems were solved using simulation. In the queueing examples, interarrival- and service-time patterns were given. In these examples, the times between arrivals and the service times were always probabilistic, which is usually the case. However, it is possible to have a constant interarrival time (as in the case of a line moving at a constant speed in the assembly of an automobile), or a constant service time (as in the case of robotized spot welding on the same assembly line). The following example illustrates how probabilistic interarrival times might occur.

EXAMPLE 4.8

Mechanics arrive at a centralized tool crib as shown in Table 4.1. Attendants check in and check out the requested tools to the mechanics. The collection of data begins at 10:00 A.M. and continues until 20 different interarrival times are recorded.

Table 4.1. ARRIVAL DATA

Arrival Number	*Arrival (Hour: Minutes:: Seconds)*	*Interarrival Time (Minutes:: Seconds)*
1	10: 05:: 03	—
2	10: 12:: 16	7:: 13
3	10: 15:: 48	3:: 32
4	10: 24:: 27	8:: 39
5	10: 32:: 19	7:: 52
6	10: 35:: 43	3:: 24
7	10: 39:: 51	4:: 08
8	10: 40:: 30	0:: 39
9	10: 41:: 17	0:: 47
10	10: 44:: 12	2:: 55
11	10: 45:: 47	1:: 33
12	10: 50:: 47	5:: 00
13	11: 00:: 05	9:: 18
14	11: 04:: 58	4:: 53
15	11: 06:: 12	1:: 14
16	11: 11:: 23	5:: 11
17	11: 16:: 31	5:: 08
18	11: 17:: 18	0:: 47
19	11: 21:: 26	4:: 08
20	11: 24:: 43	3:: 17
21	11: 31:: 19	6:: 36

Rather than record the actual time of day, the absolute time from a given origin could have been computed. Thus, the first mechanic could have arrived at time zero, the second mechanic at time 7: 13 (7 minutes, 13 seconds), and so on.

EXAMPLE 4.9

Another way of presenting interarrival data is to determine the number of arrivals per time period. Since these arrivals occur over approximately $1\frac{1}{2}$ hours, it is convenient to look at 10-minute time intervals for the first 20 mechanics. That is, in the first 10-minute time period, one arrival occurred at 10: 05:: 03. In the second time period, two mechanics arrived, and so on. The results are summarized in Table 4.2. This data could then be plotted in a histogram as shown in Figure 4.5.•

Table 4.2. ARRIVALS IN SUCCESSIVE TIME PERIODS

Time Period	*Number of Arrivals*	*Time Period*	*Number of Arrivals*
1	1	6	1
2	2	7	3
3	1	8	3
4	3	9	2
5	4	—	—

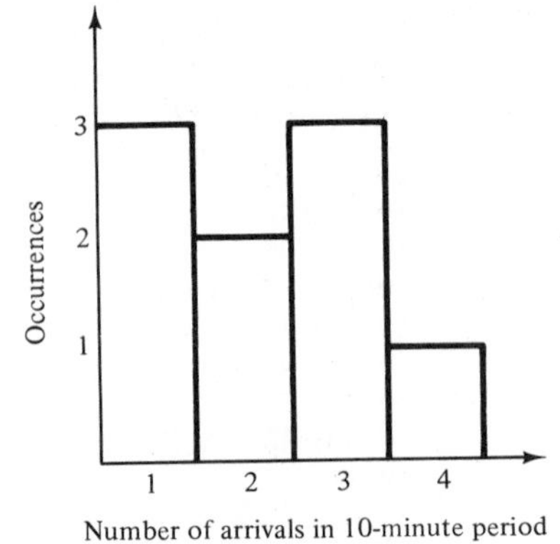

Figure 4.5. Histogram of arrivals per time period.

The distribution of time between arrivals and the distribution of the number of arrivals per time period are important in the simulation of waiting-line systems. "Arrivals" occur in numerous ways; as machine breakdowns, as jobs coming into a jobshop, as units being assembled on a line, as orders to a warehouse, and so on.

Service times may be constant or probabilistic. If service times are completely random, the exponential distribution is often used for simulation purposes. However, there are several other possibilities. It may be possible that the service times are constant, but some random variability causes fluctuations in either a positive or negative way. For example, the time it takes for a lathe to traverse a 10-centimeter shaft should always be the same. However, the material may have slight differences in hardness or the tool may wear causing different processing times. In these cases the normal distribution may describe the service time.

A special case occurs when the phenomenon of interest seems to follow the normal probability distribution, but the random variable is restricted to be greater than and/or less than a certain value. In this case, the truncated normal distribution can be utilized.

The gamma and Weibull distributions are also used to model interarrival and service times. (Actually, the exponential distribution is a special case of both the gamma and the Weibull distributions.) The differences between the exponential, gamma, and Weibull distributions involve the location of the modes of the pdf's and the shapes of their tails for large and small times [Fishman, 1973]. The exponential distribution has its mode at the origin, but the gamma and Weibull distributions have their modes at some point (≥ 0) which is a function of the parameter values selected. The tail of the gamma distribution is long, like an exponential distribution, while the tail of the Weibull distribution may decline more rapidly or less rapidly than that of an exponential distribution. In practice, this means that if there are more large service times than an exponential distribution can account for, a Weibull distribution may provide a better model of these service times.

2. *Inventory systems.* In realistic inventory systems there are three random variables: the number of units demanded per order or per time period, the time between demands, and the lead time. (The lead time is defined as the time between placing an order for stocking the inventory system and the receipt of that order.) In very simple mathematical models of inventory systems demand is a constant over time, and lead time is zero, or a constant. However, in most realistic cases, and hence, in simulation models, demand occurs randomly in time and the number of units demanded each time a demand occurs is also random, as illustrated by Figure 4.6.

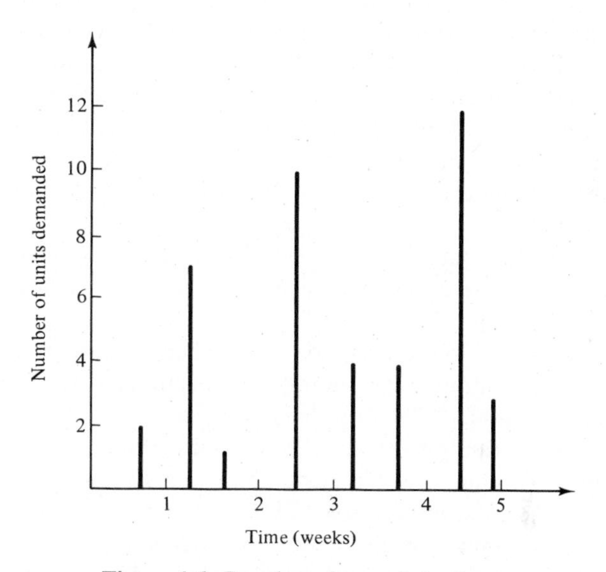

Figure 4.6. Random demands in time.

Distributional assumptions for demand and lead time in inventory theory texts are usually based on mathematical tractability, but those assumptions may not be valid in a realistic context. In practice, the lead-time distribution can often be fitted fairly well by a gamma distribution [Hadley and Whitin, 1963]. Unlike analytic models, simulation models can accommodate whatever assumptions appear most reasonable.

The geometric, Poisson and negative binomial distributions provide a range of distribution shapes that satisfy a variety of demand patterns [Fishman, 1973]. The geometric distribution, which is a special case of the negative binomial, has its mode at unity, given that at least one demand has occurred. If demand data are characterized by a long tail, the negative binomial distribution may be appropriate. The Poisson distribution is often used to model demand because it is simple, it is extensively tabulated, and it is well known. The tail of the Poisson distribution is generally shorter than that of the negative binomial, which means that fewer large demands will occur if a Poisson model is used than if a negative binomial distribution is used (assuming that both models have the same mean demand).

3. *Reliability and maintainability.* Time to failure has been modeled with numerous distributions, including the exponential, gamma and Weilbull. If only random failures occur, the time to failure distribution may be modeled as exponential. The gamma distribution arises from modeling standby redundancy where each component has an exponential time to failure. The Weibull distribution has been extensively used to represent time to failure, and its nature is such that it may be made to approximate many observed phenomena [Hines and Montgomery, 1980]. When there are a number of components in a system and failure is due to the most serious of a large number of defects, or possible defects, the Weibull distribution seems to do particularly well as a model. In situations where most failures are due to wear, the normal distribution may very well be appropriate [Hines and Montgomery, 1980]. The log-normal distribution has been found to be applicable in describing time to failure for some types of components, and the literature seems to indicate increased use of this distribution in reliability models.

4. *Limited data.* In many instances simulations begin before data collection has been completed. There are three distributions which have application to incomplete or limited data. These are the uniform, triangular, and beta distributions. The uniform distribution can be used when an interarrival or service time is known to be random, but no information is immediately available about the distribution [Gordon, 1975]. It is only necessary to specify the continuous interval in which the random variable may occur. The triangular distribution can be used when assumptions are made about the minimum, maximum, and modal values of the random variable. Finally, the beta distribution provides a variety of distributional forms on the unit interval which with appropriate modification can be shifted to any desired interval. The uniform distribution is a special case of the beta distribution.

5. *Other distributions.* Several other distributions may be useful in discrete system simulation. The Bernoulli and binomial distributions are two discrete distributions which may describe phenomena of interest. The hyperexponential distribution is similar to the exponential distribution, but with its greater variability, it may be useful in certain instances.

4.3. Discrete Distributions

Discrete random variables are used to describe random phenomena in which only integer values can occur. Numerous examples were given in Section 4.2, for example, demands for inventory items. Four distributions are described in the following subsections.

1. *Bernoulli trials and the Bernoulli distribution.* Consider an experiment consisting of n trials, each of which can be a success or a failure. Let $X_j = 1$ if the jth experiment resulted in a success, and let $X_j = 0$ if the jth experiment resulted in a failure. The n Bernoulli trials are called a Bernoulli process if the trials are independent, each trial has only two possible outcomes (success or failure), and the probability of a success remains constant from trial to trial. Thus,

$$p(x_1, x_2, \ldots, x_n) = p_1(x_1) \cdot p_2(x_2) \cdot \cdots \cdot p_n(x_n)$$

and

$$p_j(x_j) = p(x_j) = \begin{cases} p, & x_j = 1, j = 1, 2, \ldots, n \\ 1 - p = q, & x_j = 0, j = 1, 2, \ldots, n \\ 0, & \text{otherwise} \end{cases} \tag{4.8}$$

For one trial, the distribution given in equation (4.8) is called the Bernoulli distribution.

The mean and variance of X_j are calculated as follows:

$$E(X_j) = 0 \cdot q + 1 \cdot p = p$$

and

$$V(X_j) = [(0^2 \cdot q) + (1^2 \cdot p)] - p^2 = p(1 - p)$$

2. *Binomial distribution.* The random variable X that denotes the number of successes in n Bernoulli trials has a binomial distribution given by $p(x)$, where

$$p(x) = \begin{cases} \binom{n}{x} p^x q^{n-x}, & x = 0, 1, 2, \ldots, n \\ 0, & \text{otherwise} \end{cases} \tag{4.9}$$

Equation (4.9) is motivated by determining the probability of a particular outcome with all the successes, each denoted by S, occurring in the first x trials, followed by the $n - x$ failures, each denoted by an F. That is,

$$P(\overbrace{\text{SSS} \cdots\cdots \text{SS}}^{x \text{ of these}}\overbrace{\text{FF} \cdots\cdots \text{FF}}^{n\text{-}x \text{ of these}}) = p^x q^{n-x}$$

where $q = 1 - p$. There are

$$\binom{n}{x} = \frac{n!}{x!(n - x)!}$$

outcomes having the required number of S's and F's. Therefore, equation (4.9) results. An easy approach to determining the mean and variance of the binomial distribution is to consider X as a sum of n independent Bernoulli random variables, each with mean p and variance $p(1 - p) = pq$. Then,

$$X = X_1 + X_2 + \cdots + X_n$$

and the mean, $E(X)$, is given by

$$E(X) = p + p + \cdots + p = np \tag{4.10}$$

and the variance $V(X)$ is given by

$$V(X) = pq + pq + \cdots + pq = npq \tag{4.11}$$

EXAMPLE 4.10

A production process manufactures semiconductor chips used in microprocessors, on the average at 2% fraction defective. Every day a random sample of size 50 is taken from the process. If the sample contains more than two defectives, the process will be stopped. Determine the probability that the process is stopped by the sampling scheme.

Considering the sampling process as $n = 50$ Bernoulli trials, each with $p = 0.02$, the total number of defectives in the sample, X, would have a binomial distribution given by

$$p(x) = \begin{cases} \binom{50}{x}(0.02)^x(0.98)^{50-x}, & x = 0, 1, 2, \ldots, 50 \\ 0, & \text{otherwise} \end{cases}$$

It is much easier to compute the right-hand side of the following identity to determine the probability that more than two defectives are found in a sample:

$$P(X > 2) = 1 - P(X \leq 2)$$

The probability $P(X \leq 2)$ is calculated from

$$\begin{aligned} P(X \leq 2) &= \sum_{x=0}^{2} \binom{50}{x}(0.02)^x(0.98)^{50-x} \\ &= (0.98)^{50} + 50(0.02)(0.98)^{49} + 1225(0.02)^2(0.98)^{48} \\ &\doteq 0.92 \end{aligned}$$

Thus, the probability that the production process is stopped on any day, based on the sampling process, is approximately 0.08. The mean number of defectives in a random sample of size 50 is given by

$$E(X) = np = 50(0.02) = 1$$

and the variance is given by

$$V(X) = npq = 50(0.02)(0.98) = 0.98$$

The cdf for the binomial distribution has been tabulated by Romig [1953] and others. The tables decrease the effort considerably for computing probabilities such as $P(a < X \leq b)$. Under certain conditions on n and p, both the Poisson distribution and the normal distribution may be used to approximate the binomial distribution [Hines and Montgomery, 1980].

3. *Geometric distribution.* The geometric distribution is related to a sequence of Bernoulli trials; the random variable of interest, X, is defined to be the number of trials to achieve the first success. The distribution of X is given by

$$p(x) = \begin{cases} q^{x-1}p, & x = 1, 2, \ldots \\ 0, & \text{otherwise} \end{cases} \tag{4.12}$$

The event $\{X = x\}$ occurs when there are $x - 1$ failures followed by a success. Each of the failures has an associated probability of $q = 1 - p$, and each success has probability p. Thus,

$$P(\text{FFF}\cdots\text{FS}) = q^{x-1}p$$

The mean and variance are given by

$$E(X) = \frac{1}{p} \tag{4.13}$$

and

$$V(X) = \frac{q}{p^2} \tag{4.14}$$

Example 4.11

Forty percent of the assembled microprocessors are rejected at the inspection station. Find the probability that the first acceptable microprocessor is the third one inspected. Considering each inspection as a Bernoulli trial with $q = 0.4$ and $p = 0.6$,

$$p(3) = 0.4^2(0.6) = 0.096$$

Thus, in only about 10% of the cases is the first acceptable microprocessor the third one from any arbitrary starting point.

4. *Poisson distribution.* The Poisson distribution describes many random processes quite well and is mathematically quite simple. The Poisson distribution was introduced in 1837 by S. D. Poisson in a book concerning criminal and civil justice matters. The title of this rather old text is "Recherches sur la probabilité des jugements en matière criminelle et en matière civile." Evidently, the rumor handed down through generations of probability theory professors concerning the origin of the Poisson distribution is just not true. Rumor has it that the Poisson distribution was first used to model deaths from the kicks of horses in the Prussian Army.

The Poisson probability mass function is given by

$$p(x) = \begin{cases} \dfrac{e^{-\alpha}\alpha^x}{x!}, & x = 0, 1, \cdots \\ 0, & \text{otherwise} \end{cases} \tag{4.15}$$

where $\alpha > 0$. One of the important properties of the Poisson distribution is that the mean and variance are both equal to α, that is,

$$E(X) = \alpha = V(X)$$

The cumulative distribution function is given by

$$F(x) = \sum_{i=0}^{x} \frac{e^{-\alpha}\alpha^i}{i!} \tag{4.16}$$

Many introductory probability theory texts contain the tabulated values of the cdf. The pmf and cdf for a Poisson distribution with $\alpha = 2$ are shown in Figure 4.7.

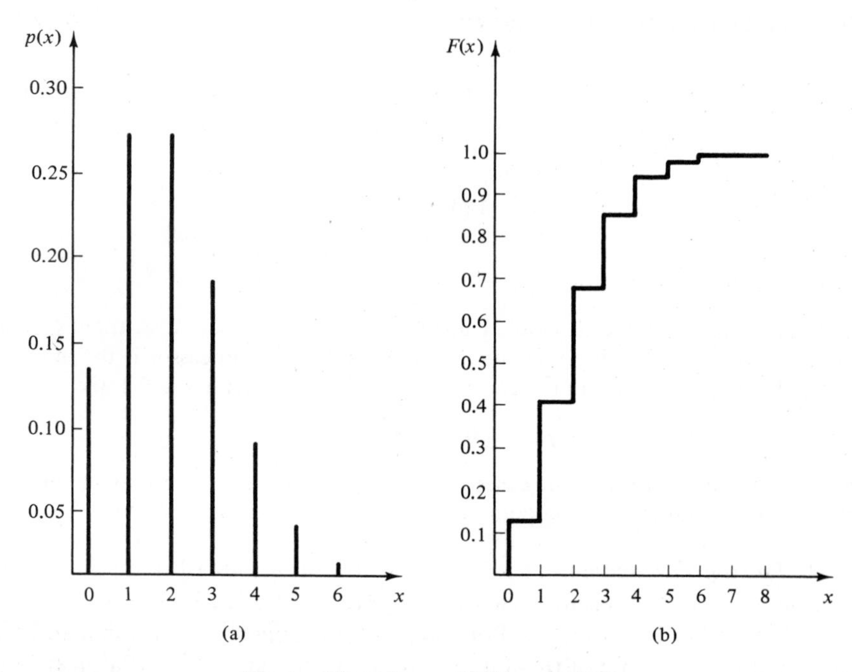

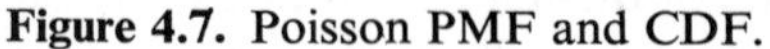

Figure 4.7. Poisson PMF and CDF.

EXAMPLE 4.12

A computer terminal repair person is "beeped" each time there is a call for service. The number of beeps per hour is known to occur in accordance with a Poisson distribution with a mean of $\alpha = 2$ per hour. The probability of three beeps in the next hour is given by equation (4.15) with $x = 3$, as follows:

$$p(3) = \frac{e^{-2}2^3}{3!} = \frac{(0.135)(8)}{6} = 0.18$$

This same result can be read from the left side of Figure 4.7.

EXAMPLE 4.13

In Example 4.12 determine the probability of two or more beeps in a 1-hour period.

$$P(2 \text{ or more}) = 1 - p(0) - p(1) = 1 - F(1)$$
$$= 1 - 0.406 = 0.594$$

The cumulative probability, $F(1)$, can be read from the right side of Figure 4.7.

EXAMPLE 4.14

The lead-time demand in an inventory system is the accumulation of demand for an item from the point at which an order is placed until the order is received. That is,

$$L = \sum_{i=1}^{T} D_i \tag{4.17}$$

where L is the lead-time demand, D_i is the demand during the ith time period, and T is the number of time periods during the lead time. Both D_i and T may be random variables.

An inventory manager desires that the probability of a stockout not exceed a certain fraction during the lead time. For example, it may be stated that the probability of a shortage during the lead time not exceed 5%.

If the lead-time demand is Poisson distributed, the determination of the reorder point is greatly facilitated. The reorder point is the level of inventory at which a new order is placed.

Assume that the lead-time demand is Poisson distributed with a mean of $\alpha = 10$ units and that 95% protection from a stockout is desired. Thus, it is desired to find the smallest value of x such that the probability that the lead-time demand does not exceed x is greater than or equal to 0.95. Using equation (4.16) requires finding the smallest x such that

$$F(x) = \sum_{i=0}^{x} \frac{e^{-10}10^i}{i!} \geq 0.95$$

The desired result occurs at $x = 15$, which can be found by using a table of cdf values, or by computation of $p(0), p(1), \ldots$.

4.4. Continuous Distributions

Continuous random variables can be used to describe random phenomena in which the variable of interest can take on any value in some interval: for example, the time to failure or the length of a rod. Seven distributions are described in the following subsections.

1. *Uniform distribution.* A random variable X is uniformly distributed on the interval $[a, b]$ if its pdf is given by

$$f(x) = \begin{cases} \dfrac{1}{b-a}, & a < x \le b \\ 0, & \text{otherwise} \end{cases} \tag{4.18}$$

The cdf is given by

$$F(x) = \begin{cases} 0, & x < a \\ \dfrac{x-a}{b-a}, & a \le x \le b \\ 1, & x > b \end{cases} \tag{4.19}$$

Note that

$$P(x_1 < X < x_2) = F(x_2) - F(x_1) = \frac{x_2 - x_1}{b-a}$$

is proportional to the length of the interval, for all x_1 and x_2 satisfying $a \le x_1 \le x_2 \le b$. The mean and variance of the distribution are given by

$$E(X) = \frac{a+b}{2} \tag{4.20}$$

and

$$V(X) = \frac{(b-a)^2}{12} \tag{4.21}$$

The pdf and cdf when $a = 1$ and $b = 6$ are shown in Figure 4.8.

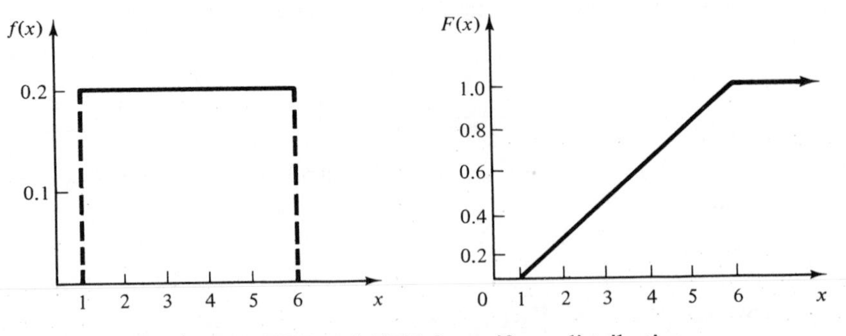

Figure 4.8. PDF and CDF for uniform distribution.

The uniform distribution plays a vital role in simulation. Random numbers, uniformly distributed between zero and 1, provide the means to generate random events. Numerous methods for generating uniformly distributed random numbers have been devised, as discussed in Chapter 7. Uniformly distributed random numbers are then used to generate samples of random variates from all other distributions, as discussed in Chapter 8.

EXAMPLE 4.15

A simulation of a warehouse operation is being developed. About every 3 minutes, a call comes for a forklift truck operator to proceed to a certain location. An initial assumption is made that the time between calls (arrivals) is uniformly distributed with a mean of 3 minutes. By equation (4.21), the uniform distribution with a mean of 3 and the greatest possible variability would have parameter values of $a = 0$ and $b = 6$ minutes. With very limited data (such as a mean of approximately 3 minutes) plus the knowledge that the quantity of interest is variable in a random fashion, the uniform distribution with greatest variance is usually the safest assumption to make, at least until more data are available.

EXAMPLE 4.16

A bus arrives every 20 minutes at a specified stop beginning at 6:40 A.M. and continuing until 8:40 A.M. A certain passenger does not know the schedule, but arrives randomly (uniformly distributed) between 7:00 A.M. and 7:30 A.M. every morning. What is the probability that the passenger waits more than 5 minutes for a bus?

The passenger has to wait more than 5 minutes only if the arrival time is between 7:00 A.M. and 7:15 A.M. or between 7:20 A.M. and 7:30 A.M. If X is a random variable that denotes the number of minutes past 7:00 A.M. that the passenger arrives, the desired probability is

$$P(0 < X < 15) + P(20 < X < 30)$$

Now, X is a uniform random variable on (0, 30). Therefore, the desired probability is given by

$$F(15) + F(30) - F(20) = \frac{15}{30} + 1 - \frac{20}{30} = \frac{5}{6}$$

2. *Exponential distribution.* A random variable X is said to be exponentially distributed with parameter $\lambda > 0$ if its pdf is given by

$$f(x) = \begin{cases} \lambda e^{-\lambda x}, & x \geq 0 \\ 0, & \text{elsewhere} \end{cases} \tag{4.22}$$

The density function is shown in Figures 4.9 and 4.3. Figure 4.9 also shows the cdf.

The exponential distribution has been used to model interarrival times when arrivals are completely random and to model service times which are highly variable. In these instances, λ is a rate: arrivals per hour or services per minute. The exponential distribution has also been used to model the lifetime of a com-

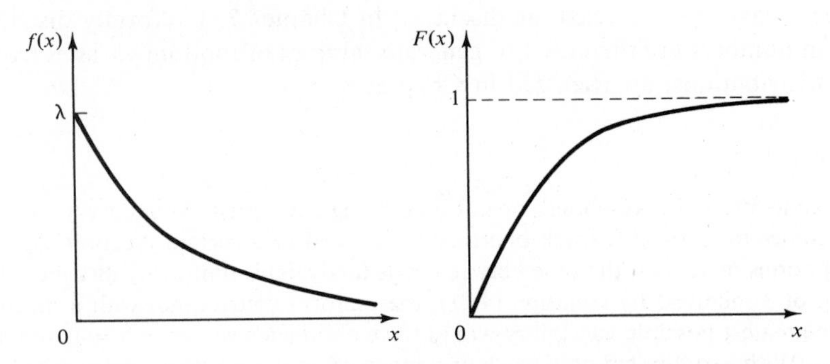

Figure 4.9. Exponential density function and cumulative distribution function.

ponent that fails catastrophically (instantaneously), such as a light bulb. Then, λ is the failure rate.

Several different exponential pdf's are shown in Figure 4.10. The value of the intercept on the vertical axis is always equal to the value of λ. Note also that all pdf's eventually intersect. (Why?)

The exponential distribution has mean and variance given by

$$E(X) = \frac{1}{\lambda} \quad \text{and} \quad V(X) = \frac{1}{\lambda^2} \tag{4.23}$$

Thus, the mean and standard deviation are equal. The cdf can be determined by integrating equation (4.22) to obtain

$$F(x) = \begin{cases} 0, & x < 0 \\ \int_0^x \lambda e^{-\lambda t}\, dt = 1 - e^{-\lambda x}, & x \geq 0 \end{cases} \tag{4.24}$$

EXAMPLE 4.17

Suppose that the life of an industrial lamp, in thousands of hours, is exponentially distributed with failure rate $\lambda = 1/3$ (one failure every 3000 hours, on the average). The probability that the lamp will last longer than its mean life of 3000 hours is given by $P(X > 3) = 1 - P(X \leq 3) = 1 - F(3)$. Equation (4.24) is used to compute $F(3)$, obtaining

$$P(X > 3) = 1 - (1 - e^{-3/3}) = e^{-1} = 0.368$$

Regardless of the value of λ, this result will always be the same! That is, the probability that the random variable is greater than its mean is 0.368, for any value of λ.

The probability that the industrial lamp will last between 2000 and 3000 hours is

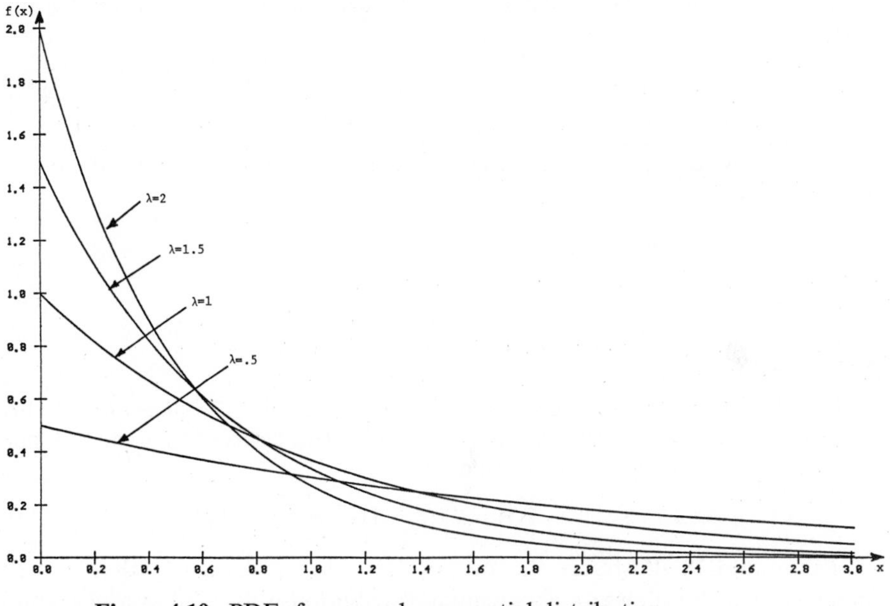

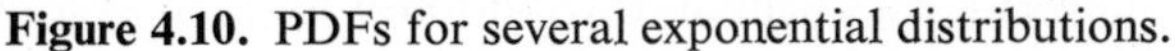

Figure 4.10. PDFs for several exponential distributions.

determined by

$$P(2 \leq X \leq 3) = F(3) - F(2)$$

Again, using the cdf given by equation (4.24),

$$\begin{aligned} F(3) - F(2) &= (1 - e^{-3/3}) - (1 - e^{-2/3}) \\ &= -0.368 + 0.513 = 0.145 \end{aligned}$$

One of the most important properties of the exponential distribution is that it is "memoryless," which means that for all $s \geq 0$ and $t \geq 0$,

$$P(X > s + t \mid X > s) = P(X > t) \tag{4.25}$$

Let X represent the life of a component (a battery, light bulb, semiconductor, cathode ray tube, etc.) and assume that X is exponentially distributed. Equation (4.25) states that the probability that the component lives for at least $s + t$ hours, given that it has survived s hours, is the same as the initial probability that it lives for at least t hours. If the component is alive at time s (if $X > s$), then the distribution of the remaining amount of time that it survives, namely $X - s$, is the same as the original distribution of a new component. That is, the component does not "remember" that it has already been in use for a time s. A used component is as good as new.

That equation (4.25) holds is shown by examining the conditional probability

$$P(X > s + t \mid X > s) = \frac{P(X > s + t)}{P(X > s)} \tag{4.26}$$

Equation (4.24) can be used to determine the numerator and denominator of equation (4.26), yielding

$$\begin{aligned} P(X > s + t \mid X > s) &= \frac{e^{-\lambda(s+t)}}{e^{-\lambda s}} = e^{-\lambda t} \\ &= P(X > t) \end{aligned}$$

EXAMPLE 4.18

Find the probability that the industrial lamp in Example 4.17 will last for another 1000 hours given that it is operating after 2500 hours. This determination can be found using equations (4.25) and (4.24), as follows:

$$P(X > 3.5 \mid X > 2.5) = P(X > 1) = e^{-1/3} = 0.717$$

Example 4.18 illustrates the memoryless property, namely that a used component which follows an exponential distribution is as good as a new component. The probability that a new component will have a life greater than 1000 hours is also equal to 0.717. Stated in general, suppose that a component which has a lifetime that follows the exponential distribution with parameter λ is observed and found to be operating at an arbitrary time. Then, the distribution of the remaining lifetime is also exponential with parameter λ. The exponential distribution is the only continuous distribution which has the memoryless property. (The geometric distribution is the only discrete distribution that possesses the memoryless property.)

3. *Gamma distribution.* A function used in defining the gamma distribution is the gamma function, which is defined for all $\beta > 0$ as

$$\Gamma(\beta) = \int_0^\infty x^{\beta-1} e^{-x}\, dx \tag{4.27}$$

By integrating equation (4.27) by parts, it can be shown that

$$\Gamma(\beta) = (\beta - 1)\Gamma(\beta - 1) \tag{4.28}$$

If β is an integer, then using $\Gamma(1) = 1$ and equation (4.28), it can be seen that

$$\Gamma(\beta) = (\beta - 1)! \tag{4.29}$$

The gamma function can be thought of as a generalization of the factorial notion which applies to all positive numbers, not just integers.

A random variable X is gamma distributed with parameters β and θ if its pdf is given by

$$f(x) = \begin{cases} \dfrac{\beta\theta}{\Gamma(\beta)}(\beta\theta x)^{\beta-1}e^{-\beta\theta x}, & x > 0 \\ 0, & \text{otherwise} \end{cases} \tag{4.30}$$

The parameter β is called the shape parameter and θ is called the scale parameter. Several gamma distributions for $\theta = 1$ and various values of β are shown in Figure 4.11.

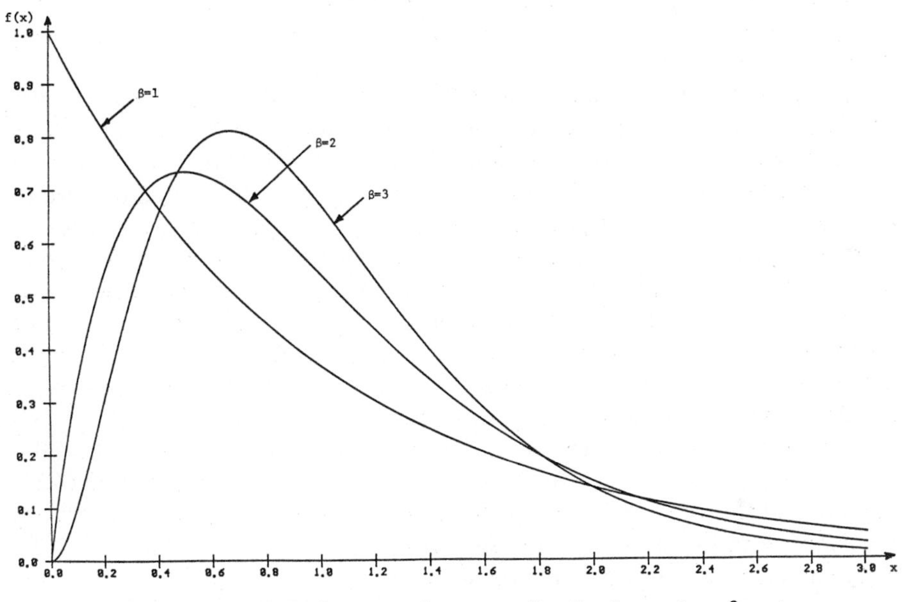

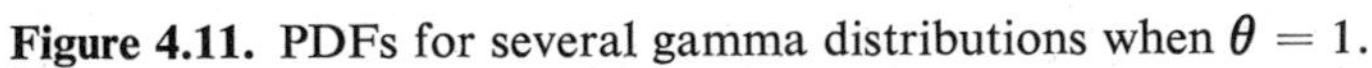

Figure 4.11. PDFs for several gamma distributions when $\theta = 1$.

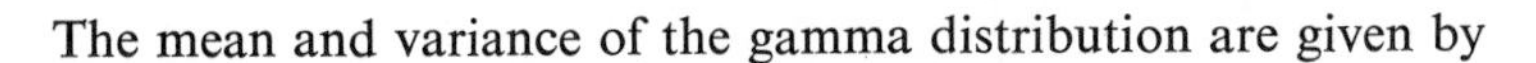

The mean and variance of the gamma distribution are given by

$$E(X) = \frac{1}{\theta} \tag{4.31}$$

and

$$V(X) = \frac{1}{\beta\theta^2} \tag{4.32}$$

The cdf of X is given by

$$F(x) = \begin{cases} 1 - \displaystyle\int_x^\infty \frac{\beta\theta}{\Gamma(\beta)}(\beta\theta t)^{\beta-1}e^{-\beta\theta t}\,dt, & x > 0 \\ 0, & x \leq 0 \end{cases} \tag{4.33}$$

When β is an integer, the gamma distribution is related to the exponential distribution in the following manner: If the random variable, X, is the sum of β independent, exponentially distributed random variables, each with parameter $\beta\theta$, then X has a gamma distribution with parameters β and θ. Thus, if

$$X = X_1 + X_2 + \cdots + X_\beta \tag{4.34}$$

where the pdf of X_j is given by

$$g(x_j) = \begin{cases} (\beta\theta)e^{-\beta\theta x_j}, & x_j \geq 0 \\ 0, & \text{otherwise} \end{cases}$$

and the X_j are mutually independent, X has the pdf given in equation (4.30). Note that when $\beta = 1$, an exponential distribution results. This result follows from equation (4.34) or from letting $\beta = 1$ in equation (4.30).

4. *Erlang distribution.* The pdf given by equation (4.30) is often referred to as the Erlang distribution of order k when $\beta = k$, an integer. Erlang was a Danish telephone engineer who was an early developer of queueing theory. The Erlang distribution could arise in the following context: Consider a series of k stations that must be passed through in order to complete the servicing of a customer. An additional customer cannot enter the first station until the customer in process has negotiated all the stations. Each station has an exponential distribution of service time with parameter $k\theta$. Equations (4.31) and (4.32), to determine the mean and variance of a gamma distribution, are valid regardless of the value of β. However, when $\beta = k$, an integer, equation (4.34) may be used to derive the mean of the distribution in a fairly straightforward manner. The expected value of the sum of random variables is the sum of the expected value of each random variable. Thus,

$$E(X) = E(X_1) + E(X_2) + \cdots + E(X_k)$$

The expected value of the exponentially distributed X_j are each given by $1/k\theta$. Thus,

$$E(X) = \frac{1}{k\theta} + \frac{1}{k\theta} + \cdots + \frac{1}{k\theta} = \frac{1}{\theta}$$

If the random variables X_j are independent, the variance of their sum is the sum of the variances, or

$$V(X) = \frac{1}{(k\theta)^2} + \frac{1}{(k\theta)^2} + \cdots + \frac{1}{(k\theta)^2} = \frac{1}{k\theta^2}$$

When $\beta = k$, a positive integer, the cdf given by equation (4.33) may be integrated by parts, giving

$$F(x) = \begin{cases} 1 - \sum_{i=0}^{k-1} \dfrac{e^{-k\theta x}(k\theta x)^i}{i!}, & x > 0 \\ 0, & x \leq 0 \end{cases} \tag{4.35}$$

which is the sum of Poisson terms with mean $\alpha = k\theta x$. Tables of the cumulative Poisson distribution may be used to evaluate the cdf when the shape parameter is an integer.

EXAMPLE 4.19

A college professor is leaving home for the summer, but would like to have a light burning at all times to discourage burglars. The professor rigs up a device that will hold two light bulbs. The device will switch the current to the second bulb if the first bulb fails. The box in which the light bulbs are packaged says, "Average life 1000 hours, exponentially distributed." The professor will be gone 90 days (2160 hours). What is the probability that a light will be burning when the summer is over and the professor returns?

The probability that the system will operate at least x hours is called the reliability function $R(x)$, where

$$R(x) = 1 - F(x)$$

In this case, the total system lifetime is given by equation (4.34) with $\beta = k = 2$ bulbs and $k\theta = 1/1000$ per hour, so $\theta = 1/2000$ per hour. Thus, $F(2160)$ can be determined from equation (4.35) as follows:

$$F(2160) = 1 - \sum_{i=0}^{1} \frac{e^{-(2)(1/2000)(2160)}[(2)(1/2000)(2160)]^i}{i!}$$

$$= 1 - e^{-2.16} \sum_{i=0}^{1} \frac{(2.16)^i}{i!} = 0.636$$

Therefore, the chances are about 36% that the light will be burning when the professor returns.

EXAMPLE 4.20

A medical examination is given in three stages by a physician. Each stage is exponentially distributed with a mean service time of 20 minutes. Find the probability that the exam will take 50 minutes or less. Also, determine the expected length of the exam. In this case, $k = 3$ stages and $k\theta = 1/20$, so that $\theta = 1/60$ per minute. Thus, $F(50)$ can be determined from equation (4.35) as follows:

$$F(50) = 1 - \sum_{i=0}^{2} \frac{e^{-(3)(1/60)(50)}[(3)(1/60)(50)]^i}{i!}$$

$$= 1 - \sum_{i=0}^{2} \frac{e^{-5/2}(5/2)^i}{i!}$$

The cumulative Poisson distribution, readily tabulated elsewhere, can be used to determine that

$$F(50) = 1 - 0.543 = 0.457$$

The probability is 0.457 that the exam will take 50 minutes or less. The expected length of the exam is determined from equation (4.31) as

$$E(X) = \frac{1}{\theta} = \frac{1}{1/60} = 60 \text{ minutes}$$

In addition, the variance of X is $V(X) = 1/\beta\theta^2 = 1200$ minutes^2. In general, a gamma distribution is less variable than an exponential distribution with the same mean. Incidentally, the mode of the Erlang distribution is given by

$$\text{Mode} = \frac{k-1}{k\theta} \tag{4.36}$$

Thus, the modal value in this example is

$$\text{Mode} = \frac{3-1}{3(1/60)} = 40 \text{ minutes}$$

5. *Normal distribution.* A random variable X with mean μ ($-\infty < \mu < \infty$) and variance $\sigma^2 > 0$ has a normal distribution if it has the pdf

$$f(x) = \frac{1}{\sigma\sqrt{2\pi}} \exp\left[-\frac{1}{2}\left(\frac{x-\mu}{\sigma}\right)^2\right], \qquad -\infty < x < \infty \tag{4.37}$$

The normal distribution is used so often that the notation $X \sim N(\mu, \sigma^2)$ has been adopted by many authors to indicate that the random variable X is normally distributed with mean μ and variance σ^2. The normal pdf is shown in Figure 4.12.

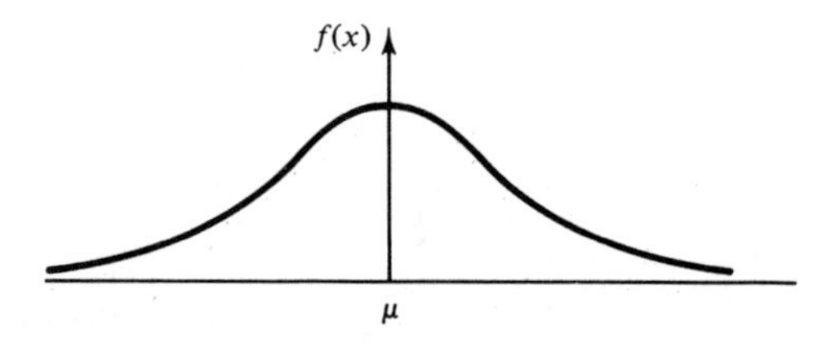

Figure 4.12. PDF of the normal distribution.

Some of the special properties of the normal distribution are listed here:

a. $\lim_{x\to-\infty} f(x) = 0$ and $\lim_{x\to\infty} f(x) = 0$; the value of $f(x)$ approaches zero as x approaches negative infinity and, similarly, as x approaches positive infinity.
b. $f(\mu - x) = f(\mu + x)$; the pdf is symmetric about μ.

c. The maximum value of the pdf occurs at $x = \mu$. (Thus, the mean and mode are equal.)

The cdf for the normal distribution is given by

$$F(x) = P(X \leq x) = \int_{-\infty}^{x} \frac{1}{\sigma\sqrt{2\pi}} \exp\left[\frac{-1}{2}\left(\frac{t-\mu}{\sigma}\right)^2\right] dt \tag{4.38}$$

It is not possible to evaluate equation (4.38) in closed form. Numerical methods could be used, but it appears that it would be necessary to evaluate the integral for each pair (μ, σ^2). However, a transformation of variables, $z = (t - \mu)/\sigma$, allows the evaluation to be independent of μ and σ. If $X \sim N(\mu, \sigma^2)$, let $Z = (X - \mu)/\sigma$ to obtain

$$\begin{aligned} F(x) = P(X \leq x) &= P\left(Z \leq \frac{x-\mu}{\sigma}\right) \\ &= \int_{-\infty}^{(x-\mu)/\sigma} \frac{1}{\sqrt{2\pi}} e^{-z^2/2}\, dz \\ &= \int_{-\infty}^{(x-\mu)/\sigma} \phi(z)\, dz = \Phi\left(\frac{x-\mu}{\sigma}\right) \end{aligned} \tag{4.39}$$

The pdf

$$\phi(z) = \frac{1}{\sqrt{2\pi}} e^{-z^2/2}, \qquad -\infty < z < \infty \tag{4.40}$$

is the pdf of a normal distribution with mean zero and variance 1. Thus, $Z \sim N(0, 1)$ and it is said that Z has a standard normal distribution. The standard normal distribution is shown in Figure 4.13. The cdf for the standard normal distribution is given by

$$\Phi(z) = \int_{-\infty}^{z} \frac{1}{\sqrt{2\pi}} e^{-t^2/2}\, dt \tag{4.41}$$

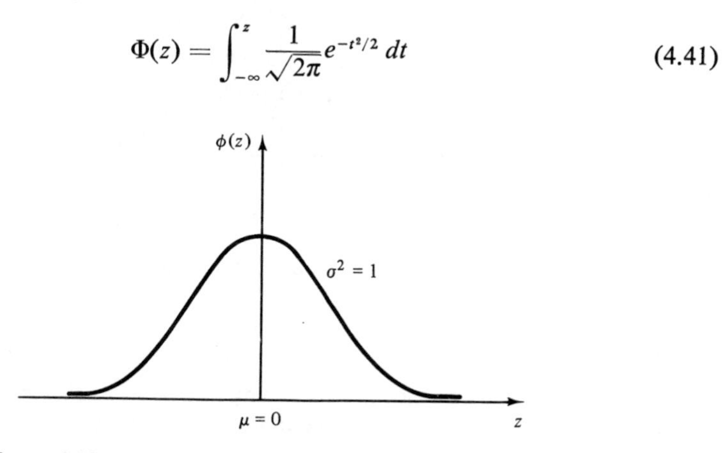

Figure 4.13. PDF of the standard normal distribution.

Equation (4.41) has been widely tabulated. The probabilities $\Phi(z)$ for $z \geq 0$ are given in Table A.3. Several examples are now given that indicate how equation (4.39) and Table A.3 are used.

EXAMPLE 4.21

It is known that $X \sim N(50, 9)$. Determine $F(56) = P(X \leq 56)$. Using equation (4.39), we get

$$F(56) = \Phi\left(\frac{56 - 50}{3}\right) = \Phi(2) = 0.9772$$

from Table A.3. The intuitive interpretation is shown in Figure 4.14. Figure 4.14(a) shows the pdf of $X \sim N(50, 9)$ with the specific value, $x_0 = 56$, marked. The shaded portion is the desired probability. Figure 4.14(b) shows the standard normal distribution or $Z \sim N(0, 1)$ with the value 2 marked since $x_0 = 56$ is 2σ (where $\sigma = 3$) greater than the mean. It is helpful to make both sketches such as those in Figure 4.14 to avoid confusion in determining required probabilities.

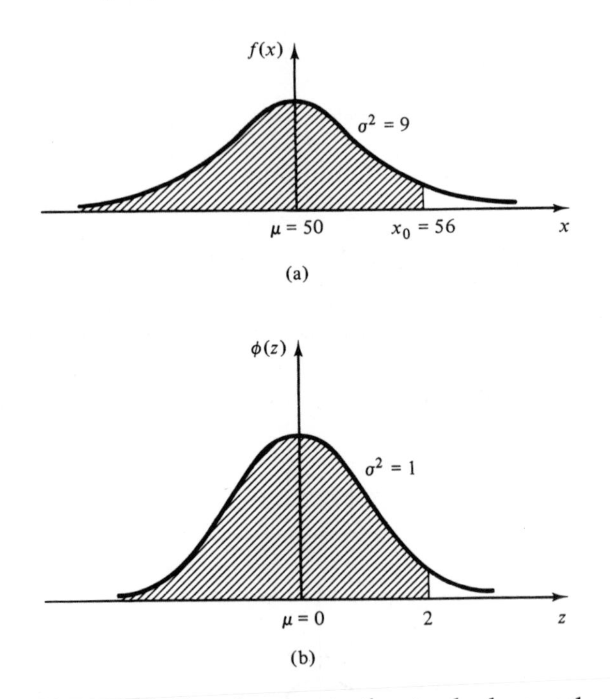

Figure 4.14. Transforming to the standard normal distribution.

EXAMPLE 4.22

The time required to load an oceangoing vessel, X, is distributed $N(12, 4)$. The probability that the vessel will be loaded in less than 10 hours is given by $F(10)$, where

$$F(10) = \Phi\left(\frac{10 - 12}{2}\right) = \Phi(-1) = 0.1587$$

The value of $\Phi(-1) = 0.1587$ is determined from Table A.3 using the symmetry property of the normal distribution. Note that $\Phi(1) = 0.8413$. The complement of 0.8413, or 0.1587, is contained in the tail, the shaded portion of the standard normal distribution shown in Figure 4.15(a). In Figure 4.15(b), the symmetry property is used to determine the shaded region to be $\Phi(-1) = 1 - \Phi(1) = 0.1587$. [Using this logic, the student can ascertain that $\Phi(2) = 0.9772$ and $\Phi(-2) = 1 - \Phi(2) = 0.0228$. In general, $\Phi(-x) = 1 - \Phi(x)$.]

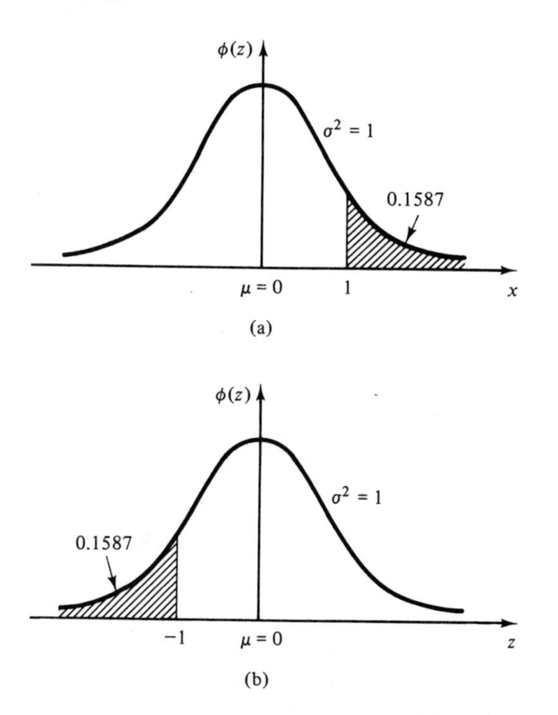

Figure 4.15. Using the symmetry property of the normal distribution.

The probability that 12 or more hours will be required to load the ship can also be determined by inspection, using the symmetry property of the normal pdf and the mean as shown by Figure 4.16. The shaded portion of Figure 4-16(a) shows the problem as originally stated [i.e., determine $P(X < 12)$]. Now, $P(X > 12) = 1 - F(12)$. The standardized normal in Figure 4.16(b) is used to determine $F(12) = \Phi(0) = 0.50$. Thus, $P(X > 12) = 1 - 0.50 = 0.50$. [The shaded portions in both Figure 4.16(a) and (b) contain 0.50 of the area under the normal pdf.]

The probabiiity that between 10 and 12 hours will be required to load a ship is given by

$$P(10 \leq X \leq 12) = F(12) - F(10) = 0.5000 - 0.1587 = 0.3413$$

using earlier results presented in this example. The desired area is shown in the shaded portion of Figure 4.17(a). The equivalent problem shown in terms of the standardized normal distribution is shown in Figure 4.17(b). The probability statement $F(12) - F(10) = \Phi(0) - \Phi(-1) = 0.5000 - 0.1587 = 0.3413$, using Table A.3.

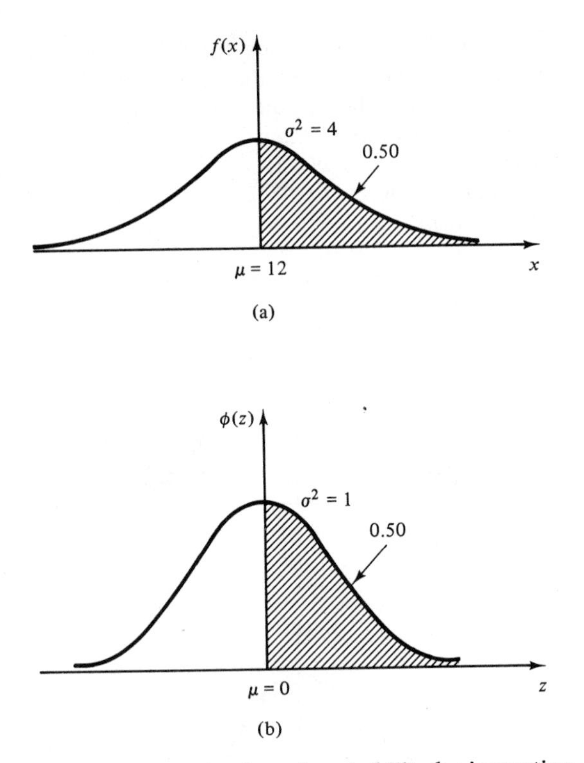

Figure 4.16. Determination of probability by inspection.

EXAMPLE 4.23

The time to pass through a queue to begin self-service at a cafeteria has been found to be $N(10, 9)$. The probability that an arriving customer waits between 9 and 12 minutes is determined as follows:

$$P(9 \le X \le 12) = F(12) - F(9) = \Phi\left(\frac{12-10}{3}\right) - \Phi\left(\frac{9-10}{3}\right)$$
$$= \Phi(0.667) - \Phi(-0.333)$$

The shaded area shown in Figure 4.18(a) represents the probability $F(12) - F(9)$. The shaded area shown in Figure 4.18(b) represents the equivalent probability, $\Phi(0.667) - \Phi(-0.333)$, for the standardized normal distribution. Using Table A.3, $\Phi(0.667) = 0.7476$. Now, $\Phi(-0.333) = 1 - \Phi(0.333) = 1 - 0.6304 = 0.3696$. Thus, $\Phi(0.667) - \Phi(-0.333) = 0.3780$. The probability is 0.3780 that the customer will pass through the queue in a time between 9 and 12 minutes.

EXAMPLE 4.24

Lead-time demand, X, for an item is approximated by a normal distribution with a mean of 25 and a variance of 9. It is desired to determine a value for lead time that will only be exceeded 5% of the time. Thus, the problem is to find x_0 such that

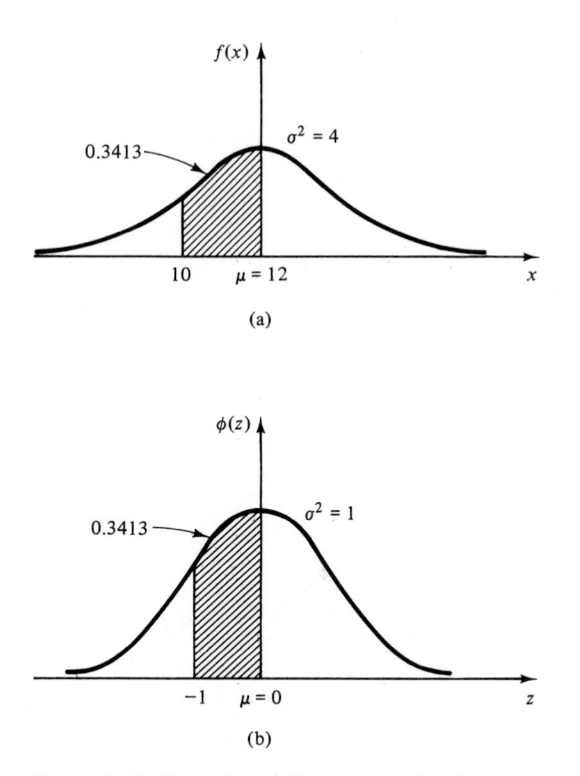

Figure 4.17. Transformation to standard normal for vessel loading problem.

$P(X > x_0) = 0.05$, as shown by the shaded area in Figure 4.19(a). The equivalent problem is shown as the shaded area in Figure 4.19(b). Now,

$$P(X > x_0) = P\left(Z > \frac{x_0 - 25}{3}\right) = 1 - \Phi\left(\frac{x_0 - 25}{3}\right) = 0.05$$

or, equivalently,

$$\Phi\left(\frac{x_0 - 25}{3}\right) = 0.95$$

From Table A.3 it can be seen that $\Phi(1.645) = 0.95$. Thus, x_0 can be determined by solving

$$\frac{x_0 - 25}{3} = 1.645$$

or

$$x_0 = 29.935$$

Therefore, in only 5% of the cases will demand during lead time exceed available inventory if an order to purchase is made when the stock level reaches 30.

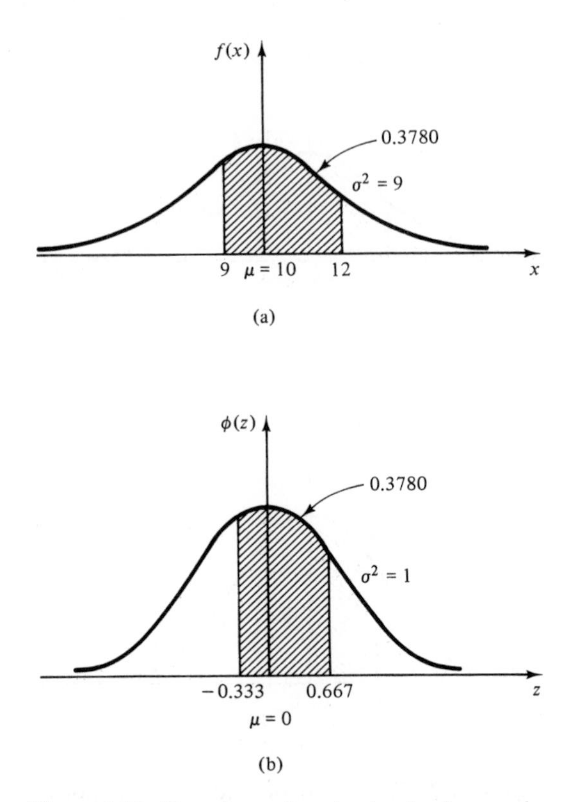

Figure 4.18. Transformation to standard normal for cafeteria problem.

EXAMPLE 4.25 (THE TRUNCATED NORMAL DISTRIBUTION)

In many practical situations phenomena exist which seem to follow the normal pdf, but the random variable is restricted to a certain segment of the real line. For example, the random variable may not assume negative values, (i.e., the life of a component, the length of a shaft, the price of a stock, the weight of foods packaged in a container, etc.). For certain values of μ and σ^2, the nonnegativity constraint is not serious, as when $\mu > 0$ and $\mu > 3\sigma$ since $\Phi(-3) = 0.0014$; that is, in only about 1 of 1000 cases would the random variable be less than zero.

In some instances the truncation may be to the right. Suppose that X is a random variable denoting the diameter of a shaft in centimeters, and it is known that $X \sim N(5.0, 0.04)$. The diameter of the shaft must not exceed 5.4 centimeters or it will not fit into the housing. All shafts exceeding 5.4 centimeters are discarded, resulting in a new random variable, Y, which follows a truncated normal distribution.

The cumulative distribution function for a random variable X having a normal distribution truncated to the right at $x = r$ may be shown to be

$$F(y) = \begin{cases} c\,\Phi\left(\dfrac{y-\mu}{\sigma}\right), & y \leq r \\ 1, & y > r \end{cases} \tag{4.42}$$

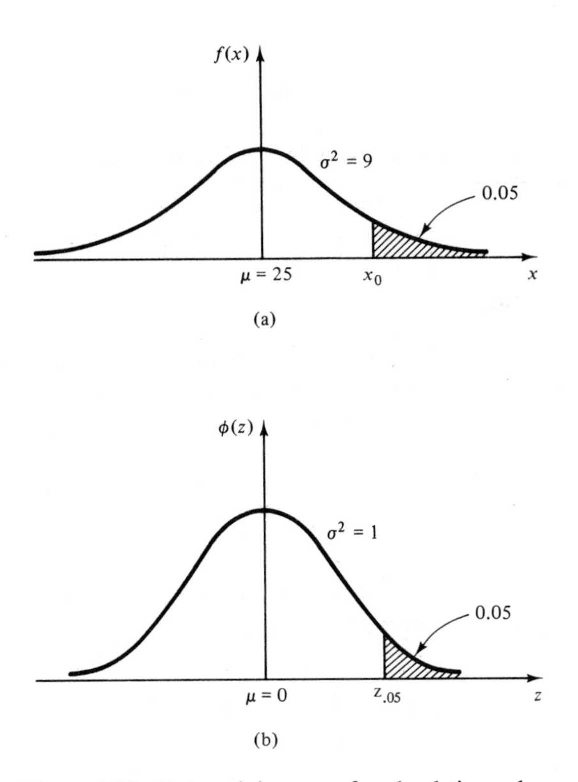

Figure 4.19. Determining x_0 for lead-time demand problem.

where

$$c = \left[\Phi\left(\frac{r - \mu}{\sigma}\right)\right]^{-1} \tag{4.43}$$

In the example of the shaft, $r = 5.4$ centimeters and c can be determined from equation (4.43) as

$$c = \left[\Phi\left(\frac{5.4 - 5.0}{0.2}\right)\right]^{-1} = \frac{1}{0.9772} = 1.023$$

Thus, $P(Y < 5.2)$ is determined as follows:

$$\begin{aligned} P(Y < 5.2) &= F(5.2) = 1.023\Phi\left(\frac{5.2 - 5.0}{0.2}\right) \\ &= 1.023\Phi(1) = (1.023)(0.8413) = 0.8606 \end{aligned}$$

If the random variable $X \sim N(\mu, \sigma^2)$ is truncated to the left at $x = l$, then the cdf of the truncated random variable is

$$F(y) = \begin{cases} c\,\Phi\left(\dfrac{y - \mu}{\sigma}\right), & y \geq l \\ 0, & y < l \end{cases} \tag{4.44}$$

where

$$c = \left[1 - \Phi\left(\frac{l - \mu}{\sigma}\right)\right]^{-1}$$

6. *Weibull distribution.* The random variable X has a Weibull distribution if its pdf has the form

$$f(x) = \begin{cases} \dfrac{\beta}{\alpha}\left(\dfrac{x - \nu}{\alpha}\right)^{\beta - 1} \exp\left[-\left(\dfrac{x - \nu}{\alpha}\right)^{\beta}\right], & x \geq \nu \\ 0, & \text{otherwise} \end{cases} \tag{4.45a}$$

The three parameters of the Weibull distribution are ν ($-\infty < \nu < \infty$), which is the location parameter; α ($\alpha > 0$), which is the scale parameter; and β ($\beta > 0$), which is the shape parameter. When $\nu = 0$, the Weibull pdf becomes

$$f(x) = \begin{cases} \dfrac{\beta}{\alpha}\left(\dfrac{x}{\alpha}\right)^{\beta - 1} \exp\left[-\left(\dfrac{x}{\alpha}\right)^{\beta}\right], & x \geq 0 \\ 0, & \text{otherwise} \end{cases} \tag{4.45b}$$

Figure 4.20 shows several Weibull densities when $\nu = 0$ and $\alpha = 1$. Letting $\beta = 1$, the Weibull distribution is reduced to

$$f(x) = \begin{cases} \dfrac{1}{\alpha} e^{-x/\alpha}, & x \geq 0 \\ 0, & \text{otherwise} \end{cases}$$

which is an exponential distribution with parameter $\lambda = 1/\alpha$.

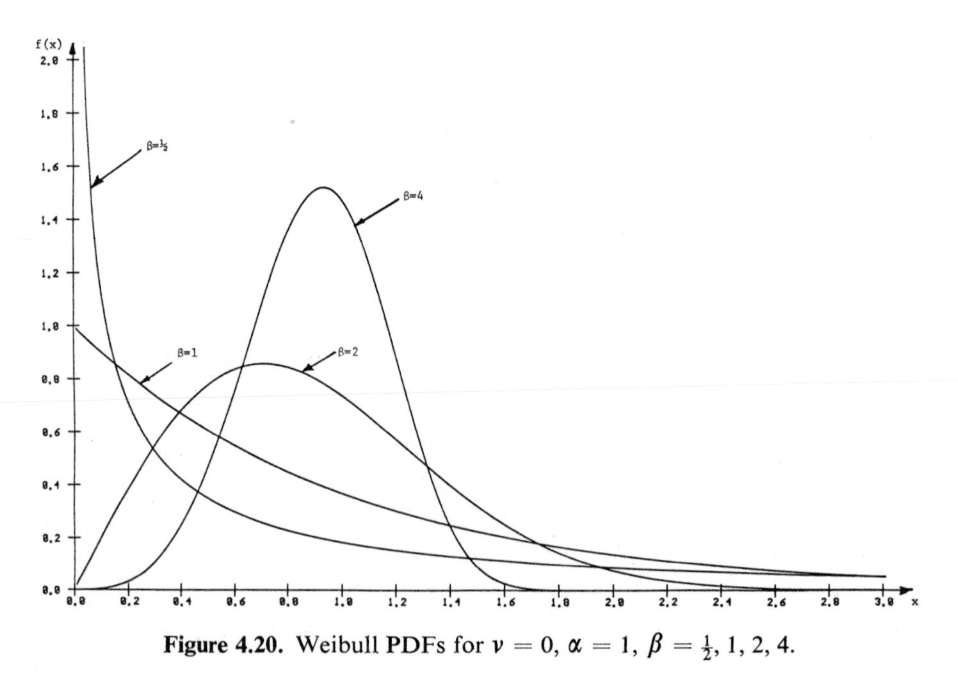

Figure 4.20. Weibull PDFs for $\nu = 0$, $\alpha = 1$, $\beta = \frac{1}{2}, 1, 2, 4$.

The mean and variance of the Weibull distribution are given by the following expressions:

$$E(X) = \nu + \alpha\Gamma\left(\frac{1}{\beta} + 1\right) \tag{4.46}$$

$$V(X) = \alpha^2\left[\Gamma\left(\frac{2}{\beta} + 1\right) - \left[\Gamma\left(\frac{1}{\beta} + 1\right)\right]^2\right] \tag{4.47}$$

where $\Gamma(\cdot)$ is defined by equation (4.27). Thus, the location parameter, ν, has no effect on the variance; however, the mean is increased or decreased by ν. The cdf of the Weibull distribution is given by

$$F(x) = \begin{cases} 0, & x < \nu \\ 1 - \exp\left[-\left(\frac{x - \nu}{\alpha}\right)^\beta\right], & x \geq \nu \end{cases} \tag{4.48}$$

EXAMPLE 4.26

The time to failure for an electronic component is known to have a Weibull distribution with $\nu = 0$, $\beta = 1/3$, and $\alpha = 200$ hours. The mean time to failure is given by equation (4.46) as

$$E(X) = 200\Gamma(3 + 1) = 200(3!) = 1200 \text{ hours}$$

The probability that a unit fails before 2000 hours is determined from equation (4.48) as

$$F(2000) = 1 - \exp\left[-\left(\frac{2000}{200}\right)^{1/3}\right]$$
$$= 1 - e^{-\sqrt[3]{10}} = 1 - e^{-2.15} = 0.884$$

EXAMPLE 4.27

The time it takes for an aircraft to land and clear the runway at a major international airport has a Weibull distribution with $\nu = 1.34$ minutes, $\beta = 0.5$, and $\alpha = 0.04$ minute. Determine the probability that an incoming airplane will take more than 1.5 minutes to land and clear the runway. In this case $P(X > 1.5)$ is determined as follows:

$$P(X \leq 1.5) = F(1.5)$$
$$= 1 - \exp\left[-\left(\frac{1.5 - 1.34}{0.04}\right)^{0.5}\right]$$
$$= 1 - e^{-2} = 1 - 0.135 = 0.865$$

Therefore, the probability that an aircraft will require more than 1.5 minutes to land and clear the runway is 0.135.

7. *Triangular distribution.* A random variable X has a triangular distribu-

tion if its pdf is given by

$$f(x) = \begin{cases} \dfrac{2(x-a)}{(b-a)(c-a)}, & a \leq x \leq b \\ \dfrac{2(c-x)}{(c-b)(c-a)}, & b < x \leq c \\ 0, & \text{elsewhere} \end{cases} \tag{4.49}$$

where $a \leq b \leq c$. The mode occurs at $x = b$. A triangular pdf is shown in Figure 4.21. The parameters (a, b, c) can be related to other measures, such as the mean and the mode, as follows:

$$E(X) = \frac{a + b + c}{3} \tag{4.50}$$

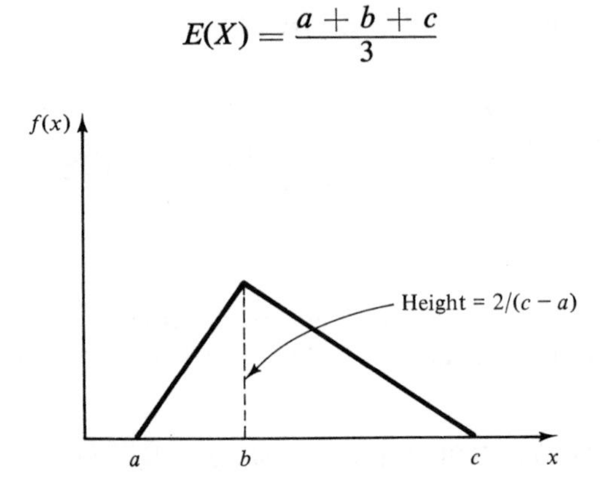

Figure 4.21. PDF of the triangular distribution.

From equation (4.50) the mode can be determined as

$$\text{Mode} = b = 3E(X) - (a + c) \tag{4.51}$$

Since $a \leq b \leq c$, it follows that

$$\frac{2a + c}{3} \leq E(X) \leq \frac{a + 2c}{3}$$

The mode is used more often than the mean to characterize the triangular distribution. As shown in Figure 4.21 its height is $2/(c - a)$ above the x axis. The variance, $V(X)$, of the triangular distribution is little used. Its determination is left as an exercise for the student. The cdf for the triangular distribution is given by

$$F(x) = \begin{cases} 0, & x \leq a \\ \dfrac{(x-a)^2}{(b-a)(c-a)}, & a < x \leq b \\ 1 - \dfrac{(c-x)^2}{(c-b)(c-a)}, & b < x \leq c \\ 1, & x > c \end{cases} \tag{4.52}$$

EXAMPLE 4.28

The central processing requirements, for programs that will execute, have a triangular distribution with $a = 0.05$ second, $b = 1.1$ seconds, and $c = 6.5$ seconds. Determine the probability that the cpu requirement for a random program is 2.5 seconds or less. The value of $F(2.5)$ is from the portion of the cdf in the interval (0.05, 1.1) plus that portion in the interval (1.1, 2.5). Using equation (4.52), both portions can be determined at one time to yield

$$F(2.5) = 1 - \frac{(6.5 - 2.5)^2}{(6.5 - 0.05)(6.5 - 1.1)} = 0.541$$

Thus, the probability is 0.541 that the cpu requirement is 2.5 seconds or less.

EXAMPLE 4.29

An electronic sensor determines the quality of semiconductor chips, rejecting those that fail. Upon demand, the sensor will give the minimum and maximum number of rejects during each hour of production over the past 24 hours. The mean is also given. Without further information, the quality control department has assumed that the number of rejected chips can be approximated by a triangular distribution. The current dump of data indicates that the minimum number of rejected chips during any hour was zero, the maximum was 10, and the mean was 4. Knowing that $a = 0$, $c = 10$, and $E(X) = 4$, the value of b can be determined from equation (4.51) as

$$b = 3(4) - (0 + 10) = 2$$

The height of the mode is $2/(10 - 0) = 0.2$. Thus, Figure 4.22 can be drawn.

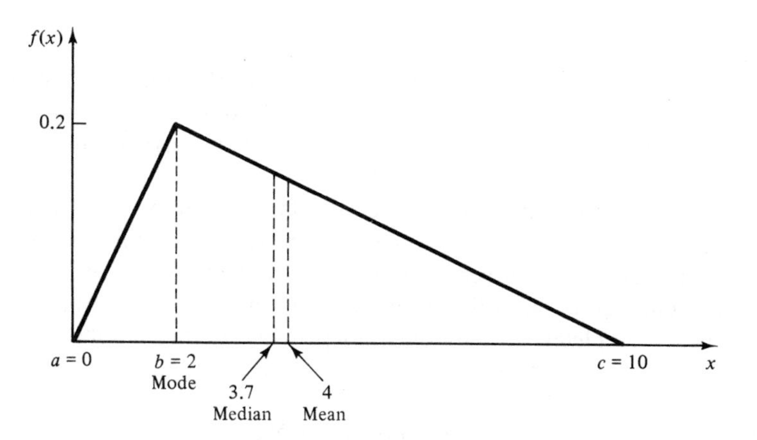

Figure 4.22. Mode, median, and mean for triangular distribution.

The median is the point at which 0.5 of the area is to the left and 0.5 is to the right. The median in this example is 3.7, also shown on Figure 4.22. Determining the median of the triangular distribution requires an initial location of the value to the left or to the right of the mode. The area to the left of the mode is determined from equation (4.52) as

$$F(2) = \frac{2^2}{20} = 0.2$$

Thus, the median is between b and c. Setting $F(x) = 0.5$ in equation (4.52), and solving for $x =$ median yields

$$0.5 = 1 - \frac{(10 - x)^2}{(10)(8)}$$

with

$$x = 3.7$$

This example clearly shows that the mean, mode, and median are not necessarily equal.

4.5. Poisson Process

Consider random events such as the arrival of jobs at a job shop, the arrival of aircraft at a runway, the arrival of ships at a port, the arrival of calls at a switchboard, the breakdown of machines in a large factory, and so on. These events may be described by a counting function $N(t)$ defined for all $t \geq 0$. This counting function will represent the number of events that occurred in $[0, t]$. Time zero is the point at which the observation began, whether or not an arrival occurred at that instant. For each interval $[0, t]$, the value $N(t)$ is an observation of a random variable where the only possible values that can be assumed by $N(t)$ are the integers $0, 1, 2, \ldots$.

The counting process, $\{N(t), t \geq 0\}$, is said to be a Poisson process with mean rate λ if the following assumptions are fulfilled.

1. Arrivals occur one at a time.
2. $\{N(t), t \geq 0\}$ has stationary increments: The distribution of the number of arrivals between t and $t + s$ depends only on the length of the interval s, and not on the starting point t. Thus, arrivals are completely at random without rush or slack periods.
3. $\{N(t), t \geq 0\}$ has independent increments: The number of arrivals during nonoverlapping time intervals are independent random variables. Thus, a large or small number of arrivals in one time interval has no effect on the number of arrivals in subsequent time intervals. Future arrivals occur completely at random, independent of the number of arrivals in past time intervals.

If arrivals occur according to a Poisson process, meeting the three assumptions above, it can be shown that the probability that $N(t)$ is equal to n is given by

$$P[N(t) = n] = \frac{e^{-\lambda t}(\lambda t)^n}{n!} \qquad \text{for } t \geq 0 \text{ and } n = 0, 1, 2, \ldots \tag{4.53}$$

Comparing equation (4.53) to equation (4.15), it can be seen that $N(t)$ has the Poisson distribution with parameter $\alpha = \lambda t$. Thus, its mean and variance are given by

$$E[N(t)] = \alpha = \lambda t = V[N(t)]$$

For any times s and t such that $s < t$, the assumption of stationary increments implies that the random variable $N(t) - N(s)$, representing the number of arrivals in the interval s to t, is also Poisson distributed with mean $\lambda(t - s)$. Thus,

$$P[N(t) - N(s) = n] = \frac{e^{-\lambda(t-s)}[\lambda(t - s)]^n}{n!} \qquad \text{for } n = 0, 1, 2, \ldots$$

and

$$E[N(t) - N(s)] = \lambda(t - s) = V[N(t) - N(s)]$$

Now, consider the time at which arrivals occur in a Poisson process. Let the first arrival occur at time A_1, the second occur at time $A_1 + A_2$, and so on as shown in Figure 4.23. Thus, $A_1, A_2, \ldots$ are successive interarrival times.

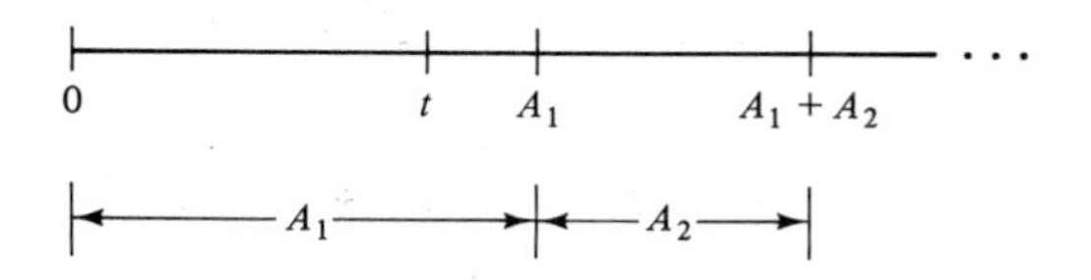

Figure 4.23. Arrival process.

Since the first arrival occurs after time t if and only if there are no arrivals in the interval $[0, t]$, it is seen that

$$\{A_1 > t\} = \{N(t) = 0\}$$

and, therefore,

$$P(A_1 > t) = P[N(t) = 0] = e^{-\lambda t}$$

the last equality following from equation (4.53). Thus, the probability that the first arrival will occur in $[0, t]$ is given by

$$P(A_1 \leq t) = 1 - e^{-\lambda t}$$

which is the cdf for an exponential distribution with parameter λ. Hence, A_1 is distributed exponentially with mean $E(A_1) = 1/\lambda$. It can also be shown that all interarrival times, $A_1, A_2, \ldots,$ are exponentially distributed and independent with mean $1/\lambda$. As an alternative definition of a Poisson process, it can be shown that if interarrival times are distributed exponentially and independently, then the number of arrivals by time t, say $N(t)$, meets the three assumptions above and, therefore, is a Poisson process.

Recall that the exponential distribution is memoryless; that is, the probability of a future arrival in a time interval of length s is independent of the time of the last arrival. The probability of the arrival depends only on the length of the time interval, s. Thus, the memoryless property is related to the properties of independent and stationary increments of the Poisson process.

Additional readings concerning the Poisson process may be obtained from many sources, including Parzen [1962], Feller [1968], and Ross [1981].

EXAMPLE 4.30

The jobs at a machine shop arrive according to a Poisson process with a mean of $\lambda = 2$ jobs per hour. Therefore, the interarrival times are distributed exponentially with the expected time between arrivals, $E(A) = 1/\lambda = \frac{1}{2}$ hour.

Properties of a Poisson process. Several properties of the Poisson process, discussed by Ross [1981] and others, are useful in discrete system simulation. The first of these properties concerns random splitting. Consider a Poisson process $\{N(t), t \geq 0\}$ having rate λ, as represented by the left side of Figure 4.24.

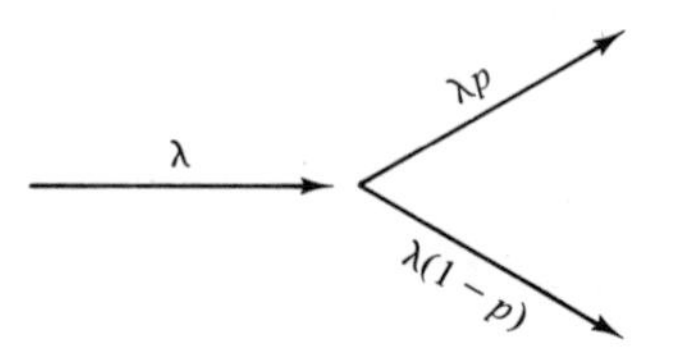

Figure 4.24. Random splitting.

Suppose that each time an event occurs it is classified as either a type I or a type II event. Suppose further that each event is classified as a type I event with probability p and type II event with probability $1 - p$, independently of all other events.

Let $N_1(t)$ and $N_2(t)$ be random variables which denote respectively the number of type I and type II events occurring in $[0, t]$. Note that $N(t) = N_1(t) + N_2(t)$. It can be shown that $N_1(t)$ and $N_2(t)$ are both Poisson processes having rates λp and $\lambda(1 - p)$ as shown in Figure 4.24. Furthermore, it can be shown that the two processes are independent.

EXAMPLE 4.31 (RANDOM SPLITTING)

Suppose that jobs arrive at a shop in accordance with a Poisson process having rate λ. Suppose further that each arrival is marked high priority with probability of 1/3 and low priority with probability of 2/3. Then a type I event would correspond to a high-priority arrival and a type II event would correspond to a low-priority arrival. If $N_1(t)$ and $N_2(t)$ are as defined above, both variables follow the Poisson process with rates $\lambda/3$ and $2\lambda/3$, respectively.

EXAMPLE 4.32

The rate in Example 4.31 is $\lambda = 3$ per hour. The probability that no high priority jobs will arrive in a 2-hour period is given by the Poisson distribution with parameter $\alpha = \lambda pt = 2$. Thus,

$$p(0) = \frac{e^{-2}2^0}{0!} = 0.135$$

Now, consider the opposite situation from random splitting, namely the pooling of two arrival streams. The process of interest is shown in Figure 4.25. It can be shown that if $N_i(t)$ are random variables representing independent Poisson processes with rates λ_i, for $i = 1$ and 2, then $N(t) = N_1(t) + N_2(t)$ is a Poisson process with rate $\lambda_1 + \lambda_2$.

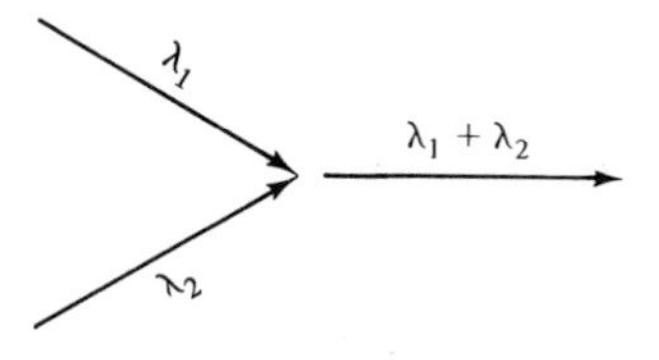

Figure 4.25. Pooled process.

EXAMPLE 4.33 (POOLED PROCESS)

A Poisson arrival stream with $\lambda_1 = 10$ arrivals per hour is combined (or pooled) with a Poisson arrival stream with $\lambda_2 = 17$ arrivals per hour. The combined process is a Poisson process with $\lambda = 27$ arrivals per hour.

4.6. Empirical Distributions

An empirical distribution may be either continuous or discrete in form. It is used when it is impossible or unnecessary to establish that a random variable has any particular known distribution. One advantage of using a known distribution in simulation is the facility with which parameters can be modified to conduct a sensitivity analysis.

EXAMPLE 4.34 (DISCRETE)

Customers at a local restaurant arrive at lunchtime in groups ranging from one to eight persons. The number of persons per party in the last 300 groups has been observed, with the results as summarized in Table 4.3. The relative frequencies appear in Table 4.3 and again in Figure 4.26, which provides a histogram of the data that were gathered. Figure 4.27 provides a cdf of the data. The cdf in Figure 4.27 is called the empirical distribution of the given data.

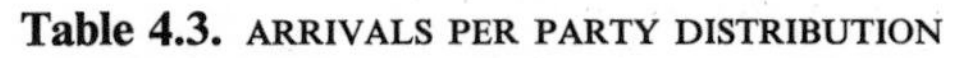

Table 4.3. ARRIVALS PER PARTY DISTRIBUTION

Arrivals per Party	*Frequency*	*Relative Frequency*	*Cumulative Relative Frequency*
1	30	0.10	0.10
2	110	0.37	0.47
3	45	0.15	0.62
4	71	0.24	0.86
5	12	0.04	0.90
6	13	0.04	0.94
7	7	0.02	0.96
8	12	0.04	1.00

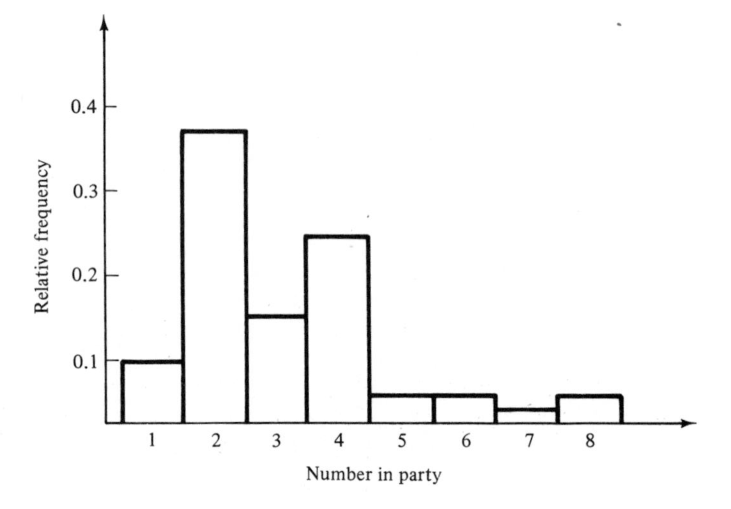

Figure 4.26. Histogram of party size.

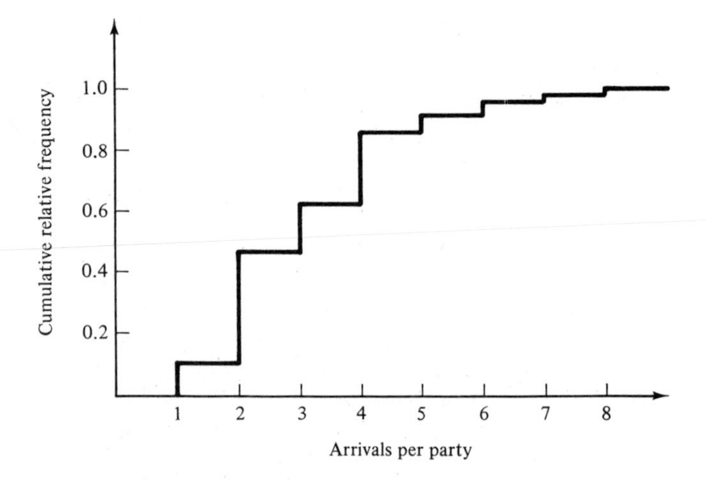

Figure 4.27. Empirical CDF of party size.

EXAMPLE 4.35 (CONTINUOUS)

The time required to repair a conveyor system which has suffered a failure has been collected for the last 100 instances with the results shown in Table 4.4. There were 21 instances in which the repair took between zero and 0.5 hour, and so on. The empirical cdf is shown in Figure 4.28. A piecewise linear curve is formed by the connection of the points of the form $[x, F(x)]$. The points are connected by a straight line. The first connected pair is (0, 0) and (0.5, 0.21); then the points (0.5, 0.21) and (1.0, 0.33) are connected; and so on.

Table 4.4. REPAIR TIMES FOR CONVEYOR

Interval (Hours)	*Frequency*	*Relative Frequency*	*Cumulative Frequency*
$0 < x \leq 0.5$	21	0.21	0.21
$0.5 < x \leq 1.0$	12	0.12	0.33
$1.0 < x \leq 1.5$	29	0.29	0.62
$1.5 < x \leq 2.0$	19	0.19	0.81
$2.0 < x \leq 2.5$	8	0.08	0.89
$2.5 < x \leq 3.0$	11	0.11	1.00

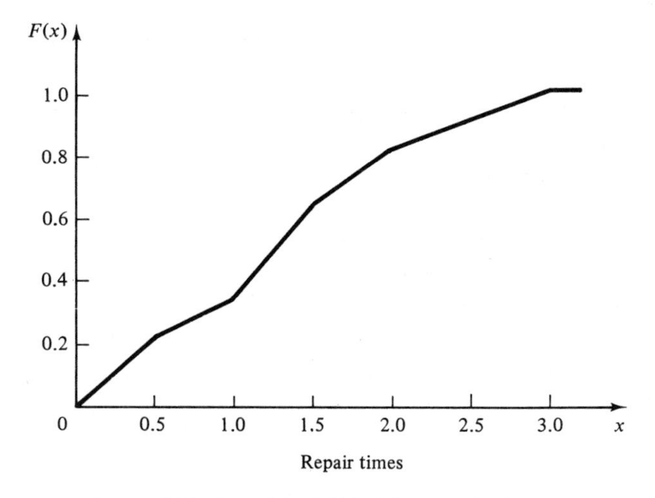

Figure 4.28. Empirical CDF for repair times.

4.7. Summary

In many instances, the world the simulator sees is probabilistic rather than deterministic. The purposes of this chapter are to review several important probability distributions, to familiarize the reader with the notation used in the remainder of the text, and to show applications of the probability distributions in a simulation context.

A major task in simulation is the collection and analysis of input data. One of the first steps in this task is hypothesizing a distributional form for the input data. This is accomplished by comparing the shape of the probability density function or mass function to a histogram of the data, and by an understanding that certain physical processes give rise to specific distributions. This chapter is intended to reinforce the properties of various distributions and to give insight into how these distributors arise in practice. In addition, probabilistic models of input data are used in generating random events in a simulation.

Several features which should have made a strong impression on the reader include the differences between discrete, continuous, and empirical distributions; the Poisson process and its properties; and the versatility of the gamma and the Weibull distributions.

REFERENCES

FELLER, WILLIAM [1968], *An Introduction to Probability Theory and Its Applications*, Vol. I, 3rd ed., Wiley, New York.

FISHMAN, GEORGE S. [1973], *Concepts and Methods in Discrete Event Digital Simulation*, Wiley, New York.

GORDON, GEOFFREY [1975], *The Application of GPSS V to Discrete System Simulation*, Prentice-Hall, Englewood Cliffs, N.J.

HADLEY, G., AND T. M. WHITIN [1963], *Analysis of Inventory Systems*, Prentice-Hall, Englewood Cliffs, N.J.

HINES, W. W., AND D. C. MONTGOMERY [1980], *Probability and Statistics in Engineering and Management Science*, 2nd ed., Wiley, New York.

MOOD, A. M., AND F. A. GRAYBILL [1963], *Introduction to the Theory of Statistics*, 2nd ed., McGraw-Hill, New York.

PARZEN, EMANUEL [1962], *Stochastic Processes*, Holden-Day, San Francisco.

ROMIG, H. G. [1953], *50–100 Binominal Tables*, Wiley, New York.

ROSS, SHELDON M. [1981], *Introduction to Probability Models*, 2nd ed., Academic Press, New York.

EXERCISES

1. A production process manufactures alternators for outboard engines used in recreational boating. On the average, 1% of the alternators will not perform up to the required standards when tested at the engine assembly plant. When a shipment of 100 alternators is received at the plant, they are tested and if more than two are defective, the shipment is returned to the alternator manufacturer. What is the probability of returning a shipment?

2. An industrial chemical that will retard the spread of fire in paint has been devel-

oped. The local sales representative has determined, from past experience, that 48% of the sales calls will result in an order.

(a) What is the probability that the first order will come on the fourth sales call of the day?

(b) If eight sales calls are made in a day, what is the probability of receiving exactly six orders?

(c) If four sales calls are made before lunch, what is the probability that one or less results in an order?

3. For the random variables, X_1 and X_2, which are exponentially distributed with parameter $\lambda = 1$, compute $P(X_1 + X_2 > 2)$.

4. Show that the geometric distribution is memoryless.

5. The number of hurricanes hitting the coast of Florida annually has a Poisson distribution with a mean of 0.8.

(a) What is the probability that more than two hurricanes will hit the Florida coast in a year?

(b) What is the probability that exactly one hurricane will hit the coast of Florida in a year?

6. Arrivals at a bank teller's cage are Poisson distributed at the rate of 1.2 per minute.

(a) What is the probability of zero arrivals in the next minute?

(b) What is the probability of zero arrivals in the next 2 minutes?

7. Lead time demand for condenser units is Poisson distributed with a mean of 6 units. Prepare a table for the inventory manager which will indicate the order level to achieve protection of the following levels: 50%, 80%, 90%, 95%, 97%, 97.5%, 99%, 99.5%, and 99.9%. For $\alpha = 6$, some values of x and $F(x)$ are as follows: [(5, 0.445), (6, 0.606), (7, 0.743), (8, 0.847), (9, 0.916), (10, 0.957), (11, 0.979), (12, 0.991), (13, 0.996), (14, 0.998), (15, 0.999), (16, 0.999)].

8. A random variable X which has pmf given by $p(x) = 1/(n + 1)$ over the range $R_X = \{0, 1, 2, \ldots, n\}$ is said to have a discrete uniform distribution.

(a) Find the mean and variance of this distribution. (*Hint:*

$$\sum_{i=1}^{n} i = \frac{n(n+1)}{2} \quad \text{and} \quad \sum_{i=1}^{n} i^2 = \frac{n(n+1)(2n+1)}{6}.\Big)$$

(b) If $R_X = \{a, a + 1, a + 2, \ldots, b\}$, determine the mean and variance of X.

9. The lifetime, in years, of a satellite placed in orbit is given by the following pdf:

$$f(x) = \begin{cases} 0.4e^{-0.4x}, & x \geq 0 \\ 0, & \text{otherwise} \end{cases}$$

(a) What is the probability that this satellite is still "alive" after 5 years?

(b) What is the probability that the satellite dies between 3 and 6 years from the time it is placed in orbit?

10. (The Poisson distribution can be used to approximate the binomial distribution when n is large and p is small, say p less than 0.1. In utilizing the Poisson approximation, let $\lambda = np$.) In the production of ball bearings, bubbles or depressions occur, rendering the ball bearing unfit for sale. It has been determined that, on the average, one in every 800 of the ball bearings has one or more of these defects.

What is the probability that a random sample of 4000 will yield fewer than three ball bearings with bubbles or depressions?

11. For an exponentially distributed random variable X, find the value of λ that satisfies the following relationship:

$$P(X \leq 3) = 0.9P(X \leq 4)$$

12. A component has an exponential time to failure distribution with mean of 10,000 hours.
    (a) The component has already been in operation for its mean life. What is the probability that it will fail by 15,000 hours?
    (b) At 15,000 hours the component is still in operation. What is the probability that it will operate for another 5000 hours?

13. Suppose that a Die-Hardly Ever battery has an exponential time to failure distribution with a mean of 48 months. At 60 months, the battery is still operating.
    (a) What is the probability that this battery is going to die in the next 12 months?
    (b) What is the probability that the battery dies in an odd year of its life?
    (c) If the battery is operating at 60 months, compute the expected additional months of life.

14. The time to service customers at a bank teller's cage is exponentially distributed with a mean of 50 seconds.
    (a) What is the probability that the two customers in front of an arriving customer will each take less than 60 seconds to complete their transactions?
    (b) What is the probability that the two customers in front will finish their transactions so that an arriving customer can reach the teller's cage within 2 minutes?

15. Determine the variance, $V(X)$, of the triangular distribution.

16. The daily use of water, in thousands of liters, at the Hardscrabble Tool and Die Works follows a gamma distribution with a shape parameter of 2 and a scale parameter of 1/4. What is the probability that the demand exceeds 4000 liters on any given day?

17. When Admiral Byrd went to the North Pole he wore battery-powered thermal underwear. The batteries failed instantaneously rather than gradually. The batteries had a life that was exponentially distributed with a mean of 12 days. The trip took 30 days. Admiral Byrd packed three batteries. What is the probability that three batteries would be a sufficient number to keep the Admiral warm?

18. Suppose that an average of 30 customers per hour arrive at the Sticky Donut Shop in accordance with a Poisson process. What is the probability that more than 5 minutes will elapse before both of the next two customers walk through the door?

19. Professor Dipsy Doodle gives six problems on each exam. Each problem requires an average of 30 minutes grading time for the entire class of 15 students. The grading time for each problem is exponentially distributed, and the problems are independent of each other.
    (a) What is the probability that the Professor will finish the grading in $2\frac{1}{2}$ hours or less?
    (b) What is the most likely grading time?
    (c) What is the expected grading time?

20. An aircraft has dual hydraulic systems. The aircraft switches to the standby system automatically if the first system fails. If both systems have failed, the plane will crash. Assume that the life of a hydraulic system is exponentially distributed with a mean of 2000 air hours.

(a) If the hydraulic systems are inspected every 2500 hours, what is the probability that an aircraft will crash before that time?

(b) What danger would there be in moving the inspection point to 3000 hours?

21. A random variable X is beta distributed if its pdf is given by

$$f(x) = \begin{cases} \dfrac{(\alpha + \beta + 1)!}{\alpha!\beta!} x^{\alpha}(1 - x)^{\beta}, & 0 < x < 1 \\ 0, & \text{otherwise} \end{cases}$$

Show that the beta distribution becomes the uniform distribution over the unit interval when $\alpha = \beta = 0$.

22. Many states have license tags which are of the following format:

letter letter letter number number number

The letters indicate the weight of the automobile, but the numbers are at random, ranging from 100 to 999.

(a) What is the probability that the next two tags seen (at random) will have numbers of 500 or higher?

(b) What is the probability that the sum of the next two tags seen (at random) will have a total of 1000 or higher? (*Hint:* Approximate the discrete uniform distribution with a continuous uniform distribution. The sum of two independent uniform distributions is a triangular distribution.)

23. Let X be a random variable that is normally distributed with a mean of 10 and a variance of 4. Find the values a and b such that $P(a < X < b) = 0.90$ and $|\mu - a| = |\mu - b|$.

24. IQ scores are normally distributed throughout society with a mean of 100 and a standard deviation of 15.

(a) A person with an IQ of 140 or higher is called a "genius." What proportion of society is in the genius category?

(b) What proportion of society will miss the genius category by 5 or less points?

(c) An IQ of 110 or better is required to make it through an accredited college or university. What proportion of society could be eliminated from completing a higher education based on a low IQ score?

25. (If $\{X_i\}$ are n independent normal random variables, and if X_i has mean μ_i and variance σ_i^2, then the sum

$$Y = X_1 + X_2 + \cdots + X_n$$

is normal with mean $\sum_{i=1}^{n} \mu_i$ and variance $\sum_{i=1}^{n} \sigma_i^2$.) Three shafts are made and assembled in a linkage. The length of each shaft, in centimeters, is distributed as follows:

Shaft 1: $N \sim (60, 0.09)$
Shaft 2: $N \sim (40, 0.05)$
Shaft 3: $N \sim (50, 0.11)$

(a) What is the distribution of the linkage?
(b) What is the probability that the linkage will be longer than 150.2 centimeters?
(c) The tolerance limits for the assembly are (149.83, 150.21). What proportion of assemblies are within the tolerance limits?

26. The circumference of battery posts in a nickel–cadmium battery are Weibull distributed with $\nu = 3.25$ centimeters, $\beta = 1/3$, and $\alpha = 0.005$ centimeters.
(a) Determine the probability that a battery post chosen at random will have a circumference larger than 3.40 centimeters.
(b) If battery posts are larger than 3.50 centimeters, they will not go through the hole provided; if they are smaller than 3.30 centimeters, the clamp will not tighten sufficiently. What proportion of posts will have to be scrapped for one of these reasons?

27. The time to failure of a nickel–cadmium battery is Weibull distributed with parameters $\nu = 0$, $\beta = 1/4$, and $\alpha = 1/2$ years.
(a) Find the fraction of batteries that are expected to fail prior to $1\frac{1}{2}$ years.
(b) What fraction of batteries are expected to last longer than the mean life?
(c) What fraction of batteries are expected to fail between $1\frac{1}{2}$ and $2\frac{1}{2}$ years?

28. The gross weight of three-axle trucks that have been checked at the Hahira Inspection Station on Interstate Highway 85 follows a Weibull distribution with parameters $\nu = 6.8$ tons, $\beta = 1.5$, and $\alpha = 1/2$ ton. Determine the appropriate weight limit such that 0.01 of the trucks will be cited for traveling overweight.

29. The current reading on Sag Revas's gas mileage indicator is an average of 25.3 miles per gallon. Assume that gas mileage on Sag's car follows a triangular distribution with a minimum value of zero and a maximum value of 50 miles per gallon. What is the value of the median?

30. A postal letter carrier has a route consisting of five segments with the time in minutes to complete each segment being normally distributed with mean and variance as shown

High Brook Drive	$N(38, 16)$
High Point Road	$N(99, 29)$
Knob Hill Appartments	$N(85, 25)$
Franklin Road	$N(73, 20)$
Chastain Shopping Center	$N(52, 12)$

In addition to the above times, the letter carrier must organize the mail at the central office which requires a time that is distributed by $N(90, 25)$. The drive to the starting point of the route requires a time that is distributed $N(10, 4)$. The return from the route requires a time that is distributed $N(15, 4)$. The letter carrier then performs administrative tasks with a time that is distributed $N(30, 9)$.
(a) What is the expected length of the letter carrier's work day?
(b) Overtime occurs after eight hours of work on a given day. What is the probability that the letter carrier works overtime on any given day?
(c) What is the probability that the letter carrier works overtime on two or more days in a six day week?
(d) What is the probability that the route will be completed within ± 24 minutes of eight hours on any given day?
(*Hint*: See Exercise 25.)

31. The time to failure of a WD-1 transistor is known to be Weibull distributed with parameters $\nu = 0$, $\beta = 1/2$ and $\alpha = 400$ days. Find the fraction expected to survive 600 days.

32. The TV sets on display at Schocker's Department Store are hooked up such that, when one fails, an exact model like the one which failed will switch on. Three such units are hooked up in this series arrangement. The lives of these TVs are independent of one another. Each TV has a life which is exponentially distributed with a mean of 10,000 hours. Find the probability that the combined life of the system is greater than 32,000 hours.

33. High temperature in Biloxi, Mississippi on July 21, denoted by the random variable X, has the following probability density function where X is in degrees F.

$$f(x) = \begin{cases} \dfrac{2(x - 85)}{119}, & 85 \leq x \leq 92 \\ \dfrac{2(102 - x)}{170}, & 92 < x \leq 102 \\ 0, & \text{elsewhere} \end{cases}$$

(a) What is the variance of the temperature, $V(X)$?
(If you worked Exercise 15, this is quite easy.)
(b) What is the median temperature?
(c) What is the modal temperature?

34. The time to failure of Eastinghome light bulbs is Weibull distributed with $\nu = 1.8 \times 10^3$ hours, $\beta = 1/2$ and $\alpha = 1/3 \times 10^3$ hours.
(a) What fraction of bulbs are expected to last longer than the mean lifetime?
(b) What is the median lifetime of a light bulb?

35. Lead time demand is gamma distributed in 100's of units with a shape parameter of 2 and a scale parameter of 1/4. What is the probability that the lead time exceeds 4 (hundred) units during an upcoming cycle?

QUEUEING MODELS

Simulation is often used in the analysis of queueing models. In a simple but typical queueing model, shown in Figure 5.1, customers arrive from time to time and join a queue, or waiting line, are eventually served, and finally leave the system. The term "customer" refers to any type of entity that can be viewed as requesting "service" from a system. Therefore, many service facilities, production systems, repair and maintenance facilities, communications and computer systems, and transport and material handling systems can be viewed as queueing systems.

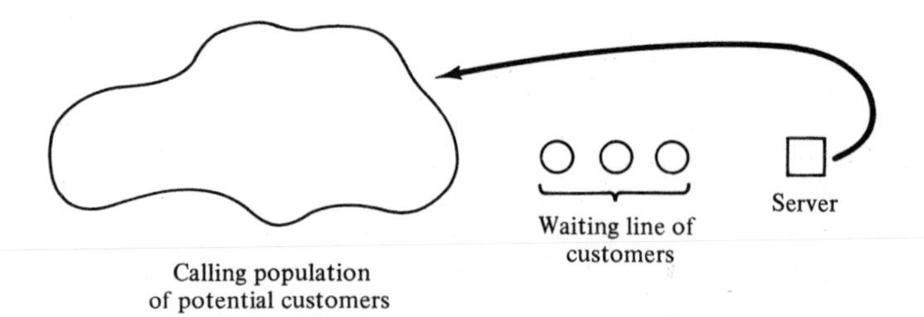

Figure 5.1. Simple queueing model.

Queueing models, whether solved mathematically or analyzed through simulation, provide the analyst with a powerful tool for designing and evaluating the performance of queueing systems. Typical measures of system performance

5

include server utilization (percentage of time a server is busy), length of waiting lines, and delays of customers. Quite often, when designing or attempting to improve a queueing system, the analyst (or decision maker) is involved in trade-offs between server utilization and customer satisfaction in terms of line lengths and delays. Queueing theory and/or simulation analysis is used to predict these measures of system performance as a function of the input parameters. The input parameters include the arrival rate of customers, the service demands of customers, the rate at which a server works, and the number and arrangement of servers. To a certain degree, some of the input parameters are under management's direct control. Thus, the performance measures are under their indirect control, provided that the relationship between the performance measures and the input parameters is adequately understood for the given system.

For relatively simple systems, these performance measures can be computed mathematically at great savings in time and expense compared to the use of a simulation model. But for realistic models of complex systems, simulation is usually required. Nevertheless, analytically tractable models, although usually requiring many simplifying assumptions, are valuable for rough but quick estimates of system performance. These rough estimates may then be refined by use of a detailed and more realistic simulation model. Simple models are also useful for developing an understanding of the dynamic behavior of queueing systems and the relationships between various performance measures.

This chapter will not develop the mathematical theory of queues but instead will discuss some of the well-known models For an elementary treatment of queueing theory, the reader is referred to the survey chapters in Wagner [1975] or Hillier and Lieberman [1980]. More extensive treatments with a view toward applications are given by Gross and Harris [1974], Cooper [1981], and White et al. [1975]. The latter text especially emphasizes engineering and management applications.

This chapter discusses the dynamic behavior and general characteristics of queues, the meaning and relationships of the important performance measures, estimation of the mean measures of performance from a simulation, the effect of varying the input parameters, and the mathematical solution of a small number of important and basic queueing models.

5.1. Characteristics of Queueing Systems

The key elements of a queueing system are the customers and servers. The term "customer" can refer to people, machines, trucks, mechanics, patients, pallets, airplanes, cases, orders, or dirty clothes—anything that arrives at a facility and requires service. The term "server" might refer to receptionists, repairpersons, mechanics, tool crib clerks, medical personnel, automatic storage and retrieval machines (e.g., cranes), runways at an airport, automatic packers, order pickers, or washing machines—any person or machine which provides the requested service. Although the terminology employed will be that of a customer arriving to a service facility, sometimes, in fact, the server moves to the customer; for example, a repairperson moving to a broken machine. This in no way invalidates the models but is merely a matter of terminology. Table 5.1 lists a number of different systems together with a subsystem consisting of "arriving customers" and one or more "servers." The remainder of this section describes the elements of a queueing system in more detail.

5.1.1. The Calling Population

The population of potential customers, referred to as the calling population, may be assumed to be finite or infinite. For example, consider a bank of five machines which are curing tires. After an interval of time, a machine automatically opens and must be attended by a worker who removes the tire and puts an uncured tire into the machine. The machines are the "customers," who "arrive" at the instant they automatically open. The worker is the "server," who "serves" an open machine as soon as possible. The calling population is finite and consists of the five machines.

In systems with a large population of potential customers, the calling population is usually assumed to be infinite. For such systems, this assumption is usually innocuous and, furthermore, it may simplify the model. Examples of infinite populations include the potential customers of a restaurant, bank, or

Table 5.1. EXAMPLES OF QUEUEING SYSTEMS

System	*Customers*	*Server(s)*
Reception desk	People	Receptionist
Repair facility	Machines	Repairperson
Garage	Trucks	Mechanic
Tool crib	Mechanics	Tool crib clerk
Hospital	Patients	Nurses
Warehouse	Pallets	Crane
Airport	Airplanes	Runway
Production line	Cases	Case packer
Warehouse	Orders	Order picker
Road network	Cars	Traffic light
Grocery	Shoppers	Checkout station
Laundry	Dirty linen	Washing machines/dryers
Job shop	Jobs	Machines/workers
Lumberyard	Trucks	Overhead crane
Saw mill	Logs	Saws
Computer	Jobs	CPU, disk, tapes
Telephone	Calls	Exchange
Ticket office	Football fans	Clerk
Mass transit	Riders	Buses, trains

other similar service facility, and a very large group of machines serviced by a technician. Even though the actual population may be finite but large, it is generally safe to use infinite population models provided that the number of customers being served or waiting for service at any given time is a small proportion of the population of potential customers.

The main difference between finite and infinite population models is in how the arrival rate is defined. In an infinite population model, the arrival rate (i.e., the average number of arrivals per unit of time) is not affected by the number of customers who have left the calling population and joined the queueing system. When the arrival process is homogeneous over time (e.g., there are no "rush hours"), the arrival rate is usually assumed to be constant. On the other hand, for finite calling population models, the arrival rate to the queueing system does depend on the number of customers being served and waiting. To take an extreme case, suppose that the calling population has one member, for example, Air Force One. When Air Force One is being serviced by the team of mechanics who are on duty 24 hours per day, the arrival rate is zero, because there are no other potential customers (airplanes) who can arrive at the service facility (team of mechanics). A more typical example is that of the five tire curing machines serviced by a single worker. When all five are closed and curing a tire, the worker is idle and the arrival rate is at a maximum, but the instant a machine opens and requires service, the arrival rate decreases. At those times when all five are open (so four machines are waiting for service while the worker is attending the other one), the arrival rate is zero, that is, no arrival is possible until the worker finishes with a machine, in which case it returns to the calling population and becomes a potential arrival.

5.1.2. System Capacity

In many queueing systems there is a limit to the number of customers that may be in the waiting line or system. For example, an automatic car wash may have room for only 10 cars to wait in line to enter the mechanism. It may be too dangerous or illegal for cars to wait in the street. An arriving customer who finds the system full does not enter but returns immediately to the calling population. Some systems, such as the registration process for students, may be considered as having unlimited capacity. There are no limits on the number of students allowed to wait to register. As will be seen later, when a system has limited capacity, a distinction is made between the arrival rate (i.e., the number of arrivals per time unit) and the effective arrival rate (i.e., the number who arrive and enter the system per time unit).

5.1.3. The Arrival Process

The arrival process for infinite-population models is usually characterized in terms of interarrival times of successive customers. Arrivals may occur at scheduled times or at random times. When at random times, the interarrival times are usually characterized by a probability distribution. In addition, customers may arrive one at a time or in batches. The batch may be of constant size or of random size.

The most important model for random arrivals is the Poisson arrival process. If A_n represents the interarrival time between customer $n - 1$ and customer n (A_1 is the actual arrival time of the first customer), then for a Poisson arrival process, A_n is exponentially distributed with mean $1/\lambda$ time units. The arrival rate is λ customers per time unit. The number of arrivals in a time interval of length t, say $N(t)$, has the Poisson distribution with mean λt customers. For further discussion of the relationship between the Poisson distribution and the exponential distribution, the reader is referred to Section 4.5.

The Poisson arrival process has been successfully employed as a model of the arrival of people to restaurants, drive-in banks, and other service facilities; the arrival of telephone calls to a telephone exchange; the arrival of demands, or orders for a service or product; and the arrival of failed components or machines to a repair facility.

A second important class of arrivals is the scheduled arrivals, say patients to a physician's office, or raw material entering a production process. In this case, the interarrival times $\{A_n, n = 1, 2, \ldots\}$ may be constant, or constant plus or minus a small random amount to represent early or late arrivals.

A third situation occurs when at least one customer is assumed to always be present in the queue, so that the server is never idle because of a lack of customers. For example, the "customers" may represent raw material for a product, and sufficient raw material is assumed to be always available.

For finite-population models, the arrival process is characterized in a completely different fashion. Define a customer as *running* when that customer is outside the queueing system and a member of the potential calling population. For example, a tire curing machine is "running" when it is closed and curing a tire, and it becomes "not running" the instant it opens and demands service from the worker. Define a *runtime* of a given customer as the length of time from departure from the queueing system until that customer's next arrival to the queue. Let $A_1^{(i)}, A_2^{(i)}, \ldots$ be the successive runtimes of customer i, and let $S_1^{(i)}, S_2^{(i)}, \ldots$ be the corresponding successive system times; that is, $S_n^{(i)}$ is the total time spent in system by customer i during the nth visit. Figure 5.2 illustrates these concepts for machine 3 in the tire curing example. The total arrival process is the superposition of the arrival points of all customers. Figure 5.2 shows the first and second arrival of machine 3, but these two times are not necessarily two successive arrivals to the system. For instance, if it is assumed that all machines are running at time 0, the first arrival to the system occurs at time $A_1 = \min\{A_1^{(1)}, A_1^{(2)}, A_1^{(3)}, A_1^{(4)}, A_1^{(5)}\}$. If $A_1 = A_1^{(2)}$, then machine 2 is the first arrival (i.e., the first to open) after time 0. As discussed earlier, the arrival rate is not constant but is a function of the number of running customers.

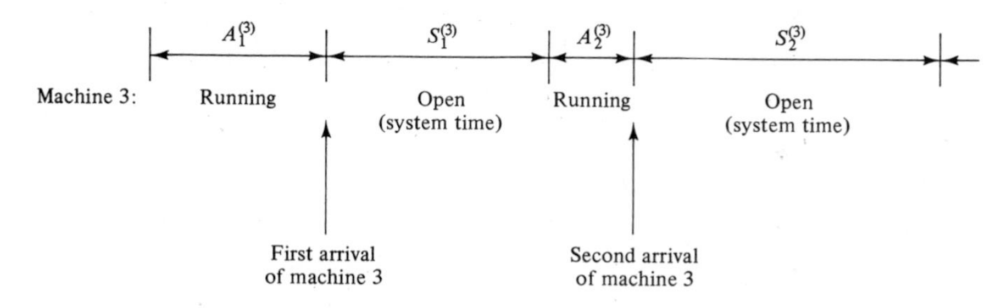

Figure 5.2. Arrival process for a finite-population model.

One important application of finite-population models is the machine repair problem. The machines are the customers and a runtime is also called time to failure. When a machine fails, it "arrives" at the queueing system (the repair facility) and remains there until it is "served" (repaired). Times to failure for a given class of machine have been characterized by the exponential, the Weibull, and the gamma distributions. Models with an exponential runtime are sometimes analytically tractable; an example is given in Section 5.6. Successive times to failure are usually assumed to be statistically independent, but they could depend on other factors, such as the age of a machine since its last major overhaul.

5.1.4. Queue Behavior and Queue Discipline

Queue behavior refers to customer actions while in a queue waiting for service to begin. In some situations, there is a possibility that incoming customers may balk (leave when they see that the line is too long), renege (leave after being in the line when they see that the line is moving too slowly), or jockey (move from one line to another if they think they have chosen a slow line).

Queue discipline refers to the logical ordering of customers in a queue and determines which customer will be chosen for service when a server becomes free. Common queue disciplines include first in, first out (FIFO); last in, first out (LIFO); service in random order (SIRO); shortest processing time first (SPT); and service according to priority (PR). In a job shop, queue disciplines are sometimes based on due dates and on expected processing time for a given type of job. Note that a FIFO queue discipline implies that services begin in the same order as arrivals, but that customers may leave the system in a different order because of different-length service times.

5.1.5. Service Times and the Service Mechanism

The service times of successive arrivals are denoted by $S_1, S_2, S_3, \ldots$. They may be constant or of random duration. In the latter case, $\{S_1, S_2, \ldots\}$ is usually characterized as a sequence of independent and identically distributed random variables. The exponential, Weibull, gamma, and truncated normal distributions have all been used successfully as models of service times in different situations. Sometimes services may be identically distributed for all customers of a given type or class or priority, while customers of different types may have completely different service-time distributions. In addition, in some systems, service times depend upon time of day, or the length of the waiting line. For example, servers may work faster than usual when the waiting line is long, thus effectively reducing the service times.

A queueing system consists of a number of service centers and interconnecting queues. Each service center consists of some number of servers, c, working in parallel; that is, upon getting to the head of the line, a customer takes the first available server. Parallel service mechanisms are either single server ($c = 1$), multiple server ($1 < c < \infty$), or unlimited servers ($c = \infty$). (A self-service facility is usually characterized as having an unlimited number of servers.)

Example 5.1

Consider a discount warehouse where customers may either serve themselves or wait for one of three clerks, and finally leave after paying a single cashier. The system is represented by the flow diagram in Figure 5.3. The subsystem consisting of queue 2 and service center 2 is shown in more detail in Figure 5.4. Other variations of service mechanisms include batch service (a server serving several customers simultaneously), or a customer requiring several servers simultaneously. In the discount warehouse, a clerk may pick several small orders at the same time, but it may take two of the clerks to handle one heavy item.

Example 5.2

A candy manufacturer has a production line which consists of three machines separated by inventory-in-process buffers. The first machine makes and wraps the individual pieces of candy, the second packs 50 pieces in a box, and the third machine

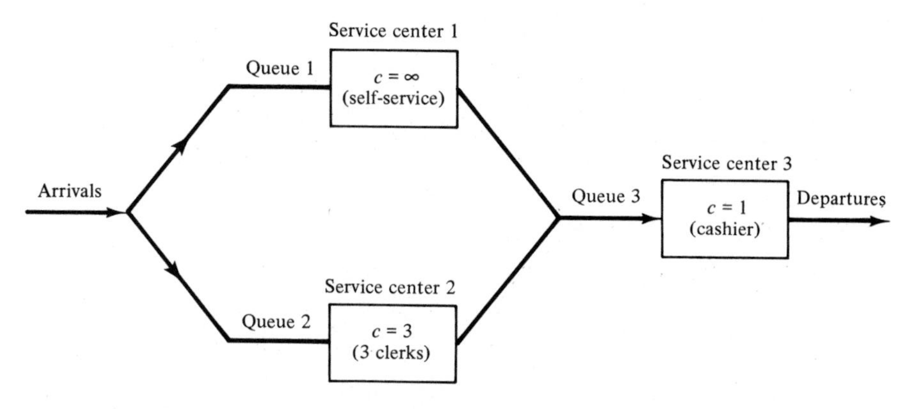

Figure 5.3. Discount warehouse with three service centers.

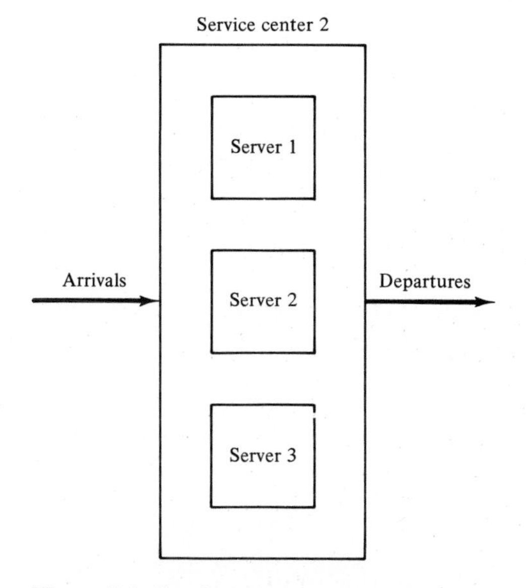

Figure 5.4. Service center 2, with $c = 3$ parallel servers.

seals and wraps the box. The two inventory buffers have capacities of 1000 boxes each. As illustrated by Figure 5.5, the system is modeled as having three service centers, each center having $c = 1$ server (a machine), with queue capacity constraints between machines. It is assumed that a sufficient supply of raw material is always available at the first queue. Because of the queue capacity constraints, machine 1 shuts down whenever the inventory buffer (queue 2) fills to capacity, while machine 2 shuts down whenever the buffer empties. In brief, the system consists of three single-server queues in series with queue capacity constraints and a continuous arrival stream at the first queue.

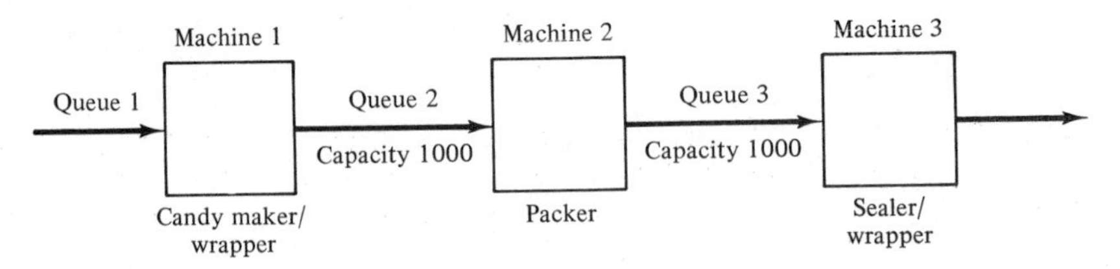

Figure 5.5. Candy production line.

5.2. Queueing Notation

Recognizing the diversity of queueing systems, Kendall (1953) proposed a notational system for parallel server systems which has been widely adopted. An abridged version of this convention is based on the format $A/B/c/N/K$. These letters represent the following system characteristics:

A represents the interarrival-time distribution.
B represents the service-time distribution.
[Common symbols for A and B include M (exponential), D (constant or deterministic), E_k (Erlang of order k), and G (arbitrary or general).]
c represents the number of parallel servers.
N represents the system capacity.
K represents the size of the population.

For example, $M/M/1/\infty/\infty$ indicates a single-server system that has unlimited queue capacity and an infinite population of potential arrivals. The interarrival times and service times are exponentially distributed. When N and K are infinite, they may be dropped from the notation. For example, $M/M/1/\infty/\infty$ is often shortened to $M/M/1$. The tire curing system can be initially represented by $G/G/1/5/5$.

Additional notation used throughout the remainder of this chapter for parallel server systems is listed in Table 5.2. The meanings may vary slightly from system to system. All systems will be assumed to have a FIFO queue discipline.

5.3. Transient and Steady-State Behavior of Queues

Consider an $M/M/1/1/\infty$ queue which at time 0 is empty and idle. That is, consider a single-server queue with Poisson arrivals having mean λ arrivals per time unit and exponential service times with mean μ^{-1} time units. The system capacity is one, so there is never a waiting line, and arrivals from the infinite

Table 5.2. QUEUEING NOTATION FOR PARALLEL SERVER SYSTEMS

P_n	Steady-state probability of having n customers in system
$P_n(t)$	Probability of n customers in system at time t
λ	Arrival rate
λ_e	Effective arrival rate
μ	Service rate of one server
μ_e	Effective service rate of one server
ρ	Server utilization
A_n	Interarrival time between customers $n-1$ and n
S_n	Service time of the nth arriving customer
W_n	Total time spent in system by the nth arriving customer
W_n^Q	Total time spent in the waiting line by customer n
$L(t)$	The number of customers in system at time t
$L_Q(t)$	The number of customers in queue at time t
L	Long-run time-average number of customers in system
L_Q	Long-run time-average number of customers in queue
w	Long-run average time spent in system per customer
w_Q	Long-run average time spent in queue per customer

calling population who find the server busy are turned away. At time 0, the server is idle. Although this system can be solved mathematically, it will also be simulated to illustrate the transient behavior of queues.

One possible history of this $M/M/1/1/\infty$ queue is illustrated in Figure 5.6, where $L(t)$, the number of customers in the system at time t, is plotted versus time t. For this system, $L(t)$ is always 0 or 1. Figure 5.6 is derived from the data in the following example.

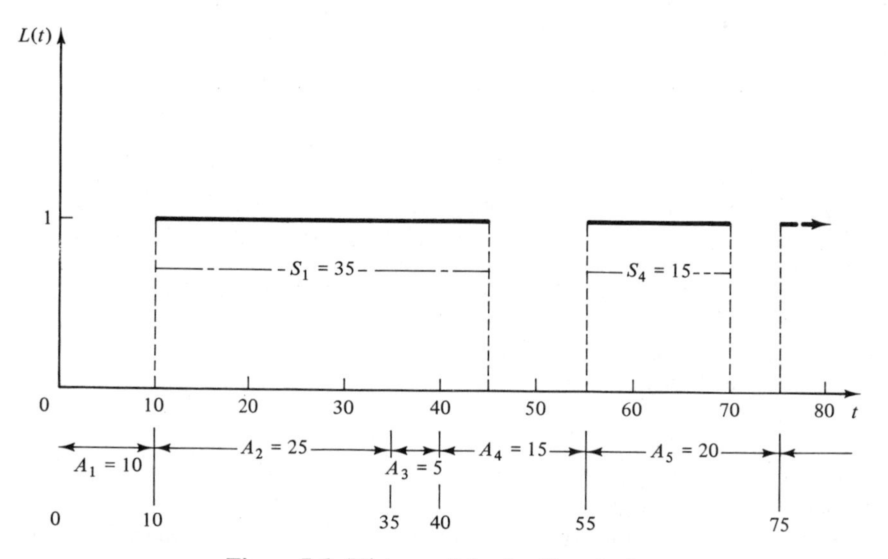

Figure 5.6. History of the loading dock.

Example 5.3

Consider a loading dock that has room for one truck and no places for trucks to wait. If a truck is at the dock, all other arriving trucks go to other loading docks. When a truck arrives, it is either turned away or begins unloading immediately. Trucks arrive in Poisson fashion at a mean arrival rate λ of 2 per hour, while loading and/or unloading is modeled as a random variable exponentially distributed with mean $\mu^{-1} = 120$ minutes, so that the service rate is $\mu = 0.5$ per hour. Since there is a very large population of potential arriving trucks, the system is modeled as an $M/M/1/1/\infty$ queue. The system will be studied by a simulation model.

Suppose that at time 0 the dock is empty, that the interarrival times in minutes are generated as $A_1 = 10$, $A_2 = 25$, $A_3 = 5$, $A_4 = 15$, $A_5 = 20$, and that the service times in minutes are generated as $S_1 = 35$, $S_2 = 20$, $S_3 = 60$, $S_4 = 15$, $S_5 = 134$, ... A graph of $L(t)$, the number of trucks present at time t, is presented in Figure 5.6. This plot may be regarded as a simulation of the truck dock from time 0 to time 75. Note that the second and third trucks are turned away at times 35 and 40 since the first truck is still present, and thus S_2 and S_3 play no role in Figure 5.6.

The observed server utilization, or proportion of time the server is busy, over the time interval 0 to $T = 75$ minutes is denoted by $\hat{\rho}$ and estimated as follows from the simulation data:

$$\hat{\rho} = \frac{35 + 15}{10 + 25 + 5 + 15 + 20} = \frac{50}{75} = 0.67$$

The statistic $\hat{\rho}$ can also be expressed as

$$\hat{\rho} = 0\left(\frac{T_0}{T}\right) + 1\left(\frac{T_1}{T}\right) = 0\left(\frac{25}{75}\right) + 1\left(\frac{50}{75}\right) = 0.67$$

where T_i/T is the fraction of time there are i customers being served; $\hat{\rho}$ is called a time-weighted, or time-integrated average. In many systems the analyst is interested in long-run, or steady-state, characteristics of the system. For an $M/M/1/1/\infty$ queue the long-run proportion of time the server is busy can be derived mathematically and is given by

$$\rho = \frac{\lambda}{\lambda + \mu} = \frac{2}{2 + 0.5} = 0.80$$

In addition, ρ is independent of the initial conditions. Note that long-run utilization ρ is a constant between 0 and 1, while observed utilization $\hat{\rho}$ is a random variable (or statistic) because it depends on the random arrivals and random service times. In addition, $\hat{\rho}$ depends on T, the length of time the system is observed, and on I, the initial conditions (i.e., the state of the system at time 0). However, as T becomes large, the effect of the initial conditions lessens and becomes negligible and $\hat{\rho}$ approaches ρ; that is, with probability 1,

$$\hat{\rho} \longrightarrow \rho \qquad \text{as } T \longrightarrow \infty \tag{5.1}$$

regardless of the initial conditions I. In other words, condition (5.1) remains true whether I is the empty and idle condition, or the busy condition, or even if I is chosen by the flip of a coin. Condition (5.1) implies that the statistic $\hat{\rho}$ is strongly consistent;

if the simulation run length T is sufficiently long, the estimate $\hat{\rho}$ can be made arbitrarily close to long-run utilization ρ.

For this system, the estimate $\hat{\rho}$ is biased low because of the initial empty and idle condition. That is, $E(\hat{\rho}) < \rho$, which means that if an observer recorded system behavior over an unlimited number of 75-minute time intervals, each interval beginning at time 0 under identical conditions, namely, the empty and idle condition, then the ensemble average utilization $E(\hat{\rho})$ would be less than the long-run average ρ. This bias causes problems when estimating long-run measures of system performance of queueing systems, and will be discussed in that context in Chapter 11.

Using the mathematical theory of queues, it can be shown for the $M/M/1/1/\infty$ queue that $L(t)$, the number in system at time t, has the probability distribution given by

$$P(L(t) = 0) = P_0(t) = \frac{\mu}{\lambda + \mu} + a_0 e^{-(\lambda+\mu)t}$$

and (5.2)

$$P(L(t) = 1) = P_1(t) = \frac{\lambda}{\lambda + \mu} + a_1 e^{-(\lambda+\mu)t}$$

where a_0 and a_1 are constants which are independent of t but do depend on the initial conditions I:

$$a_0 = P_0(0) - \frac{\mu}{\lambda + \mu}$$

$$a_1 = P_1(0) - \frac{\lambda}{\lambda + \mu}$$

As $t \longrightarrow \infty$, $e^{-(\lambda+\mu)t} \longrightarrow 0$ since $\lambda > 0$ and $\mu > 0$, and therefore

$$P_0(t) \longrightarrow P_0 = \frac{\mu}{\lambda + \mu}$$

and (5.3)

$$P_1(t) \longrightarrow P_1 = \frac{\lambda}{\lambda + \mu}$$

regardless of the initial conditions I. Note that $P_0(t) + P_1(t) = 1$ for any time $t \geq 0$, and $P_0 + P_1 = 1$. The probabilities $\{P_i(t)\}$ are called the transient probabilities, as they give the system state probabilities for time t given that the system was in initial state I at time $t = 0$, where I is defined by $P_0(0)$ and $P_1(0)$. The probabilities P_0 and P_1 are called the steady-state probabilities. After a sufficiently long period of time, the system is said to be in steady state, or more accurately, in statistical equilibrium. For a system in statistical equilibrium, an observer who notes the system state at an arbitrary point of time t would have probability $P_0 = \mu/(\lambda + \mu)$ of finding the system empty, and probability $P_1 = \lambda/(\lambda + \mu)$ of finding it occupied.

EXAMPLE 5.4

Consider the truck dock of Example 5.3, which was modeled as an $M/M/1/1/\infty$ queue with $\lambda = 2$ arrivals per hour and mean service time $\mu^{-1} = 2$ hours. Assume that at time 0 the system is empty; therefore,

$$P_0(0) = 1, \qquad P_1(0) = 0$$

and by equations (5.2),

$$P_0(t) = \frac{0.5}{2 + 0.5} + \left(1 - \frac{0.5}{2 + 0.5}\right)e^{-2.5t}$$
$$= 0.2 + 0.8e^{-2.5t}$$

and

$$P_1(t) = \frac{2}{2 + 0.5} + \left(0 - \frac{2}{2 + 0.5}\right)e^{-2.5t}$$
$$= 0.8 - 0.8e^{-2.5t}$$

Using equations (5.3) (or letting $t \longrightarrow \infty$) yields the steady-state probabilities:

$$P_0 = 0.2, \qquad P_1 = 0.8$$

Note that $P_0(t) > P_0$ and $P_1(t) < P_1$ for all times $t \geq 0$, which is another indication of the downward bias in the observed utilization $\hat{\rho}$ due to the empty and idle condition at time 0. Figure 5.7 depicts the approach to statistical equilibrium given the empty and idle initial conditions. Figure 5.7 shows clearly what would happen if a number of

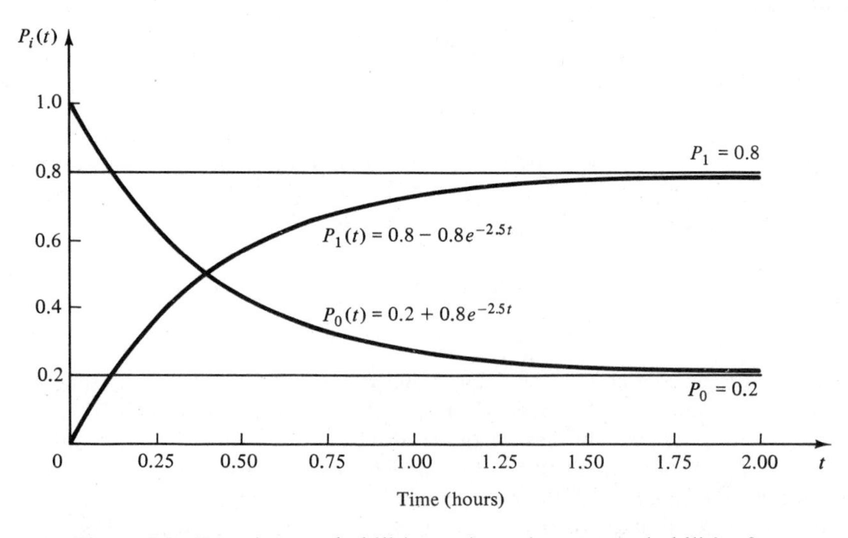

Figure 5.7. Transient probabilities and steady-state probabilities for the $M/M/1/1/\infty$ Queue.

simulation runs using different random numbers were made over a short interval of time, say from time 0 to time $T = 0.50$ hour, all with the system idle initially. For $t = 0.50$, $P_0(t) = P_0(0.5) = 0.429$ and $P_1(t) = P_1(0.5) = 0.571$. Thus, the proportion of runs in which the system will be empty at time $t = 0.50$ hour is 42.9%, considerably greater than the steady-state proportion of 20%. This transient effect is due entirely to the empty and idle initial condition I.

Although the arrival rate to the truck dock is $\lambda = 2$ arrivals per hour, this potential is not realized due to system capacity constraints. The effective arrival rate, defined as the average number of arrivals who enter the system per hour, is given by

$$\lambda_e = \lambda(1 - P_1) = 0.4 \text{ arrival per hour}$$

Similarly, the effective service rate, defined as the number of customers served per time unit, is given by

$$\mu_e = \mu(1 - P_0) = (0.5)(0.8) = 0.4 \text{ per hour}$$

which can be compared to the service rate, $\mu = 0.5$ per hour when the server is busy. In general, for systems with Poisson arrivals, if the system capacity is N, the effective arrival rate is $\lambda_e = \lambda(1 - P_N)$ since arrivals can enter with probability $1 - P_N$, and for single-server queues, the effective service rate is $\mu_e = \mu(1 - P_0)$. If the system is in statistical equilibrium, the effective service rate and effective arrival rate are equal, as illustrated by the truck dock of Example 5.4.

The concepts of transient and steady-state probabilities, long-run measures of performance, and statistical equilibrium are applicable to a wide class of queueing and other models. Except for the simplest types of models, it is usually impossible to compute the transient probabilities $\{P_i(t), t \geq 0, i = 0, 1, 2, \ldots\}$ or to estimate how rapidly convergence to statistical equilibrium occurs. For a fairly large class of queueing models, the Markovian queues, it is possible to derive the steady-state probabilities $\{P_i, i = 0, 1, 2, \ldots\}$. Some examples of Markovian queues are given in Sections 5.5 and 5.6. Nevertheless, complex and realistic models that contain such features as nonexponential services and non-Poisson arrivals, routing based on priorities, and capacity constraints on subsystems, continue to defy mathematical analysis, and thus simulation is an attractive tool for analyzing such systems. When estimating long-run, or steady-state, measures of performance from a simulation run, the analyst should consider the following:

1. Estimators such as $\hat{\rho}$ are random variables. For fixed run length T, different random number streams will lead to different values of $\hat{\rho}$.
2. An estimator, such as $\hat{\rho}$, depends on run length T and initial conditions I. As T is increased, the estimator becomes more accurate (its variance decreases) and simultaneously the bias created by I is reduced. If the run length T is too short, estimators will be highly variable and biased, and the simulation results may be misleading. Chapter 11 will discuss ways of reducing the bias due to I and of estimating the variability of the estimator.

5.4. Long-Run Measures of Performance of Queueing Systems

The major long-run mean measures of performance of queueing systems are the long-run time average number of customers in the system (L) and in the queue (L_Q), the long-run average time spent in system (w) and in the queue (w_Q) per customer, and the server utilization, or proportion of time that a server is busy (ρ). The term "system" usually refers to the waiting line plus the service mechanism but in general can refer to any subsystem of the queueing system, whereas the term "queue" refers to the waiting line alone. Other measures of performance of interest include the long-run proportion of customers who are delayed in queue longer than t_0 time units, the long-run proportion of customers turned away because of capacity constraints, and the long-run proportion of time the waiting line contains more than k_0 customers.

This section defines the major measures of performance for a general $G/G/c/N/P$ queueing system, discusses their relationships, and shows how they can be estimated from a simulation run. There are two types of estimators: an ordinary sample average and a time-integrated (or time-weighted) sample average.

5.4.1. Time-Average Number in System, L

Consider a queueing system over a period of time T, and let $L(t)$ denote the number of customers in the system at time t. A simulation of such a system is shown in Figure 5.8. Let T_i denote the total time during $[0, T]$ in which the system

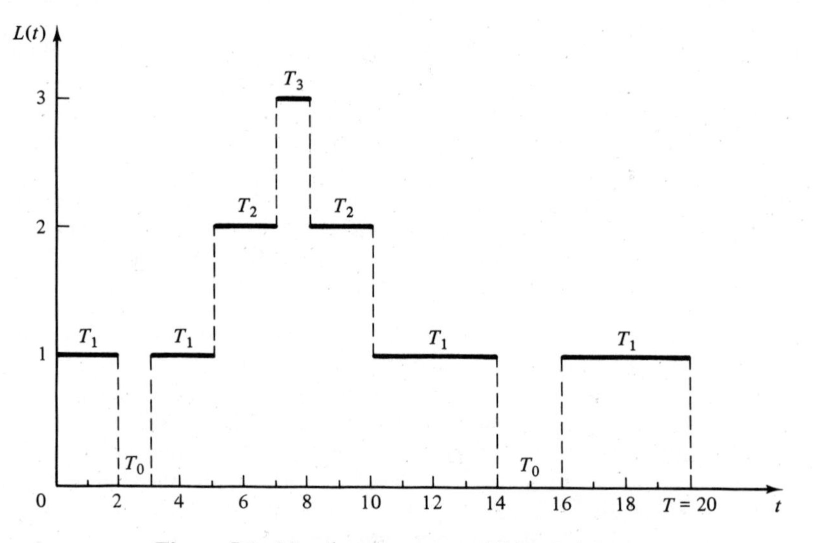

Figure 5.8. Number in system, $L(t)$, at time t.

contained exactly i customers. In Figure 5.8, it is seen that $T_0 = 3$, $T_1 = 12$, $T_2 = 4$, and $T_3 = 1$. (The line segments whose lengths total $T_1 = 12$ are labelled "T_1" in Figure 5.8, etc.) In general, $\sum_{i=0}^{\infty} T_i = T$. The time-weighted-average number in system is defined by

$$\hat{L} = \frac{\sum_{i=0}^{\infty} iT_i}{T} = \sum_{i=0}^{\infty} i\left(\frac{T_i}{T}\right) \tag{5.4}$$

For Figure 5.8, $\hat{L} = [0(3) + 1(12) + 2(4) + 3(1)]/20 = 23/20 = 1.15$. Note that T_i/T is the proportion of time the system contains exactly i customers. The estimator, $\hat{L}$, is an example of a time-weighted average.

By considering Figure 5.8 it can be seen that the total area beneath the function $L(t)$ can be decomposed into rectangles of height i and length T_i. For example, the rectangle of area $3 \times T_3$ has base running from $t = 7$ to $t = 8$ (thus $T_3 = 1$); however, most of the rectangles are broken into parts, such as the rectangle of area $2 \times T_2$ which has part of its base between $t = 5$ and $t = 7$ and the remainder from $t = 8$ to $t = 10$ (thus $T_2 = 2 + 2 = 4$). It follows that the total area is given by $\sum_{i=0}^{\infty} iT_i = \int_0^T L(t)\,dt$ and therefore

$$\hat{L} = \frac{1}{T}\sum_{i=0}^{\infty} iT_i = \frac{1}{T}\int_0^T L(t)\,dt \tag{5.5}$$

The expressions in equations (5.4) and (5.5) are always equal for any queueing system, regardless of the number of servers, the queue discipline, or any other special circumstances. Equation (5.5) justifies the terminology *time integrated average*.

Many queueing systems exhibit a certain kind of long-run stability in terms of their average performance. For such systems, as time T gets large, the observed time-average number in the system, $\hat{L}$, approaches a limiting value, say L, which is called the long-run time-average number in system. That is, with probability 1

$$\hat{L} = \frac{1}{T}\int_0^T L(t)\,dt \longrightarrow L \qquad \text{as } T \longrightarrow \infty \tag{5.6}$$

The estimator $\hat{L}$ is said to be strongly consistent for L. If simulation run length T is sufficiently long, the estimator $\hat{L}$ becomes arbitrarily close to L. In addition, $\hat{L}$ depends on the initial conditions, I, just as did $\hat{\rho}$ in Section 5.3.

Equations (5.5) and (5.6) can be applied to any subsystem of a queueing system as well as to the whole system. If $L_Q(t)$ denotes the number of customers in the waiting line, and T_i^Q denotes the total time during $[0, T]$ in which exactly i customers are in the waiting line, then

$$\hat{L}_Q = \frac{1}{T}\sum_{i=0}^{\infty} iT_i^Q = \frac{1}{T}\int_0^T L_Q(t)\,dt \longrightarrow L_Q \qquad \text{as } T \longrightarrow \infty$$

where $\hat{L}_Q$ is the observed time average number of customers in the waiting line from time 0 to time T, and L_Q is the long-run time-average number in the waiting line.

EXAMPLE 5.5

Suppose that Figure 5.8 represents a single-server queue, that is, a $G/G/1/N/K$ queueing system ($N \geq 3$, $K \geq 3$). Then the number of customers in the waiting line alone is given by $L_Q(t)$, defined by

$$L_Q(t) = \begin{cases} 0 & \text{if } L(t) = 0 \\ L(t) - 1 & \text{if } L(t) \geq 1 \end{cases}$$

and shown in Figure 5.9. Thus, $T_0^Q = 5 + 10 = 15$, $T_1^Q = 2 + 2 = 4$, and $T_2^Q = 1$. Therefore,

$$\hat{L}_Q = \frac{0(15) + 1(4) + 2(1)}{20} = 0.3$$

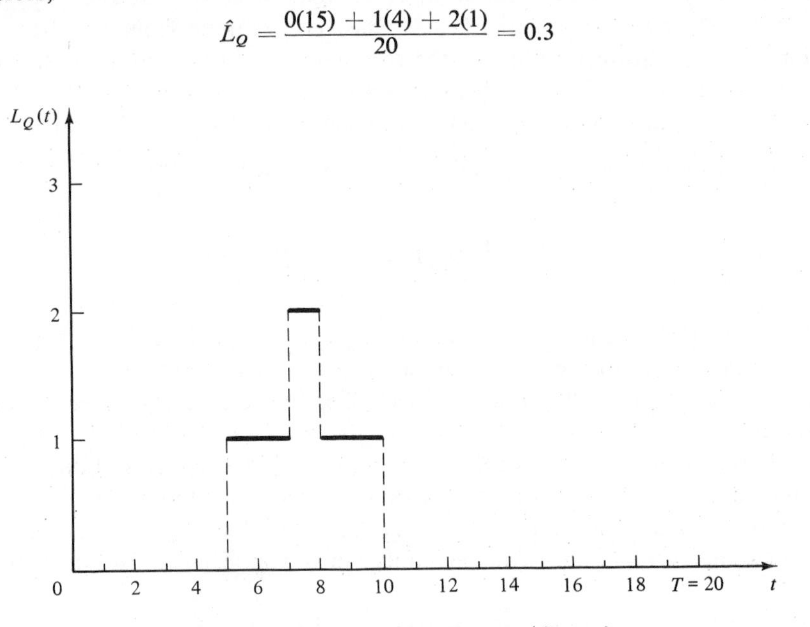

Figure 5.9. Number in waiting line, $L_Q(t)$, at time t.

5.4.2. AVERAGE TIME SPENT IN SYSTEM PER CUSTOMER, w

Simulate a queueing system for some period of time, say T, and record the time each customer spends in the system during $[0, T]$, say $W_1, W_2, \ldots, W_N$, where N is the number of arrivals during $[0, T]$. Then the average time spent in system per customer, called the average system time, is given by the ordinary sample average

$$\hat{w} = \frac{1}{N} \sum_{i=1}^{N} W_i \tag{5.7}$$

For stable systems, as $N \longrightarrow \infty$, with probability 1,

$$\hat{w} \longrightarrow w \tag{5.8}$$

where w is called the long-run average system time.

If the system under consideration is a waiting line alone, equations (5.7) and (5.8) are written as

$$\hat{w}_Q = \frac{1}{N} \sum_{i=1}^{N} W_i^Q \longrightarrow w_Q \qquad \text{as } N \longrightarrow \infty \tag{5.9}$$

where W_i^Q is the total time customer i spends in the waiting line, $\hat{w}_Q$ the observed average time spent in queue (called delay), and w_Q the long-run average delay per customer. The estimators $\hat{w}$ and $\hat{w}_Q$ are influenced by initial conditions I and run length T, analogously to $\hat{L}$ and $\hat{\rho}$.

EXAMPLE 5.6

For the system history shown in Figure 5.8, $N = 5$, $W_1 = 2$, and $W_5 = 20 - 16 = 4$, but W_2, W_3, and W_4 cannot be computed unless more is known about the system. Assume that the system has a single-server and a FIFO queue discipline. This implies that customers will depart from the system in the same order in which they arrived. Each jump upward of $L(t)$ in Figure 5.8 represents an arrival. Arrivals occur at times 0, 3, 5, 7, and 16. Similarly, departures occur at times 2, 8, 10, and 14. (A departure may or may not have occurred at time 20.) Under these assumptions, it is seen that $W_2 = 8 - 3 = 5$, $W_3 = 10 - 5 = 5$, $W_4 = 14 - 7 = 7$, and therefore

$$\hat{w} = \frac{2 + 5 + 5 + 7 + 4}{5} = \frac{23}{5} = 4.6$$

Thus, on the average, an arbitrary customer spends 4.6 time units in the system. As for time spent in the waiting line, it can be seen that $W_1^Q = 0$, $W_2^Q = 0$, $W_3^Q = 8 - 5 = 3$, $W_4^Q = 10 - 7 = 3$, and $W_5^Q = 0$; thus,

$$\hat{w}_Q = \frac{0 + 0 + 3 + 3 + 0}{5} = 1.2$$

5.4.3. THE CONSERVATION EQUATION: $L = \lambda w$

For the system exhibited in Figure 5.8, there were $N = 5$ arrivals in $T = 20$ time units, and thus the observed arrival rate was $\hat{\lambda} = N/T = 1/4$ customer per time unit. Recall that $\hat{L} = 1.15$ and $\hat{w} = 4.6$; hence, it follows that

$$\hat{L} = \hat{\lambda}\hat{w} \tag{5.10}$$

This relationship between L, λ, and w is not coincidental; it holds for almost all queueing systems or subsystems regardless of the number of servers, the queue discipline, or any other special circumstances. Allowing $T \longrightarrow \infty$ and $N \longrightarrow \infty$, equation (5.10) becomes

$$L = \lambda w \tag{5.11}$$

where $\hat{\lambda} \longrightarrow \lambda$, and λ is the long-run average arrival rate. Equation (5.11) is called a conservation equation and is usually attributed to Little [1961]. It says that the average number of customers in the system at an arbitrary point in time is equal to the average number of arrivals per time unit times the average time spent in the system. For Figure 5.8, there is one arrival every 4 time units (on the average) and each arrival spends 4.6 time units in the system (on the average), so at an arbitrary point in time there will be $(1/4)(4.6) = 1.15$ customers present (on the average).

Equation (5.10) can be derived by reconsidering Figure 5.8 in the following manner: Figure 5.10 shows system history, $L(t)$, exactly as in Figure 5.8, with

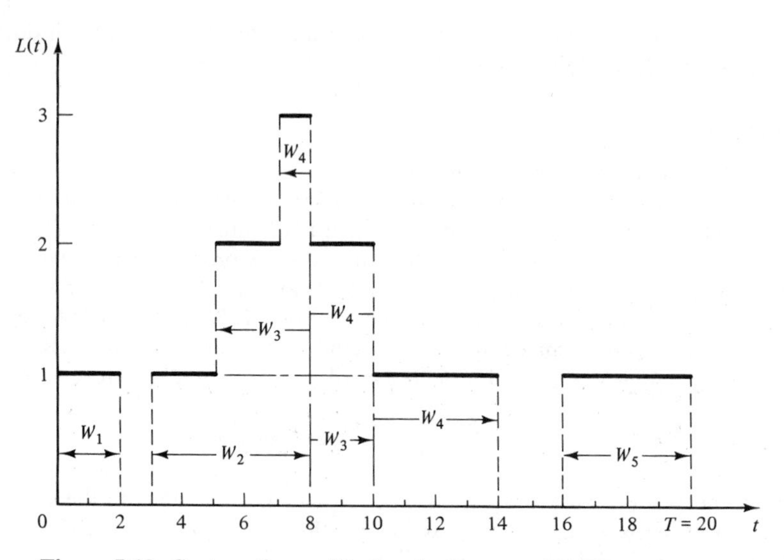

Figure 5.10. System times, W_i, for single-server FIFO system.

each customer's time in the system, W_i, represented by a rectangle. This representation again assumes a single-server system with a FIFO queue discipline. The rectangles for the third and fourth customers are in two and three separate pieces, respectively. The ith rectangle has a height of 1 and a length of W_i for each $i = 1, 2, \ldots, N$. It follows that the total system time of all customers is given by the total area beneath the number in system function, $L(t)$; that is,

$$\sum_{i=1}^{N} W_i = \int_0^T L(t)\,dt \tag{5.12}$$

Therefore, by combining equations (5.5) and (5.7) with $\hat{\lambda} = N/T$, it follows that

$$\hat{L} = \frac{1}{T}\int_0^T L(t)\,dt = \frac{N}{T}\frac{1}{N}\sum_{i=1}^{N} W_i = \hat{\lambda}\hat{w}$$

which is Little's equation (5.10). The intuitive and informal derivation presented here depended on the single-server FIFO assumptions, but these assumptions are not necessary. In fact, equation (5.12), which was the key to the derivation, holds (at least approximately) in great generality, and thus so do equations (5.10) and (5.11). Exercises 11 and 12 ask the student to derive equations (5.12) and (5.10) under different assumptions.

[*Technical note:* If, as defined in Section 5.4.2, W_i is the system time for customer i during $[0, T]$, then equation (5.12) and hence equation (5.10) hold exactly. Some authors choose to define W_i as total system time for customer i; this change will only affect the value of W_i for those customers i who arrive before time T but do not depart until after time T (possibly customer 5 in Figure

5.10). With this change in definition, equations (5.12) and (5.10) only hold approximately. Nevertheless, as $T \longrightarrow \infty$ and $N \longrightarrow \infty$ the error in equation (5.10) decreases to zero and therefore the conservation equation (5.11), namely $L = \lambda w$, for long-run measures of performance holds exactly.]

5.4.4. SERVER UTILIZATION

Server utilization is defined as the proportion of time that a server is busy. Observed server utilization, denoted by $\hat{\rho}$, is defined over a specified time interval $[0, T]$. Long-run server utilization is denoted by ρ. For systems that exhibit long-run stability,

$$\hat{\rho} \longrightarrow \rho \qquad \text{as } T \longrightarrow \infty$$

EXAMPLE 5.7

Referring to Figure 5.8 or 5.10, and assuming that the system has a single server, it can be seen that the server utilization is $\hat{\rho} = (\text{total busy time})/T = (\sum_{i=1}^{\infty} T_i)/T = (T - T_0)/T = 17/20$.

Server utilization in G/G/1/∞/∞ queues. Consider any single-server queueing system with average arrival rate λ customers per time unit, average service time $E(S) = 1/\mu$ time units, and infinite queue capacity and calling population. [$E(S) = \mu^{-1}$ implies that, when busy, the server is working at rate μ customers per time unit, on the average; μ is called the service rate.] The server alone is a subsystem that can be considered as a queueing system in itself, and hence the conservation equation (5.11), $L = \lambda w$, can be applied to the server. For stable systems, the average arrival rate to the server (say λ_s) must be identical to the average arrival rate to the system (λ); for $\lambda_s \leq \lambda$ certainly, but if $\lambda_s < \lambda$, then the waiting line would tend to grow in length at an average rate of $\lambda - \lambda_s$ customers per time unit, which is an unstable system. For the server subsystem, the average system time is $w = E(S) = \mu^{-1}$. The actual number of customers in the server subsystem is either 0 or 1, as shown in Figure 5.11 for the system represented by Figure 5.8. Hence, the average number, $\hat{L}_s$, is given by

$$\hat{L}_s = \frac{1}{T}\int_0^T (L(t) - L_Q(t))\,dt = \frac{T - T_0}{T}$$

In this case, $\hat{L}_s = 17/20 = \hat{\rho}$. In general, for a single-server queue, the average number of customers being served at an arbitrary point in time is equal to server utilization. As $T \longrightarrow \infty$, $\hat{L}_s = \hat{\rho} \longrightarrow L_s = \rho$. Combining these results into $L = \lambda w$ for the server subsystem yields

$$\rho = \lambda E(S) = \frac{\lambda}{\mu} \tag{5.13}$$

or, that long-run server utilization in a single-server queue is equal to the average arrival rate divided by the average service rate. For a single-server queue to be

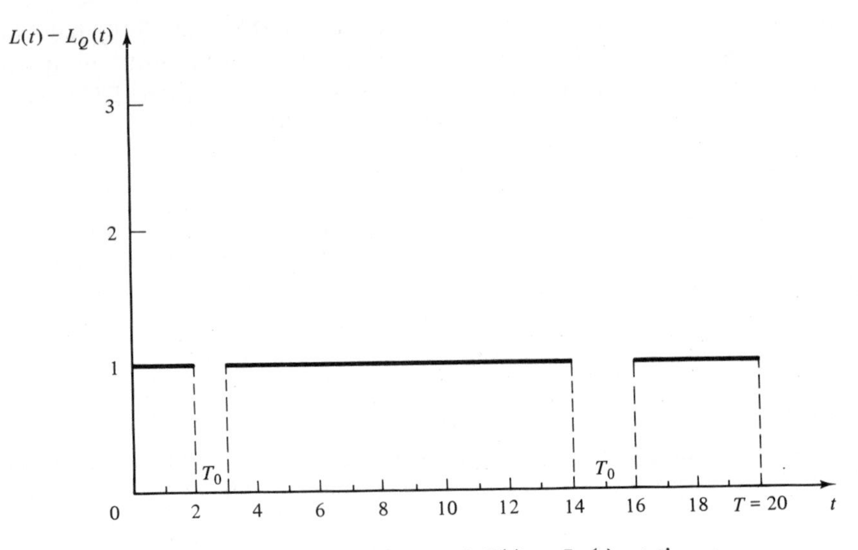

Figure 5.11. Number being served, $L(t) - L_Q(t)$, at time t.

stable, the arrival rate λ must be less than the service rate μ; that is,

$$\lambda < \mu$$

or

$$\rho = \frac{\lambda}{\mu} < 1 \tag{5.14}$$

If the arrival rate is greater than the service rate ($\lambda > \mu$), the server will eventually get further and further behind. After a time the server will be always busy and the waiting line will tend to grow in length at an average rate of $(\lambda - \mu)$ customers per time unit, because departures will be occurring at rate μ per time unit. For stable single-server systems ($\lambda < \mu$ or $\rho < 1$), long-run measures of performance such as average queue length L_Q (and also L, w, and w_Q) are well defined and have meaning. For unstable systems ($\lambda > \mu$), long-run server utilization is 1, and long-run average queue length is infinite; that is,

$$\frac{1}{T}\int_0^T L_Q(t)\,dt \longrightarrow L_Q = +\infty \qquad \text{as } T \longrightarrow \infty$$

Similarly, $L = w = w_Q = \infty$, and therefore these long-run measures of performance are meaningless for unstable queues. The quantity λ/μ is called the traffic intensity and is a measure of the work load offered to the system.

Server utilization in $G/G/c/\infty/\infty$ queues. Consider a queueing system with c identical servers in parallel. If an arriving customer finds more than one server idle, the customer chooses a server without favoring any particular server. (For example, the choice of server might be made at random.) Arrivals occur at rate λ from an infinite calling population, and each server works at rate μ customers per time unit. Using equation (5.11), $L = \lambda w$, applied to the server subsystem alone, an argument similar to the one given for a single-server leads to the result that, for systems in statistical equilibrium, the average number of busy servers, say L_s, is given by

$$L_s = \lambda E(S) = \frac{\lambda}{\mu} \tag{5.15}$$

Clearly, $0 \leq L_s \leq c$. The long-run average server utilization is defined by

$$\rho = \frac{L_s}{c} = \frac{\lambda}{c\mu} \tag{5.16}$$

so that $0 \leq \rho \leq 1$. The utilization ρ can be interpreted as the proportion of time an arbitrary server is busy in the long run.

The maximum service rate of the $G/G/c/\infty/\infty$ system is $c\mu$, which occurs when all servers are busy. For the system to be stable, the average arrival rate λ must be less than the maximum service rate $c\mu$; that is, the system is stable if and only if

$$\lambda < c\mu \tag{5.17}$$

or equivalently, the traffic intensity λ/μ is less than the number of servers c. If $\lambda > c\mu$, then arrivals are occurring, on the average, faster than the system can handle them, all servers will be continuously busy, and the waiting line will grow in length at an average rate of $(\lambda - c\mu)$ customers per time unit. Such a system is unstable and the long-run performance measures (L, L_Q, w, and w_Q) are again meaningless for such systems.

Note that condition (5.17) generalizes condition (5.14), and the equation for utilization for stable systems, equation (5.16), generalizes equation (5.13).

Equations (5.15) and (5.16) can also be applied when some servers work more than others, for example, when customers favor one server over others, or when certain servers only serve customers if all other servers are busy. In this case, L_s given by equation (5.15) is still the average number of busy servers, but ρ as given by equation (5.16) cannot be applied to an individual server. Instead, ρ must be interpreted as the average utilization of all servers.

EXAMPLE 5.8

Customers arrive at random to a license tag office at a rate of $\lambda = 50$ customers per hour. Presently there are 20 clerks, each serving $\mu = 5$ customers per hour on the average. Therefore the long-run, or steady-state, average utilization of a server, given by equation (5.16), is

$$\rho = \frac{\lambda}{c\mu} = \frac{50}{20(5)} = 0.5$$

and the average number of busy servers is

$$L_s = c\rho = 20(0.5) = 10$$

Thus, in the long run, a typical clerk is busy serving customers only 50% of the time. The office manager asks if the number of servers can be decreased. By equation (5.17), it follows that for the system to be stable, it is necessary for the number of servers to satisfy

$$c > \frac{\lambda}{\mu}$$

or $c > 50/5 = 10$. Thus, possibilities for the manager to consider include $c = 11$, or $c = 12$, or $c = 13, \ldots$. Note that $c \geq 11$ only guarantees long-run stability in the sense that all servers when busy can handle the incoming work load (i.e., $c\mu > \lambda$) on the average. The office manager may well desire to have more than the minimum number of servers ($c = 11$) because of other factors, such as customer delays and length of the waiting line. A stable queue can still have very long lines on the average.

Server utilization and system performance. As will be seen here and in later sections, system performance can vary widely for a given value of utilization, ρ. Consider a $G/G/1/\infty/\infty$ queue: that is, a single-server queue with arrival rate λ and service rate μ, and utilization $\rho = \lambda/\mu < 1$.

At one extreme, consider the $D/D/1$ queue, which has deterministic arrival and service times. Then all interarrival times $\{A_1, A_2, \ldots\}$ are equal to $E(A) = 1/\lambda$ and all service times $\{S_1, S_2, \ldots\}$ are equal to $E(S) = 1/\mu$. Assuming that a customer arrives to an empty system at time 0, the system evolves in a completely deterministic and predictable fashion, as shown in Figure 5.12. It

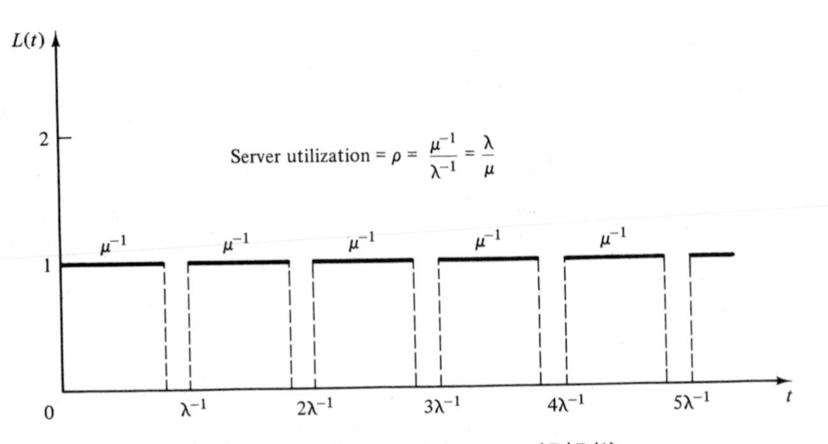

Figure 5.12. Deterministic queue ($D/D/1$).

can be seen that $L = \rho = \lambda/\mu$, $w = E(S) = \mu^{-1}$, and $L_Q = w_Q = 0$. Server utilization can assume any value between 0 and 1, and there is never any line whatsoever.

What, then, causes lines to build, if not a high server utilization? In general, it is the variability of interarrival and/or service times that causes lines to fluctuate in length.

EXAMPLE 5.9

Consider a physician who schedules patients every 10 minutes and who spends S_i minutes with the ith patient, where

$$S_i = \begin{cases} 9 \text{ minutes with probability } 0.9 \\ 12 \text{ minutes with probability } 0.1 \end{cases}$$

Thus, arrivals are deterministic ($A_1 = A_2 = \cdots = \lambda^{-1} = 10$) but services are stochastic (or probabilistic), with mean and variance given by

$$E(S_i) = 9(0.9) + 12(0.1) = 9.3 \text{ minutes}$$

and

$$\begin{aligned} V(S_i) &= E(S_i^2) - [E(S_i)]^2 \\ &= 9^2(0.9) + 12^2(0.1) - (9.3)^2 \\ &= 0.81 \end{aligned}$$

Since $\rho = \lambda/\mu = E(S)/E(A) = 9.3/10 = 0.93 < 1$, the system is stable and the physician will be busy 93% of the time in the long run. In the short run, lines will not build up as long as patients require only 9 minutes of service, but because of the variability in the service times, 10% of the patients will require 12 minutes, which in turn will cause a temporary line to form. The system is simulated with service times, $S_1 = 9$, $S_2 = 12$, $S_3 = 9$, $S_4 = 9$, $S_5 = 9, \ldots$; assuming that at time 0 a patient arrived to find the doctor idle and subsequent patients arrived at precisely times 10, 20, 30, . . . , the system evolves as in Figure 5.13. The delays in queue are $W_1^Q = W_2^Q = 0$, $W_3^Q = 22 - 20 = 2$, $W_4^Q = 31 - 30 = 1$, $W_5^Q = 0$. The occurrence of a relatively long service time (here $S_2 = 12$) caused a waiting line to temporarily form. In general, because of the variability of the interarrival and service distributions, relatively small interarrival times and/or relatively large service times occasionally do occur, and these in turn cause lines to lengthen. Conversely, the occurrence of a large interarrival time and/or a small service time will tend to shorten an existing waiting line. The relationship between utilization, service and interarrival variability, and system performance will be explored in more detail for a few parallel server queues in Section 5.5.

5.4.5. COSTS IN QUEUEING PROBLEMS

In many queueing situations, costs can be associated with various aspects of the waiting line or servers. Suppose that the system incurs a cost for each customer in the queue, say at a rate of \$10 per hour per customer. If customer j spends W_j^Q hours in the queue, then $\sum_{j=1}^{N} (\$10 \cdot W_j^Q)$ is the total cost of the N

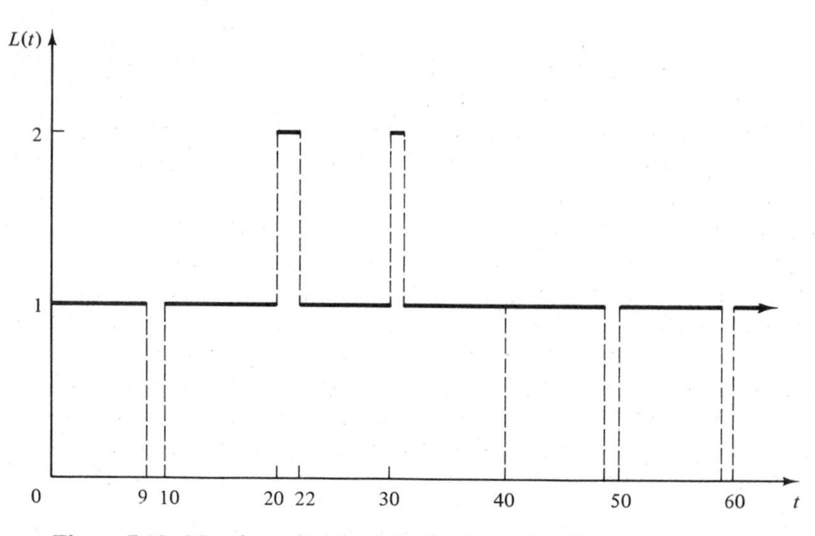

Figure 5.13. Number of patients in the doctor's office at time t.

customers who arrive during the simulation. Thus, the average cost per customer is

$$\sum_{j=1}^{N} \frac{\$10 \cdot W_j^Q}{N} = \$10 \cdot \hat{w}_Q$$

by equation (5.9). If $\hat{\lambda}$ customers per hour arrive (on the average), the average cost per hour is

$$\left(\hat{\lambda}\frac{\text{customers}}{\text{hour}}\right)\left(\frac{\$10 \cdot \hat{w}_Q}{\text{customer}}\right) = \$10 \cdot \hat{\lambda}\hat{w}_Q = \$10 \cdot \hat{L}_Q/\text{hour}$$

the last equality following by Little's equation (5.10). An alternative way to derive the average cost per hour is to consider equation (5.5). If T_i^Q is the total time over the interval $[0, T]$ that the system contains exactly i customers, then $\$10.00\ iT_i^Q$ is the cost incurred by the system during the time exactly i customers are present. Thus, the total cost is $\sum_{i=1}^{\infty} (\$10 \cdot iT_i^Q)$, and the average cost per hour is

$$\sum_{i=1}^{\infty} \frac{\$10 \cdot iT_i^Q}{T} = \$10 \cdot \hat{L}_Q/\text{hour}$$

by equation (5.5). In these cost expressions, $\hat{L}_Q$ may be replaced by L_Q (if the long-run number in queue is known), or by L or $\hat{L}$ (if costs are incurred while the customer is being served in addition to being delayed).

The server may also impose costs on the system. If a group of c parallel servers $(1 \leq c < \infty)$ have utilization ρ, and each server imposes a cost of \$5 per hour while busy, the total server cost per hour is

$$\$5 \cdot c\rho$$

since $c\rho$ is the average number of busy servers. If server cost is only imposed when the servers are idle, the server cost per hour would be

$$\$5 \cdot c(1 - \rho)$$

since $c(1 - \rho) = c - c\rho$ is the average number of idle servers. In many problems, two or more of these various costs are combined into a total cost. Such problems are illustrated by Examples 5.16 and 5.17, and Exercises 1, 9 and 18. In most cases, the objective is to minimize total costs (given certain constraints) by varying those parameters under management's control, such as the number of servers, the arrival rate, the service rate, and/or system capacity.

5.5. Steady-State Behavior of Infinite-Population Markovian Models

This section presents the steady-state solution of a number of queueing models that can be solved mathematically. For the infinite-population models, the arrival process is assumed to be a Poisson process with mean λ arrivals per time unit; that is, the interarrival times are assumed to be exponentially distributed with rate parameter λ. Service times may be exponentially distributed (M), Erlang of order k (E_k), constant (D) or arbitrary (G). The queue discipline will be FIFO. Because of the exponential distributional assumptions, these models are called Markovian models.

A queueing system is said to be in statistical equilibrium, or steady state, provided the probability that the system is in a given state is not time dependent; that is,

$$P(L(t) = n) = P_n(t) = P_n$$

is independent of time t. An example was given in Section 5.3 of a system, the $M/M/1/1/\infty$, which approached statistical equilibrium over time. In Exercise 14 the student is asked to show that if the $M/M/1/1/\infty$ system is in statistical equilibrium at time $t = 0$, then it remains in statistical equilibrium for all time. These two characteristics—approaching statistical equilibrium given any starting state, and remaining in statistical equilibrium once it is reached—are characteristic of many stochastic models, and in particular of all the systems studied in the following subsections. On the other hand, if an analyst were interested in the transient behavior of a queue over a relatively short period of time and given some specific initial conditions (such as idle and empty), the results to be presented here are inappropriate. A transient mathematical analysis such as the one in Section 5.3 or, more likely, a simulation model would be the chosen tool of analysis.

The mathematical models whose solutions are shown in the following subsections may be used to obtain approximate results even when the assumptions of the model do not strictly hold. These results may be considered as a rough guide to the behavior of the system. A simulation may then be used for a more refined analysis. However, it should be remembered that a mathematical analysis (when it is applicable) provides the true value of the model parameter (e.g., L), whereas a simulation analysis delivers a statistical estimate (e.g., $\hat{L}$) of the parameter. On the other hand, for complex systems, a simulation model generally is more accurate than a mathematical model.

For the simple models studied here, the steady-state parameter L, the time-average number of customers in the system, can be computed as

$$L = \sum_{n=0}^{\infty} nP_n \tag{5.18}$$

where $\{P_n\}$ are the steady-state probabilities (as defined in Table 5.2). As discussed in Section 5.4, and expressed in equation (5.6), L can also be interpreted as a long-run measure of performance of the system. Once L is given, the other steady-state parameters can be readily computed by Little's equation (5.11) applied to the whole system and the queue alone:

$$\begin{aligned} w &= \frac{L}{\lambda} \\ w_Q &= w - \mu^{-1} \\ L_Q &= \lambda w_Q \end{aligned} \tag{5.19}$$

where λ is the arrival rate and μ the service rate per server.

For the $G/G/c/\infty/\infty$ queues considered in this section to have a statistical equilibrium, a necessary and sufficient condition is that $\lambda/c\mu < 1$, where λ is the arrival rate, μ is the service rate of one server, and c is the number of parallel servers. For these unlimited capacity, infinite-calling-population models, it shall be assumed that the theoretical server utilization, $\rho = \lambda/c\mu$, satisfies $\rho < 1$. For models with finite system capacity or finite calling population, the quantity $\lambda/c\mu$ can assume any positive value.

5.5.1. Single-Server Queues with Poisson Arrivals and Unlimited Capacity: $M/G/1$

Suppose that service times have mean μ^{-1} and variance σ^2, and there is one server. If $\rho = \lambda/\mu < 1$, then the $M/G/1$ queue has a steady-state probability distribution with steady-state characteristics as given in Table 5.3. In general, there is no simple expression for the steady-state probabilities $P_1, P_2, \ldots$. When

$\lambda < \mu$, the quantity $\rho = \lambda/\mu$ is the server utilization, or long-run proportion of time the server is busy. As seen in Table 5.3, $1 - P_0 = \rho$ can also be interpreted as the steady-state probability that the system contains one or more customers. Note also that $L - L_Q = \rho$ is the time-average number of customers being served.

Table 5.3. STEADY-STATE PARAMETERS OF THE $M/G/1$ QUEUE

ρ	$\dfrac{\lambda}{\mu}$
L	$\rho + \dfrac{\lambda^2(\mu^{-2} + \sigma^2)}{2(1-\rho)} = \rho + \dfrac{\rho^2(1 + \sigma^2\mu^2)}{2(1-\rho)}$
w	$\mu^{-1} + \dfrac{\lambda(\mu^{-2} + \sigma^2)}{2(1-\rho)}$
w_Q	$\dfrac{\lambda(\mu^{-2} + \sigma^2)}{2(1-\rho)}$
L_Q	$\dfrac{\lambda^2(\mu^{-2} + \sigma^2)}{2(1-\rho)} = \dfrac{\rho^2(1 + \sigma^2\mu^2)}{2(1-\rho)}$
P_0	$1 - \rho$

EXAMPLE 5.10

Widget-making machines malfunction apparently at random and require a mechanic's attention. It is assumed that malfunctions occur according to a Poisson process at rate $\lambda = 1.5$ per hour. Observation over several months has found that repair times by the single mechanic take an average time of 30 minutes with a standard deviation of 20 minutes. Thus the service rate is $\mu = 2$ per hour and $\sigma^2 = (20)^2$ minutes^2 = 1/9 hour^2. The "customers" are the widget makers, and the appropriate model is the $M/G/1$ queue, since only the mean and variance of service times is known, not their distribution. The proportion of time the mechanic is busy is $\rho = \lambda/\mu = 1.5/2 = 0.75$, and by Table 5.3, the steady-state time average number of broken machines is

$$L = 0.75 + \frac{(1.5)^2[(0.5)^2 + 1/9]}{2(1 - 0.75)}$$
$$= 0.75 + 1.625 = 2.375 \text{ machines}$$

Thus, an observer who notes the state of the repair system at arbitrary times would find an average of 2.375 broken machines (in the long run).

A close look at the formulas in Table 5.3 reveals the source of the waiting lines and delays in an $M/G/1$ queue. For example, L_Q may be rewritten as

$$L_Q = \frac{\rho^2}{2(1-\rho)} + \frac{\lambda^2\sigma^2}{2(1-\rho)}$$

The first term involves only the ratio of mean arrival rate λ, to mean service rate μ. As shown by the second term, if λ and μ are held constant, the average length of the waiting line (L_Q) depends on the variability, σ^2, of the service times. If two systems have identical mean service times and mean interarrival times, the one with the more variable service times (larger σ^2) will tend to have longer lines, and will have longer lines on the average. Intuitively, if service times are highly variable, there is a high probability of a large service time occurring (say much larger than the mean service time), and when large service times do occur, there is a higher than usual tendency for lines to form and delays of customers to increase. (The reader should not confuse "steady state" with low variability or short lines; a system in steady-state or statistical equilibrium can be highly variable and can have long waiting lines.)

EXAMPLE 5.11

There are two workers competing for a job. Able claims an average service time which is faster than Baker's, but Baker claims to be more consistent, if not as fast. The arrivals occur according to a Poisson process at a rate of $\lambda = 2$ per hour (1/30 per minute). Able's statistics are an average service time of 24 minutes with a standard deviation of 20 minutes. Baker's service statistics are an average service time of 25 minutes, but a standard deviation of only 2 minutes. If the average length of the queue is the criterion for hiring, which worker should be hired? For Able, $\lambda = 1/30$ per minute, $\mu^{-1} = 24$ minutes, $\sigma^2 = 20^2 = 400$ minutes^2, $\rho = \lambda/\mu = 24/30 = 4/5$, and the average queue length is computed as

$$L_Q = \frac{(1/30)^2[24^2 + 400]}{2(1 - 4/5)} = 2.711 \text{ customers}$$

For Baker, $\lambda = 1/30$ per minute, $\mu^{-1} = 25$ minutes, $\sigma^2 = 2^2 = 4$ minutes^2, $\rho = 25/30 = 5/6$, and the average queue length is

$$L_Q = \frac{(1/30)^2[25^2 + 4]}{2(1 - 5/6)} = 2.097 \text{ customers}$$

Although working faster on the average, Able's greater service variability results in an average queue length about 30% greater than Baker's. On the other hand, the proportion of arrivals who would find Able idle and thus experience no delay is $P_0 = 1 - \rho = 1/5 = 20\%$, while the proportion who would find Baker idle and thus experience no delay is $P_0 = 1 - \rho = 1/6 = 16.7\%$. On the basis of average queue length, L_Q, Baker wins.

Several cases of the $M/G/1$ queue are of special note, namely, when service times are exponential, Erlang of order k, or deterministic.

The M/M/1 queue. Suppose that service times in an $M/G/1$ queue are exponentially distributed with mean μ^{-1}. Then the variance as given by equation (4.23) is $\sigma^2 = \mu^{-2}$. Since the mean and standard deviation of the exponential distribution are equal, the $M/M/1$ queue may often be a useful approximate model when service times have standard deviations approximately equal to their means. The steady-state parameters, given in Table 5.4, may be computed by substituting $\sigma^2 = \mu^{-2}$ into the formulas in Table 5.3. Alternatively, L may be computed by equation (5.18) from the steady-state probabilities, P_n, given in Table 5.4, and then w, w_Q, and L_Q computed by equations (5.19). The student can show that the two expressions for each parameter are equivalent by substituting $\rho = \lambda/\mu$ into the right-hand side of each equation in Table 5.4.

Table 5.4. STEADY-STATE PARAMETERS OF THE $M/M/1$ QUEUE

L	$\dfrac{\lambda}{\mu - \lambda} = \dfrac{\rho}{1 - \rho}$
w	$\dfrac{1}{\mu - \lambda} = \dfrac{1}{\mu(1 - \rho)}$
w_Q	$\dfrac{\lambda}{\mu(\mu - \lambda)} = \dfrac{\rho}{\mu(1 - \rho)}$
L_Q	$\dfrac{\lambda^2}{\mu(\mu - \lambda)} = \dfrac{\rho^2}{1 - \rho}$
P_n	$\left(1 - \dfrac{\lambda}{\mu}\right)\left(\dfrac{\lambda}{\mu}\right)^n = (1 - \rho)\rho^n$

EXAMPLE 5.12

The interarrival times as well as the service times at a single-chair unisex barbershop have been shown to be exponentially distributed. The values of λ and μ are 2 per hour and 3 per hour, respectively. That is, the time between arrivals averages 1/2 hour, exponentially distributed, and the service time averages 20 minutes, also exponentially distributed. The server utilization and the probabilities for zero, one, two, three, and four or more customers in the shop are computed as follows:

$$\rho = \frac{\lambda}{\mu} = \frac{2}{3}$$

$$P_0 = 1 - \frac{\lambda}{\mu} = \frac{1}{3}$$

$$P_1 = \left(\frac{1}{3}\right)\left(\frac{2}{3}\right) = \frac{2}{9}$$

$$P_2 = \left(\frac{1}{3}\right)\left(\frac{2}{3}\right)^2 = \frac{4}{27}$$

$$P_3 = \left(\frac{1}{3}\right)\left(\frac{2}{3}\right)^3 = \frac{8}{81}$$

$$P_{\geq 4} = 1 - \sum_{n=0}^{3} P_n = 1 - \frac{1}{3} - \frac{2}{9} - \frac{4}{27} - \frac{8}{81} = \frac{16}{81}$$

From the calculations it can be seen that the probability that the barber is busy is $1 - P_0 = \rho = 0.67$, and thus the probability that the barber is idle is 0.33. The time-average number of customers in the system is given by Table 5.4 as

$$L = \frac{\lambda}{\mu - \lambda} = \frac{2}{3 - 2} = 2 \text{ customers}$$

The average time an arrival spends in the system can be obtained from Table 5.4 or equation (5.19) as

$$w = \frac{L}{\lambda} = \frac{2}{2} = 1 \text{ hour}$$

The average time the customer spends in the queue can be obtained using equation (5.19) as

$$w_Q = w - \mu^{-1} = 1 - \frac{1}{3} = \frac{2}{3} \text{ hour}$$

Using Table 5.4, the time-average number in the queue is given by

$$L_Q = \frac{\lambda^2}{\mu(\mu - \lambda)} = \frac{4}{3(1)} = \frac{4}{3} \text{ customers}$$

Finally, note multiplying $w = w_Q + \mu^{-1}$ through by λ and using Little's equation (5.11) yields

$$L = L_Q + \frac{\lambda}{\mu} = \frac{4}{3} + \frac{2}{3} = 2 \text{ customers}$$

EXAMPLE 5.13

For the $M/M/1$ queue with service rate $\mu = 10$ customers per hour, consider how L and w increase as the arrival rate, λ, increases from 5 to 8.64 by increments of 20%, and then to $\lambda = 10$.

λ	5.0	6.0	7.2	8.64	10.0
ρ	0.500	0.600	0.720	0.864	1.00
L	1.00	1.50	2.57	6.35	∞
w	0.20	0.25	0.36	0.73	∞

For any $M/G/1$ queue, if $\lambda/\mu \geq 1$, waiting lines tend to continually grow in length; the long-run measures of performance, L, w, w_Q, and L_Q are all infinite ($L = w = w_Q = L_Q = \infty$); and a steady-state probability distribution does not exist. As shown here for $\lambda < \mu$, if ρ is close to 1, waiting lines and delays will tend to be long. Note that the increase in average system time, w, and average number in system, L, is highly nonlinear as a function of ρ. For example, as λ increases by 20%, L increases first by 50% (from 1.00 to 1.50), then by 71% (to 2.57), and by 147% (to 6.35).

EXAMPLE 5.14

If arrivals are occurring at rate $\lambda = 10$ per hour, and management has a choice of two servers, one who works at rate $\mu_1 = 11$ customers per hour and the second at rate $\mu_2 = 12$ customers per hour, the respective utilizations are $\rho_1 = \lambda/\mu_1 = 10/11 = 0.909$ and $\rho_2 = \lambda/\mu_2 = 10/12 = 0.833$. If the $M/M/1$ queue is used as an approximate model, then with the first server the average number in the system would be, by Table 5.4,

$$L_1 = \frac{\rho_1}{1 - \rho_1} = 10$$

and with the second server, the average number in the system would be

$$L_2 = \frac{\rho_2}{1 - \rho_2} = 5$$

Thus, a decrease in service rate from 12 to 11 customers per hour, a mere 8.3% decrease, would result in an increase in average number in system from 5 to 10, which is a 100% increase.

The $M/E_k/1$ queue. Suppose that service times have the Erlang distribution of order k with mean $1/\mu$, and variance $\sigma^2 = 1/k\mu^2$, as given by equations (4.31) and (4.32) by replacing θ by μ and β by k. The probability density function of an Erlang distribution is shown in Figure 4.11. The standard deviation of an Erlang distribution of order k is equal to its mean divided by $\sqrt{k}$, and thus the Erlang model for service times may be useful for approximating service times which have a skewed distribution with the mean greater than the standard deviation. For $k \geq 10$, an Erlang random variable is approximately normally distributed, and as $k \longrightarrow \infty$, an Erlang random variable approaches a constant value of $1/\mu$.

The steady-state parameters for the $M/E_k/1$ are given in Table 5.5. They may be computed by substituting $\sigma^2 = 1/k\mu^2$ into the formulas for the $M/G/1$ queue in Table 5.3.

Table 5.5. STEADY-STATE PARAMETERS FOR THE $M/E_k/1$ QUEUE

L	$\dfrac{\lambda}{\mu} + \dfrac{1+k}{2k}\dfrac{\lambda^2}{\mu(\mu-\lambda)} = \rho + \dfrac{1+k}{2k}\dfrac{\rho^2}{1-\rho}$
w	$\dfrac{1}{\mu} + \dfrac{1+k}{2k}\dfrac{\lambda}{\mu(\mu-\lambda)} = \mu^{-1} + \dfrac{1+k}{2k}\dfrac{\rho\mu^{-1}}{1-\rho}$
w_Q	$\dfrac{1+k}{2k}\dfrac{\lambda}{\mu(\mu-\lambda)} = \dfrac{1+k}{2k}\dfrac{\rho\mu^{-1}}{1-\rho}$
L_Q	$\dfrac{1+k}{2k}\dfrac{\lambda^2}{\mu(\mu-\lambda)} = \dfrac{1+k}{2k}\dfrac{\rho^2}{1-\rho}$

EXAMPLE 5.15

Patients arrive for a physical examination according to a Poisson process at the rate of one per hour. The physical examination requires three stages, each one independently and exponentially distributed with a service time of 15 minutes. A patient must go through all three stages before the next patient is admitted to the treatment facility. Determine the average number of delayed patients, L_Q, for this system.

If patients follow this treatment pattern, the service-time distribution will be Erlang of order $k = 3$. The necessary parameters are $\lambda = 1/60$ per minute and $\mu = 1/45$ per minute; thus,

$$L_Q = \frac{1+3}{2(3)}\,\frac{(1/60)^2}{(1/45)(1/45 - 1/60)} = \frac{2}{3}\left(\frac{135}{60}\right) = 1\tfrac{1}{2} \text{ patients}$$

EXAMPLE 5.16

Suppose that mechanics arrive randomly at a tool crib at Poisson rate $\lambda = 10$ per hour. It is known that the single tool clerk serves a mechanic in 4 minutes on the average with a standard deviation of approximately 2 minutes. It is assumed that service times follow an Erlang distribution of order k. The mean is $E(X) = 1/\mu = 4$ minutes and the variance is $V(X) = 1/k\mu^2 = 2^2$ minutes^2, which implies that $k = [E(X)]^2/V(X) = 4^2/2^2 = 4$. Suppose that mechanics make \$15.00 per hour. Then wages for non-productive waiting in line amount to \$15w_Q per mechanic's visit to the tool crib. Since there are $\lambda = 10$ visits per hour on the average, the average cost per hour of having mechanics delayed is $\lambda(\$15w_Q) = \$15L_Q$, using $L_Q = \lambda w_Q$. By Table 5.5, since the service rate is $\mu = 1/4$ per minute $= 15$ per hour, the mean delay per mechanic is

$$w_Q = \frac{1+k}{2k}\frac{\lambda}{\mu(\mu-\lambda)} = \frac{5}{8}\frac{10}{15(15-10)} = 0.0833 \text{ hour}$$

and the mean number of mechanics in the waiting line is

$$L_Q = \lambda w_Q = 0.833$$

Thus, the average cost per mechanic's visit is \$1.25 and the average cost per hour is \$12.50.

In Table 5.5, note that $k = 1$ yields the steady-state parameters for the $M/M/1$ queue, since an Erlang distribution of order $k = 1$ is an exponential distribution.

The M/D/1 queue. Assume now that service times have no variability, that is, $\sigma^2 = 0$, which means that all service times assume the constant value $1/\mu$. The steady-state parameters are given in Table 5.6. They may be derived

Table 5.6. STEADY-STATE PARAMETERS FOR THE $M/D/1$ QUEUE

L	$\dfrac{\lambda}{\mu} + \dfrac{1}{2}\dfrac{\lambda^2}{\mu(\mu-\lambda)} = \rho + \dfrac{1}{2}\dfrac{\rho^2}{1-\rho}$
w	$\dfrac{1}{\mu} + \dfrac{1}{2}\dfrac{\lambda}{\mu(\mu-\lambda)} = \mu^{-1} + \dfrac{1}{2}\dfrac{\rho\mu^{-1}}{1-\rho}$
w_Q	$\dfrac{1}{2}\dfrac{\lambda}{\mu(\mu-\lambda)} = \dfrac{1}{2}\dfrac{\rho\mu^{-1}}{1-\rho}$
L_Q	$\dfrac{1}{2}\dfrac{\lambda^2}{\mu(\mu-\lambda)} = \dfrac{1}{2}\dfrac{\rho^2}{1-\rho}$

from Table 5.3 by substituting $\sigma^2 = 0$, or from Table 5.5 by letting $k \to \infty$ with λ and μ held constant. Note that the average line length L_Q for the $M/D/1$ queue is exactly one-half the average line length of the $M/M/1$ queue. This shows how a decrease in service variability can drastically reduce lengths of waiting lines. (Waiting times are similarly affected by the variance of service times.)

EXAMPLE 5.17

Arrivals at a large jetport are all directed to the same runway. At a certain time of the day, these arrivals are Poisson distributed at a rate of 30 per hour. The time to land an aircraft is a constant 90 seconds. Determine L_Q, w_Q, L, and w for this airport.

In this case $\lambda = 0.5$ per minute, and $1/\mu = 1.5$ minutes, or $\mu = 2/3$ per minute. The runway utilization is

$$\rho = \frac{\lambda}{\mu} = \frac{1/2}{2/3} = \frac{3}{4}$$

The steady-state parameters are given by

$$L_Q = \frac{(3/4)^2}{2(1 - 3/4)} = \frac{9}{8} = 1.125 \text{ aircraft}$$

$$w_Q = \frac{L_Q}{\lambda} = \frac{9/8}{1/2} = \frac{9}{4} = 2.25 \text{ minutes}$$

$$w = w_Q + \frac{1}{\mu} = 2.25 + 1.5 = 3.75 \text{ minutes}$$

and

$$L = L_Q + \frac{\lambda}{\mu} = 1.125 + 0.75 = 1.875 \text{ aircraft}$$

If a delayed aircraft burns \$5000 worth of fuel per hour on the average, the average fuel cost due to delay is $\$5000 \cdot L_Q = \$5625$ per hour. The average fuel cost due to delay per aircraft is $\$5000 \cdot w_Q = \$5000(2.25/60) = \$187.50$.

The effect of utilization and service variability. For any $M/G/1$ queue, if lines are too long, they can be reduced by decreasing the server utilization ρ, or by decreasing the service time variability, σ^2. These remarks hold for almost all queues, not just the $M/G/1$ queue. The utilization factor ρ can be reduced by decreasing the arrival rate λ, increasing the service rate μ, or by increasing the number of servers, since in general, $\rho = \lambda/c\mu$, where c is the number of parallel servers. The effect of additional servers will be studied in the following subsections. Figure 5.14 illustrates the effect of service variability. The mean steady-state number in the queue, L_Q, is plotted versus utilization ρ for a number of different coefficients of variation. The coefficient of variation (cv) of a positive random variable X is defined by

$$(\text{cv})^2 = \frac{V(X)}{[E(X)]^2}$$

and is a measure of the variability of a distribution. The larger its value, the more variable is the distribution. For deterministic service times, $V(X) = 0$, so $\text{cv} = 0$. For Erlang service times of order k, $V(X) = 1/k\mu^2$ and $E(X) = 1/\mu$, so that $\text{cv} = 1/\sqrt{k}$. For exponential service times at service rate μ, the mean service time is $E(X) = 1/\mu$ and the variance is $V(X) = 1/\mu^2$, so that $\text{cv} = 1$. If service times have standard deviation greater than their mean (i.e., if $\text{cv} > 1$), then the hyperexponential distribution, which can achieve any desired coefficient of variation greater than 1, provides a good model. One occasion where it arises is given in Exercise 13.

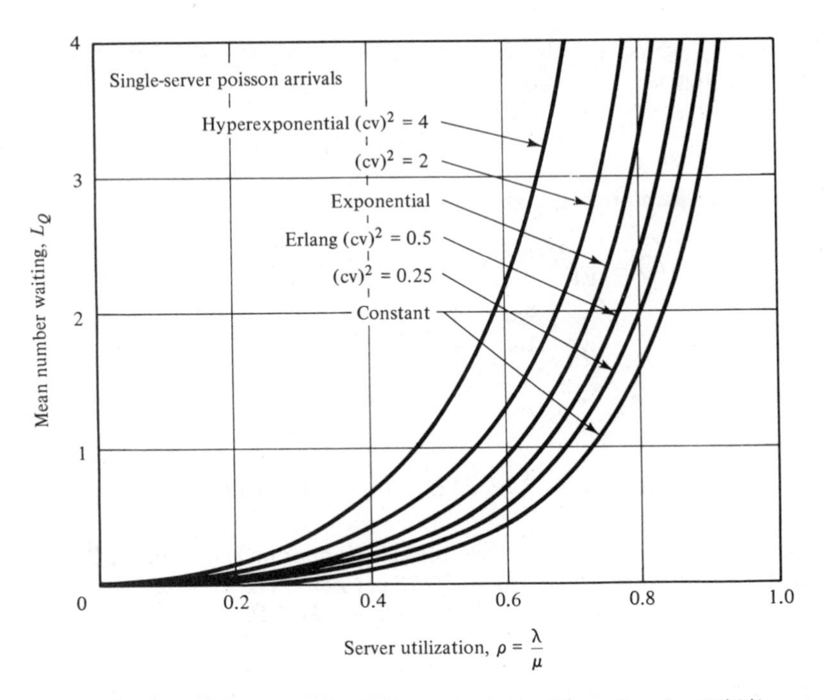

Figure 5.14. Mean number of customers waiting, L_Q, in $M/G/1$ queue having service distributions with given CV. (Adapted from Geoffrey Gordon, *System Simulation*, 2nd ed., Prentice-Hall, Englewood Cliffs, N.J., 1978.)

5.5.2. Single-Server Queues with Poisson Arrivals and Limited Capacity: $M/M/1/N/\infty$

Suppose that service times are exponentially distributed at rate μ, there is a single server, and the total system capacity is N customers. If an arrival occurs when the system is full, that arrival is turned away and does not enter the system. As in the preceding section, suppose that arrivals occur randomly according to a Poisson process with rate λ arrivals per time unit. For any values of λ and μ, the $M/M/1/N$ queue has a statistical equilibrium with steady-state characteristics as given in Table 5.7. When the arrival rate λ equals the service rate μ, the mean number in system L equals $N/2$; that is, the system is half-full (or half-empty) on the average. In Table 5.7, the quantity $a = \lambda/\mu$ is used to define the steady-state parameters.

The effective arrival rate, λ_e, is defined as the mean number of arrivals per time unit who enter and remain in the system. For all systems, $\lambda_e \leq \lambda$; for the unlimited capacity systems, $\lambda_e = \lambda$; but for systems such as the present one which turn customers away when full, $\lambda_e < \lambda$. The effective arrival rate is computed by

$$\lambda_e = \lambda(1 - P_N)$$

Table 5.7. STEADY-STATE PARAMETERS FOR THE $M/M/1/N$ QUEUE (N = SYSTEM CAPACITY, $a = \lambda/\mu$)

Parameter	Value	Condition
L	$\dfrac{a[1-(N+1)a^N + Na^{N+1}]}{(1-a^{N+1})(1-a)}$	$\lambda \neq \mu$
	$\dfrac{N}{2}$	$\lambda = \mu$
$1 - P_N$	$\dfrac{1-a^N}{1-a^{N+1}}$	$\lambda \neq \mu$
	$\dfrac{N}{N+1}$	$\lambda = \mu$
λ_e	$\lambda(1-P_N) = \mu(1-P_0) = \mu_e$	
ρ	$\dfrac{\lambda_e}{\mu} = 1 - P_0$	
w	$\dfrac{L}{\lambda_e}$	
w_Q	$w - \dfrac{1}{\mu}$	
L_Q	$\lambda_e w_Q = L - (1 - P_0)$	
P_n	$\dfrac{(1-a)a^n}{1-a^{N+1}}$ $\quad \lambda \neq \mu$	$n = 0, 1, 2, \ldots, N$
	$\dfrac{1}{N+1}$ $\quad \lambda = \mu$	

since $1 - P_N$ is the probability that a customer, upon arrival, will find space and be able to enter the system. When using Little's equations (5.19) to compute mean time spent in system w and in queue w_Q, λ must be replaced by λ_e.

By Table 5.7, for $\lambda \neq \mu$, the probability the system is full is given by

$$P_N = \frac{(1-a)a^N}{1-a^{N+1}}$$

where $a = \lambda/\mu$, and the probability the system is empty is

$$P_0 = \frac{1-a}{1-a^{N+1}}$$

When $a = 1$, or $\lambda = \mu$, all states are equally likely, so $P_0 = P_N = 1/(N+1)$. As expected, it can be shown that P_N increases and P_0 decreases as a increases, that is, as the arrival rate becomes large compared to the service rate.

EXAMPLE 5.18

The unisex barbershop described in Example 5.12 can hold only three customers, one in service and two waiting. Additional customers are turned away when the system is full. Determine the measures of effectiveness for this system. The traffic intensity is as previously determined, namely $\lambda/\mu = 2/3$. The probability that there are three customers in the system is computed by

$$P_N = P_3 = \frac{(1-2/3)(2/3)^3}{1-(2/3)^4} = \frac{8}{65} = 0.123$$

The expected number of customers in the shop is given by

$$L = \frac{2/3[1 - 4(2/3)^3 + 3(2/3)^4]}{[1 - (2/3)^4](1 - 2/3)} = \frac{66}{65} = 1.015 \text{ customers}$$

Now, the effective arrival rate, λ_e, is given by

$$\lambda_e = 2\left(1 - \frac{8}{65}\right) = 2\left(\frac{57}{65}\right) = \frac{114}{65} = 1.754 \text{ customers per hour}$$

Then, w can be calculated as

$$w = \frac{1.015}{1.754} = 0.579 \text{ hour}$$

In order to calculate L_Q, first determine P_0 as

$$P_0 = \frac{(1 - 2/3)(2/3)^0}{1 - (2/3)^4} = \frac{1/3}{65/81} = \frac{27}{65} = 0.415$$

Then, the average length of the queue is given by

$$L_Q = L - (1 - P_0) = 1.015 - (1 - .415) = 0.43 \text{ customer}$$

Note that $1 - P_0 = 0.585$ is the average number of customers being served, or equivalently, the probability that the single server is busy. Thus, the server utilization, or proportion of time the server is busy in the long run, is given by

$$\rho = 1 - P_0 = \frac{\lambda_e}{\mu} = 0.585$$

Finally, the waiting time in the queue is determined by Little's equation as

$$w_Q = \frac{0.43}{1.754} = 0.245 \text{ hour}$$

The reader should compare these results to those of the unisex barbershop before the capacity constraint was placed on the system. Specifically, in systems with limited capacity, the traffic intensity λ/μ can assume any positive value and no longer equals the server utilization $\rho = \lambda_e/\mu$. Note that server utilization decreases from 67% to 58.5% when the system imposes a capacity constraint.

Since P_0 and P_3 have been computed, it is easy to check the value of L using equation (5.17), $L = \sum_{n=1}^{N} nP_n$. To make the check requires computation of P_1 and P_2:

$$P_1 = \frac{(1 - 2/3)(2/3)}{1 - (2/3)^4} = \frac{18}{65} = 0.277$$

Since $P_0 + P_1 + P_2 + P_3 = 1$,

$$P_2 = 1 - P_0 - P_1 - P_3$$
$$= 1 - \frac{27}{65} - \frac{18}{65} - \frac{8}{65} = \frac{12}{65} = 0.185$$

Then, using equation (5.17) yields

$$L = 0\left(\frac{27}{65}\right) + 1\left(\frac{18}{65}\right) + 2\left(\frac{12}{65}\right) + 3\left(\frac{8}{65}\right)$$

$$= \frac{0 + 18 + 24 + 24}{65} = \frac{66}{65} = 1.015 \text{ customers}$$

which verifies the previous result.

5.5.3. MULTISERVER QUEUE: $M/M/c/\infty/\infty$

Suppose that there are c channels operating in parallel. Each of these channels has an independent and identical exponential service time distribution with mean $1/\mu$. The arrival process is Poisson with rate λ. Arrivals will join a single queue and enter the first available service channel. The queueing system is shown in Figure 5.15. If the number in system is $n < c$, an arrival will enter an available channel. However, when $n \geq c$, a queue will build if arrivals occur.

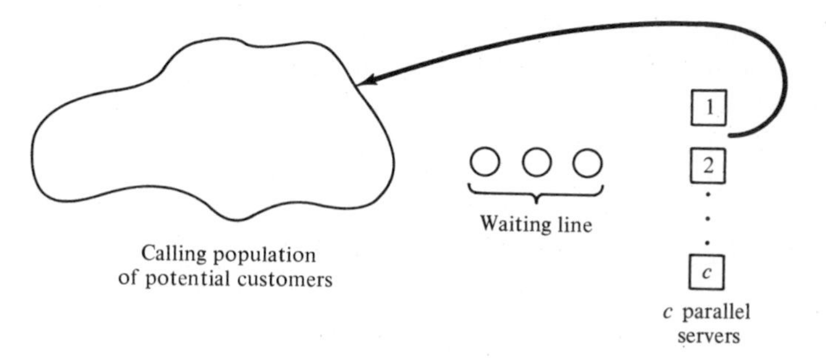

Figure 5.15. Multiserver queueing system.

The traffic intensity is defined by λ/μ. If $\lambda \geq c\mu$, the arrival rate is greater than or equal to the maximum service rate of the system, the service rate when all servers are busy. Thus, the system cannot handle the load put upon it, and therefore it has no statistical equilibrium. If $\lambda > c\mu$, the waiting line grows in length at rate $(\lambda - c\mu)$ customers per time unit, on the average. Customers are entering the system at rate λ per time unit but are leaving the system at a maximum rate of $c\mu$ per time unit.

For the $M/M/c$ to have statistical equilibrium, the traffic intensity must satisfy $\lambda/\mu < c$, in which case $\lambda/c\mu = \rho$, the server utilization. The steady-state parameters are listed in Table 5.8. Most of the measures of performance can be expressed fairly simply in terms of P_0, the probability that the system is empty, or $\sum_{n=c}^{\infty} P_n$, the probability that all servers are busy, denoted by $P(L(t) \geq c)$, where $L(t)$ is a random variable representing the number in system in statistical equilibrium. Thus, $P(L(t) = n) = P_n$, $n = 0, 1, 2, \ldots$. The value of P_0 is

Table 5.8. STEADY-STATE PARAMETERS FOR THE $M/M/c$ QUEUE

ρ	$\dfrac{\lambda}{c\mu}$
P_0	$\left\{\left[\sum_{n=0}^{c-1} \frac{(\lambda/\mu)^n}{n!}\right] + \left[\left(\frac{\lambda}{\mu}\right)^c\left(\frac{1}{c!}\right)\left(\frac{c\mu}{c\mu-\lambda}\right)\right]\right\}^{-1}$ $= \left\{\left[\sum_{n=0}^{c-1} \frac{(c\rho)^n}{n!}\right] + \left[(c\rho)^c\left(\frac{1}{c!}\right)\frac{1}{1-\rho}\right]\right\}^{-1}$
$P(L(t) \geq c)$	$\dfrac{(\lambda/\mu)^c P_0}{c!\,(1-\lambda/c\mu)} = \dfrac{(c\rho)^c P_0}{c!\,(1-\rho)}$
L	$c\rho + \dfrac{(c\rho)^{c+1}P_0}{c(c!)(1-\rho)^2} = c\rho + \dfrac{\rho P(L(t) \geq c)}{1-\rho}$
w	$\dfrac{L}{\lambda}$
w_Q	$w - \dfrac{1}{\mu}$
L_Q	$\lambda w_Q = \dfrac{(c\rho)^{c+1}P_0}{c(c!)(1-\rho)^2} = \dfrac{\rho P(L(t) \geq c)}{1-\rho}$
$L - L_Q$	$\dfrac{\lambda}{\mu} = c\rho$

necessary for computing all the measures of effectiveness, and the equation for P_0 is somewhat more complex than in the previous cases. However, P_0 depends only on c and ρ. A good approximation to P_0 can be obtained by using Figure 5.16, where P_0 is plotted versus ρ on semilog paper for various values of c. Figure 5.17 is a plot of L versus ρ for different values of c.

The results in Table 5.8 simplify to those in Table 5.4 when $c = 1$, the case of a single server. Note that the average number of busy servers, or the average number of customers being served, is given by the simple expression, $L - L_Q = \lambda/\mu = c\rho$.

EXAMPLE 5.19

Many early examples of queueing theory applied to practical problems concerning tool cribs. Attendants manage the tool cribs while mechanics, assumed to be from an infinite calling population, arrive for service. Assume Poisson arrivals at rate 2 mechanics per minute and exponentially distributed service times with mean 40 seconds.

Now, $\lambda = 2$ per minute, and $\mu = 60/40 = 3/2$ per minute. Since the traffic intensity is greater than 1, that is, since

$$\frac{\lambda}{\mu} = \frac{2}{3/2} = \frac{4}{3} > 1$$

more than one server is needed if the system is to have a statistical equilibrium. The requirement for steady state is that $c > \lambda/\mu = 4/3$. Thus at least $c = 2$ attendants are needed. The quantity 4/3 is the expected number of busy servers, and for $c \geq 2$, $\rho = 4/(3c)$ is the long-run proportion of time each server is busy. (What would happen if there were only $c = 1$ server?)

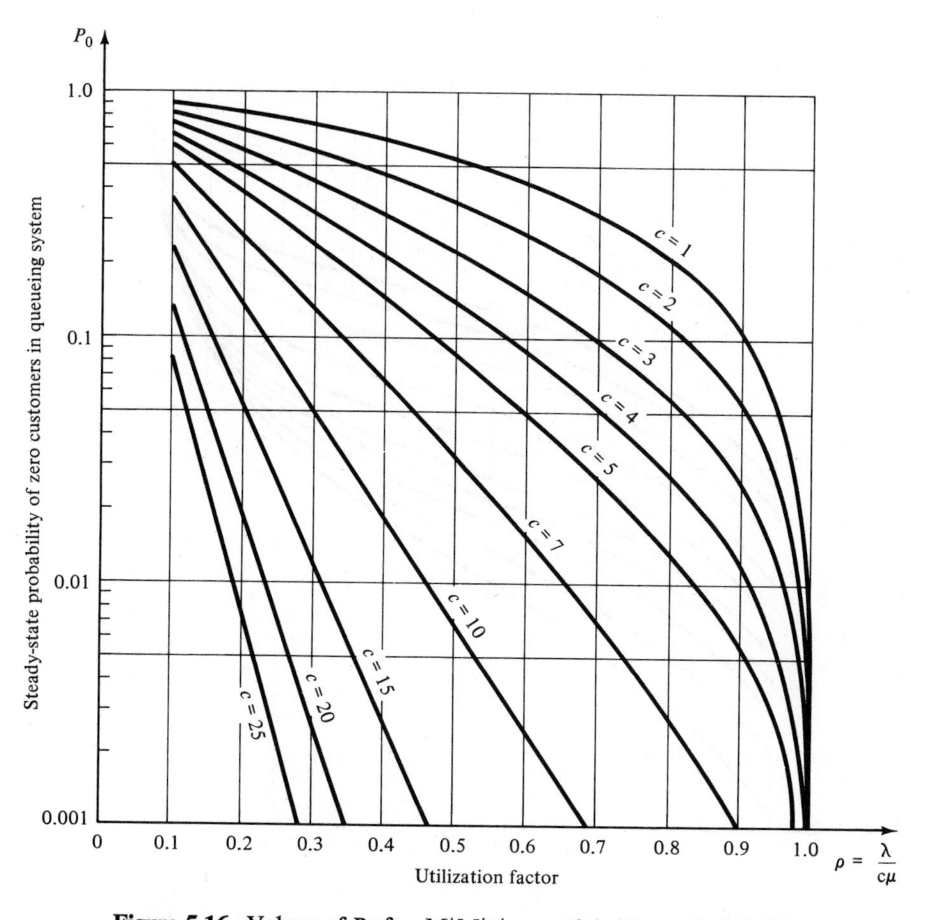

Figure 5.16. Values of P_0 for $M/M/c/\infty$ model. (From F. S. Hillier and G. J. Lieberman, *Introduction to Operations Research*, 3rd ed., © 1980, p. 422. Adapted with permission of Holden-Day, Inc., San Francisco.)

Let there be $c = 2$ attendants. First, P_0 is calculated as

$$P_0 = \left\{\sum_{n=0}^{1} \frac{(4/3)^n}{n!} + \left(\frac{4}{3}\right)^2\left(\frac{1}{2!}\right)\left[\frac{2(3/2)}{2(3/2) - 2}\right]\right\}^{-1}$$

$$= \left\{1 + \frac{4}{3} + \left(\frac{16}{9}\right)\left(\frac{1}{2}\right)(3)\right\}^{-1} = \left(\frac{15}{3}\right)^{-1} = \frac{1}{5} = 0.2$$

Proceeding, the probability that all servers are busy is given by

$$P(L(t) \geq 2) = \frac{(4/3)^2}{2!\,(1 - 2/3)}\left(\frac{1}{5}\right) = \left(\frac{8}{3}\right)\left(\frac{1}{5}\right) = \frac{8}{15} = 0.533$$

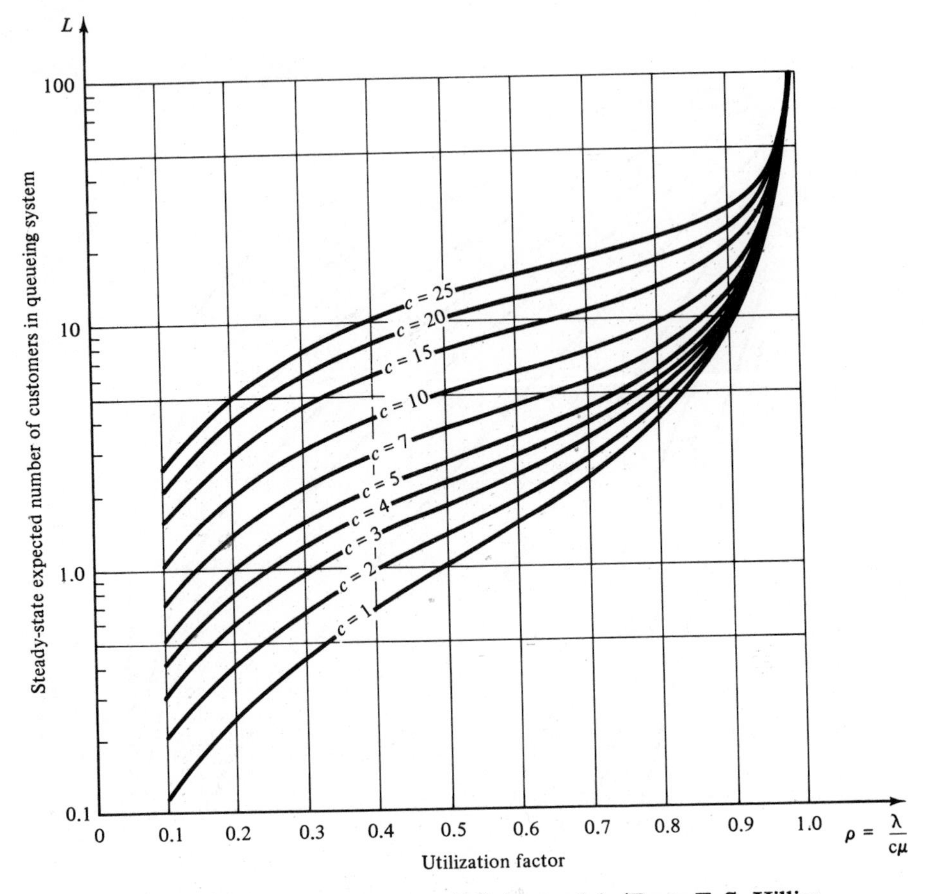

Figure 5.17. Values of L for $M/M/c/\infty$ model. (From F. S. Hillier and G. J. Lieberman, *Introduction to Operations Research*, 3rd ed., © 1980, p. 423. Adapted with permission of Holden-Day, Inc., San Francisco.)

Thus, the time-average length of the waiting line of mechanics is

$$L_Q = \frac{(2/3)(8/15)}{1 - 2/3} = 1.07 \text{ mechanics}$$

and the time-average number in system is given by

$$L = L_Q + \frac{\lambda}{\mu} = \frac{16}{15} + \frac{4}{3} = \frac{12}{5} = 2.4 \text{ mechanics}$$

Using Little's relationships, the average time a mechanic spends at the tool crib is

$$w = \frac{L}{\lambda} = \frac{2.4}{2} = 1.2 \text{ minutes}$$

while the average time spent waiting for an attendant is

$$w_Q = w - \frac{1}{\mu} = 1.2 - \frac{2}{3} = 0.533 \text{ minute}$$

EXAMPLE 5.20

Using the data of Example 5.19, compute P_0 and L from Figures 5.16 and 5.17. First compute

$$\rho = \frac{\lambda}{c\mu} = \frac{2}{2(3/2)} = \frac{2}{3} = 0.667$$

Entering the utilization factor of 0.667 on the horizontal axis of Figure 5.16 gives a value for P_0 of 0.2 on the vertical axis. Similarly, a value of $L = 2.4$ is read from the vertical axis of Figure 5.17.

5.6. Steady-State Behavior of Finite-Population Models ($M/M/c/K/K$)

In many practical problems, the assumption of an infinite calling population leads to invalid results because the calling population is, in fact, small. When the calling population is small, the presence of one or more customers in the system has a strong effect on the distribution of future arrivals, and the use of an infinite-population model can be misleading. Typical examples include a small group of machines that break down from time to time and require repair, or a small group of mechanics who line up at a counter for parts or tools. In the extreme case, if all the machines are broken, no new "arrivals" (breakdowns) of machines can occur; similarly, if all the mechanics are in line, no arrival is possible to the tool and parts counter. Contrast this to the infinite-population models in which the arrival rate, λ, of customers to the system is assumed independent of the state of the system.

Consider a finite-calling-population model with K customers. The runtime between calls for service for each member of the population is assumed to be exponentially distributed with mean $1/\lambda$ time units; service times are also exponentially distributed with mean $1/\mu$ time units; there are c parallel servers, and system capacity is K, so that all arrivals remain for service. All performance measures can be expressed in terms of K, λ/μ, and c. Because of the complexity of the formulas, finite queueing tables have been tabulated to ease the computations [Descloux, 1962; Hillier and Yu, 1981; Peck and Hazelwood, 1958]. Table 5.9 is a typical page from *Finite Queueing Tables* by Peck and Hazelwood. For each population size K and number of servers c, the user enters the tables with a "service factor" X, and exits with two performance measures:

D = probability that an arrival has to wait for service
F = efficiency factor, which is defined as the long-run proportion of customers either running or being served

Table 5.9. SELECTED PAGE FROM *Finite Queueing Tables* FOR POPULATION SIZE $K = 10$

X	c	D	F	X	c	D	F	X	c	D	F
.064	2	.119	.995	.125	3	.100	.994	.180	2	.614	.890
	1	.547	.940		2	.369	.962		1	.975	.549
.066	2	.126	.995		1	.878	.737	.190	5	.016	.999
	1	.562	.936	.130	4	.022	.999		4	.078	.995
.068	3	.020	.999		3	.110	.994		3	.269	.973
	2	.133	.994		2	.392	.958		2	.654	.873
	1	.577	.931		1	.893	.718		1	.982	.522
.070	3	.022	.999	.135	4	.025	.999	.200	5	.020	.999
	2	.140	.994		3	.121	.993		4	.092	.994
	1	.591	.926		2	.415	.952		3	.300	.968
.075	3	.026	.999		1	.907	.699		2	.692	.854
	2	.158	.992	.140	4	.028	.999		1	.987	.497
	1	.627	.913		3	.132	.991	.210	5	.025	.999
.080	3	.031	.999		2	.437	.947		4	.108	.992
	2	.177	.990		1	.919	.680		3	.333	.961
	1	.660	.899	.145	4	.032	.999		2	.728	.835
.085	3	.037	.999		3	.144	.990		1	.990	.474
	2	.196	.988		2	.460	.941	.220	5	.030	.998
	1	.692	.883		1	.929	.662		4	.124	.990
.090	3	.043	.998	.150	4	.036	.998		3	.366	.954
	2	.216	.986		3	.156	.989		2	.761	.815
	1	.722	.867		2	.483	.935		1	.993	.453
.095	3	.049	.998		1	.939	.644	.230	5	.037	.998
	2	.237	.984	.155	4	.040	.998		4	.142	.988
	1	.750	.850		3	.169	.987		3	.400	.947
.100	3	.056	.998		2	.505	.928		2	.791	.794
	2	.258	.981		1	.947	.627		1	.995	.434
	1	.776	.832	.160	4	.044	.998	.240	5	.044	.997
.105	3	.064	.997		3	.182	.986		4	.162	.986
	2	.279	.978		2	.528	.921		3	.434	.938
	1	.800	.814		1	.954	.610		2	.819	.774
.110	3	.072	.997	.165	4	.049	.997		1	.996	.416
	2	.301	.974		3	.195	.984	.250	6	.010	.999
	1	.822	.795		2	.550	.914		5	.052	.997
.115	3	.081	.996		1	.961	.594		4	.183	.983
	2	.324	.971	.170	4	.054	.997		3	.469	.929
	1	.843	.776		3	.209	.982		2	.844	.753
.120	4	.016	.999		2	.571	.906		1	.997	.400
	3	.090	.995		1	.966	.579	.260	6	.013	.999
	2	.346	.967	.180	5	.013	.999		5	.060	.996
	1	.861	.756		4	.066	.996		4	.205	.980
.125	4	.019	.999		3	.238	.978		3	.503	.919

Source: L. G. Peck and R. N. Hazelwood, *Finite Queueing Tables*, © 1958. Reproduced with permission of John Wiley & Sons, Inc., New York.

The service factor is computed by one of the formulas

$$X = \frac{\lambda/\mu}{\lambda/\mu + 1} = \frac{1/\mu}{1/\mu + 1/\lambda} = \frac{\lambda}{\lambda + \mu} \tag{5.20}$$

All other mean performance measures can be computed from the definition of F and the use of the conservation equation $L = \lambda w$ applied to each subsystem of the overall system. The complete model is illustrated in Figure 5.18.

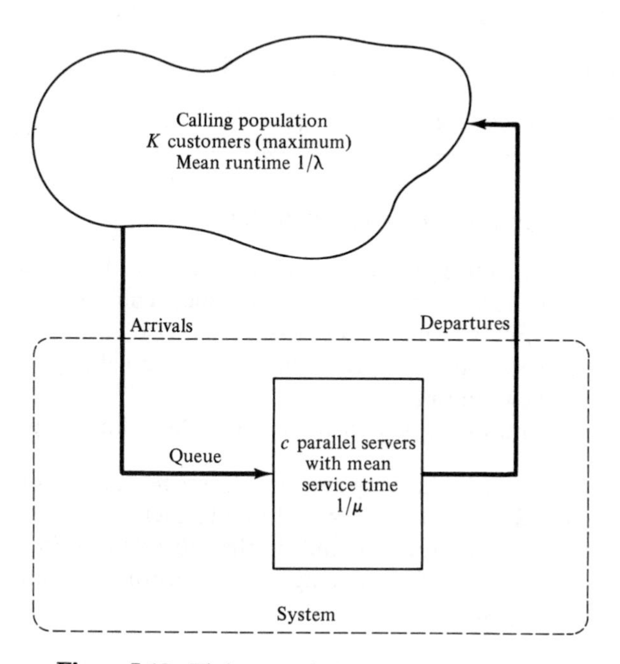

Figure 5.18. Finite-population queueing model.

First, by definition, F is the long-run proportion of customers either running or being served. Thus,

$$F = \frac{K - L_Q}{K} \tag{5.21}$$

Solving for L_Q, the mean number of customers waiting for service, yields

$$L_Q = K(1 - F) \tag{5.22}$$

Second, use $L = \lambda w$ to compute the long-run effective arrival rate to the queue and to compute the other performance measures. The results of applying $L = \lambda w$ to a number of subsystems are given in Table 5.10. Note that the effective

Table 5.10. APPLICATION OF LITTLE'S EQUATION $L = \lambda w$ TO THE SUBSYSTEMS OF A FINITE-POPULATION MODEL

Subsystem	$L = \lambda w$
Servers only	$L - L_Q = \lambda_e\left(\frac{1}{\mu}\right)$
Queue only	$L_Q = \lambda_e w_Q$
Queue and server (the system)	$L = \lambda_e\left(w_Q + \frac{1}{\mu}\right)$
Calling population	$K - L = \lambda_e\left(\frac{1}{\lambda}\right)$
All customers	$K = \lambda_e\left(w + \frac{1}{\lambda}\right)$
Servers and calling population	$K - L_Q = \lambda_e\left(\frac{1}{\mu} + \frac{1}{\lambda}\right)$

arrival rate λ_e has several valid interpretations:

λ_e = long-run effective arrival rate of customers to the queue
= long-run effective arrival rate of customers entering service
= long-run rate at which customers exit service
= long-run rate at which customers enter the calling population (and begin a new runtime)
= long-run rate at which customers exit the calling population

For each subsystem, L is the long-run time average number of customers in the subsystem; λ is λ_e, the effective arrival rate to each subsystem; and w is the long-run average time a customer spends in the subsystem per visit. In addition to the relations in Table 5.10, the average time a customer spends in the queue and in service, w, is given by

$$w = w_Q + \frac{1}{\mu}$$

where w_Q is the average time a customer spends in the queue alone.

From the last equation in Table 5.10, plus equation (5.22), λ_e can be computed. Using the other $L = \lambda w$ relationships, the results for L_Q, λ_e, w_Q, w, and L are computed and summarized in Table 5.11. The steady-state probability of n customers in the system (i.e., in queue or in service) is also given as P_n. Theoretically, L can be computed from equation (5.18) or, equivalently, L_Q can be computed from

$$L_Q = \sum_{n=c+1}^{K} (n - c)P_n$$

These equations, in fact, are used to compute the finite queueing tables, such as Table 5.9.

Table 5.11. PERFORMANCE MEASURES FOR FINITE-POPULATION MODEL, BASED ON THE EFFICIENCY FACTOR F

L_Q	$K(1-F)$
λ_e	$\dfrac{K-L_Q}{1/\mu+1/\lambda}=(K-L_Q)X\mu=KFX\mu$
w_Q	$\dfrac{L_Q}{\lambda_e}=\dfrac{L_Q(1/\mu+1/\lambda)}{K-L_Q}$
$L-L_Q$	$\dfrac{\lambda_e}{\mu}=(K-L_Q)X=KFX$
w	$w_Q+\dfrac{1}{\mu}$
L	$\lambda_e w=L_Q+\dfrac{\lambda_e}{\mu}=L_Q+(K-L_Q)X$
$K-L$	$\dfrac{\lambda_e}{\lambda}=KF(1-X)$
ρ	$\dfrac{L-L_Q}{c}=\dfrac{\lambda_e}{c\mu}=\dfrac{KFX}{c}$
P_0	$\left[\sum_{n=0}^{c-1}\binom{K}{n}\left(\dfrac{\lambda}{\mu}\right)^n+\sum_{n=c}^{K}\binom{K}{n}\dfrac{n!\,c^c(\lambda/c\mu)^n}{c!}\right]^{-1}$
P_n	$\begin{cases}\binom{K}{n}\left(\dfrac{\lambda}{\mu}\right)^n P_0, & n=0,1,\ldots,c-1\\ \binom{K}{n}\dfrac{n!\,c^c(\lambda/c\mu)^n}{c!}P_0, & n=c,c+1,\ldots,K\end{cases}$

EXAMPLE 5.21

There are two workers that are responsible for 10 milling machines. The machines run on the average for 20 minutes, then require an average 5-minute service period, both times exponentially distributed. Determine the various measures of performance for this system using Table 5.9.

First, the value of the service factor X is computed by equation (5.20) as

$$X=\frac{5}{5+20}=0.200$$

since $1/\lambda=20$ and $1/\mu=5$. Then, the values of D and F are read from the tables as 0.692 and 0.854, respectively. The probability is $D=0.692$ that an arrival will have to wait. The average number of machines waiting for service is given by equation (5.22) as

$$L_Q=10(1-0.854)=1.46 \text{ machines}$$

The average waiting time in the queue can be computed as

$$w_Q=\frac{1.46(5+20)}{10-1.46}=4.27 \text{ minutes}$$

Now, the average number of machines being serviced is given by

$$L-L_Q=(10-1.46)(0.200)=1.708 \text{ machines}$$

Since the machines must be running, waiting to be serviced, or in service, the average number of running machines is given by

$$K - L = 10 - (1.708 + 1.46) = 6.832 \text{ machines}$$

A frequently asked question is: What will happen if the number of servers is increased or decreased? If the number of workers in this example increases to three ($c = 3$), then $D = 0.300$ and $F = 0.968$. It follows that the average number of running machines increases to

$$K - L = 10(0.968)(0.8) = 7.744 \text{ machines}$$

an increase of 0.912 machine, on the average. In addition, the probability that a broken machine has to wait for service decreases from $D = 0.692$ (with two servers) to $D = 0.300$ (with three servers).

Conversely, what happens if the number of servers decreases to one? The value of D becomes 0.987 and the value of F is 0.497. Then, the time-average number of running machines, $K - L$, is given by

$$K - L = 10(0.497)(0.8) = 3.976 \text{ machines}$$

The decrease from two to one server has resulted in a drop of nearly three machines running, on the average. Exercise 15 asks the reader to examine the effect on server utilization of adding or deleting one server.

Example 5.21 illustrates several general relationships that have been found to hold for almost all queues. If the number of servers is decreased, delays, server utilization, and the probability of an arrival having to wait to begin service all increase.

5.7. Summary

Queueing models have found widespread use in the analysis of service facilities, production and material handling systems, telephone and communications systems, and many other situations where congestion or competition for scarce resources may occur. This chapter has introduced the basic concepts of queueing models, and shown how simulation, and in some cases a mathematical analysis, can be used to estimate the performance measures of a system.

A simulation can be used to generate one or more artificial histories of a complex system. This simulation generated data in turn can be used to estimate desired performance measures of the system. Commonly used performance measures, including L, L_Q, w, w_Q, ρ and λ_e, were introduced and formulas given for their estimation from data.

When simulating any system which evolves over time, the analyst must decide whether transient behavior or steady-state performance is to be studied. The differences between the transient behavior and the steady-state behavior of systems was illustrated by means of a simple example, the $M/M/1/1$ queue. The considerations discussed in this chapter apply to the simulation of any complex, dynamic system. Most importantly, when estimating steady-state performance measures from simulation generated data, it is necessary to recognize and to deal with the possibly deleterious effect of the initial conditions on the estimators of steady-state performance. As was seen, these estimators may be severely biased (either high or low) if the initial conditions, I, are unrepresentative of steady state, and/or if simulation run length, T, is too short. These estimation problems are discussed at greater length in Chapter 11.

Whether the analyst is interested in transient or steady-state performance of a system, it should be recognized that the estimates obtained from a simulation of a stochastic queue are exactly that—estimates. That is, an estimate contains random error, and thus a proper statistical analysis is required to assess the accuracy of the estimate. Methods for conducting such a statistical analysis are discussed in Chapters 11 and 12.

In the last two sections, it was shown that a number of simple models can be solved mathematically. Although the assumptions of such models may not be met exactly in a practical application, these models can still be useful in providing a rough estimate of a performance measure. In many cases, models with exponentially distributed interarrival and service times will provide a conservative estimate of system behavior; for example, if the model predicts that average waiting time, w, will be 12.7 minutes, then average waiting time in the real system is likely to be less than 12.7 minutes. This conservative nature of exponential models arises because (1) performance measures, such as w and L, are generally increasing functions of the variance of interarrival times and service times (recall the $M/G/1$ queue), and (2) the exponential distribution is fairly highly variable, having its standard deviation always equal to its mean. Thus, if the arrival process or service mechanism of the real system is less variable than exhibited by exponentially distributed interarrival or service times, then it is likely that the average number in the system, L, and the average time spent in system, w, will be less than what is predicted by the exponential model.

Another application of the mathematical models is to the design of systems when the number of needed servers at a work station or service center is not known. Quite often, if the arrival rate λ_e and the service rate μ are known or can be estimated, then the simple inequality $\lambda_e/c\mu < 1$ can be used to provide an initial estimate for the number of servers, c, at a work station. For a large system with many work stations, it could be quite time consuming to have to simulate every possibility $(c_1, c_2, \ldots)$ for the number of servers, c_i, at work station i. Thus, a bit of mathematical analysis using rough estimates may save a great deal of computer time and analyst's time.

Finally, the qualitative behavior of the simple exponential models of queues carries over to more complex systems. In general, it is the variability of service times and the variability of the arrival process that causes waiting lines to build up and congestion to occur. For most systems, if the arrival rate increases, if the service rate decreases, or if the variance of service times or interarrival times increases, then the system will become more congested. Congestion can be decreased by adding more servers or by reducing the mean value and variability of service times. Simulation can be a great aid in quantifying these relationships and evaluating alternative system designs.

REFERENCES

COOPER, ROBERT B. [1981], *Introduction to Queueing Theory*, 2nd ed., North-Holland, New York.

DESCLOUX, A. [1962], *Delay Tables for Finite- and Infinite-Source Systems*, McGraw-Hill, New York.

GROSS, DONALD, AND CARL HARRIS [1974], *Fundamentals of Queueing Theory*, Wiley, New York.

HILLIER, FREDERICK S. AND GERALD J. LIEBERMAN [1980], *Introduction to Operations Research*, 3rd ed., Holden-Day, San Francisco.

HILLIER, FREDERICK S., AND OLIVER S. YU [1981], *Queueing Tables and Graphs*, Elsevier North-Holland, New York.

KENDALL, D. G. [1953], "Stochastic Processes Occurring in the Theory of Queues and Their Analysis by the Method of Imbedded Markov Chains," *Annals of Mathematical Statistics*, Vol. 24, pp. 338–54.

LITTLE, J. D. C. [1961], "A Proof for the Queueing Formula $L = \lambda w$," *Operations Research*, Vol. 16, pp. 651–65.

PECK, L. G., AND R. N. HAZELWOOD [1958], *Finite Queueing Tables*, Wiley, New York.

WAGNER, HARVEY M. [1975], *Principles of Operations Research*, 2nd ed., Prentice-Hall, Englewood Cliffs, N.J.

WHITE, J. A., J. W. SCHMIDT, AND G. K. BENNETT [1975], *Analysis of Queueing Systems*, Academic Press, New York.

EXERCISES

1. A tool crib has exponential interarrival and service times, and serves a very large group of mechanics. The mean time between arrivals is 4 minutes. It takes 3 minutes on the average for a tool crib attendant to service a mechanic. The attendant is paid $6 per hour and the mechanic is paid $10 per hour. Would it be advisable to have a second tool crib attendant?
2. A two-runway (one runway for landing, one runway for taking off) airport is being designed for propeller-driven aircraft. The time to land an airplane is known to be exponentially distributed with a mean of $1\frac{1}{2}$ minutes. If airplane arrivals are assumed to occur at random, what arrival rate can be tolerated if the average wait in the sky is not to exceed 3 minutes?
3. The Port of Trop can service only one ship at a time. However, there is mooring space for three more ships. Trop is a favorite port of call, but if no mooring space is available, the ships have to go to the Port of Poop. An average of seven ships arrive each week, according to a Poisson process. The Port of Trop has the capacity to handle an average of eight ships a week, with service times exponentially distributed. What is the expected number of ships waiting or in service at the Port of Trop?
4. At Metropolis City Hall, two workers "pull strings" every day. Strings arrive to be

pulled on an average of one every 10 minutes throughout the day. It takes an average of 15 minutes to pull a string. Both times between arrivals and service times are exponentially distributed. What is the probability that there are no strings to be pulled in the system at a random point in time? What is the expected number of strings waiting to be pulled? What is the probability that both string pullers are busy? What is the effect on performance if a third string puller, working at the same speed as the first two, is added to the system?

5. At Tony and Cleo's bakery, one kind of birthday cake is offered. It takes 15 minutes to decorate this particular cake and the job is performed by one particular baker. In fact, this is all this baker does. What mean time between arrivals (exponentially distributed) can be accepted if the mean length of the queue for decorating is not to exceed five cakes?

6. A machine shop repairs small electric motors which arrive according to a Poisson process at a rate of 12 per week (5-day, 40-hour workweek). An analysis of past data indicates that engines can be repaired, on the average, in 2.5 hours, with a variance of 1 hour. How many working hours should a customer expect to leave a motor at the repair shop (not knowing the status of the system)? If the variance of the repair time could be controlled, what variance would reduce the expected waiting time to 6.5 hours?

7. Arrivals at a self-service gasoline pump occur in a Poisson fashion at a rate of 12 per hour. Service time has a distribution which averages 4 minutes with a standard deviation of $1\frac{1}{3}$ minutes. What is the expected number of vehicles in the system?

8. Classic Car Care has one worker who washes cars in a four-step method—soap, rinse, dry, vacuum. The time to complete each step is exponentially distributed with a mean of 9 minutes. Every car goes through every step before another car begins the process. On the average one car every 45 minutes arrives for a wash job, according to a Poisson process. What is the average time a car waits to begin the wash job? What is the average number of cars in the car wash system? What is the average time required to wash a car?

9. A room has 10 cotton spinning looms. Once the looms are set up, they run automatically. The setup time is exponentially distributed with a mean of 10 minutes. The machines run for an average of 40 minutes, also exponentially distributed. Loom operators are paid $10 an hour and looms not running incur a cost of $40 an hour. How many loom operators should be employed to minimize the total cost of the loom room? If the objective becomes "on the average, no loom should wait more than 1 minute for an operator," how many persons should be employed? How many operators should be employed to ensure that an average of at least 7.5 looms are running at all times?

10. Given the following information for a finite calling population problem with exponentially distributed runtimes and service times:

$$K = 10$$

$$\frac{1}{\mu} = 15$$

$$\frac{1}{\lambda} = 82$$

$$c = 2$$

Compute L_Q, and w_Q. Determine the value of λ such that $L_Q = L/2$.

11. Suppose that Figure 5.8 represents the number in system for a last in, first out (LIFO) single-server system. Customers are not preempted (i.e., kicked out of service) but upon service completion the most recent arrival next begins service. For this LIFO system apportion the total area under $L(t)$ to each individual customer, as was done in Figure 5.10 for the FIFO system. Using the figure, show that Equations (5.12), (5.10), and (5.11) hold for the single-server LIFO system.

12. Repeat Exercise 11 assuming that:
(a) Figure 5.8 represents a FIFO system with $c = 2$ servers.
(b) Figure 5.8 represents a LIFO system with $c = 2$ servers.

13. Consider a $M/G/1$ queue with the following type of service distribution. Customers request one of two types of service in the proportions p and $1 - p$. Type i service is exponentially distributed at rate μ_i, $i = 1, 2$. Let X_i denote a type i service time and X an arbitrary service time. Then $E(X_i) = 1/\mu_i$, $V(X_i) = 1/\mu_i^2$ and

$$X = \begin{cases} X_1 & \text{with probability } p \\ X_2 & \text{with probability } (1 - p) \end{cases}$$

The random variable X is said to have a hyperexponential distribution with parameters (μ_1, μ_2, p).
(a) Show that $E(X) = p/\mu_1 + (1 - p)/\mu_2$ and $E(X^2) = 2p/\mu_1^2 + 2(1 - p)/\mu_2^2$.
(b) Use $V(X) = E(X^2) - [E(X)]^2$ to show $V(X) = 2p/\mu_1^2 + 2(1 - p)/\mu_2^2 - [p/\mu_1 + (1 - p)/\mu_2]^2$.
(c) For any hyperexponential random variable, if $\mu_1 \neq \mu_2$ and $0 < p < 1$, show that its coefficient of variation is greater than 1; that is, $(\text{cv})^2 = V(X)/[E(X)]^2 > 1$. Thus, the hyperexponential distribution provides a family of statistical models for service times which are more variable than exponentially distributed service times. [*Hint:* The algebraic expression for $(\text{cv})^2$, using parts (a) and (b), can be manipulated into the form $(\text{cv})^2 = 2p(1 - p)(1/\mu_1 - 1/\mu_2)^2/[E(X)]^2 + 1$.]
(d) Many choices of μ_1, μ_2, and p lead to the same overall mean $E(X)$ and $(\text{cv})^2$. If a distribution with mean $E(X) = 1$ and coefficient of variation $\text{cv} = 2$ is desired, find values of μ_1, μ_2, and p to achieve this. [*Hint:* Choose $p = 1/4$ arbitrarily; then solve the following equations for μ_1 and μ_2.

$$\frac{1}{4\mu_1} + \frac{3}{4\mu_2} = 1$$

$$\frac{3}{8}\left(\frac{1}{\mu_1} - \frac{1}{\mu_2}\right)^2 + 1 = 4.]$$

14. (a) Using Equation (5.2), show that if the $M/M/1/1/\infty$ queue is in statistical equilibrium at time $t = 0$; that is, if $P_0(0) = \mu/(\lambda + \mu)$ and $P_1(0) = \lambda/(\lambda + \mu)$, then the system is in statistical equilibrium for all times t. (This result holds for a wide class of stochastic models.)
(b) Suppose that the $M/M/1/1/\infty$ queue is busy at time $t = 0$ (instead of empty and idle as in the truck dock of Example 5.4). How does this affect the transient probabilities $P_i(t)$? Make a plot of $P_i(t)$ versus t for $i = 0, 1$, as in Figure 5.7. What is the effect of these new initial conditions on the estimator $\hat{\rho}$ of long-run utilization ρ?
(c) For the $M/M/1/1/\infty$ queue, show that $E(\hat{\rho}) = \int_0^T P_1(t)\,dt/T$. Use this to justify

the assertion that $\hat{p}$ is biased when initial conditions are the empty state, and when initial conditions are the busy state. [*Hint:* Evaluate $E(\hat{p})$ explicitly, using Equation (5.2).]

15. In Example 5.21, compare the systems with $c = 1$, $c = 2$, and $c = 3$ servers on the basis of server utilization ρ (the proportion of time a typical server is busy). The equation for ρ is in Table 5.11.

16. (a) Derive the equation $K - L = KF(1 - X)$ for the number of "running" customers in a finite-population model. Any of the $L = \lambda w$ conservation equations in Table 5.10 plus Equation (5.20) may be used.

(b) Use Table 5.10 and equations (5.20) and (5.21) to derive the additional relationships:

$$L - L_Q = \frac{1/\mu}{1/\mu + 1/\lambda + w_Q} K$$

$$L_Q = \frac{w_Q}{1/\mu + 1/\lambda + w_Q} K$$

$$K - L = \frac{1/\lambda}{1/\mu + 1/\lambda + w_Q} K$$

$$F = \frac{1/\mu + 1/\lambda}{1/\mu + 1/\lambda + w_Q}$$

Interpret each of these results.

17. A small lumberyard is supplied by a fleet of 10 trucks. One overhead crane is available to unload the long logs from the trucks. It takes an average of 1 hour to unload a truck. After unloading, a truck takes an average of 3 hours to get the next load of logs and return to the lumberyard.

(a) Certain distributional assumptions are needed to analyze this problem with the models of this chapter. State them and discuss their reasonableness.

(b) With one crane, what is the average number of trucks waiting to be unloaded? On the average, how many trucks arrive at the yard each hour? What percentage of trucks upon arrival find the crane busy? Is this the same as the long-run proportion of time the crane is busy?

(c) Suppose that a second crane is installed at the lumberyard. Answer the same questions as in part (b). Make a chart comparing one crane to two cranes.

(d) If the value of the logs brought to the yard is approximately $200 per truck load and long-run crane costs are $50 per hour per crane (whether busy or not), determine the optimal number of cranes on the basis of cost per hour.

(e) In addition to the costs in part (d), if management decides to consider the cost of idle trucks and drivers, what is the optimal number of cranes? A truck and its driver is estimated to cost approximately $40 per hour, and is considered to be idle when it is waiting for a crane.

18. A tool crib with one attendant serves a group of 10 mechanics. Mechanics work for an exponentially distributed amount of time with mean 20 minutes, then go to the crib to request a special tool. Service times by the attendant are exponentially distributed with a mean of 3 minutes. If the attendant is paid $6 per hour and the mechanic is paid $10 per hour, would it be advisable to have a second attendant? (Compare to Exercise 1.)

INVENTORY SYSTEMS

Inventory systems are frequently encountered in practice, and simulation has often proven the only method of analysis. However, if an inventory problem fits into the structure of a mathematical model that can be optimized directly, there is no need for simulation. An additional benefit of studying mathematical models of inventory systems is the understanding of relationships between the various costs and parameters and the policy decisions that result. Although the contents of this chapter are incomplete compared to the plethora of literature on inventory theory and applications, several of the well-known models are introduced.

The first area presented is the deterministic economic order quantity model, with numerous extensions. Three different areas of probabilistic models are briefly introduced. The first of these is the classical "newspaper seller's problem," a stochastic single-period model. The second probabilistic model is the reorder point (ROP) model. In this text only the heuristic approximate treatment is described. The last probabilistic model is the case of periodic review. Only the simple approximate model is introduced.

For full treatment of mathematical inventory models the reader is referred to the classic text by Hadley and Whitin [1963]. Also, the text written by Peterson and Silver [1979] is useful because of its practical considerations, including case studies.

6

6.1. Measures of Effectiveness

Revenues and costs in a firm can be affected by inventory decisions. This section sets the stage for the remainder of the chapter by briefly discussing some of the costs that are commonly considered in analyzing an inventory system.

The objective of an inventory system can either be the maximization of profit, or the minimization of cost. Consider the latter case. Then, the major classes of cost can be the item cost, the ordering or procurement cost, the holding or carrying cost, and the shortage cost.

The item cost C does not usually vary as a function of inventory policy. However, there are cases where it must be considered. In the procurement mode, where inventories are obtained from a vendor, there may be quantity discounts. When larger amounts (order quantities) are purchased, the price may be decreased. The reduction in price can apply to all units purchased, or to incremental units. If the reduction is incremental, succeeding quantities of units within the order can be reduced in price. (It is also possible for succeeding quantities to increase in price, as with the purchase of electricity!) The logic behind quantity discounts, from the vendor's point of view, is that the cost of sales is nonlinear. The vendor may as well reduce the price on higher-order quantities to induce greater purchasing, and if it takes a price cut to do it, that price cut can be more than offset by increased profits.

In manufacturing there is an analogy to quantity discounts. Manufacturing progress functions have been developed, initially in the aircraft industry, which indicate that worker learning can significantly reduce the time required to construct an item. Thus, the cost of subsequent items can be reduced as the production lot size increases. In the case where the decision of whether to make or buy is made on item cost alone, manufacturing progress must be considered. Fabrycky and Banks [1967] treated this subject in some detail.

Ordering cost, in both the purchasing and manufacturing environments, is usually considered as independent of the purchase quantity or production lot size. When purchasing, there are certain ordering costs incurred, such as processing the purchase order, receiving the reports, accounting for the order, and so on. These costs are incurred whether there is only one unit being purchased, or whether there are many units. In modeling inventory systems in this chapter, the symbol A is given to the ordering cost. The dimensions of A are dollars per order.

If the lot is to be manufactured, the fixed cost, A, is known as the "setup cost." This is a function of processing the manufacturing order, preparing the machines for the new task, training the workers, and then tearing down the machinery after the job has been completed.

When holding an inventory, there are opportunity costs that occur. These inventories represent sums of money that could be invested in other revenue-producing activities. At the least, the equivalent cost of the inventory could be placed in government bonds, or perhaps in the money market. Additionally, inventory holding charges result from out-of-pocket costs from taxes, insurance, pilferage, deterioration, damage, materials handling, and the cost of storage space.

The inventory carrying cost, iC, is a fraction, i, of the item cost, C, over a specified time period. For example, $i = 0.25$ on an annual basis is taken to mean that the inventory carrying charge is 25% of the item cost over the period of a year, or, $0.25C$. (A period may be a day, a week, a month, a year, etc.)

If a demand occurs when the desired item is out of stock, an economic loss known as a shortage cost can occur. The economic loss depends on whether the units short are backordered, provided via substitution, or a lost sale (cancellation of the order) results. In every case, an added information processing cost will result. In the backorder and lost sales cases, a loss of goodwill may occur because of customer dissatisfaction. In the substitution case, goodwill may or may not be lost since the substitution of a more valuable item may actually please the customer. However, the seller usually charges the original price in this instance, resulting in an opportunity cost equal to the difference in the costs of the requested item and the substitute.

In the backorder case, special handling and shipping costs may also occur.

If the item is used in manufacturing, a backorder could result in a production stoppage.

In the lost-sales case, lost profit on the sale must be considered. There may be other costs in the lost-sales case, including dissatisfaction by the customer resulting in future orders being placed elsewhere.

Consider the backorder case, in which demands occurring when no units are in stock are supplied, if possible, when an order is received. It is rather difficult to determine the exact nature of this cost. However, Hadley and Whitin [1963] describe it generally as

$$\pi(t) = \pi + \pi' t$$

where t is the length of time which the backorder exists. The term π is a fixed cost for each unit backordered and π' is a variable cost which is linear in the length of time for which the backorder exists.

In the lost-sales case, demands are lost if they occur when the system is out of stock. The cost of a lost sale is not a function of time. Each lost sale would result in a constant cost π. In this chapter, models for the lost sales case are not treated. The interested reader is referred to Hadley and Whitin [1963].

Johnson and Montgomery [1974] indicate that assuming a constant cost π is also appropriate for the substitution case. Note that models for the substitution case do not appear in this brief survey chapter.

6.2. Inventory Policies

Consider an inventory control policy where the inventory level is observed at discrete time points which are spaced N time units apart. Then, N is the length of the review period. At every N time periods, the status of the inventory position, I_j, is checked, and if it is at or below the reorder point L, enough units are ordered to bring the inventory up to a target level M. The order quantity Q_j, at the jth review point, is determined as either

$$Q_j = \begin{cases} 0 & \text{if } I_j > L \\ M - I_j & \text{if } I_j \leq L \end{cases}$$

The resulting three-parameter policy is denoted as periodic review, or a (M, L, N) policy.

A special case of the (M, L, N) policy occurs when $L = M$. Then, an order for $M - I_j$ units is placed at every review point j. This (M, N) policy is called an order up to M policy. It is a useful policy when the ordering cost is very low or not to be considered.

Continuous review policies result if inventories are monitored at all times and when the inventory position reaches a specified point an order results. A continuous review (M, L) policy operates such that at any time t, the inventory $I(t)$ drops to the reorder point L, or below the reorder point, an order for $Q(t) = M - I(t)$ units is placed. If units are demanded one at a time, the order size will always be $Q = M - L$. This results in what is usually called a (Q, L) policy, or fixed reorder quantity policy. Then, an order for Q units is placed each time the inventory level reaches L units.

6.3. Deterministic Systems

The discussion of inventory systems begins with a collection of models in which both demands and lead times are known with certainty. In the real world, demands do not usually occur with certainty, nor do lead times. However, the deterministic models provide a framework for understanding more complex probabilistic models. Additionally, some extensions of the most basic model aid in understanding what complexities may occur in an inventory system. Finally, there are some real-world situations which are solvable by the models presented.

6.3.1. Lot-Size (EOQ) Model with No Stockouts and Zero Lead Time

In this section a model is derived for the case where shortages are not allowed and where the replenishment of inventory occurs instantaneously. If no shortages are allowed, the cost of a shortage must be very large—infinitely large. When replenishment occurs instantaneously, the lead time (the time between placing and receiving an order) is zero.

The lot-size formula is traced back to F. Harris [1915] of the Westinghouse Corporation, who derived the result. Because R. H. Wilson aggressively sold the idea as part of an inventory management system, the lot-size model is sometimes referred to as the Wilson formula. Historically speaking, F. E. Raymond [1931] prepared the first full-length volume dealing with inventory problems. His book shows how the lot-size model could be extended for practical purposes.

When shortage cost is infinite, the total variable cost per period will be the sum of the ordering or procurement cost per period and the holding cost per period. That is, $C_T = C_P + C_H$. Intuitively, if a high inventory is maintained, procurement will be infrequent, but the holding cost per period may be excessive. Conversely, frequent procurements will decrease the holding cost, but will increase the ordering cost per period. If total cost per period is the criterion, it can be minimized by balancing ordering cost per period and holding cost per period.

The total cost for the period, C_T, is determined by examination of the geometry of Figure 6.1. The total variable cost per period is made up of two components. The first of these, the ordering cost per period, will be the ordering cost per order, A, divided by the number of periods per cycle, or $C_P = A/N$. From the geometry of Figure 6.1, it is evident that the number of periods per

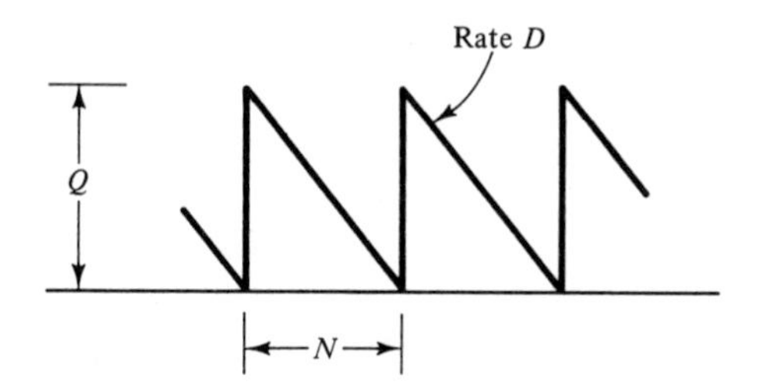

Figure 6.1. Economic order quantity.

inventory cycle is $N = Q/D$, where Q is the order quantity and D is the demand per period. Therefore,

$$C_P = \frac{AD}{Q} \tag{6.1}$$

During an inventory cycle, the average inventory on hand is given by $Q/2$. The holding cost per period for each unit on hand is iC, the inventory carrying cost as a proportion of the item cost, i, multiplied by the item cost, C. Thus, the holding cost component is as follows:

$$C_H = \frac{QiC}{2} \tag{6.2}$$

Equations (6.1) and (6.2) combine to form the total cost, or

$$C_T = \frac{AD}{Q} + \frac{QiC}{2} \tag{6.3}$$

The only variable in equation (6.3) is the order quantity Q. The optimal value of Q, the economic order quantity (EOQ), can be determined by taking the derivative of equation (6.3) with respect to Q, setting the result to zero, and solving for Q, calling the optimal value Q^*. This procedure is as follows:

$$\frac{dC_T}{dQ} = \frac{-AD}{Q^2} + \frac{iC}{2} = 0$$

$$Q^* = \sqrt{\frac{2AD}{iC}} \tag{6.4}$$

The minimum total variable cost per period, C_T^*, may be found by substituting equation (6.4) into equation (6.3) to obtain

$$C_T^* = \frac{AD}{\sqrt{2AD/iC}} + \frac{(\sqrt{2AD/iC})iC}{2}$$

which can be simplified to yield

$$C_T^* = \sqrt{2AiCD} \tag{6.5}$$

EXAMPLE 6.1

Suppose that an inventory manager purchases an item having the following demand and cost data:

D = 4 units per month
C = \$500 per unit
i = 0.02 (expressed as a fraction of item cost on a monthly basis)
A = \$80 per order

The optimum order quantity is determined from equation (6.4) as

$$Q^* = \sqrt{\frac{2(\$80)(4)}{(0.02)(\$500)}} = 8 \text{ units}$$

Also, C_T^* can be determined from equation (6.5) as

$$C_T^* = \sqrt{2(\$80)(0.02)(\$500)(4)} = \$80$$

The components of total cost are individually given by equations (6.1) and (6.2). The ordering cost amount in the optimal C_T^* is

$$C_P = \frac{\$80(4)}{8} = \$40$$

The holding cost component, by subtraction, or by equation (6.2) is

$$C_H = \frac{8(0.02)(\$500)}{2} = \$40$$

Thus, the costs are balanced at the optimal solution.

Table 6.1 displays the costs associated with purchasing different quantities. Notice that there is little difference whether the optimal value of $Q^* = 8$ is purchased or whether it is seven or nine. Figure 6.2 displays the total cost graphically and shows its relatively insensitive nature near the optimal value.

6.3.2. EOQ MODEL WITH DETERMINISTIC LEAD TIME

In many real-life situations there is some positive lead time associated with the procurement process. Lead time is the time between the placing of an order and the receipt of that order. In the simplest case, where the lead time is less

Table 6.1. C_T ASSOCIATED WITH SEVERAL VALUES OF Q

Q	C_P	C_H	C_T
0	$ ∞	$ 0	$ ∞
1	320.00	5.00	325.00
.	.	.	.
.	.	.	.
.	.	.	.
7	45.71	35.00	80.71
8	40.00	40.00	80.00
9	35.56	45.00	80.56
.	.	.	.
.	.	.	.
.	.	.	.
20	16.00	100.00	116.00

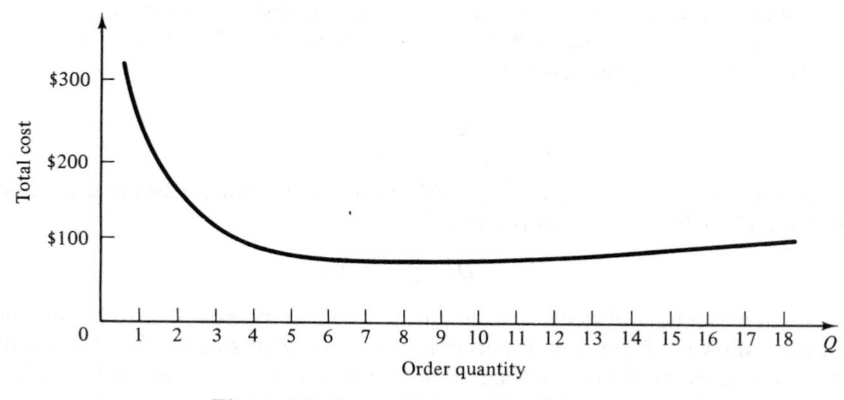

Figure 6.2. Insensitivity of the TC curve.

than the length of the inventory cycle, not more than one order can be outstanding. Such an inventory system is depicted in Figure 6.3. Figure 6.3 contains several elements that did not appear in Figure 6.1. First, the symbol L is used to identify the order level, or reorder point as it is often called. When the inventory on hand plus that on order reaches L, it is time to place another order. A deterministic lead time of T time periods will be incurred each time an order is placed. The dashed line in Figure 6.3 represents the amount on hand plus that on order.

The optimal order quantity for this model is the same as that for the model with zero lead time. There is an additional policy variable in the model of Figure 6.3: the optimal order level, which is determined geometrically as follows:

$$L^* = DT$$

Thus, the order level is just the lead-time demand.

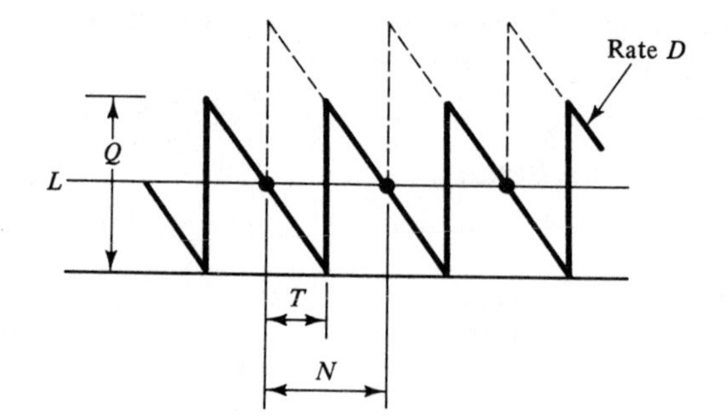

Figure 6.3. EOQ model with lead time.

EXAMPLE 6.2

To illustrate what happens when more than one order is outstanding, let the parameters of Example 6.1 be used, but augment these with a lead time of 3 months. Now, the length of a cycle is given by

$$N = \frac{Q}{D}$$

With $Q = 8$ and $D = 4$, a cycle length of 2 months is indicated. (An order is placed every 2 months.) The order level is given by

$$L^* = DT = 12 \text{ units}$$

Thus, when the amount of stock on hand plus the amount on order falls to 12, an order is placed as shown in Figure 6.4. An order is placed when the stock on hand is actually 4 units. At that point in time, placing an order will result in two outstanding orders. Twenty units will be on hand plus on order. After replenishment stock arrives, there is only one order for 8 units outstanding. This condition will exist for 1 month, at which

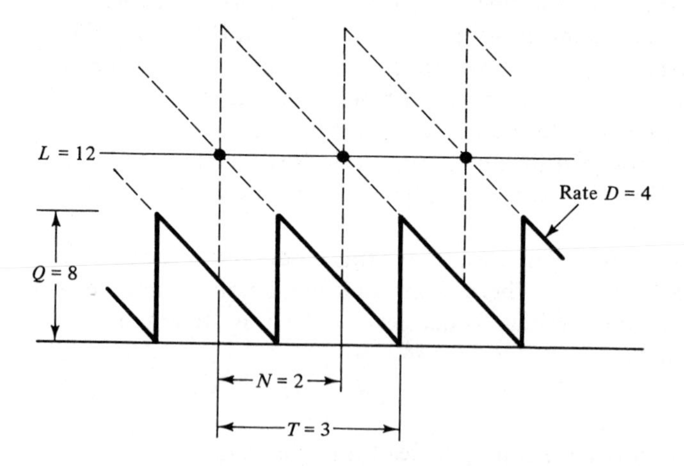

Figure 6.4. Inventory system with orders outstanding.

time another order is placed. The student is encouraged to study Figure 6.4. Exercises at the conclusion of this chapter require drawing an analogous figure for different lead times.

6.3.3. EOQ Model with Backorders and Deterministic Lead Time

The models presented in Sections 6.3.1 and 6.3.2 were based on the assumption that no shortages were to be allowed (i.e., there is infinite shortage cost). Thus, the models involved a trade-off of ordering and holding cost. When shortage cost is finite, an economic advantage occurs when a certain amount of shortage is allowed. For instance, the holding cost component C_H of total variable cost could be reduced by allowing a shortage condition to exist, thereby incurring a shortage cost component C_S of total variable cost.

As indicated in Section 6.1, the shortage cost consists of a term π which is independent of time and a term π' which is linear in the length of time for which the shortage occurs. Also, recall that a shortage may result in a backorder, lost sale, or substitution. In this chapter, the backorder case is described in the models where shortages are allowed and the resulting C_S will be a function of time, or $\pi' t$, where t is the length of time the backorder exists.

The inventory process is illustrated by Figure 6.5. The maximum shortage during the cycle, s, occurs just before the replenishment quantity, Q, is received. The unit periods of holding, H, are shown shaded above the positive inventory axis just as the unit periods of shortage, S, are shown below the axis.

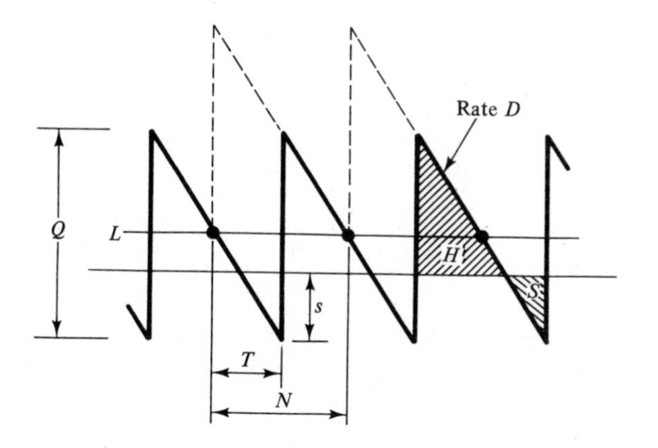

Figure 6.5. EOQ model with shortage cost.

Following an earlier development by Fabrycky and Banks [1967], the number of periods per cycle may be expressed as

$$N = T + \frac{Q - DT}{D} = \frac{Q}{D}$$

The total number of unit periods of stock on hand during the cycle is given by the area of the triangle, H, which is the product of 1/2 its base multiplied by its height. The height of H is $Q + L - DT$ and the length of the base is $(Q + L - DT)/D$. Thus,

$$H = \frac{Q + L - DT}{2}\frac{Q + L - DT}{D}$$

$$= \frac{(Q + L - DT)^2}{2D}$$

and the total number of unit periods of shortage during the cycle is given by the area of the triangle, S, which is also 1/2 its base, s, multiplied by its height. Thus,

$$S = \frac{s}{2}\frac{s}{D} = \frac{DT - L}{2}\frac{DT - L}{D}$$

$$= \frac{(DT - L)^2}{2D}$$

The total variable cost for the period is given by

$$C_T = C_P + C_H + C_S$$

where C_S is the shortage cost for the period. The C_P term is, as determined previously,

$$C_P = \frac{AD}{Q}$$

The holding cost for the period will be the holding cost per unit per period, iC, multiplied by the average number of units in stock for the period, or

$$C_H = iC\frac{H}{N} = \frac{iC(Q + L - DT)^2}{2Q}$$

Shortage cost for the period will be the shortage cost per unit short per period, π', multiplied by the average number of units short for the period, or

$$C_S = \frac{\pi' S}{N} = \frac{\pi'(DT - L)^2}{2Q}$$

The total variable cost for the period is a summation of the three cost components, or

$$C_T = \frac{AD}{Q} + \frac{iC(Q + L - DT)^2}{2Q} + \frac{\pi'(DT - L)^2}{2Q} \tag{6.6}$$

Total variable cost is a function of both Q and L. Taking the partial derivative of equation (6.6) with respect to Q and then L, and setting the result equal to zero, gives

$$Q^* = \sqrt{\frac{2AD}{iC} + \frac{2AD}{\pi'}} \tag{6.7}$$

and

$$L^* = DT - \sqrt{\frac{2AiCD}{\pi'(iC + \pi')}} \tag{6.8}$$

and then substituting equations (6.7) and (6.8) back into equation (6.6) gives

$$C_T^* = \sqrt{\frac{2AiC\pi' D}{iC + \pi'}} \tag{6.9}$$

The derivation of equations (6.7), (6.8), and (6.9) requires a considerable amount of mathematical maneuvering. This is left as an optional exercise for the interested reader.

EXAMPLE 6.3

Suppose that an inventory manager purchases an item having the following parameters:

$D = 10$ units per day
$T = 16$ days
$C = \$200$ per unit
$A = \$16$ per order
$\pi' = \$0.10$ per unit per day
$i = 0.001$ per unit per day, as a fraction of item cost

Then, using equations (6.7) and (6.8) yields

$$Q^* = \sqrt{\frac{2(\$16)(10)}{0.001(\$200)} + \frac{2(\$16)(10)}{\$0.10}} = 69.281 \doteq 69$$

$$L^* = 10(16) - \sqrt{\frac{2(\$16)(0.001)(\$200)(10)}{\$.10[(0.001)(\$200) + \$0.10]}} = 113.810 \doteq 114$$

In this example Q^* and L^* are not integers. A round-off rule could be developed to determine Q^* and L^*. However, rounding to the nearest integer should have little impact on the total variable cost. In this example, the total variable cost computed from equation (6.9) gives the following result:

$$C_T^* = \sqrt{\frac{2(\$16)(0.001)(\$200)(\$0.10)(10)}{(0.001)(\$200) + \$0.10}} = \$4.618$$

Table 6.2 shows the total variable cost based on equation (6.6) when values in the neighborhood of (Q^*, L^*) are used. The differences in Table 6.2 are certainly insignificant from C_T^* and from each other.

Table 6.2. TOTAL VARIABLE COST SURROUNDING THE OPTIMAL VALUES

	Q	
L	69	70
113	\$4.621	\$4.619
114	\$4.618	\$4.619

6.3.4. MANUFACTURING LOT-SIZE MODEL

This section presents a model for the case where stock is manufactured. In this situation there is a finite replenishment or production rate R, where $R > D$. The model for the case where shortages are not allowed is developed. Extension to the case with a finite shortage cost is left as an exercise to the interested reader.

The determination of the economic lot size Q^*, in a manufacturing environment, is similar to the determination of the minimum-cost order quantity for the purchase alternative. The costs for the model presented will be setup (rather than ordering) cost and holding cost. If few setups are made, the holding cost can be excessive. However, increased setups, while lowering the holding cost, will increase the setup cost as a component of total variable cost.

Figure 6.6 shows the geometry of the inventory process under discussion.

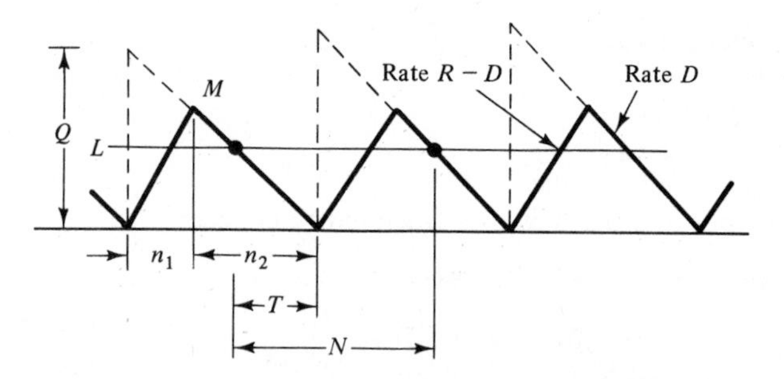

Figure 6.6. Manufacturing alternative inventory geometry.

The ordering (setup) cost for the period will be the same as in previous models. That is,

$$C_P = \frac{AD}{Q}$$

The cycle of length N can be equated to the sum of n_1, the time during which inventory is increasing, plus n_2, the time during which inventory is decreasing, or

$$N = n_1 + n_2$$

Next, n_1 and n_2 are determined as

$$n_1 = \frac{M}{R - D}$$

and

$$n_2 = \frac{M}{D}$$

where M is the maximum inventory value. Now

$$N = \frac{Q}{D} = M\left(\frac{1}{R - D} + \frac{1}{D}\right)$$

so that

$$M = Q\left(1 - \frac{D}{R}\right)$$

The total number of unit periods of stock on hand during the cycle, H, is given by

$$H = \frac{NM}{2}$$

The holding-cost component then becomes

$$C_H = \frac{HiC}{2} = \frac{Q(1 - D/R)iC}{2}$$

The total variable cost is given by the sum of the ordering cost and the holding cost, or

$$C_T = \frac{AD}{Q} + \frac{Q(1 - D/R)iC}{2} \tag{6.10}$$

The procurement quantity resulting in a minimum total variable cost may be found by differentiating equation (6.10) with respect to Q, setting the result equal to zero, and solving for Q, to obtain the following:

$$Q^* = \sqrt{\frac{2AD}{iC(1 - D/R)}} \tag{6.11}$$

The point at which setup is commenced, L, is given by

$$L^* = DT \tag{6.12}$$

The minimum total variable cost can be found by substituting equation (6.11) into equation (6.10) to obtain

$$C_T^* = \sqrt{2A(1 - D/R)iCD} \tag{6.13}$$

EXAMPLE 6.4

Suppose that an item costs \$100 to manufacture and has a deterministic demand of 400 per year. Each time a setup occurs results in a cost of \$1000. The carrying cost fraction is 4/15 on an annual basis. These items can be manufactured at a rate of 1600 per year. The lead time to prepare for manufacturing is $1\frac{1}{2}$ months.

From equation (6.11), the optimal lot size is given by

$$Q^* = \sqrt{\frac{2(\$1000)(400)}{(4/15)(\$100)(1 - 400/1600)}} = 200 \text{ units}$$

and from equation (6.12) the point at which setup should begin is $L^* = 400(1.5/12) = 50$ units. Finally, the optimal total variable cost is

$$C_T^* = \sqrt{2(\$1000)\left(\frac{1 - 400}{1600}\right)(\$100)\left(\frac{4}{15}\right)(400)} = \$4000$$

6.3.5 QUANTITY DISCOUNT MODEL

Often, vendors offer a price discount as the number of items purchased increases. Price discounts may take the form of a reduced per unit cost when the next price break is reached, and there may be several price breaks.

The introduction of a price discount schedule requires that the cost equation include item cost in the form $C_I = CD$. Since the item cost is different for each price break, several different total costs must be calculated to determine the optimal procurement quantity.

To demonstrate the methodology, the total cost will be based on item cost, ordering cost, and holding cost only. The resulting C_T is given by

$$C_T = CD + \frac{AD}{Q} + \frac{QiC}{2} \tag{6.14}$$

The procedure described can easily be extended to include shortage cost.

If there are b price breaks, there will be b total cost curves, one for each price break. The total cost curves when there are three price breaks are shown in Figure 6.7. The first price break occurs at Q_1 and the second occurs at Q_2. The portion of the total cost curve associated with each price break is represented by the solid portion. The dashed portion is not realizable. The problem then is to find the lowest point on the discontinuous curve. The following iterative procedure[1] will yield the least-cost procurement quantity:

1. Calculate the total cost at each price break using equation (6.14).
2. Compute trial minimum-cost order quantities for each price interval

[1] It should be noted that a more efficient procedure which may require less calculations can be used. However, the suggested procedure is favored by students and always yields the optimal result. The interested reader may refer to Johnson and Montgomery [1974] for the more efficient procedure.

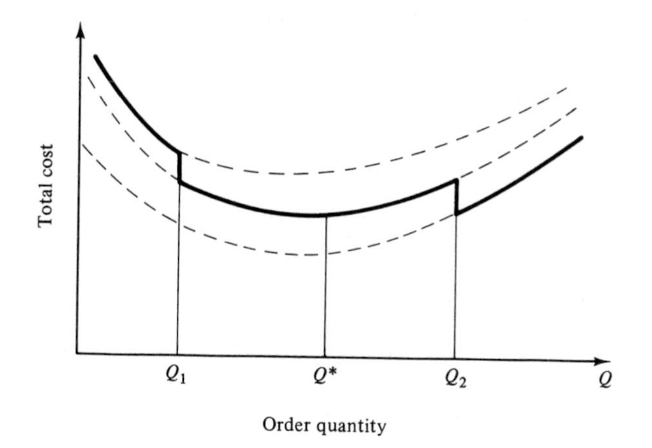

Figure 6.7. Total cost curves with price discounts.

using equation (6.4). Discard any that are not realizable. Calculate the total cost of any that are realizable using equation (6.14).

3. Determine the minimum of the total costs that have been determined. The optimal order quantity is associated with the minimum total cost.
4. Set $L^* = DT$.

EXAMPLE 6.5

As an example of the price discount procedure just described, consider the case of a purchasing manager confronted by the following situation:

$D = 10$ units per week
$T = 2$ weeks
$A = \$20$ per procurement
$i = 0.52$ as a fraction of item cost on an annual basis (0.01 on a weekly basis)

The price discount schedule is as follows:

Order Quantity	*Price per Unit*
$1 \leq Q < 100$	\$10.00
$100 \leq Q < 200$	9.50
$200 \leq Q$	9.25

First compute the total cost at the order quantities where the price breaks occur:

$$C_T(100) = \$9.50(10) + \frac{\$20(10)}{100} + \frac{100(0.01)(\$9.50)}{2} = \$101.75$$

$$C_T(200) = \$9.25(10) + \frac{\$20(10)}{200} + \frac{200(0.01)(\$9.25)}{2} = \$102.75$$

Next, compute the trial order quantities that would be required for each interval:

First interval:

$$\text{Trial } Q^* = \sqrt{\frac{2(\$20)(10)}{(0.01)(\$10)}} = 63 \text{ units}$$

This is realizable, so compute the associated total cost as follows:

$$C_T(63) = \$10(10) + \frac{\$20(10)}{63} + \frac{63(0.01)(\$10)}{2} = \$106.32$$

Second interval:

$$\text{Trial } Q^* = \sqrt{\frac{2(\$20)(10)}{0.01(\$9.50)}} = 65 \text{ units}$$

This is not realizable; therefore, discard it from further consideration.

Third interval:

$$\text{Trial } Q^* = \sqrt{\frac{2(\$20)(10)}{0.01(\$9.25)}} = 66 \text{ units}$$

This is not realizable; therefore, discard it from further consideration.

Next, find the minimum and optimal C_T of the remaining candidates $C_T =$ min (\$101.75, \$102.75, \$106.32) $=$ \$101.75, where $Q^* = 100$ units and $L^* = DT = 10(2) = 20$ units.

6.3.6. Model for Price Inflation

In this section a model is presented for an inventory system in which the item price is expected to change by a known amount. For simplicity, assume that the lead time is zero. The model involves item cost, ordering cost, and carrying cost only. An early formulation of the model was given by Naddor [1966]. A summary of models that have been developed to deal with inflationary conditions has been published [Mangiameli et al., 1981].

Knowing that the price will increase, a large amount would be purchased just before the new price goes into effect. However, if too great an amount is purchased, the holding cost will be excessive. The objective of this section is to determine just what the optimal order quantity should be.

In previous sections, once the order quantity was determined, it would be ordered repetitively every $N = Q/D$ time periods. However, in this section, after the large order is placed, future orders will revert to an order quantity that reflects the new system parameters. In this system, the decision of whether to buy at the current price is based on a cost model which has the dimension of dollars only and this dollar value will be that which would occur over the time period T_0 to T_1, where T_0 is the date at which the price increase will take effect and T_1 is the date when the next purchase will be made. The approach to the model presented in this section is to compare the cost of not taking advantage of the price change versus purchasing an increased amount at the old price right before the price changes.

The student is encouraged to question the reality of this model, particularly with respect to the constancy of future prices. Prices may change repeatedly and frequently in amounts and at times that are unknown at present.

At some date T_0, the price will be increased by an amount k. Purchases before T_0 will cost C, but after that date the price will be $C + k$. The amount, Q_0, of the one time quantity purchased just before T_0, is to be determined.

If an amount Q_0 is purchased just before time T_0, the next purchase will occur at time T_1 after an elapse of $N_0 = Q_0/D$ periods. The next purchase will be made at the new price, and the EOQ will then revert to that given by the following:

$$Q_1^* = \sqrt{\frac{2AD}{i(C + k)}} \tag{6.15}$$

After time T_1, purchases of Q_1^* will continue to be made every Q_1^*/D periods and the total variable cost per period will be given by

$$C_T^* = \sqrt{2Ai(C + k)D} \tag{6.16}$$

If no special purchase is made just before T_0, then $Q_0 = 0$ and $T_0 = T_1$. The cost per period, C_T^* from equation (6.16) commences at T_0 rather than at T_1. The dynamics of the system are shown in Figure 6.8. The dashed lines show the case when $Q_0 = 0$.

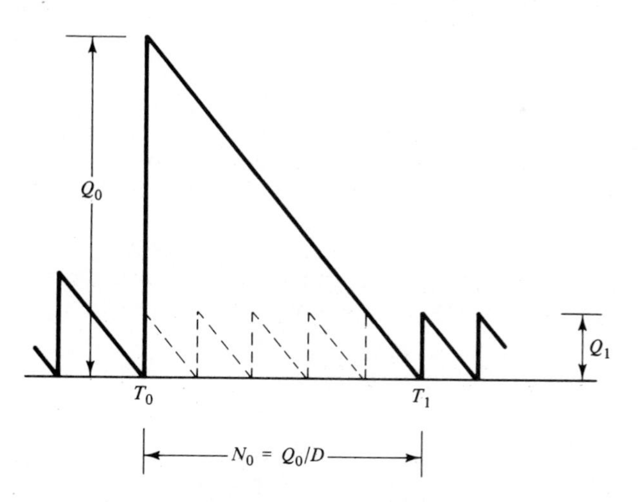

Figure 6.8. Inventory system with a known price increase.

Let V' designate the total cost of the system during the period T_0 to T_1 when an amount $Q_0 > 0$ is purchased at the old price of C per item. The total cost during this time will be the sum of the item cost, the ordering cost, and the holding cost, or

$$V' = CQ_0 + A + \frac{Q_0}{2}N_0 iC = CQ_0 + A + \frac{Q_0{}^2 iC}{2D}$$

Let V designate the total cost of the system during the period T_0 to T_1 when no purchase is made just before T_0, but when several purchases of Q_1^* are made at the new price $(C + k)$. (There will be an average of Q_0/Q_1^* such purchases during T_0 to T_1.) The total cost over the period T_0 to T_1 will be given by

$$V = (C + k)Q_0 + \frac{Q_0}{Q_1^*}A + \frac{Q_1^*}{2}i(C + k)\frac{Q_0}{D}$$

The difference between V' and V can be maximized if

$$Q_0^* = Q_1^* + \frac{k}{C}\left(Q_1^* + \frac{D}{i}\right) \tag{6.17}$$

with saving G_0^* given by

$$G_0^* = \frac{k}{C}\left[\frac{k}{2}\frac{D}{i} + Q_1^*(C + k) + A\right] \tag{6.18}$$

The details of the derivation of equation (6.17) can be determined from Naddor [1966]. [What happens to equations (6.17) and (6.18) as $k \longrightarrow 0$?] An example will aid in understanding this section.

EXAMPLE 6.6

The current price of Bunker C fuel is \$0.27 per liter. As of January 1, upcoming shortly, the price will be raised to \$0.30 per liter. The fuel is being consumed at the rate of 10,000 liters per month. The carrying cost fraction is 0.25 expressed on an annual basis. The ordering cost is \$20. What amount should be purchased on December 31?

Using the symbols in the model, $C = \$0.27$ per liter, $k = \$0.03$ per liter, $D =$ 120,000 liters per year, $i = 0.25$ per year, and $A = \$20$ per order. Q_1^* is determined from equation (6.15), after which Q_0^* can be determined from equation (6.17) as follows:

$$Q_1^* = \sqrt{\frac{2(\$20)(120{,}000)}{(0.25)(\$0.30)}} = 8000 \text{ liters}$$

Then,

$$Q_0^* = 8000 + \frac{\$0.03}{\$0.27}\left(8000 + \frac{120{,}000}{0.25}\right) = 48{,}889 \text{ liters}$$

The resulting savings if Q_0^* is purchased are computed from equation (6.18) to be

$$G_0^* = \frac{\$0.03}{\$0.27}\left[\frac{0.03}{2}\frac{120{,}000}{0.25} + 8000(\$0.30) + \$20\right] = \$1069$$

This solution requires that storage space be available for nearly 41,000 extra liters of fuel! Constraints on storage space may result in purchasing less than Q_0^* but more than Q_1^*. Exercise 13 asks that the reader determine the resulting savings when the order quantity is in this range.

6.4. Probabilistic Systems

In reality, demand in an inventory system is usually a random variable and so is the lead time. Thus, models that reflect this probabilistic environment are needed. Of the many possible models that have been developed, three are presented here, and the most elementary form of each model is shown. The first model is for a single period, where demand is considered to occur at one point in time, and there is no opportunity to reorder. The second model is for a periodic review with backorders allowed. In this model the optimal period length and target inventory are determined. The last probabilistic inventory model presented is for the case of continuous review with backorders allowed. In this instance the inventory level is monitored after every transaction, and if the level has dropped to or below the reorder point, a specified quantity of units is ordered.

6.4.1. Single-Period Model

Consider an inventory process in which the cycle length is fixed. There is only one opportunity to order. Demand is probabilistic with distribution $p(x)$, if discrete, and $f(x)$ if continuous. The relevant costs are holding cost and shortage cost only.

The problem is often called the newspaper seller's problem—a version of which was introduced in Chapter 2 and repeated here. A newspaper seller orders papers in advance of each day's news and has no opportunity to reorder papers on the day of sale. The cost of a newspaper is C and the selling price is P, $P > C$. Any newspapers not sold at the end of the day are sold for scrap at a value of B, $C > B$.

Consider a situation in which the demand is continuous. The expected holding cost, $E(C_H)$, is given by

$$E(C_H) = (C - B) \int_{x=0}^{M} (M - x) f(x)\, dx$$

where M is the inventory purchased at the beginning of the period. The expected shortage cost, $E(C_S)$, is given by

$$E(C_S) = (P - C) \int_{x=M}^{\infty} (x - M) f(x)\, dx$$

Then, the expected total relevant cost, $E(C_T)$, is given by

$$E(C_T) = (C - B)\int_{x=0}^{M} (M - x)f(x)\,dx + (P - C)\int_{x=M}^{\infty} (x - M)f(x)\,dx \qquad (6.19)$$

To find the optimal value of M, namely M^*, differentiate equation (6.19) with respect to M, set the result equal to zero and solve for M^*, as follows:

$$\frac{dE(C_T)}{dM} = (C - B)\int_{x=0}^{M} f(x)\,dx - (P - C)\int_{x=M}^{\infty} f(x)\,dx = 0 \qquad (6.20)$$

Now,

$$\int_{x=0}^{M} f(x)\,dx + \int_{x=M}^{\infty} f(x)\,dx = 1$$

so

$$\int_{x=M}^{\infty} f(x)\,dx = 1 - \int_{x=0}^{M} f(x)\,dx$$

Multiplying all terms by $(P - C)$ gives

$$(P - C)\int_{x=M}^{\infty} f(x)\,dx = (P - C) - (P - C)\int_{x=0}^{M} f(x)\,dx \qquad (6.21)$$

Substituting the right-hand side of equation (6.21) into equation (6.20) yields

$$(C - B)\int_{x=0}^{M} f(x)\,dx - (P - C) + (P - C)\int_{x=0}^{M} f(x)\,dx = 0$$

Then,

$$[(C - B) + (P - C)]\int_{x=0}^{M} f(x)\,dx = P - C$$

Finally,

$$\int_{x=0}^{M^*} f(x)\,dx = \frac{P - C}{P - B} \qquad (6.22)$$

where M^* is the optimal value of M.

EXAMPLE 6.7

A store selling Mother's Day cards must decide 6 months in advance on the number of cards to stock. Reordering is not allowed. Cards cost \$0.25 and sell for \$0.60. Any cards not sold by Mother's Day will be sold for \$0.20, and all of them can be sold at this price. The card shop owner is not sure how many cards can be sold, but thinks it is somewhere between 200 and 400.

Treating demand as uniform, $f(x) = 1/200$, $200 \leq x \leq 400$, the following costs are deciphered:

$$P = \$0.60$$
$$C = \$0.25$$
$$B = \$0.20$$

Then, using equation (6.22) yields

$$\int_{x=200}^{M} \frac{1}{200}\,dx = \frac{\$0.60 - \$0.25}{\$0.60 - \$0.20} = \frac{\$0.35}{\$0.40}$$

$$\left.\frac{x}{200}\right|_{200}^{M*} = 0.875$$

The optimal value of M^* is then $M^* = 375$.

The newspaper seller's problem was posed in a discrete format in Chapter 2. It can be shown that M^* for such a problem is given by the following:

$$F(M^* - 1) < \frac{P - C}{P - B} \leq F(M^*) \tag{6.23}$$

where F is the cumulative distribution function of demand. Also, for the discrete system, $E(C_T)$ is given by the following:

$$E(C_T) = (C - B)\sum_{x=0}^{M}(M - x)p(x) + (P - C)\sum_{x=M+1}^{\infty}(x - M)p(x) \tag{6.24}$$

EXAMPLE 6.8 (THE NEWS SELLER'S PROBLEM)

Reference is made to Example 2.3, in which the paper seller buys for 13 cents and sells for 20 cents. The unsold papers have a scrap value of 2 cents. The ratio in equation (6.23), used in making the decision, is given by

$$\frac{P - C}{P - B} = \frac{\$0.20 - \$0.13}{\$0.20 - \$0.02} = 0.39$$

The cumulative distribution must now be calculated as follows:

$$F(40) = 0.35(0.03) + 0.45(0.10) + 0.20(0.44) = 0.1435$$
$$F(50) = 0.1435 + 0.35(0.05) + 0.45(0.18) + 0.20(0.22) = 0.2860$$
$$F(60) = 0.2860 + 0.2645 = 0.5505$$
$$F(70) = 0.5505 + 0.1840 = 0.7345$$
$$F(80) = 0.7345 + 0.1705 = 0.9050$$
$$F(90) = 0.9050 + 0.0705 = 0.9755$$
$$F(100) = 0.9755 + 0.0245 = 1.0000$$

Note that equation (6.23) is satisfied when

$$F(50) < 0.39 \leq F(60)$$

Thus, the optimal number of papers (in batches of 10) for the newspaper seller to purchase each day, M^*, is 60. Then, from equation (6.24), $E(C_T^*)$, the optimal expected relevant total cost, can be calculated as

$$E(C_T^*) = (\$0.13 - \$0.02) \sum_{x=40}^{60} (60 - x)p(x) + (\$0.20 - \$0.13) \sum_{x=70}^{100} (x - 60)p(x)$$
$$= \$0.11[(20)(0.1435) + (10)(0.1425)] + \$0.07[(10)(0.1840) + (20)(0.1705) + (30)(0.0705) + (40)(0.0245)] = \$1.06$$

It is left as an exercise to the reader to verify that purchasing any quantity other than 60 newspapers will result in a higher expected relevant total cost.

6.4.2. Periodic Review with Backorders

The order up to M policy may be used when both the optimal review period (cycle length), N, and the optimal target inventory, M, are both of interest. The formulation of Section 6.4.1 does not provide for simultaneous determination of both of these policy variables. In this section, an approximate treatment of the periodic-review case with constant lead time is presented. The inventory system has both a fixed cost of ordering, A, as well as a fixed review cost, J. Also, the cost of a backorder, π, is high in comparison to the cost of carrying inventory. This will serve to keep the number of backorders rather low.

The expected total variable cost per period is given by

$$E(C_T) = \frac{A + J}{N} + iC\left(M - \mu - \frac{\bar{D}N}{2}\right) + \pi E(s) \tag{6.25}$$

In equation (6.25), μ is the expected lead time demand and $\bar{D}$ denotes the average demand rate per period. Also, $E(s)$ is the expected number of backorders per period and is defined as

$$E(s) = \frac{1}{N} \int_M^\infty (x - M) f(x; T + N)\, dx \tag{6.26}$$

Now, $f(x; T + N)$ is the pdf of demand X over the review period, N, plus the lead time T. (For probabilistic lead time, the reader is referred to Hadley and Whitin [1963].) For a given N, the optimal value of M (M^*) is a solution to

$$\int_M^\infty f(x; T + N)\, dx = \frac{iCN}{\pi} \tag{6.27}$$

Usually, the method used to determine N^* requires the tabulation of the cost, $E(C_T)$, as a function of N, using M^* for the given N to compute $E(C_T)$. An example with normally distributed lead time will clarify this rather brief sketch of the model. The reader is referred to the classic text by Hadley and Whitin [1963] for the full development and extensions.

EXAMPLE 6.9

(This example follows a similar example by Johnson and Montgomery [1974].) A very large mining operation has decided to control the inventory of high-pressure piping by a policy involving periodic review and ordering up to M. The mean demand rate for this item is 600 units per year. (A unit is a 10-meter section.) The lead time is nearly constant at 4 months. The demand in the time $T + N$ can be represented by a normal distribution with mean $600(T + N)$ and variance $800(T + N)$. The cost of each unit is \$400, and the inventory carrying charge as a proportion of item cost on an annual basis is 0.25. The cost of making a review and placing an order is \$200, and the cost of a backorder is estimated to be \$500.

It is desired to find the optimal M and N. If N is specified, M^* may be found easily. Suppose that the inventory level is reviewed every 2 months. Then $T + N = 0.5$, the expected demand in $T + N$ is $600(0.5) = 300$ units, and the variance of demand in this time is $800(0.5) = 400$ units^2. Thus, M^* is the solution to

$$1 - \Phi\left(\frac{M - 300}{20}\right) = \frac{0.25(\$400)(0.1667)}{\$500} = 0.03334$$

where $\Phi(z)$ is the cumulative distribution function of the standard normal random variable given by equation (4.41). From Table A.3,

$$\frac{M - 300}{20} = 1.833$$

Thus,

$$M^* = 300 + 20(1.833) = 337 \text{ units}$$

If N is not specified, the average annual total cost becomes

$$E(C_T) = \frac{200}{N} + (0.25)(\$400)\left[M - 200 - \frac{(600)(N)}{2}\right] + \frac{\$500}{N}\int_M^\infty (x - M)f(x; T + N)\,dx$$

The density $f(x; T + N)$ is normal with mean $600(0.1667 + N)$ and variance $800(0.3333 + N)$.

It can be shown that for a normal random variable X, having mean μ and variance σ^2,

$$\int_a^\infty (x - a)N(x; \mu, \sigma^2)\,dx = \sigma\phi\left(\frac{a - \mu}{\sigma}\right) + (\mu - a)\Phi'\left(\frac{a - \mu}{\sigma}\right)$$

where $\phi(z)$ is the standard normal density, given by equation (4.40), and $\Phi'(z) = 1 - \Phi(z)$. Also, $N(x; \mu, \sigma^2)$, the pdf of a $N(\mu, \sigma^2)$ distribution, is given by

$$N(x; \mu, \sigma^2) = \frac{1}{\sigma}\phi\left(\frac{x - \mu}{\sigma}\right)$$

Using this result, the expected annual cost may be written as

$$E(C_T) = \frac{200}{N} + (100)(M - 200 - 300N) + \frac{500}{N}\left\{\sqrt{800(0.3333 + N)}\phi\left[\frac{M - 600(0.3333 + N)}{\sqrt{800(0.3333 + N)}}\right]\right.$$
$$\left. + [600(0.3333 + N) - M]\Phi'\left[\frac{M - 600(0.3333 + N)}{\sqrt{800(0.3333 + N)}}\right]\right\}$$

Using this expression, the values of $E(C_T)$ as a function of N are computed, where the optimal M^* depends on the chosen N. The results are shown in Table 6.3. The optimal

Table 6.3. $E(C_T)$ AS A FUNCTION OF N

N (*Years*)	M^* (*Units*)	$E(C_T)$
0.0417	273	$11,024
0.0833	293	9,536
0.1667	337	10,652
0.2500	383	12,783
0.3333	430	15,182

N occurs at approximately 0.0833 year (1 month). Increments of N between 0.0417 and 0.1667 would result in greater accuracy, if such precision is warranted.

6.4.3. Continuous Review with Backorders

A process in which the inventory position is monitored after every transaction with the policy of ordering the quantity Q when the level drops to the reorder point of L units is called a continuous review system. The demand in any interval of time is a random variable whose mean is $\bar{D}$.

In this section an approximate treatment of the continuous review case is presented. Suppose that units are demanded in such small quantities that there is no severe overshoot past the reorder point. The fixed cost of ordering is A. The cost of a backorder is π per unit short, independent of the length of the shortage. The procurement lead time, T, is assumed to be constant. Variability in lead time increases the complexity of the model. Hadley and Whitin provide a thorough discussion of these effects [1963].

Hadley and Whitin [1963] provide the formulation for this case with the expected total cost per period equation given by

$$E(C_T) = \frac{A\bar{D}}{Q} + iC\left(\frac{Q}{2} + L - \mu\right) + \frac{\pi\bar{D}}{Q}E(s) \tag{6.28}$$

In equation (6.28), the terms represent ordering cost, holding cost, and shortage cost, in that order. The term μ is the mean of lead-time demand. The expected number of backorders per cycle, $E(s)$, is given by

$$E(s) = \int_L^\infty (x - L)f(x)\,dx \tag{6.29}$$

where x denotes the demand during a lead time and $f(x)$ is its probability density function. The optimal value of Q is given by

$$Q^* = \sqrt{\frac{2\bar{D}(A + \pi E(s))}{iC}} \tag{6.30}$$

and the optimal value of L is given by

$$F'(L) = \frac{QiC}{\pi \bar{D}} \tag{6.31}$$

where

$$F'(L) = \int_L^{\infty} f(x)\,dx \tag{6.32}$$

for the continuous case. Thus, $F'(L)$ is the complementary cumulative distribution of x evaluated at L. To find the optimal pair (Q^*, L^*) that minimizes $E(C_T)$, the following iterative procedure is used:

1. Assume that $E(s) = 0$ and compute Q with equation (6.30). Call this value Q_1.
2. Use equation (6.31) to find the corresponding reorder level, L, with $Q = Q_1$. Call this value L_1.
3. Use equation (6.29) to find $E(s)$ at L_1.
4. Use equation (6.30) with $L = L_1$ to compute Q_2.
5. Repeat step 2 with $Q = Q_2$. Continue this process. Convergence occurs when at an iteration the value of Q or L is repeated from the previous iteration.

EXAMPLE 6.10

(This example follows a similar example by Johnson and Montgomery [1974].) A firm uses gasoline at an expected annual rate of 12,000 metric gallons. The expected rate is constant throughout the year. However, the actual demand may vary randomly. The demand during the lead time is estimated to be normally distributed with a mean of 400 metric gallons and a standard deviation of 50 metric gallons. A procurement costs \$100 per order and the gasoline is purchased on a contract basis at \$1.60 per metric gallon. The inventory carrying charge, on an annual basis, is 0.25. Shortages result in purchases at retail and added paperwork with a resulting loss estimated to be \$1.00 per metric gallon. A fixed-reorder-quantity system is to be used. In this case $\bar{D} = 12{,}000$, $A = \$100$, $iC = (0.25)(\$1.60) = \$0.40$, $\mu = 400$, $\sigma = 50$, $\pi = \$1.00$, and

$$E(s) = \int_L^{\infty} (x - L)N(x; 400, 2500)\,dx$$

where $N(x; \mu, \sigma^2)$ is the normal density with mean μ and variance σ^2 (not to be confused with the cycle length, N). Substituting the known values in equation (6.30) gives

$$Q = \sqrt{\frac{2(12{,}000)[100 + E(s)]}{0.40}} = 245\sqrt{100 + E(s)}$$

and from equations (6.32) and (6.31)

$$F'(L) = \Phi'\left(\frac{L - 400}{50}\right) = \frac{0.4Q}{12{,}000} = \frac{Q}{30{,}000}$$

To evaluate $E(s)$, the result given in the preceding example, modified for the current usage,

$$E(s) = \int_L^\infty (x - L)N(x; \mu, \sigma^2)\,dx = \sigma\phi\left(\frac{L - \mu}{\sigma}\right) - (L - \mu)\Phi'\left(\frac{L - \mu}{\sigma}\right) \qquad (6.33)$$

can be used.

Initially, assume that $E(s) = 0$ and compute $Q_1 = 2450$. Using an order quantity of 2450, the optimal reorder point L_1 satisfies

$$\Phi'\left(\frac{L_1 - 400}{50}\right) = \frac{2{,}450}{30{,}000} = 0.0817$$

Then,

$$\frac{L_1 - 400}{50} = 1.394$$

so that

$$L_1 = 469.7 \doteq 470$$

Using $L_1 = 470$ in equation (6.33) yields

$$E(s) = 50\phi\left(\frac{470 - 400}{50}\right) - (470 - 400)(0.0817) = 1.85$$

Then Q_2 is given by

$$Q_2 = 245\sqrt{100 + 1.85} = 2473$$

The resulting reorder level is given by

$$\Phi'\left(\frac{L_2 - 400}{50}\right) = \frac{2{,}473}{30{,}000} = 0.0824$$

Then, using Table A.3,

$$\frac{L_2 - 400}{50} = 1.390$$

so that

$$L_2 = 469.5 \doteq 470$$

Since $L_1 \doteq L_2$, convergence has been reached with $Q^* = 2473$ metric gallons and $L^* = 470$ metric gallons. This policy yields a probability of shortage during any cycle equal to 0.0824. The expected number of backorders per cycle is 1.85 metric gallons. Convergence usually occurs very rapidly. However, if at any point in the algorithm, the value of $F'(L)$, using equation (6.31), is greater than unity, the cost of a shortage is so small that it is desirable to incur a large number of backorders. In that case a more exact model is needed.

6.5. Simulation in Inventory Analysis

Of all the models in this chapter, the (M, N) system with probabilistic demand in Section 6.4.2 and the (Q, L) system with probabilistic demand in Section 6.4.3 are the closest to reality. However, when there are dynamic and/or sequential effects with uncertainty present it may not be possible to obtain sound policies from these models. The dynamic realities are nonstationary demand and nonstationary lead time. The (M, N) and (Q, L) systems are not designed for this circumstance. In reality, inventory systems are usually sequential, one of many stages in a production and distribution system. Simulation may be the only methodology to develop realistic policies in these instances.

The (M, L) policy was mentioned in Section 6.2, but it was not discussed in Section 6.4. The (M, L) policy can be shown to have total cost no larger than the best (Q, L) model. Why was it not presented? The computational effort to find the best (M, L) policy is prohibitive. Incidentally, (M, L) systems are often found in practice, but the policy variables are usually arbitrary. In addition to the dynamic and/or sequential effects mentioned above, the computational difficulties of formulating (M, L) policies make such determinations good candidates for simulation.

Peterson and Silver [1979] give a number of illustrative problems which are candidates for solution by simulation. For example, these authors describe problems involving products with interdependent demand (hot dogs and buns) and problems of coordinated replenishment which have been solved using simulation. Many other examples are given in Peterson and Silver [1979] with references.

6.6. Summary

Inventory problems occur frequently in practice and discrete-event system simulation is often used as the solution methodology. Faced with an inventory problem, the analyst should initially determine if a closed-form solution might accomplish the result with much less expenditure of resources. A vast amount of literature concerning mathematical models of inventory systems has appeared. This chapter contains several deterministic models, with extensions, and three probabilistic models to give the reader some insight into the types of inventory problems that can be solved mathematically.

If the inventory system is complex, then a model with closed-form solution is probably unavailable. If simulation is to be used as the solution methodology, it is necessary that the analyst understand the relationships between the parameters and decision variables in an inventory system. By studying the models in this chapter, and by solving the exercises at the end of the chapter, valuable insight should be obtained.

REFERENCES

FABRYCKY, W. J., AND J. BANKS [1967], *Procurement and Inventory Systems: Theory and Analysis*, Reinhold, New York.

HADLEY, G., AND T. M. WHITIN [1963], *Analysis of Inventory Systems*, Prentice-Hall, Englewood Cliffs, N.J.

HARRIS, F. [1915], *Operations and Cost* (Factory Management Series), A. W. Shaw and Company, Chicago, pp. 48–52.

JOHNSON, L. A., AND D. C. MONTGOMERY [1974], *Operations Research in Production Planning, Scheduling, and Inventory Control*, Wiley, New York.

MANGIAMELI, P. M., J. BANKS, AND H. SCHWARZBACH [1981], "Static Inventory Models and Inflationary Cost Increases," *The Engineering Economist*, Vol. 26.

NADDOR, E. [1966], *Inventory Systems*, Wiley, New York.

PETERSON, R., AND E. SILVER [1979], *Decision Systems for Inventory Management and Production Planning*, Wiley, New York.

RAYMOND, F. E. [1931], *Quantity and Economy in Manufacturing*, McGraw-Hill, New York.

EXERCISES

1. In Example 6.2, change T to the following values, then prepare an analogy to Figure 6.4 with the new value.
(a) $T = 4$
(b) $T = 5$

2. Using equation (6.4) can result in a nonintegral value. Q^* could be rounded to $Q^* - u$ or $Q^* + v$, where u and v are the values that yield discrete integers below and above Q^*. Develop a round-off rule based on the associated costs and parameters.

3. An item that costs \$200 has a demand rate of 30 units per month. The ordering cost per order is \$60. The inventory carrying cost fraction is 0.24 on an annual basis. If the lead time for the item is 1 month:
(a) Determine Q^* and L^*.
(b) Construct the inventory geometry analogous to Figure 6.3.

4. Given equation (6.6), derive equations (6.7) and (6.8). Then, having obtained equations (6.7) and (6.8), obtain equation (6.9).

5. What happens to equations (6.7), (6.8), and (6.9) as $\pi' \longrightarrow \infty$?

6. An inventory manager operates by intuition. For instance, faced with the data given in Example 6.3, this inventory manager would set $Q = 60$ and $L = 100$. How far from the optimal total variable cost is such a policy?

7. Show that equation (6.11) reduces to equation (6.4) as $R \longrightarrow \infty$.

8. Given equation (6.10), derive equation (6.11). Then, given equation (6.11), derive equation (6.13).

9. Extend the model of Section 6.3.4 to include a finite shortage cost.

10. Using the data of Example 6.4, determine C_T if Q is too low ($Q = 150$) or too high ($Q = 250$). Does a Q exist such that

$$C_T(Q - 10) = C_T(Q + 10)$$

where $C_T(Q)$ is the total variable cost for purchasing a quantity equal to Q? If so, find Q.

11. Suppose that the price per unit in Exercise 3 is quoted in accordance with the following schedule:

Purchase Quantity	*Price per Unit*
$1 \leq Q < 20$	\$200
$20 \leq Q < 100$	190
$100 \leq Q$	180

Find the minimum cost Q^* and L^*.

12. The current price of an item is \$20, but an announcement has been made that as of 1 month from today, the price is going up 10%. It cost \$40 to place an order. The annual demand for this item is 800 units. The annual inventory carrying cost as a fraction of item cost is 0.20. The shortage cost is infinitely high.

(a) If an amount Q_0 is purchased just before the price rises, what should the continuing order quantity be after Q_0 is used?

(b) What should the order quantity be immediately before the price increase?

(c) Draw the inventory process, to scale, for the decisions in parts (a) and (b).

(d) What is the gain G_0^* using the policy determined?

13. Determine the savings G_0 in Example 6.6 if a maximum of 24,000 liters of fuel could be stored.

14. Demand for an item is exponentially distributed as $f(x) = \frac{1}{4}e^{-x/4}$, $x \geq 0$. There is only one opportunity to order. The purchase price of the item is \$1000 and each item has a salvage value of \$400. The items are sold for \$1800. How many items should be purchased?

15. A polar voyage is being planned. Special-purpose heavy-duty 12-volt batteries are required for this trip. Demand for these batteries, for a trip of the scheduled duration, is Poisson with mean 4. Each battery costs \$250. They can be sold for their lead content for \$50. If there is a need for a battery, and it is unavailable, an aircraft will have to fly the battery to the site at a cost of \$1000 each. How many batteries should be taken?

16. Show that purchasing 50 or 70 newspapers in Example 6.8 will result in a higher relevant total cost.

17. In Example 6.9, show that an N of 3 months results in an M^* of 383 and a $E(C_T)$ of \$12,783.

18. A bakery uses flour at an expected annual rate of 20,000 kilograms. The expected rate of usage is essentially constant throughout the year. However, the actual demand may vary randomly. The lead time demand is normally distributed with a mean of 800 kilograms and a variance of 3600 kilograms^2. The procurement cost is $200 per order and a kilogram of flour costs $2. The inventory carrying fraction, on an annual basis, is 0.30. A shortage results in rescheduling production and buying flour at retail with the resulting cost proportional to the size of the shortage. The associated shortage cost is $3 per kilogram. Determine the optimal values of Q and L using a continuous review model.

part three

RANDOM NUMBERS

RANDOM NUMBER GENERATION

Random numbers are a necessary basic ingredient in the simulation of almost all discrete systems. Most computer languages have a subroutine that will generate a random number for the asking. Similarly, simulation languages generate random numbers that are used to generate event times and other random variables. In this chapter, the generation of random numbers and their subsequent testing for randomness is described. Chapter 8 will show how random numbers are used to generate a random variable with any desired probability distribution.

7.1. Properties of Random Numbers

A sequence of random numbers, $R_1, R_2, \ldots$, must have two important statistical properties, uniformity and independence. Each random number R_i is an independent sample drawn from a continuous uniform distribution between zero and 1. That is, the pdf is given by

$$f(x) = \begin{cases} 1, & 0 \leq x \leq 1 \\ 0, & \text{otherwise} \end{cases}$$

This density function is shown in Figure 7.1. The expected value of each R_i is given by

7

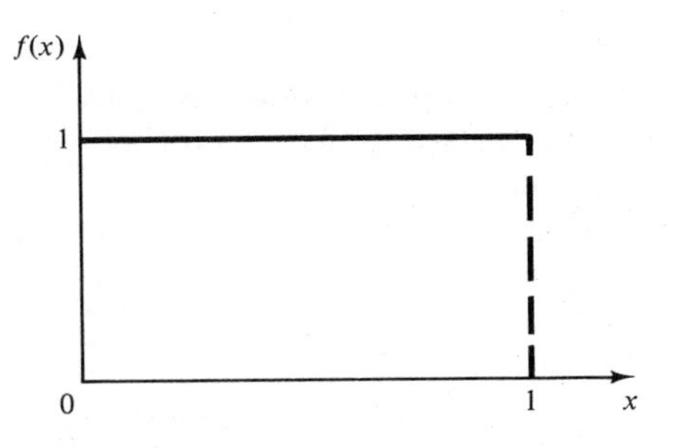

Figure 7.1. PDF for random numbers.

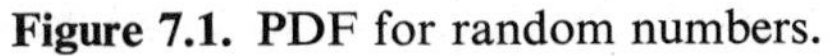

$$E(R) = \int_0^1 x\,dx = \frac{x^2}{2}\Big|_0^1 = \frac{1}{2}$$

and, the variance is given by

$$V(R) = \int_0^1 x^2\,dx - [E(R)]^2 = \frac{x^3}{3}\Big|_0^1 - \left(\frac{1}{2}\right)^2 = \frac{1}{3} - \frac{1}{4} = \frac{1}{12}$$

Some consequences of the uniformity and independence property are the following:

1. If the interval (0, 1) is divided into n classes, or subintervals of equal length, the expected number of observations in each interval is N/n, where N is the total number of observations.

2. The probability of observing a value in a particular interval is independent of the previous values drawn.

7.2. Generation of Pseudo-Random Numbers

Notice that the title of this section has the word "pseudo" in it. "Pseudo" means false, so false random numbers are being generated! However, such is not necessarily the case. In this instance, "pseudo" is used to imply that the very act of generating random numbers by a known method removes the potential for true randomness. If the method is known, the set of random numbers can be replicated. Then an argument can be made that the numbers are not truly random. The goal of any generation scheme, however, is to produce a sequence of numbers between zero and 1 which simulates, or imitates, the ideal properties of uniform distribution and independence as closely as possible.

To be sure, when generating pseudo-random numbers certain problems or errors can occur. These errors, or departures from ideal randomness, are all related to the properties stated previously. Some examples of such departures include the following:

1. The generated numbers may not be uniformly distributed.
2. The generated numbers may be discrete instead of continuous.
3. The mean of the generated numbers may be too high or too low.
4. The variance of the generated numbers may be too high or too low.
5. There may be cyclic variations. The following are examples:
   a. Autocorrelation between numbers.
   b. Numbers successively higher or lower than adjacent numbers.
   c. Several numbers above the mean followed by several numbers below the mean.

Departures from uniformity and/or independence for a particular generation scheme may be detected by tests such as those described in Section 7.4. If such departures are detected, the generation scheme would probably be dropped in search of an acceptable generator. Generators that pass all the tests in Section 7.4, and even more stringent tests, have been developed; thus, there is no excuse for using a generator that has been found to be bad.

Usually, random numbers will be generated by a digital computer, as part of the simulation. There are numerous methods that can be used to generate the values. Before describing some of these methods, or routines, there are a number of important considerations:

1. The routine should be fast. Individual computations are inexpensive, but simulation may require many computations. The total cost can be managed by selecting a cost-effective method of random generation.

2. The routine should not require a lot of core storage. Core storage is relatively very expensive, and may be needed for the remaining parts of the simulation. Virtually all the routines in the paragraphs below require only minimal amounts of core storage.
3. The routine should have a sufficiently long cycle. The cycle length, or period, represents the length of the random number sequence before previous numbers begin to repeat themselves in an earlier order. Thus, if 100 events are to be generated, the period should be at least that long. However, if only a few random numbers are needed, say 10, it would be just as simple to read these values in as data obtained from a table of random numbers.

   A special case of cycling is degeneration. A routine degenerates when the same random numbers appear repeatedly. Such an occurrence is certainly unacceptable. This can happen rapidly with some methods, as attested by Example 7.3 below.
4. The random numbers should be replicable. Given the starting point (or conditions) it should be possible to generate the same set of random numbers, completely independent of the status of the system that is being simulated. This is necessary since many runs of the alternative systems with the same event list may be required, as a means of facilitating comparisons between systems.
5. Most important, and as indicated previously, the generated random numbers should closely approximate the ideal statistical properties of uniformity and independence.

7.3. Techniques for Generating Random Numbers

The more common methods for generating random numbers are discussed in the following paragraphs. The linear congruential method of Section 7.3.3 is the most widely used technique today. The other techniques are mostly of historical interest. With some techniques, the number of decimal places generated is a function of the requirements of the application and is controlled by the user.

7.3.1. Midsquare Method

The midsquare method was proposed by von Neumann and Metropolis in the mid-1940s. The technique starts with an initial number, or seed. This number is squared, and the middle digits of this square become the random number after placement of the decimal. The middle digits are then squared to generate the second random number. The technique continues in this fashion. An example will show how the technique operates.

EXAMPLE 7.1

Suppose that a sequence of four-digit random numbers is needed. Begin with a seed of 5497. Let X_i be the ith value to be squared and R_i be the ith random number. Then,

$$
\begin{aligned}
X_0 &= 5497 \\
X_0^2 &= (5497)^2 = 30{,}217{,}009 \Longrightarrow X_1 = 2170 \\
R_1 &= 0.2170 \\
X_1^2 &= (2170)^2 = 04{,}708{,}900 \Longrightarrow X_2 = 7089 \\
R_2 &= 0.7089 \\
X_2^2 &= (7089)^2 = 50{,}253{,}921 \Longrightarrow X_3 = 2539 \\
R_3 &= 0.2539 \\
&\vdots
\end{aligned}
$$

This technique is more of historical than of practical interest. There are a number of drawbacks to using this technique. It is difficult to state simple general conditions for choosing a seed that guarantees that the sequence will not degenerate and will have a long period. Additionally, zeros, once they appear, are carried in subsequent numbers, as the next example will show.

EXAMPLE 7.2

Suppose that the seed in Example 7.1 is changed slightly to 5197. Then,

$$
\begin{aligned}
X_0 &= 5197 \\
X_0^2 &= (5197)^2 = 27{,}008{,}809 \Longrightarrow X_1 = 0088 \\
R_1 &= 0.0088 \\
X_1^2 &= 00{,}007{,}744 \Longrightarrow X_2 = 0077 \\
R_2 &= 0.0077 \\
X_2^2 &= 00{,}005{,}929 \Longrightarrow X_3 = 0059 \\
R_3 &= 0.0059 \\
&\vdots
\end{aligned}
$$

The leading zeros will appear in every succeeding R_i.

A third problem is that the technique can degenerate, either by going to an X_i value of zero, or an X_i value that is repeated. The next example shows how this can occur.

EXAMPLE 7.3

Suppose that in the process of generating four-digit random numbers, an X_i value of 6500 appeared. Then,

$$X_i = 6500$$
$$X_i^2 = (6500)^2 = 42{,}250{,}000 \Longrightarrow X_{i+1} = 2500$$
$$R_i = 0.2500$$
$$(X_{i+1})^2 = 06{,}250{,}000 \Longrightarrow X_{i+2} = 2500$$
$$R_{i+1} = 0.2500$$

This is a degenerate condition since all subsequent values of X_i will be 2500.

7.3.2. Other Techniques of Historical Interest

Numerous other methods have been developed for generating random numbers. These methods are described here for completeness rather than because of their current utility.

1. *The midproduct technique.* Similar to the midsquare technique, the midproduct technique starts by selecting two seeds, X_0 and X'_0, each containing the same number of digits, D. Now, multiply X_0 by X'_0 to get a number U_1. Set X_1 equal to the middle D digits of U_1. With placement of the decimal, obtain R_1. Next, multiply X_1 by X_0 to obtain U_2. Set X_2 equal to the middle D digits of U_2. Place the decimal to obtain R_2. An example illustrates the method.

Example 7.4

Use the midproduct technique to generate a sequence of four-digit random numbers with $X'_0 = 2938$ and $X_0 = 7229$.

$$U_1 = X'_0 X_0 = (2938)(7229) = 21{,}238{,}802 \Longrightarrow X_1 = 2388$$
$$R_1 = 0.2388$$
$$U_2 = X_0 X_1 = (7229)(2388) = 17{,}262{,}852 \Longrightarrow X_2 = 2628$$
$$R_2 = 0.2628$$
$$U_3 = X_1 X_2 = (2388)(2628) = 6{,}275{,}664 \Longrightarrow X_3 = 2756$$
$$R_3 = 0.2756$$
$$\vdots$$

This technique is very similar to the midsquare technique. However, the period is longer and the results are more uniformly distributed. The technique does have a major shortcoming in that it can eventually degenerate.

2. *Constant multiplier technique.* A slight variation of the midproduct technique is to use a constant multiplier, K. The constant is multiplied by the random number seed X_0. Both the constant and the random number seed have D digits. The result is a value V_1. The middle four digits are taken to become X_1. With placement of the decimal obtain R_1, and so on.

EXAMPLE 7.5

Use the constant multiplier technique with $K = 3987$ and $X_0 = 7223$ to obtain a sequence of four-digit random numbers.

$$
\begin{aligned}
V_1 &= (K)(X_0) \\
V_1 &= (3987)(7223) = 28{,}798{,}101 \Longrightarrow X_1 = 7981 \\
R_1 &= 0.7981 \\
V_2 &= (K)(X_1) \\
V_2 &= (3987)(7981) = 31{,}820{,}247 \Longrightarrow X_2 = 8202 \\
R_2 &= 0.8202 \\
&\vdots
\end{aligned}
$$

This technique has the same problems as the midproduct technique. Also, the success of the method is highly dependent on the selection of the constant.

3. *Additive congruential method.* The behavior of the additive congruential method is not well known, but it employs a somewhat different approach than the previously discussed methods. As will be seen in the example below, the method is quite fast, since no multiplication is required. What is required is a sequence of n numbers $X_1, X_2, \ldots, X_n$. The generator then produces an extension of the sequence, or $X_{n+1}, X_{n+2}, \ldots$. The method by which the values are generated is as follows:

$$X_i = (X_{i-1} + X_{i-n}) \bmod m$$

By definition, $a = b \bmod m$ if $a - b$ is divisible by m with zero remainder. The following example illustrates modular arithmetic and the subsequent example shows how rapidly random integers can be generated using the additive congruential method.

EXAMPLE 7.6

In mod 4 arithmetic, the numbers 2, 6, 10, 14 . . . are all equivalent because $(10 - 2)$, $(10 - 6)$, . . . are all divisible by 4 with remainder zero. When using mod m, the final result should be an integer between 0 and $m - 1$. Thus, $17 \bmod 3 = 2$, $14 \bmod 5 = 4$, $37 \bmod 2 = 1$, $16 \bmod 4 = 0$, and so on.

EXAMPLE 7.7

Let the sequence of integers X_1, X_2, X_3, X_4, X_5 be 57, 34, 89, 92, and 16 (so $n = 5$). Let $m = 100$. This sequence can be extended using the additive congruential method as follows:

$$
\begin{aligned}
X_6 &= (X_5 + X_1) \bmod 100 = 73 \bmod 100 = 73 \\
X_7 &= (X_6 + X_2) \bmod 100 = 107 \bmod 100 = 7 \\
X_8 &= (X_7 + X_3) \bmod 100 = 96 \bmod 100 = 96
\end{aligned}
$$

$$X_9 = (X_8 + X_4) \bmod 100 = 188 \bmod 100 = 88$$
$$X_{10} = (X_9 + X_5) \bmod 100 = 104 \bmod 100 = 4$$
$$X_{11} = (X_{10} + X_6) \bmod 100 = 77 \bmod 100 = 77$$
$$\vdots$$

EXAMPLE 7.8

Random numbers can be generated using the additive congruential method using the relationship

$$R_{i-n} = \frac{X_i}{m} \tag{7.1a}$$

where $i = n + 1, n + 2, \ldots$. Using the results of Example 7.7 yields

$$R_1 = \frac{X_6}{100} = 0.73$$
$$R_2 = \frac{X_7}{100} = 0.07$$
$$\vdots$$

7.3.3. LINEAR CONGRUENTIAL METHOD

The linear congruential method, initially proposed by Lehmer [1951], produces a sequence of integers, $X_1, X_2, \ldots$ between zero and $m - 1$ according to the following recursive relationship:

$$X_{i+1} = (aX_i + c) \bmod m, \qquad i = 0, 1, 2, \ldots \tag{7.2}$$

The initial value X_0 is called the seed, a is called the constant multiplier, c is the increment, and m is the modulus. The selection of the values for a, c, m, and X_0 drastically affects the statistical properties and the cycle length. Variations of equation (7.2) are quite common in the computer generation of random numbers. An example will illustrate how this technique operates.

EXAMPLE 7.9

Use the linear congruential method to generate a sequence of random numbers with $X_0 = 27$, $a = 17$, $c = 43$, and $m = 100$. Here, the integer values generated will all be between zero and 99 because of the value of the modulus. Also, note that random integers are being generated rather than random numbers. These random integers will be uniformly distributed between zero and 99. Random numbers between zero and 1 can be generated using equation (7.1a) slightly modified to become

$$R_i = \frac{X_i}{m}, \qquad i = 1, 2, \ldots \tag{7.1b}$$

The sequence of X_i and subsequent R_i values is computed as follows:

$$X_0 = 27$$
$$X_1 = (17 \cdot 27 + 43) \bmod 100 = 502 \bmod 100 = 2$$
$$R_1 = \frac{2}{100} = 0.02$$
$$X_2 = (17 \cdot 2 + 43) \bmod 100 = 77 \bmod 100 = 77$$
$$R_2 = \frac{77}{100} = 0.77$$
$$X_3 = (17 \cdot 77 + 43) \bmod 100 = 1352 \bmod 100 = 52$$
$$R_3 = \frac{52}{100} = 0.52$$
$$\vdots$$

Recall that $a = b \bmod m$ provided that $(a - b)$ is divisible by m with no remainder. Thus, $X_1 = 502 \bmod 100$, but 502/100 equals 5 with a remainder of 2, so that $X_1 = 2$. In other words, $(502 - 2)$ is evenly divisible by $m = 100$, so $X_1 = 502$ "reduces" to $X_1 = 2 \bmod 100$. (A shortcut for the modulo, or reduction operation for the case $m = 10^b$, a power of 10, is illustrated in Example 7.11.)

The ultimate test of the linear congruential method, as of any generation scheme, is how closely the generated numbers $R_1, R_2, \ldots$ approximate uniformity and independence. There are, however, several secondary properties which must be considered. These include maximum density and maximum period.

First, note that the numbers generated from equation (7.1b) can only assume values from the set $I = \{0, 1/m, 2/m, \ldots, (m-1)/m\}$, since each X_i is an integer in the set $\{0, 1, 2, \ldots, m-1\}$. Thus, each R_i is discrete on I, instead of continuous on the interval [0, 1]. This approximation appears to be of little consequence provided that the modulus m is a very large integer. (Such values as $m = 2^{31} - 1$ and $m = 2^{48}$ are in common use in generators appearing in GPSS, SIMSCRIPT, FORTRAN, and other languages.) By maximum density is meant that the values assumed by R_i, $i = 1, 2, \ldots$, leave no large gaps on [0, 1].

Second, to help achieve maximum density, and to avoid cycling (i.e., recurrence of the same sequence of generated numbers) in practical applications, the generator should have the largest possible period. Maximal period can be achieved by the proper choice of a, c, m, and X_0 [Fishman, 1978]. For m a power of 2, say $m = 2^b$, and $c \neq 0$, the longest possible period is $P = m = 2^b$, which is achieved provided that c is relatively prime to m; that is, the greatest common factor of c and m is 1, and $a = 1 + 4k$, where k is an integer.

If $c \neq 0$ in equation (7.2), the form is called the mixed congruential method. When $c = 0$, the form is known as the multiplicative congruential method.

EXAMPLE 7.10

Using the multiplicative congruential method, find the period of the generator for $a = 13$, $m = 2^6 = 64$, and $X_0 = 1, 2, 3$, and 4. The solution is given in Table 7.1. When the seed is 1 and 3, the sequence has period 16. However, a period of length eight is achieved when the seed is 2 and a period of length four occurs when the seed is 4.

Table 7.1. PERIOD DETERMINATION USING VARIOUS SEEDS

i	X_i	X_i	X_i	X_i
0	1	2	3	4
1	13	26	39	52
2	41	18	59	36
3	21	42	63	20
4	17	34	51	4
5	29	58	23	
6	57	50	43	
7	37	10	47	
8	33	2	35	
9	45		7	
10	9		27	
11	53		31	
12	49		19	
13	61		55	
14	25		11	
15	5		15	
16	1		3	

For m a power of 2, say $m = 2^b$, and $c = 0$, the longest possible period is $P = m/4 = 2^{b-2}$, which is achieved provided that the seed X_0 is odd and the multiplier, a, is given by $a = \pm 3 + 8k$, where k is an integer.

In Example 7.10, $b = 6$ and $m = 2^6 = 64$. The maximal period is $P = m/4 = 16$. Note that this period is achieved using odd seeds $X_0 = 1$ and $X_0 = 3$, but even seeds, $X_0 = 2$ and $X_0 = 4$, yield periods of eight and four, both less than the maximum. Note that $a = 13$ is of the form $-3 + 8k$ with $k = 2$, as required to achieve maximal period.

When $X_0 = 1$, the generated sequence assumes values from the set $\{1, 5, 9, 13, \ldots, 53, 57, 61\}$. The "gaps" in the sequence of generated random numbers, R_i, are quite large (i.e., the gap is $5/64 - 1/64$ or 0.0625). Such a gap gives rise to concern about the density of the generated sequence.

The generator in Example 7.10 is not viable for any application—its period is too short and its density is insufficiently low. However, the example shows the importance of properly choosing a, c, m, and X_0.

The speed and efficiency in using the generator on a digital computer is also a selection consideration. Speed and efficiency are aided by use of a modulus,

m, which is either a power of 2 or close to a power of 2. Since most digital computers use a binary representation of numbers, the modulo, or remaindering, operation of equation (7.2) can be conducted efficiently when the modulo is a power of 2 (i.e., $m = 2^b$). After ordinary arithmetic yields a value for $aX_i + c$, X_{i+1} is obtained by dropping the leftmost binary digits in $aX_i + c$ and then using only the b rightmost binary digits. The following example illustrates, by analogy, this operation using $m = 10^b$, because most human beings think in decimal representation.

EXAMPLE 7.11

Let $m = 10^2 = 100$, $a = 19$, $c = 0$, and $X_0 = 63$, and generate a sequence of random integers using equation (7.1).

$$\begin{aligned}
X_0 &= 63 \\
X_1 &= (19)(63) \bmod 100 = 1197 \bmod 100 = 97 \\
X_2 &= (19)(97) \bmod 100 = 1843 \bmod 100 = 43 \\
X_3 &= (19)(43) \bmod 100 = 817 \bmod 100 = 17 \\
&\ \vdots
\end{aligned}$$

When m is a power of 10, say $m = 10^b$, the modulo operation is quite simple. Merely save the b rightmost (decimal) digits. By analogy, the modulo operation is most efficient for binary computers when $m = 2^b$ for some $b > 0$.

EXAMPLE 7.12

The last example in this section is in actual use. It has been extensively tested [Learmonth and Lewis, 1973; Lewis et al., 1969]. The values for a, c, and m have been preselected to ensure that the characteristics desired in a generator are most likely to be achieved. Although a, c, and m are fixed, the user can either specify the seed, X_0, or use the seed that is resident in the routine. By changing X_0, the user can control the repeatability of the stream. The generator in this example appears in the IMSL Scientific Subroutine Package [1978], written in FORTRAN originally for IBM 360/370 computers.

Let $a = 7^5 = 16{,}807$, $m = 2^{31} - 1 = 2{,}147{,}483{,}647$ (a prime number), and $c = 0$. Further, specify a seed, $X_0 = 123{,}457$. Since m is a prime number, the period, P, equals $m - 1$ (well over 2 billion). The first few numbers generated are as follows:

$$\begin{aligned}
X_1 &= 7^5(123{,}457) \bmod (2^{31} - 1) = 2{,}074{,}941{,}799 \bmod (2^{31} - 1) \\
X_1 &= 2{,}074{,}941{,}799 \\
R_1 &= \frac{X_1}{2^{31}} = 0.9662 \\
X_2 &= 7^5(2{,}074{,}941{,}799) \bmod (2^{31} - 1) = 559{,}872{,}160 \\
R_2 &= \frac{X_2}{2^{31}} = 0.2607
\end{aligned}$$

$$X_3 = 7^5(559{,}872{,}160) \bmod (2^{31} - 1) = 1{,}645{,}535{,}613$$

$$R_3 = \frac{X_3}{2^{31}} = 0.7662$$

$$\vdots$$

Note that this IMSL routine divides by $m + 1$ instead of m; however, for such a large value of m, the effect is negligible.

7.4. Tests for Random Numbers

The properties of random numbers—uniformity and independence—were discussed in Section 7.1. To determine if a random number generator is providing numbers that possess the desired properties, certain tests should be performed. The tests can be placed in two categories according to the properties of interest. The first entry in the list below concerns testing for uniformity. The second through fifth entries concern testing for independence. A brief description of the five types of tests discussed in this chapter is as follows:

1. *Frequency test.* Uses the Kolmogorov–Smirnov or the chi-square test to compare the distribution of the set of numbers generated to a uniform distribution.
2. *Runs test.* Tests the runs up and down or the runs above and below the mean by comparing the actual values to expected values. The statistic for comparison is the chi-square.
3. *Autocorrelation test.* Tests the correlation between numbers and compares the sample correlation to the expected correlation of zero.
4. *Gap test.* Counts the number of digits that appear between repetitions of a particular digit and then uses the Kolmogorov–Smirnov test to compare with the expected number of gaps.
5. *Poker test.* Treats numbers grouped together as a poker hand. Then the hands obtained are compared to what is expected using the chi-square test.

In testing for uniformity, the hypotheses are as follows:

$$H_0: \quad R_i \sim U[0, 1]$$
$$H_1: \quad R_i \not\sim U[0, 1]$$

The null hypothesis, H_0, reads that the numbers are distributed uniformly on the interval [0, 1]. Failure to reject the null hypothesis means that no evidence of nonuniformity has been detected on the basis of this test. This does not imply that further testing of the generator for uniformity is unnecessary.

In testing for independence, the hypotheses are as follows;

$$H_0: \quad R_i \sim \text{independently}$$
$$H_1: \quad R_i \not\sim \text{independently}$$

This null hypothesis, H_0, reads that the numbers are independent. Failure to reject the null hypothesis means that no evidence of dependence has been detected on the basis of this test. This does not imply that further testing of the generator for independence is unnecessary.

For each test, a level of significance α must be stated. Now, α is the probability of rejecting the null hypothesis given that the null hypothesis is true, or

$$\alpha = P(\text{reject } H_0 \mid H_0 \text{ true})$$

The decision maker sets the value of α for any test. Frequently, α is set to 0.01 or 0.05.

If several tests are conducted on the same set of numbers, the probability of rejecting the null hypothesis on at least one test, by chance alone [i.e., making a Type I (α) error], increases. Say that $\alpha = 0.05$ and that five different tests are conducted on a sequence of numbers. The probability of rejecting the null hypothesis on at least one test, by chance alone, may be as large as 0.25.

Similarly, if one test is conducted on many sets of numbers from a generator, the probability of rejecting the null hypothesis on at least one test by chance alone [i.e., making a Type I (α) error], increases as more sets of numbers are tested. Additionally, if 100 sets of numbers were subjected to the test, with $\alpha = 0.05$, it would be expected that five of those tests would be rejected by chance alone. If the number of rejections in 100 tests is close to 100α, then there is no compelling reason to discard the generator. The concept discussed in this and the preceding paragraph is discussed further at the conclusion of Example 7.19.

If one of the well-known simulation languages or random number generators is used, it is probably unnecessary to use the tests mentioned above and described in Sections 7.4.1 through 7.4.5. (However, a generator such as RANDU, distributed by IBM in the late 1960s and still available on some computers, has been found unreliable due to autocorrelation among triplets of random numbers.) If a new method has been developed, it is advisable to try these tests and others on many samples of numbers from the generator. Some additional tests commonly used are Good's serial test for sampling numbers [1953, 1967], the median-spectrum test [Cox and Lewis, 1966; Durbin, 1967], and a variance heterogeneity test [Cox and Lewis, 1966]. Even if a set of numbers passes all the tests, it is no guarantee of randomness. It is always possible that some underlying pattern will go undetected.

In the examples of tests that follow, the hypotheses are not restated. The hypotheses are as indicated in the paragraphs above.

7.4.1. Frequency Tests

A basic test that should always be performed to validate a new generator is the test of uniformity. Two different methods of testing are available. They are the Kolmogorov–Smirnov and the chi-square test. Both of these tests measure the degree of agreement between the distribution of a sample of generated random numbers and the theoretical uniform distribution. Both tests are based on the null hypothesis of no significant difference between the sample distribution and the theoretical distribution.

1. *The Kolmogorov–Smirnov test.* This test compares the continuous cdf, $F(x)$, of the uniform distribution to the empirical cdf, $S_N(x)$, of the sample of N observations. By definition,

$$F(x) = x, \qquad 0 \leq x \leq 1$$

If the sample from the random number generator is $R_1, R_2, \ldots, R_N$, the empirical cdf, $S_N(x)$, is defined by

$$S_N(x) = \frac{\text{number of } R_1, R_2, \ldots, R_N \text{ which are} \leq x}{N}$$

As N becomes larger, $S_N(x)$ should become a better approximation to $F(x)$, provided that the null hypothesis is true.

In Section 4.6, empirical distributions were described. The cdf of an empirical distribution is a step function with jumps at each observed value. This behavior was illustrated by Example 4.34.

The Kolmogorov–Smirnov test is based on the largest absolute deviation between $F(x)$ and $S_N(x)$ over the range of the random variable. That is, it is based on the statistic

$$D = \max |F(x) - S_N(x)| \tag{7.3}$$

The sampling distribution of D is known and is tabulated as a function of N in Table A.7. For testing against a uniform cdf, the test procedure follows these steps:

Step 1. Rank the data from smallest to largest. Let $R_{(i)}$ denote the ith smallest observation, so that

$$R_{(1)} \leq R_{(2)} \leq \cdots \leq R_{(N)}$$

Step 2. Compute

$$D^+ = \max_{1 \leq i \leq N} \left\{ \frac{i}{N} - R_{(i)} \right\}$$

$$D^- = \max_{1 \leq i \leq N} \left\{ R_{(i)} - \frac{i-1}{N} \right\}$$

Step 3. Compute $D = \max(D^+, D^-)$.

Step 4. Determine the critical value, D_α, from Table A.7 for the specified significance level α and the given sample size N.

Step 5. If the sample statistic D is greater than the critical value, D_α, the null hypothesis, that the data are a sample from a uniform distribution, is rejected. If $D \leq D_\alpha$, conclude that no difference has been detected between the true distribution of $\{R_1, R_2, \ldots, R_N\}$ and the uniform distribution.

EXAMPLE 7.13

Suppose that the five numbers 0.44, 0.81, 0.14, 0.05, 0.93 were generated and it is desired to perform a test for uniformity using the Kolmogorov–Smirnov test with a level of significance α of 0.05. First, the numbers must be ranked from smallest to largest. The calculations can be facilitated by use of Table 7.2. The top row lists the

Table 7.2. CALCULATIONS FOR KOLMOGOROV–SMIRNOV TEST

$R_{(i)}$	0.05	0.14	0.44	0.81	0.93
i/N	0.20	0.40	0.60	0.80	1.00
$i/N - R_{(i)}$	0.15	0.26	0.16	—	0.07
$R_{(i)} - (i-1)/N$	0.05	—	0.04	0.21	0.13

numbers from smallest ($R_{(1)}$) to largest ($R_{(5)}$). The computations for D^+, namely $i/N - R_{(i)}$, and for D^-, namely $R_{(i)} - (i-1)/N$, are easily accomplished using Table 7.2. The statistics are computed as $D^+ = 0.26$ and $D^- = 0.21$. Therefore, $D = \max\{0.26, 0.21\} = 0.26$. The critical value of D, obtained from Table A.7 for $\alpha = 0.05$ and $N = 5$, is 0.565. Since the computed value, 0.26, is less than the tabulated critical value, 0.565, the hypothesis of no difference between the distribution of the generated numbers and the uniform distribution is not rejected.

The calculations in Table 7.2 are illustrated in Figure 7.2, where the empirical cdf, $S_N(x)$, is compared to the uniform cdf, $F(x)$. It can be seen that D^+ is the largest deviation of $S_N(x)$ above $F(x)$, and that D^- is the largest deviation of $S_N(x)$ below $F(x)$. For example, at $R_{(3)}$ the value of D^+ is given by $3/5 - R_{(3)} = 0.60 - 0.44 = 0.16$ and D^- is given by $R_{(3)} - 2/5 = 0.44 - 0.40 = 0.04$. Although the test statistic D is defined

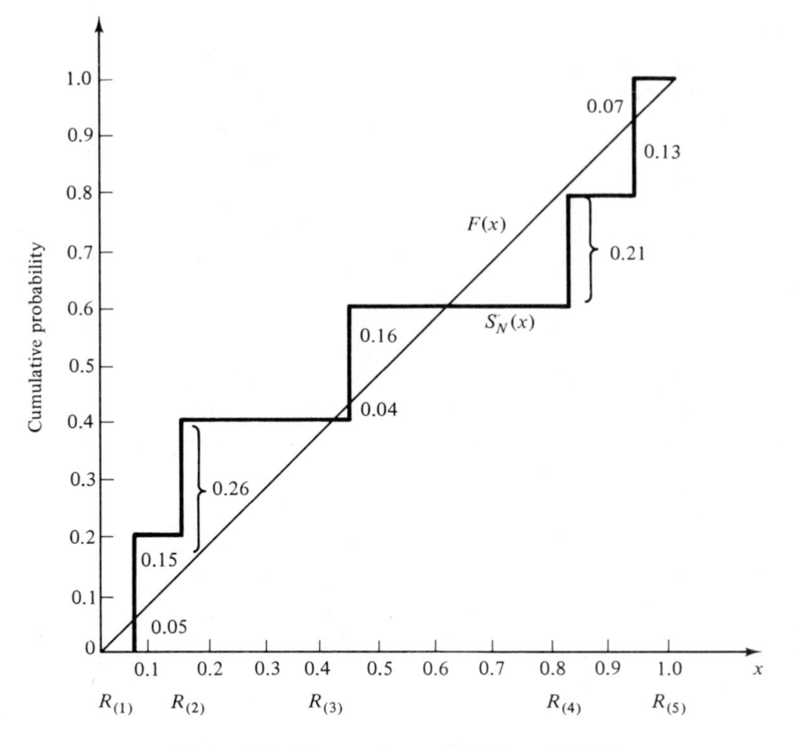

Figure 7.2. Comparison of $F(x)$ and $S_N(x)$.

by equation (7.3) as the maximum deviation over all x, it can be seen from Figure 7.2 that the maximum deviation will always occur at one of the jump points $R_{(1)}, R_{(2)}, \ldots$ and thus the deviation at other values of x need not be considered.

2. *The chi-square test.* The chi-square test uses the sample statistic

$$\chi_0^2 = \sum_{i=1}^{n} \frac{(O_i - E_i)^2}{E_i}$$

where O_i is the observed number in the ith class, E_i is the expected number in the ith class, and n is the number of classes. For the uniform distribution, E_i, the expected number in each class, is given by

$$E_i = \frac{N}{n}$$

for equally spaced classes, where N is the total number of observations. It can be shown that the sampling distribution of χ_0^2 is approximately the chi-square distribution with $n - 1$ degrees of freedom.

EXAMPLE 7.14

Use the chi-square test with $\alpha = 0.05$ to test whether the data shown below are uniformly distributed. Table 7.3 contains the essential computations. The test uses $n = 10$ intervals of equal length, namely [0, 0.1), [0.1, 0.2), . . . , [0.9, 1.0).

0.34	0.90	0.25	0.89	0.87	0.44	0.12	0.21	0.46	0.67
0.83	0.76	0.79	0.64	0.70	0.81	0.94	0.74	0.22	0.74
0.96	0.99	0.77	0.67	0.56	0.41	0.52	0.73	0.99	0.02
0.47	0.30	0.17	0.82	0.56	0.05	0.45	0.31	0.78	0.05
0.79	0.71	0.23	0.19	0.82	0.93	0.65	0.37	0.39	0.42
0.99	0.17	0.99	0.46	0.05	0.66	0.10	0.42	0.18	0.49
0.37	0.51	0.54	0.01	0.81	0.28	0.69	0.34	0.75	0.49
0.72	0.43	0.56	0.97	0.30	0.94	0.96	0.58	0.73	0.05
0.06	0.39	0.84	0.24	0.40	0.64	0.40	0.19	0.79	0.62
0.18	0.26	0.97	0.88	0.64	0.47	0.60	0.11	0.29	0.78

Table 7.3. COMPUTATIONS FOR CHI-SQUARE TEST

Interval	O_i	E_i	$O_i - E_i$	$(O_i - E_i)^2$	$\frac{(O_i - E_i)^2}{E_i}$
1	8	10	−2	4	0.4
2	8	10	−2	4	0.4
3	10	10	0	0	0
4	9	10	−1	1	0.1
5	12	10	2	4	0.4
6	8	10	−2	4	0.4
7	10	10	0	0	0
8	14	10	4	16	1.6
9	10	10	0	0	0
10	11	10	1	1	0.1
	100	100	0		3.4

The value of χ_0^2 is 3.4. This is compared with the critical value $\chi_{0.05,9}^2 = 16.9$. Since χ_0^2 is much smaller than the tabulated value of $\chi_{0.05,9}^2$, the null hypothesis of no difference between the sample distribution and the uniform distribution is not rejected.

Different authors have offered considerations concerning the application of the χ^2 test. In the application to a data set the size of that in Example 7.14, the considerations do not apply. That is, if 100 values are in the sample and from 5 to 10 intervals of equal length are used, the test will be acceptable. In general, it is recommended that n and N be chosen so that each $E_i \geq 5$.

Both the Kolmogorov–Smirnov and the chi-square test are acceptable for testing the uniformity of a sample of data, provided that the sample size is large. However, the Kolmogorov–Smirnov test is the more powerful of the two and is recommended. Furthermore, the Kolmogorov–Smirnov test can be applied to small sample sizes, whereas the chi-square is only valid for large samples, say $N \geq 50$.

Imagine a set of 100 numbers which are being tested for independence where the first 10 values are in the range 0.01–0.10, the second 10 values are in the range 0.11–0.20, and so on. This set of numbers would pass the frequency tests with ease, but the ordering of the numbers produced by the generator would not be random. The tests in the remainder of this chapter are concerned with the independence of random numbers which are generated. The presentation of the tests is similar to that by Schmidt and Taylor [1970].

7.4.2. Runs Tests

1. *Runs up and runs down.* Consider a generator that provided a set of 40 numbers which were in the following sequence:

0.08	0.09	0.23	0.29	0.42	0.55	0.58	0.72	0.89	0.91
0.11	0.16	0.18	0.31	0.41	0.53	0.71	0.73	0.74	0.84
0.02	0.09	0.30	0.32	0.45	0.47	0.69	0.74	0.91	0.95
0.12	0.13	0.29	0.36	0.38	0.54	0.68	0.86	0.88	0.91

Both the Kolmogorov–Smirnov test and the chi-square test would indicate that the numbers are uniformly distributed. However, a glance at the ordering shows that the numbers are successively larger in blocks of 10 values. If these numbers are rearranged as follows, there is far less reason to doubt their independence:

0.41	0.68	0.89	0.84	0.74	0.91	0.55	0.71	0.36	0.30
0.09	0.72	0.86	0.08	0.54	0.02	0.11	0.29	0.16	0.18
0.88	0.91	0.95	0.69	0.09	0.38	0.23	0.32	0.91	0.53
0.31	0.42	0.73	0.12	0.74	0.45	0.13	0.47	0.58	0.29

The runs test examines the arrangement of numbers in a sequence to test the hypothesis of independence.

Before defining a run, a look at a sequence of coin tosses will help with some terminology. Consider the following sequence generated by tossing a coin 10 times:

H T T H H T T T H T

There are three mutually exclusive outcomes, or events, with respect to the sequence. Two of the possibilities are rather obvious. That is, the toss can result in a head or a tail. The third possibility is "no event." The first head is preceded by no event and the last tail is succeeded by no event. Every sequence begins and ends with no event.

A run is defined as a succession of similar events preceded and followed by a different event. The length of the run is the number of events that occur in the run. In the coin-flipping example above there are six runs. The first run is of length one, the second and third of length two, the fourth of length three, and the fifth and sixth of length one.

There are two possible concerns in a runs test for a sequence of numbers. The number of runs is the first concern and the length of runs is a second concern. The types of runs counted in the first case might be runs up and runs down. An up run is a sequence of numbers each of which is succeeded by a larger number. Similarly, a down run is a sequence of numbers each of which is succeeded by a smaller number. To illustrate the concept, consider the following sequence of 15 numbers:

$$\begin{array}{ccccccccc} 0.87 & ^{-}0.15 & ^{+}0.23 & ^{+}0.45 & ^{+}0.69 & ^{-}0.32 & ^{-}0.30 & ^{-}0.19 & ^{+}0.24 \\ ^{-}0.18 & ^{+}0.65 & ^{+}0.82 & ^{+}0.93 & ^{-}0.22 & ^{+}0.81 & & & \end{array}$$

The numbers are given a "+" or a "−" depending on whether they are followed by a larger number or a smaller number. Since there are 15 numbers, and they are all different, there will be 14 +'s and −'s. The last number is followed by "no event" and hence will get neither a + nor a −. The sequence of 14 +'s and −'s is as follows:

$$- \quad + \quad + \quad + \quad - \quad - \quad - \quad + \quad - \quad + \quad + \quad + \quad - \quad +$$

Each succession of +'s and −'s form a run. There are eight runs. The first run is of length one, the second and third are of length three, and so on. Further, there are four runs up and four runs down.

There can be too few runs or too many runs. Consider the following sequence of numbers:

0.08 0.18 0.23 0.36 0.42 0.55 0.63 0.72 0.89 0.91

This sequence has one run, a run up. It is unlikely that a valid random number generator would produce such a sequence. Next, consider the following sequence:

0.08 0.93 0.15 0.96 0.26 0.84 0.28 0.79 0.36 0.57

This sequence has nine runs, five up and four down. It is unlikely that a sequence of 10 numbers will have this many runs. What is more likely is that the number of runs will be somewhere between the two extremes. These two extremes can be formalized as follows: If N is the number of numbers in a sequence, the maximum number of runs is $N - 1$ and the minimum number of runs is one.

If a is the total number of runs in a sequence, the mean and variance of a is given by

$$\mu_a = \frac{2N - 1}{3} \tag{7.4}$$

and

$$\sigma_a^2 = \frac{16N - 29}{90} \tag{7.5}$$

For $N > 20$, the distribution of a is reasonably approximated by a normal distribution, $N(\mu_a, \sigma_a^2)$. This approximation could be used to test the independence of numbers from a generator. In that case the standardized normal test statistic is developed by subtracting the mean from the observed number of runs, a, and dividing by the standard deviation. That is, the test statistic is

$$Z_0 = \frac{a - \mu_a}{\sigma_a}$$

Substituting equation (7.4) for μ_a and the square root of equation (7.5) for σ_a yields

$$Z_0 = \frac{a - [(2N - 1)/3]}{\sqrt{(16N - 29)/90}}$$

where $Z_0 \sim N(0, 1)$. Failure to reject the hypothesis of independence occurs when $-z_{\alpha/2} \leq Z_0 \leq z_{\alpha/2}$, where α is the level of significance. The critical values and rejection region are shown in Figure 7.3.

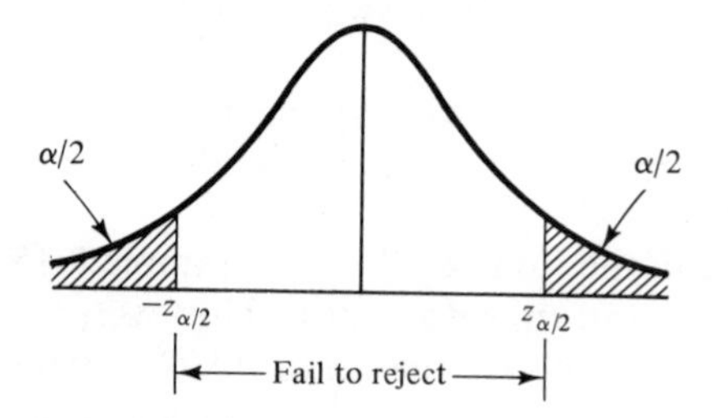

Figure 7.3. Failure to reject hypothesis.

EXAMPLE 7.15

Based on runs up and runs down, determine whether the following sequence of 40 numbers is such that the hypothesis of independence can be rejected where $\alpha = 0.05$.

0.41	0.68	0.89	0.94	0.74	0.91	0.55	0.62	0.36	0.27
0.19	0.72	0.75	0.08	0.54	0.02	0.01	0.36	0.16	0.28
0.18	0.01	0.95	0.69	0.18	0.47	0.23	0.32	0.82	0.53
0.31	0.42	0.73	0.04	0.83	0.45	0.13	0.57	0.63	0.29

The sequence of runs up and down is as follows:

+ + + − + − + − − − + + − + − − + − + − − + − − + − + +
− − + + − + − − + + −

There are 26 runs in this sequence. With $N = 40$ and $a = 26$, equations (7.4) and (7.5) yield

$$\mu_a = \frac{2(40) - 1}{3} = 26.33$$

and

$$\sigma_a^2 = \frac{16(40) - 29}{90} = 6.79$$

Then,

$$Z_0 = \frac{26 - 26.33}{\sqrt{6.79}} = -0.13$$

Now, the critical value is $z_{0.025} = 1.96$, so the independence of the numbers cannot be rejected on the basis of this test.

2. *Runs above and below the mean.* The test for runs up and runs down is not completely adequate to assess the independence of a group of numbers. Consider the following 40 numbers:

0.63	0.72	0.79	0.81	0.52	0.94	0.83	0.93	0.87	0.67
0.54	0.83	0.89	0.55	0.88	0.77	0.74	0.95	0.82	0.86
0.43	0.32	0.36	0.18	0.08	0.19	0.18	0.27	0.36	0.34
0.31	0.45	0.49	0.43	0.46	0.35	0.25	0.39	0.47	0.41

The sequence of runs up and runs down is as follows:

$$+ + + - + - + - - - + + - + - - + - + - - + - -$$
$$+ - + + - - + + - + - - + + -$$

This sequence is exactly the same as that in Example 7.15. Thus, the numbers would pass the runs up and runs down test. However, it can be observed that the first 20 numbers are all above the mean $[(0.99 + 0.00)/2 = 0.495]$ and the last 20 numbers are below the mean. Such an occurrence is highly unlikely. The previous runs analysis can be used to test for this condition, if the definition of a run is changed. Runs will be described as being above the mean or below the mean. A "+" sign will be used to denote an observation above the mean, and a "−" sign will denote an observation below the mean.

For example, consider the following sequence of 20 two-digit random numbers:

0.40	0.84	0.75	0.18	0.13	0.92	0.57	0.77	0.30	0.71
0.42	0.05	0.78	0.74	0.68	0.03	0.18	0.51	0.10	0.37

The pluses and minuses are as follows:

$$- + + - - + + + - + - - + + + - - + - -$$

In this case, there is a run of length one below the mean followed by a run of length two above the mean, and so on. In all, there are 11 runs, five of which are above the mean, and six of which are below the mean. Let n_1 and n_2 be the number of individual observations above and below the mean and let b be the total number of runs. Note that the maximum number of runs is $N = n_1 + n_2$, and the minimum number of runs is one. Given n_1 and n_2, the mean, with a continuity correction suggested by Swed and Eisenhart [1943], and the variance of b are given by

$$\mu_b = \frac{2n_1n_2}{N} + \frac{1}{2} \tag{7.6}$$

and

$$\sigma_b^2 = \frac{2n_1n_2(2n_1n_2 - N)}{N^2(N-1)} \tag{7.7}$$

For either n_1 or n_2 greater than 20, b is approximately normally distributed. The test statistic can be formed by subtracting the mean from the number of runs and dividing by the standard deviation, or

$$Z_0 = \frac{b - (2n_1n_2/N) - 1/2}{\left[\dfrac{2n_1n_2(2n_1n_2 - N)}{N^2(N-1)}\right]^{1/2}}$$

Failure to reject the hypothesis of independence occurs when $-z_{\alpha/2} \leq Z_0 \leq z_{\alpha/2}$, where α is the level of significance. The rejection region is shown in Figure 7.3.

EXAMPLE 7.16

Determine whether there is an excessive number of runs above or below the mean for the sequence of numbers given in Example 7.15. The assignment of +'s and −'s results in the following:

− + + + + + + + − − − + + − + − − − − −
− − + + − − − − − + + − − + − + − − + + −

The values of n_1, n_2, and b are as follows:

$$n_1 = 18$$
$$n_2 = 22$$
$$N = n_1 + n_2 = 40$$
$$b = 17$$

Equations (7.6) and (7.7) are used to determine μ_b and σ_b^2 as follows:

$$\mu_b = \frac{2(18)(22)}{40} + \frac{1}{2} = 20.3$$

and

$$\sigma_b^2 = \frac{2(18)(22)[(2)(18)(22) - 40]}{(40)^2(40-1)} = 9.54$$

Since n_2 is greater than 20, the normal approximation is acceptable, resulting in a Z_0 value of

$$Z_0 = \frac{17 - 20.3}{\sqrt{9.54}} = -1.07$$

Since $z_{0.025} = 1.96$, the hypothesis of independence cannot be rejected on the basis of this test.

3. *Runs test: length of runs.* Yet another concern is the length of runs. As an example of what might occur, consider the following sequence of numbers:

$$0.16, 0.27, 0.58, 0.63, 0.45, 0.21, 0.72, 0.87, 0.27, 0.15, 0.92, 0.85 \ldots$$

Assume that this sequence continues in a like fashion: two numbers below the mean followed by two numbers above the mean. A test of runs above and below the mean would detect no departure from independence. However, it is to be expected that runs other than of length two should occur.

Let Y_i be the number of runs of length i in a sequence of N numbers. For an independent sequence, the expected value of Y_i for runs up and down is given by

$$E(Y_i) = \frac{2}{(i+3)!}[N(i^2 + 3i + 1) - (i^3 + 3i^2 - i - 4)], \qquad i \leq N - 2 \tag{7.8}$$

$$E(Y_i) = \frac{2}{N!}, \qquad i = N - 1 \tag{7.9}$$

For runs above and below the mean, the expected value of Y_i is approximately given by

$$E(Y_i) = \frac{Nw_i}{E(I)}, \qquad N > 20 \tag{7.10}$$

where w_i, the approximate probability that a run has length i, is given by

$$w_i = \left(\frac{n_1}{N}\right)^i \left(\frac{n_2}{N}\right) + \left(\frac{n_1}{N}\right)\left(\frac{n_2}{N}\right)^i, \qquad N > 20 \tag{7.11}$$

and where $E(I)$, the approximate expected length of a run, is given by

$$E(I) = \frac{n_1}{n_2} + \frac{n_2}{n_1}, \qquad N > 20 \tag{7.12}$$

The approximate expected total number of runs (of all lengths) in a sequence of length N, $E(A)$, is given by

$$E(A) = \frac{N}{E(I)}, \qquad N > 20 \tag{7.13}$$

The appropriate test is the chi-square test with O_i being the observed number of runs of length i. Then the test statistic is

$$\chi_0^2 = \sum_{i=1}^{L} \frac{[O_i - E(Y_i)]^2}{E(Y_i)}$$

where $L = N - 1$ for runs up and down and $L = N$ for runs above and below the mean. If the null hypothesis of independence is true, then χ_0^2 is approximately chi-square distributed with $L - 1$ degrees of freedom.

EXAMPLE 7.17

Given the following sequence of numbers, can the hypothesis that the numbers are independent be rejected on the basis of the length of runs up and down ($\alpha = 0.05$)?

0.30	0.48	0.36	0.01	0.54	0.34	0.96	0.06	0.61	0.85
0.48	0.86	0.14	0.86	0.89	0.37	0.49	0.60	0.04	0.83
0.42	0.83	0.37	0.21	0.90	0.89	0.91	0.79	0.57	0.99
0.95	0.27	0.41	0.81	0.96	0.31	0.09	0.06	0.23	0.77
0.73	0.47	0.13	0.55	0.11	0.75	0.36	0.25	0.23	0.72
0.60	0.84	0.70	0.30	0.26	0.38	0.05	0.19	0.73	0.44

For this sequence the +'s and −'s are as follows:

+ − − + − + − + + − + − + + − + + − +
− + − − + − + − − + − − + + + − − − + +
− − − + − + − − − + − + − − − + − + + −

The length of runs in the sequence is as follows:

1, 2, 1, 1, 1, 1, 2, 1, 1, 1, 2, 1, 2, 1, 1, 1, 1, 2, 1, 1,
1, 2, 1, 2, 3, 3, 2, 3, 1, 1, 1, 3, 1, 1, 1, 3, 1, 1, 3, 1

The number of observed runs of each length is as follows:

Run Length, i	1	2	3
Observed Runs, O_i	26	9	5

The expected number of runs of lengths one, two, and three are computed from equation (7.8) as

$$E(Y_1) = \frac{2}{4!}[60(1 + 3 + 1) - (1 + 3 - 1 - 4)]$$
$$= 25.08$$
$$E(Y_2) = \frac{2}{5!}[60(4 + 6 + 1) - (8 + 12 - 2 - 4)]$$
$$= 10.77$$
$$E(Y_3) = \frac{2}{6!}[60(9 + 9 + 1) - (27 + 27 - 3 - 4)]$$
$$= 3.04$$

The mean total number of runs (up and down) is given by equation (7.4) as

$$\mu_a = \frac{2(60) - 1}{3} = 39.67$$

Thus far, the $E(Y_i)$ for $i = 1, 2$, and 3 total 38.89. The expected number of runs of length 4 or more is the difference $\mu_a - \sum_{i=1}^{3} Y_i$, or 0.78.

As observed by Hines and Montgomery [1980], there is no general agreement regarding the minimum value of expected frequencies in applying the chi-square test. Values of 3, 4, and 5 are widely used and a minimum of five was suggested earlier in this chapter. Should an expected frequency be too small, it can be combined with the expected frequency in an adjacent class interval. The corresponding observed frequencies would then be combined also, and L would be reduced by one. With the foregoing calculations and procedures in mind, Table 7.4 is constructed. The critical value

Table 7.4. LENGTH OF RUNS UP AND DOWN: χ^2 TEST

Run Length, i	*Observed Number of Runs,* O_i	*Expected Number of Runs,* $E(Y_i)$	$\frac{[O_i - E(Y_i)]^2}{E(Y_i)}$
1	26	25.08	0.03
2	9 } 14	10.77 } 14.59	} 0.02
≥3	5	3.82	
	40	39.67	0.05

$\chi^2_{0.05,1}$ is 3.84. (The degrees of freedom equals the number of class intervals minus one.) Since $\chi^2_0 = 0.05$ is less than the critical value, the hypothesis of independence cannot be rejected on the basis of this test.

EXAMPLE 7.18

Given the same sequence of numbers in Example 7.17, can the hypothesis that the numbers are independent be rejected on the basis of the length of runs above and below the mean ($\alpha = 0.05$)? For this sequence, the +'s and −'s are as follows:

$$\begin{array}{l} - - - - + - + - + + - + - + + - - + - + \\ - + - - + + + + + + + - - + + - - - - + \\ + - - + - + - - - + + + + - - - - - + - \end{array}$$

The number of runs of each length is as follows:

Run Length, i	1	2	3	≥4
Observed Runs, O_i	17	9	1	5

There are 28 values above the mean ($n_1 = 28$) and 32 values below the mean ($n_2 = 32$). The probabilities of runs of various lengths, w_i, are determined from equation (7.11) as

$$w_1 = \left(\frac{28}{60}\right)^1 \frac{32}{60} + \frac{28}{60}\left(\frac{32}{60}\right)^1 = 0.498$$

$$w_2 = \left(\frac{28}{60}\right)^2 \frac{32}{60} + \frac{28}{60}\left(\frac{32}{60}\right)^2 = 0.249$$

$$w_3 = \left(\frac{28}{60}\right)^3 \frac{32}{60} + \frac{28}{60}\left(\frac{32}{60}\right)^3 = 0.125$$

$$\vdots$$

The expected length of a run, $E(I)$, is determined from equation (7.12) as

$$E(I) = \frac{28}{32} + \frac{32}{28} = 2.02$$

Now, equation (7.10) can be used to determine the expected number of runs of various lengths as

$$E(Y_1) = \frac{60(0.498)}{2.02} = 14.79$$

$$E(Y_2) = \frac{60(0.249)}{2.02} = 7.40$$

$$E(Y_3) = \frac{60(0.125)}{2.02} = 3.71$$

The total number of runs expected is given by equation (7.13) as $E(A) = 60/2.02 = 29.7$. This indicates that approximately 3.8 runs of length four or more can be expected. Proceeding by combining adjacent cells in which $E(Y_i) < 5$ produces Table 7.5.

Table 7.5. LENGTH OF RUNS ABOVE AND BELOW THE MEAN: χ^2 TEST

Run Length, i	*Observed Number of Runs,* O_i	*Expected Number of Runs,* $E(Y_i)$	$\frac{[O_i - E(Y_i)]^2}{E(Y_i)}$
1	17	14.79	0.33
2	9	7.40	0.35
3	1 } 6	3.71 } 7.51	} 0.30
≥ 4	5	3.80	
	32	29.70	0.98

The critical value $\chi^2_{0.05,2}$ is 5.99. (The degrees of freedom equals the number of class intervals minus one.) Since $\chi^2_0 = 0.98$ is less than the critical value, the hypothesis of independence cannot be rejected on the basis of this test.

7.4.3. Tests for Autocorrelation

The tests for autocorrelation are concerned with the dependence between numbers in a sequence. As an example, consider the following sequence of numbers:

0.12	0.01	0.23	0.28	0.89	0.31	0.64	0.28	0.83	0.93
0.99	0.15	0.33	0.35	0.91	0.41	0.60	0.27	0.75	0.88
0.68	0.49	0.05	0.43	0.95	0.58	0.19	0.36	0.69	0.87

From a visual inspection, these numbers appear random, and they would probably pass all the tests presented to this point. However, an examination of the 5th, 10th, 15th (every five numbers beginning with the fifth), and so on, indicates a very large number in that position. Now, 30 numbers is a rather small sample size to reject a random number generator, but the notion is that numbers in the sequence might be related. In this particular section, a method for determining whether such a relationship exists is described. The relationship would not have to be all high numbers. It is possible to have all low numbers in the locations being examined, or the numbers may alternately shift from very high to very low.

The test to be described below requires the computation of the autocorrelation between every m numbers (m is also known as the lag) starting with the ith number. Thus, the autocorrelation ρ_{im} between the following numbers would be of interest: $R_i, R_{i+m}, R_{i+2m}, \ldots, R_{i+(M+1)m}$. The value M is the largest integer such that $i + (M + 1)m \leq N$, where N is the total number of values in the sequence. (Thus, a subsequence of length $M + 2$ is being tested.)

Since a nonzero autocorrelation implies a lack of independence, the following two-tailed test is appropriate:

$$H_0: \quad \rho_{im} = 0$$

$$H_1: \quad \rho_{im} \neq 0$$

For large values of M, the distribution of the estimator of ρ_{im}, or $\hat{\rho}_{im}$, is approximately normal if the values $R_i, R_{i+m}, R_{i+2m}, \ldots, R_{i+(M+1)m}$ are uncorrelated. Then the test statistic can be formed as follows:

$$Z_0 = \frac{\hat{\rho}_{im}}{\sigma_{\hat{\rho}_{im}}}$$

which is distributed normally with a mean of zero and a variance of 1, under the assumption of independence, for large M.

The formula for $\hat{\rho}_{im}$, in a slightly different form, and the standard deviation of the estimator, $\sigma_{\hat{\rho}_{im}}$, are given by Schmidt and Taylor [1970] as follows:

$$\hat{\rho}_{im} = \frac{1}{M+1}\left[\sum_{k=0}^{M} R_{i+km}R_{i+(k+1)m}\right] - 0.25$$

and

$$\sigma_{\hat{\rho}_{im}} = \frac{\sqrt{13M + 7}}{12(M + 1)}$$

After computing Z_0, do not reject the null hypothesis of independence if $-z_{\alpha/2} \leq Z_0 \leq z_{\alpha/2}$, where α is the level of significance. Figure 7.3, presented earlier, illustrates this test.

If $\rho_{im} > 0$, the subsequence is said to exhibit positive autocorrelation. In this case, successive values at lag m have a higher probability than expected of being close in value (i.e., high random numbers in the subsequence followed by high, and low followed by low). On the other hand, if $\rho_{im} < 0$, the subsequence is exhibiting negative autocorrelation, which means that low random numbers tend to be followed by high ones, and vice versa. The desired property of independence, which implies zero autocorrelation, means that there is no discernible relationship of the nature discussed here between successive random numbers at lag m.

EXAMPLE 7.19

Test whether the 3rd, 8th, 13th, and so on, numbers in the sequence at the beginning of this section are autocorrelated. (Use $\alpha = 0.05$.) Here, $i = 3$ (beginning with the third number), $m = 5$ (every five numbers), $N = 30$ (30 numbers in the sequence), and $M = 4$ (largest integer such that $3 + (M + 1)5 \leq 30$). Then,

$$\begin{aligned}\hat{\rho}_{35} &= \frac{1}{4+1}[(0.23)(0.28) + (0.28)(0.33) + (0.33)(0.27) + (0.27)(0.05) + (0.05)(0.36)] \\ &\quad - 0.25 \\ &= -0.1945\end{aligned}$$

and

$$\sigma_{\hat{\rho}_{35}} = \frac{\sqrt{13(4) + 7}}{12(4 + 1)} = 0.1280$$

Then, the test statistic assumes the value

$$Z_0 = \frac{-0.1945}{0.1280} = -1.516$$

Now, the critical value is

$$Z_{0.025} = 1.96$$

Therefore, the hypothesis of independence cannot be rejected on the basis of this test.

It can be observed that this test is not very sensitive for small values of M, particularly when the numbers being tested are on the low side. Imagine what would happen if each of the entries in the foregoing computation of $\hat{\rho}_{im}$ were equal to zero. Then, $\hat{\rho}_{im}$ would be equal to -0.25 and the calculated Z would have the value of -1.95, not quite enough to reject the hypothesis of independence.

There are many sequences that can be formed in a set of data, given a large value of N. For example, beginning with the first number in the sequence, possibilities include (1) the sequence of all numbers, (2) the sequence formed from the first, third, fifth, . . . , numbers, (3) the sequence formed from the first, fourth, . . . , numbers, and so on. If $\alpha = 0.05$, there is a probability of 0.05 of rejecting a true hypothesis. If 10 independent sequences are examined, the probability of finding no significant autocorrelation, by chance alone, is $(0.95)^{10}$ or 0.60. Thus, 40% of the time significant autocorrelation would be detected when it does not exist. If α is 0.10 and 10 tests are conducted, there is a 65% chance of finding autocorrelation by chance alone. In conclusion, when "fishing" for autocorrelation, upon performing numerous tests, autocorrelation may eventually be detected, perhaps by chance alone, even when there is no autocorrelation present.

7.4.4. Gap Test

The gap test is used to determine the significance of the interval between the recurrence of the same digit. A gap of length x occurs between the recurrence of some digit. The following example illustrates the length of gaps associated with the digit 3:

$$\begin{array}{l}
4,\ 1,\ \underline{3},\ 5,\ 1,\ 7,\ 2,\ 8,\ 2,\ 0,\ 7,\ 9,\ 1,\ \underline{3},\ 5,\ 2,\ 7,\ 9,\ 4,\ 1,\ 6,\ \underline{3} \\
\underline{3},\ 9,\ 6,\ \underline{3},\ 4,\ 8,\ 2,\ \underline{3},\ 1,\ 9,\ 4,\ 4,\ 6,\ 8,\ 4,\ 1,\ \underline{3},\ 8,\ 9,\ 5,\ 5,\ 7 \\
\underline{3},\ 9,\ 5,\ 9,\ 8,\ 5,\ \underline{3},\ 2,\ 2,\ \underline{3},\ 7,\ 4,\ 7,\ 0,\ \underline{3},\ 6,\ \underline{3},\ 5,\ 9,\ 9,\ 5,\ 5 \\
5,\ 0,\ 4,\ 6,\ 8,\ 0,\ 4,\ 7,\ 0,\ \underline{3},\ \underline{3},\ 0,\ 9,\ 5,\ 7,\ 9,\ 5,\ 1,\ 6,\ 6,\ \underline{3},\ 8 \\
8,\ 8,\ 9,\ 2,\ 9,\ 1,\ 8,\ 5,\ 4,\ 4,\ 5,\ 0,\ 2,\ \underline{3},\ 9,\ 7,\ 1,\ 2,\ 0,\ \underline{3},\ 6,\ \underline{3}
\end{array}$$

To facilitate the analysis, the digit 3 has been underlined. There are eighteen 3's in the list. Thus, only 17 gaps can occur. The first gap is of length 10, the second gap is of length 7, and so on. The frequency of the gaps is of interest. The probability of the first gap is determined as follows:

$$\begin{aligned}
P(\text{gap of } 10) &= \overbrace{P(\text{no } 3) \cdots P(\text{no } 3)}^{\text{there are 10 of these}} P(3) \\
&= (0.9)^{10}(0.1)
\end{aligned}$$

since the probability that any digit is not a 3 is 0.9, and the probability that any digit is a 3 is 0.1. In general,

$$P(t \text{ followed by exactly } x \text{ non-}t \text{ digits}) = (0.9)^x(0.1), \qquad x = 0, 1, 2, \ldots$$

In the example above, only the digit 3 was examined. However, to fully analyze a set of numbers for independence using the gap test, every digit, 0, 1, 2, . . . , 9, must be analyzed. The observed frequencies for all the digits are recorded and this is compared to the theoretical frequency using the Kolmogorov–Smirnov test for discretized data.

The theoretical frequency distribution for randomly ordered digits is given by

$$F(x) = 0.1 \sum_{n=0}^{x} (0.9)^n = 1 - 0.9^{x+1} \tag{7.14}$$

The procedure for the test follows the steps below:

Step 1. Specify the cdf for the theoretical frequency distribution given by equation (7.14) based on the selected class interval width.

Step 2. Arrange the observed sample of gaps in a cumulative distribution with these same classes.

Step 3. Find D, the maximum deviation between $F(x)$ and $S_N(x)$, as in equation (7.3).

Step 4. Determine the critical value, D_α, from Table A.7 for the specified value of α and the sample size N.

Step 5. If the calculated value of D is greater than the tabulated value of D_α, the null hypothesis of independence is rejected.

It should be noted that using the Kolmogorov–Smirnov test when the underlying distribution is discrete results in a reduction in the Type I error, α, and an increase in the Type II error, β. The exact value of α can be found using the methodology described by Connover [1980].

EXAMPLE 7.20

Based on the frequency with which gaps occur, analyze the 110 digits above to test whether they are independent. Use $\alpha = 0.05$. The number of gaps is given by the number of digits minus 10, or 100. The number of gaps associated with the various digits are as follows:

Digit	0	1	2	3	4	5	6	7	8	9
Number of Gaps	7	8	8	17	10	13	7	8	9	13

The gap test is presented in Table 7.6. The critical value of D is given by

$$D_{0.05} = \frac{1.36}{\sqrt{100}} = 0.136$$

Table 7.6. GAP TEST EXAMPLE

Gap Length	*Frequency*	*Relative Frequency*	*Cumulative Relative Frequency*	$F(x)$	$\lvert F(x) - S_N(x) \rvert$
0–3	35	0.35	0.35	0.3439	0.0061
4–7	22	0.22	0.57	0.5695	0.0005
8–11	17	0.17	0.74	0.7176	0.0224
12–15	9	0.09	0.83	0.8147	0.0153
16–19	5	0.05	0.88	0.8784	0.0016
20–23	6	0.06	0.94	0.9202	0.0198
24–27	3	0.03	0.97	0.9497	0.0223
28–31	0	0.0	0.97	0.9657	0.0043
32–35	0	0.0	0.97	0.9775	0.0075
36–39	2	0.02	0.99	0.9852	0.0043
40–43	0	0.0	0.99	0.9903	0.0003
44–47	1	0.01	1.00	0.9936	0.0064

Since $D = \max |F(x) - S_N(x)| = 0.0224$ is less than $D_{0.05}$, do not reject the hypothesis of independence on the basis of this test.

7.4.5. Poker Test

The poker test for independence is based on the frequency with which certain digits are repeated in a series of numbers. The following example shows an unusual amount of repetition:

$$0.255, \quad 0.577, \quad 0.331, \quad 0.414, \quad 0.828, \quad 0.909, \quad 0.303, \quad 0.001, \quad \ldots$$

In each case, a pair of like digits appears in the number that was generated. In three-digit numbers there are only three possibilities, as follows:

1. The individual numbers can all be different.
2. The individual numbers can all be the same.
3. There can be one pair of like digits.

The probability associated with each of these possibilities is given by the following:

$$\begin{aligned} P(\text{three different digits}) &= P(\text{second different from the first}) \\ &\quad \times P(\text{third different from the first and second}) \\ &= (0.9)(0.8) = 0.72 \\ P(\text{three like digits}) &= P(\text{second digit same as the first}) \\ &\quad \times P(\text{third digit same as the first}) \\ &= (0.1)(0.1) = 0.01 \\ P(\text{exactly one pair}) &= 1 - 0.72 - 0.01 = 0.27 \end{aligned}$$

Alternatively, the last result can be obtained as follows:

$$P(\text{exactly one pair}) = \binom{3}{2}(0.1)(0.9) = 0.27$$

The following example shows how the poker test (in conjunction with the chi-square test) is used to ascertain independence.

Example 7.21

A sequence of 1000 three-digit numbers has been generated and an analysis indicates that 680 have three different digits, 289 contain exactly one pair of like digits, and 31 contain three like digits. Based on the poker test, are these numbers independent? Let $\alpha = 0.05$. The test is summarized in Table 7.7.

Table 7.7. POKER TEST RESULTS

Combination, i	Observed Frequency, O_i	Expected Frequency, E_i	$\frac{(O_i - E_i)^2}{E_i}$
Three different digits	680	720	2.22
Three like digits	31	10	44.10
Exactly one pair	289	270	1.33
	1000	1000	47.65

The appropriate degrees of freedom are one less than the number of class intervals. Since $\chi^2_{0.05,2} = 5.99 < 47.65$, the independence of the numbers is rejected on the basis of this test.

7.5. Summary

This chapter describes the generation of random numbers and the subsequent testing of the generated numbers for uniformity and independence. Random numbers are used to generate random variates, the subject of Chapter 8.

Several different random number generators are described, some of which are only of historical interest. The linear congruential method is the most widely used. Of the many types of statistical tests that are used in testing random number generators, five different types are described. Some of these tests are for uniformity, and others are for testing independence.

The simulator may never work directly with a random number generator, or with the testing of random numbers from a generator. Most computers and simulation languages have routines that generate a random number, or streams of random numbers, for the asking. But even generators that have been used for years, some of which are still in use, have been found to be inadequate. So this chapter calls the simulator's attention to such possibilities, with a warning to investigate and confirm that the generator has been tested thoroughly. Some researchers have attained sophisticated expertise in developing methods for generating and testing random numbers and the subsequent application of these methods. However, this chapter provides only a basic introduction to the subject matter; more depth and breadth are required for the reader to become a specialist in the area.

One final caution is due. Even if generated numbers pass all the tests (both those covered in this chapter and those mentioned in the chapter), some underlying pattern may go undetected and the generator may not be rejected as faulty. However, there are generators available in widely used simulation languages such as SLAM and SIMSCRIPT and in the IMSL Scientific Library which have been extensively tested and validated.

REFERENCES

CONNOVER, W. J. [1980], *Practical NonParametric Statistics*, 2nd ed., Wiley, New York.

COX, D. R., AND P. A. W. LEWIS [1966], *The Statistical Analysis of Series of Events*, Barnes and Noble, New York.

DURBIN, J. [1967], "Tests of Serial Independence Based on the Cumulated Periodogram," *Bulletin of the International Institute of Statistics.*

FISHMAN, G. S. [1978], *Principles of Discrete Event Simulation*, Wiley, New York.

GOOD, I. J. [1953], "The Serial Test for Sampling Numbers and Other Tests of Randomness," *Proceedings of the Cambridge Philosophical Society*, Vol. 49, pp. 276–84.

GOOD, I. J. [1967], "The Generalized Serial Test and the Binary Expansion of $\sqrt{2}$," *Journal of the Royal Statistical Society*, Ser. A, Vol. 30, No. 1, pp. 102–7.

GRAYBEAL, W. J., AND U. W. POOCH [1980], *Simulation: Principles and Methods*, Winthrop, Cambridge, Mass.

HINES, W. W., AND D. C. MONTGOMERY [1980], *Probability and Statistics in Engineering and Management Science*, 2nd ed., Prentice-Hall, Englewood Cliffs, N.J.

International Mathematics and Statistical Libraries, Inc. (IMSL) [1978], 4th ed., Houston, Tex., April.

KNUTH, D. W. [1969], *The Art of Computer Programming: Vol. 2, Semi-numerical Algorithms*, Addison-Wesley, Reading, Mass.

LEARMONTH, G. P., AND P. A. W. LEWIS [1973], "Statistical Tests of Some Widely Used and Recently Proposed Uniform Random Number Generators," *Proceedings of the Conference on Computer Science and Statistics: Seventh Annual Symposium on the Interface*, Western Publishing, North Hollywood, Calif., pp. 163–71.

LEHMER, D. H. [1951], *Proceedings of the Second Symposium on Large-Scale Digital Computing Machinery*, Harvard University Press, Cambridge, Mass.

LEWIS, P. A. W., A. S. GOODMAN, AND J. M. MILLER [1969], "A Pseudo-Random Number Generator for the System/360," *IBM Systems Journal*, Vol. 8, pp. 136–45.

SCHMIDT, J. W., AND R. E. TAYLOR [1970], *Simulation and Analysis of Industrial Systems*, Irwin, Homewood, Ill.

SHANNON, R. E. [1975], *Systems Simulation: The Art and Science*, Prentice-Hall, Englewood Cliffs, N.J.

SWED, F. S., AND C. EISENHART [1943], "Tables for Testing Randomness of Grouping in a Sequence of Alternatives," *Annals of Mathematical Statistics*, Vol. 14, pp. 66–82.

EXERCISES

1. Beginning with a seed of 6393, generate the next three four-digit random numbers using the midsquare method.

2. Use the midproduct technique to generate a sequence of three four-digit random numbers. Let the values for X_0 and X'_0 be 4729 and 8583.

3. Use the constant multiplier technique with $K = 6787$ and $X_0 = 4129$ to obtain a sequence of three four-digit random numbers.

4. Use the linear congruential method to generate a sequence of three two-digit random integers. Let $X_0 = 27$, $a = 8$, $c = 47$, and $m = 100$.

5. Let a sequence of random numbers R_1, R_2, R_3, R_4, R_5, be 0.45, 0.37, 0.89, 0.11, and 0.66. Extend the sequence through R_{10} using the additive congruential method, where $m = 100$.

6. Use the multiplicative congruential method to generate a sequence of four three-digit random integers. Let $X_0 = 117$, $a = 43$ and $m = 1000$.

7. The sequence of numbers 0.54, 0.73, 0.98, 0.11, and 0.68 has been generated. Use the Kolmogorov–Smirnov test with $\alpha = 0.05$ to determine if the hypothesis that the numbers are uniformly distributed on the interval [0, 1] can be rejected.

8. Reverse the 100 two-digit random numbers in Example 7.14 to get a new set of random numbers. Thus, the first random number in the new set will be 0.43. Use the chi-square test, with $\alpha = 0.05$, to determine if the hypothesis that the numbers are uniformly distributed on the interval [0, 1] can be rejected.

9. Consider the first 50 two-digit values in Example 7.14. Based on runs up and runs down, determine whether the hypothesis of independence can be rejected, where $\alpha = 0.05$.

10. Consider the last 50 two-digit values in Example 7.14. Determine whether there is an excessive number of runs above or below the mean. Use $\alpha = 0.05$.

11. Consider the first 50 two-digit values in Example 7.14. Can the hypothesis that the numbers are independent be rejected on the basis of the length of runs up and down when $\alpha = 0.05$?

12. Consider the last 50 two-digit values in Example 7.14. Can the hypothesis that the numbers are independent be rejected on the basis of the length of runs above and below the mean, where $\alpha = 0.05$?

13. Consider the 60 values in Example 7.17. Test whether the 2nd, 9th, 16th, . . . , numbers in the sequence are autocorrelated, where $\alpha = 0.05$.

14. Consider the following sequence of 120 digits:

1	3	7	4	8	6	2	5	1	6	4	4	3	3	4	2	1	5	8	7
0	7	6	2	6	0	5	7	8	0	1	1	2	6	7	6	3	7	5	9
0	8	8	2	6	7	8	1	3	5	3	8	4	0	9	0	3	0	9	2
2	3	6	5	6	0	0	1	3	4	4	6	9	9	8	5	6	0	1	7
5	6	7	9	4	9	3	1	8	3	3	6	6	7	8	2	3	5	9	6
6	7	0	3	1	0	2	4	2	0	6	4	0	3	9	3	6	8	1	5

Test whether these digits can be assumed to be independent based on the frequency with which gaps occur. Use $\alpha = 0.05$.

15. Develop the poker test for:
(a) Four-digit numbers
(b) Five-digit numbers

16. A sequence of 1000 four-digit numbers has been generated and an analysis indicates the following combinations and frequencies.

Combination, i	*Observed Frequency,* O_i
Four different digits	565
One pair	392
Two pairs	17
Three like digits	24
Four like digits	2
	1000

Based on the poker test, test whether these numbers are independent. Use $\alpha = 0.05$.

17. Determine whether the linear congruential generators shown below can achieve a maximum period. Also, state restrictions on X_0 to obtain this period.

(a) The CDC version of SIMSCRIPT uses the mixed congruential method with

$$a = 2{,}814{,}749{,}767{,}109$$
$$c = 59{,}482{,}661{,}568{,}307$$
$$m = 2^{48}$$

(b) "Super Duper," developed for the IBM 360, is a multiplicative congruential generator with

$$a = 69{,}069$$
$$c = 0$$
$$m = 2^{32}$$

(c) Consider a mixed congruential generator with

$$a = 4951$$
$$c = 247$$
$$m = 256$$

(d) Consider a multiplicative congruential generator with

$$a = 6507$$
$$c = 0$$
$$m = 1024$$

18. Use the mixed congruential method to generate a sequence of three two-digit random numbers with $X_0 = 37$, $a = 7$, $c = 29$ and $m = 100$.

19. Use the mixed congruential method to generate a sequence of three two-digit random integers between 0 and 24 with $X_0 = 13$, $a = 9$ and $c = 35$.

20. Write a BASIC, FORTRAN or Pascal program that will generate four-digit random numbers using the multiplicative congruential method. Allow the user to input values of X_0, a, c and m.

21. If $X_0 = 3579$ in Exercise 17(c), generate the first random number in the sequence. Compute the random number to four place accuracy.

22. Investigate the random number generator in BASIC on a personal computer to which you have access. On many computers, random numbers are generated by a BASIC function called RND.

(a) How are the random numbers generated? Check to insure that the generator is performing in accordance with the procedure.

(b) Write routines in BASIC to conduct each of the tests described in this chapter. Generate 100 sets of random numbers, each set containing 100 random numbers. Perform each test on each set of random numbers. Draw conclusions.

23. Consider the multiplicative congruential generator under the following circumstances

(a) $a = 11, m = 16, X_0 = 7$
(b) $a = 11, m = 16, X_0 = 8$
(c) $a = 7, m = 16, X_0 = 7$
(d) $a = 7, m = 16, X_0 = 8$

Generate enough values in each case to complete a cycle. What inferences can be drawn? Is maximum period achieved?

RANDOM VARIATE GENERATION

This chapter deals with procedures for sampling from a variety of widely used continuous and discrete distributions. Previous discussions and examples of queueing and inventory systems have indicated the usefulness of statistical distributions to model activities that are generally unpredictable or uncertain. For example, interarrival times and service times at queues, and demands for a product, are quite often unpredictable in nature, at least to a certain extent. Usually, such variables are modeled as random variables with some specified statistical distribution, and standard statistical procedures exist for estimating the parameters of the hypothesized distribution and for testing the validity of the assumed statistical model. Such procedures are discussed in Chapter 9.

In this chapter it is assumed that a distribution has been completely specified, and ways are sought to generate samples from this distribution to be used as input to a simulation model. The purpose of the chapter is to explain and illustrate the most widely used techniques for generating random variates, not to give a state-of-the-art survey of the most efficient techniques. In practice, most simulation modelers will use existing routines available in FORTRAN (e.g., the IMSL library), or the routines built into the language being used, such as those routines built into SIMSCRIPT, GASP, and SLAM. Some languages such as GPSS do not have built-in routines, and some computer installations do not have random variate generators in FORTRAN, so that the modeler must

8

construct an acceptable routine. This chapter discusses the inverse transform technique, the convolution method, and more briefly, the acceptance–rejection technique. Another technique, the composition method, is discussed by Fishman [1978]. In particular, it will be shown how to generate samples from all the distributions discussed in Chapter 4.

All the techniques in this chapter assume that a source of uniform (0, 1) random numbers, $R_1, R_2, \ldots$ is readily available, where each R_i has pdf

$$f_R(x) = \begin{cases} 1, & 0 \le x \le 1 \\ 0, & \text{otherwise} \end{cases}$$

and cdf

$$F_R(x) = \begin{cases} 0, & x < 0 \\ x, & 0 \le x \le 1 \\ 1, & x > 1 \end{cases}$$

Throughout this chapter R and $R_1, R_2, \ldots$ represent random numbers uniformly distributed on [0, 1] and generated by one of the techniques in Chapter 7 or taken from a random number table such as Table A.1. The use of Table A.1 was described in Chapter 2.

8.1. Inverse Transform Technique

The inverse transform technique can be used to sample from the exponential, the Weibull and the uniform distributions, and empirical distributions. Additionally, it is the underlying principle for sampling from a wide variety of discrete distributions. The technique will be explained in detail for the exponential distribution, and then applied to other distributions. It is the most straightforward but not always the most efficient technique computationally.

8.1.1. Exponential Distribution

The exponential distribution, discussed in Section 4.4, has probability density function (pdf) given by

$$f(x) = \begin{cases} \lambda e^{-\lambda x}, & x \geq 0 \\ 0, & x < 0 \end{cases}$$

and cumulative distribution function (cdf) given by

$$F(x) = \int_{-\infty}^{x} f(t)\,dt = \begin{cases} 1 - e^{-\lambda x}, & x \geq 0 \\ 0, & x < 0 \end{cases}$$

The parameter λ can be interpreted as the mean number of occurrences per time unit. For example, if interarrival times $X_1, X_2, X_3, \ldots$ had an exponential distribution with rate λ, λ could be interpreted as the mean number of arrivals per time unit, or the arrival rate. Note that for any i

$$E(X_i) = \frac{1}{\lambda}$$

so that $1/\lambda$ is the mean interarrival time. The goal here is to develop a procedure for generating values $X_1, X_2, X_3, \ldots$ which have an exponential distribution.

The inverse transform technique can be utilized when the cdf, $F(x)$, is of such simple form that its inverse, F^{-1}, can be explicitly computed analytically. A step-by-step procedure for the inverse transform technique, illustrated by the exponential distribution, is as follows:

Step 1. Compute the cdf of the desired random variable X.
For the exponential distribution, the cdf is $F(x) = 1 - e^{-\lambda x}, x \geq 0$.

Step 2. Set $F(X) = R$ on the range of X.
For the exponential distribution, it becomes $1 - e^{-\lambda X} = R$ on the range $x \geq 0$.

Since X is a random variable (with the exponential distribution in this case), it follows that $1 - e^{-\lambda X}$ is also a random variable, here called R. As will be shown later, R has a uniform distribution over the interval (0, 1).

Step 3. Solve the equation $F(X) = R$ for X in terms of R.
For the exponential distribution, the solution proceeds as follows:

$$1 - e^{-\lambda X} = R$$

$$e^{-\lambda X} = 1 - R$$

$$-\lambda X = \ell\text{n}\,(1 - R)$$

$$X = \frac{-1}{\lambda}\,\ell\text{n}\,(1 - R) \tag{8.1}$$

Equation (8.1) is called a random variate generator for the exponential distribution. In general, equation (8.1) is written as $X = F^{-1}(R)$. Generating a sequence of values is accomplished through step 4.

Step 4. Generate (as needed) uniform random numbers $R_1, R_2, R_3, \ldots$ and compute the desired random variates by

$$X_i = F^{-1}(R_i)$$

For the exponential case, $F^{-1}(R) = (-1/\lambda)\,\ell\text{n}\,(1 - R)$ by equation (8.1), so that

$$X_i = \frac{-1}{\lambda}\,\ell\text{n}\,(1 - R_i) \tag{8.2a}$$

for $i = 1, 2, 3, \ldots$. One simplification that is usually employed in equation (8.2a) is to replace $1 - R_i$ by R_i to yield

$$X_i = \frac{-1}{\lambda}\,\ell\text{n}\,R_i \tag{8.2b}$$

which is justified since both R_i and $1 - R_i$ are uniformly distributed on (0, 1).

EXAMPLE 8.1

Table 8.1 gives a sequence of random numbers from Table A.1 and the computed exponential variates, X_i, given by equation (8.2a) with a value of $\lambda = 1$. Figure 8.1(a) is a histogram of 200 values, $R_1, R_2, \ldots, R_{200}$, from the uniform distribution, and

Table 8.1. GENERATION OF EXPONENTIAL VARIATES X_i WITH MEAN 1, GIVEN RANDOM NUMBERS R_i

i	1	2	3	4	5
R_i	0.1306	0.0422	0.6597	0.7965	0.7696
X_i	0.1400	0.0431	1.078	1.592	1.468

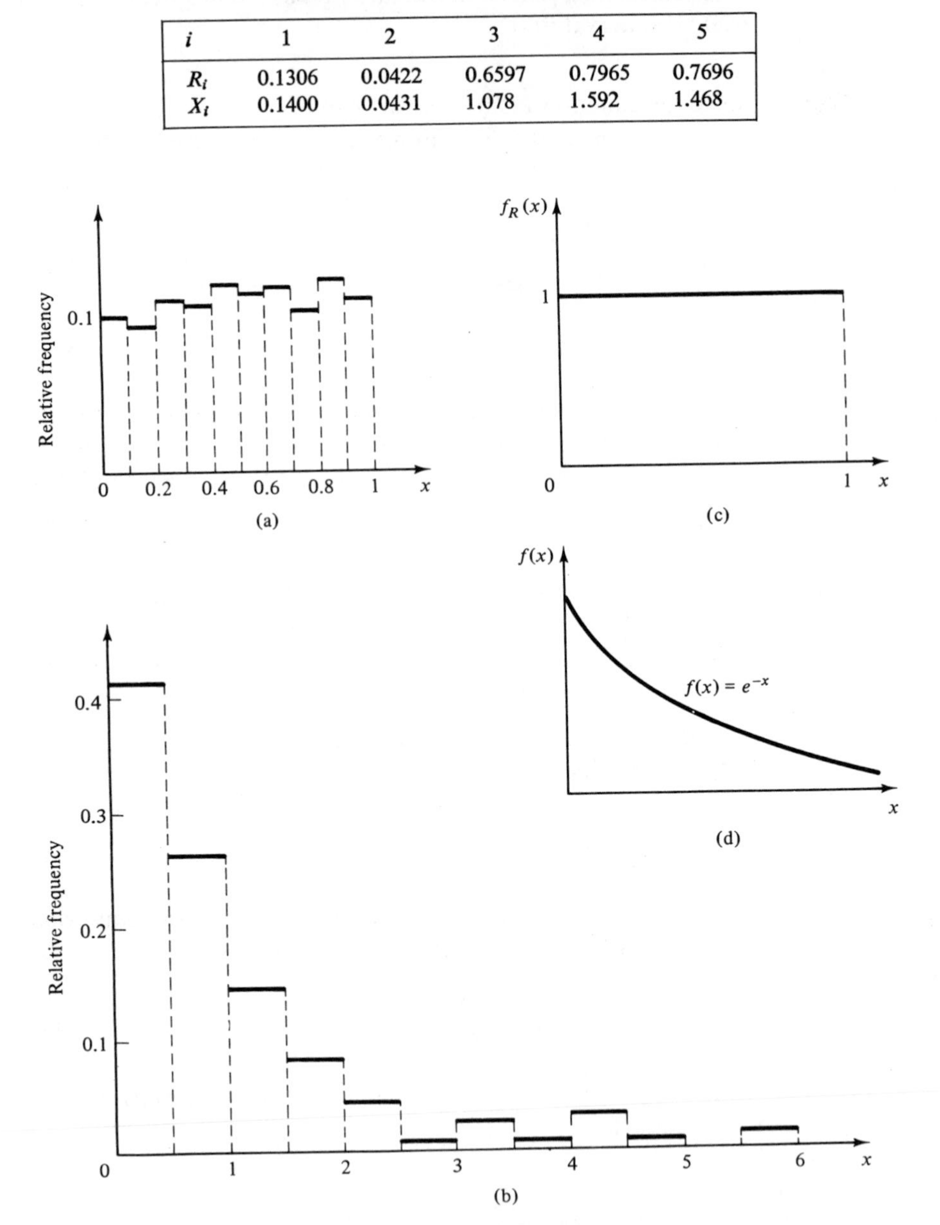

Figure 8.1. (a) Empirical histogram of 200 uniform random numbers; (b) empirical histogram of 200 exponential variates; (c) theoretical uniform density on (0, 1); (d) theoretical exponential density with mean 1.

Figure 8.1(b) is a histogram of the 200 values, $X_1, X_2, \ldots, X_{200}$, computed by equation (8.2a). Compare these empirical histograms with the theoretical density functions in Figure 8.1(c) and (d). As illustrated here, a histogram is an estimate of the underlying density function. (This fact is used in Chapter 9 as a way to identify distributions.)

Figure 8.2 gives a graphical interpretation of the inverse transform technique.

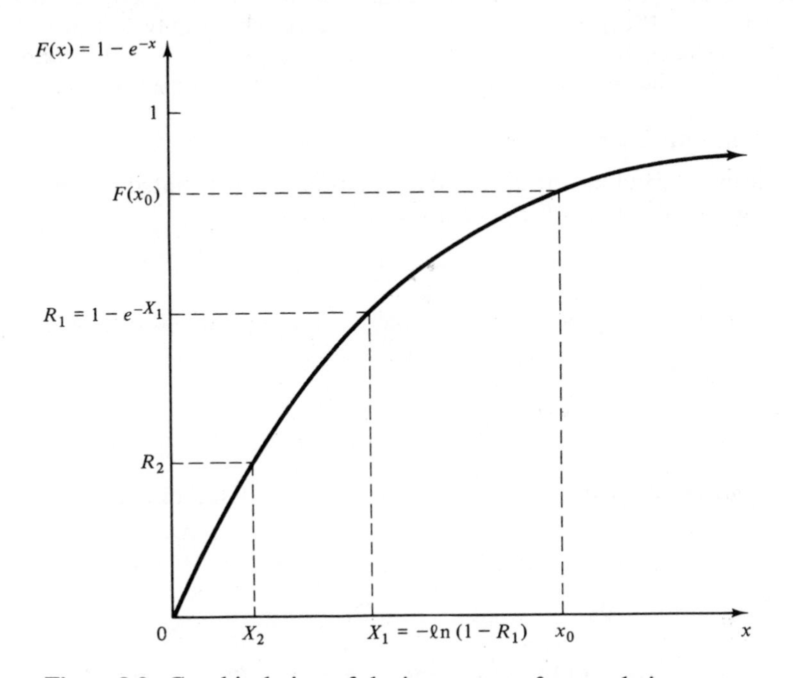

Figure 8.2. Graphical view of the inverse transform technique.

The cdf shown is $F(x) = 1 - e^{-x}$, an exponential distribution with rate $\lambda = 1$. To generate a value X_1 with cdf $F(x)$, first a random number R_1 between 0 and 1 is generated, a horizontal line is drawn from R_1 to the graph of the cdf, then a vertical line is dropped to the x axis to obtain X_1, the desired result. Note the inverse relation between R_1 and X_1, namely

$$R_1 = 1 - e^{-X_1}$$

and

$$X_1 = -\ell n\,(1 - R_1)$$

In general, the relation is written as

$$R_1 = F(X_1)$$

and

$$X_1 = F^{-1}(R_1)$$

Why does the random variable X_1 generated by this procedure have the desired distribution? Pick a value x_0 and compute the cumulative probability

$$P(X_1 \leq x_0) = P(R_1 \leq F(x_0)) = F(x_0) \tag{8.3}$$

To see the first equality in equation (8.3), refer to Figure 8.2, where the fixed numbers x_0 and $F(x_0)$ are drawn on their respective axes. It can be seen that $X_1 \leq x_0$ when and only when $R_1 \leq F(x_0)$. Since $0 \leq F(x_0) \leq 1$, the second equality in equation (8.3) follows immediately from the fact that R_1 is uniformly distributed on (0, 1). Equation (8.3) shows that the cdf of X_1 is F; hence, X_1 has the desired distribution.

8.1.2. Uniform Distribution

Consider a random variable X that is uniformly distributed on the interval $[a, b]$. A reasonable guess for generating X is given by

$$X = a + (b - a)R \tag{8.4}$$

[Recall that R is always a random number on (0, 1).] The pdf of X is given by

$$f(x) = \begin{cases} \dfrac{1}{b-a}, & a \leq x \leq b \\ 0, & \text{otherwise} \end{cases}$$

The derivation of equation (8.4) follows steps 1 through 3 of Section 8.1.1:

Step 1. The cdf is given by

$$F(x) = \begin{cases} 0, & x < a \\ \dfrac{x-a}{b-a}, & a \leq x \leq b \\ 1, & x > b \end{cases}$$

Step 2. Set $F(X) = (X - a)/(b - a) = R$.

Step 3. Solving for X in terms of R yields $X = a + (b - a)R$, which agrees with equation (8.4).

8.1.3. Weibull Distribution

The Weibull distribution was introduced in Section 4.4 as a model for "time to failure" for machines or electronic components. When the location parameter ν is set to 0, its pdf is given by equation (4.45b) as

$$f(x) = \begin{cases} \dfrac{\beta}{\alpha^\beta} x^{\beta-1} e^{-(x/\alpha)^\beta}, & x \geq 0 \\ 0, & \text{otherwise} \end{cases}$$

where $\alpha > 0$ and $\beta > 0$ are the scale and shape parameters of the distribution. To generate a Weibull variate, follow steps 1 through 3 of Section 8.1.1:

Step 1. The cdf is given by $F(x) = 1 - e^{-(x/\alpha)^\beta}$, $x \geq 0$.

Step 2. Let $F(X) = 1 - e^{-(X/\alpha)^\beta} = R$.

Step 3. Solving for X in terms of R yields

$$X = \alpha[-\ell n\,(1 - R)]^{1/\beta} \tag{8.5}$$

The derivation of equation (8.5) is left as Exercise 10 for the student. By comparing equations (8.5) and (8.1), it can be seen that if X is a Weibull variate, then X^β is an exponential variate with mean α^β. Conversely, if Y is an exponential variate with mean μ, then $Y^{1/\beta}$ is a Weibull variate with shape parameter β and scale parameter $\alpha = \mu^{1/\beta}$.

8.1.4. Triangular Distribution

Consider a random variable X which has pdf

$$f(x) = \begin{cases} x, & 0 \leq x \leq 1 \\ 2 - x, & 1 < x \leq 2 \\ 0, & \text{otherwise} \end{cases}$$

as shown in Figure 8.3. This distribution is called a triangular distribution with endpoints (0, 2) and mode at 1. Its cdf is given by

$$F(x) = \begin{cases} 0, & x \leq 0 \\ \dfrac{x^2}{2}, & 0 < x \leq 1 \\ 1 - \dfrac{(2 - x)^2}{2}, & 1 < x \leq 2 \\ 1, & x > 2 \end{cases}$$

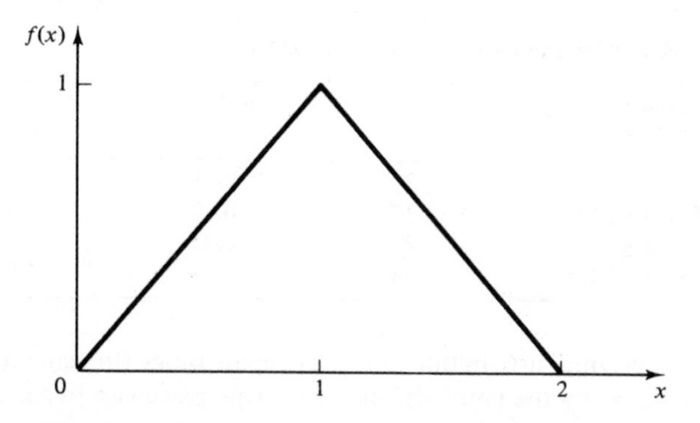

Figure 8.3. Density function for a triangular distribution.

For $0 \leq X \leq 1$,

$$R = \frac{X^2}{2} \tag{8.6}$$

and for $1 \leq X \leq 2$,

$$R = 1 - \frac{(2 - X)^2}{2} \tag{8.7}$$

By equation (8.6), $0 \leq X \leq 1$ implies that $0 \leq R \leq \frac{1}{2}$, in which case $X = \sqrt{2R}$. By equation (8.7), $1 \leq X \leq 2$ implies that $\frac{1}{2} \leq R \leq 1$, in which case $X = 2 - \sqrt{2(1 - R)}$. Thus, X is generated by

$$X = \begin{cases} \sqrt{2R}, & 0 \leq R \leq \frac{1}{2} \\ 2 - \sqrt{2(1 - R)}, & \frac{1}{2} < R \leq 1 \end{cases} \tag{8.8}$$

Exercises 2, 3, and 4 give the student practice in dealing with other triangular distributions. Note that if the pdf and cdf of the random variable X come in parts (i.e., require different formulas over different parts of the range of X), then the application of the inverse transform technique for generating X will result in separate formulas over different parts of the range of R, as in equation (8.8). A general form of the triangular distribution was discussed in Section 4.4.

8.1.5. Empirical Continuous Distributions

If the modeler has been unable to find a theoretical distribution that provides a good model for the input data, it may be necessary to use the empirical distribution of the data.

Example 8.2

Suppose that 100 broken widget repair times have been collected. The data are summarized in Table 8.2 in terms of the number of observations in various intervals. For example, there were 31 observations between 0 and 0.5 hour, 10 between 0.5 and 1 hour, and so on.

Table 8.2. SUMMARY OF REPAIR-TIME DATA

Interval (Hours)	*Frequency*	*Relative Frequency*	*Cumulative Frequency*
$0 \leq x \leq 0.5$	31	0.31	0.31
$0.5 < x \leq 1.0$	10	0.10	0.41
$1.0 < x \leq 1.5$	25	0.25	0.66
$1.5 < x \leq 2.0$	34	0.34	1.00

The true underlying distribution, $F(x)$, of repair times (the curved line in Figure 8.4) can be estimated by the empirical cdf, $\hat{F}(x)$ (the piecewise linear curve in Figure

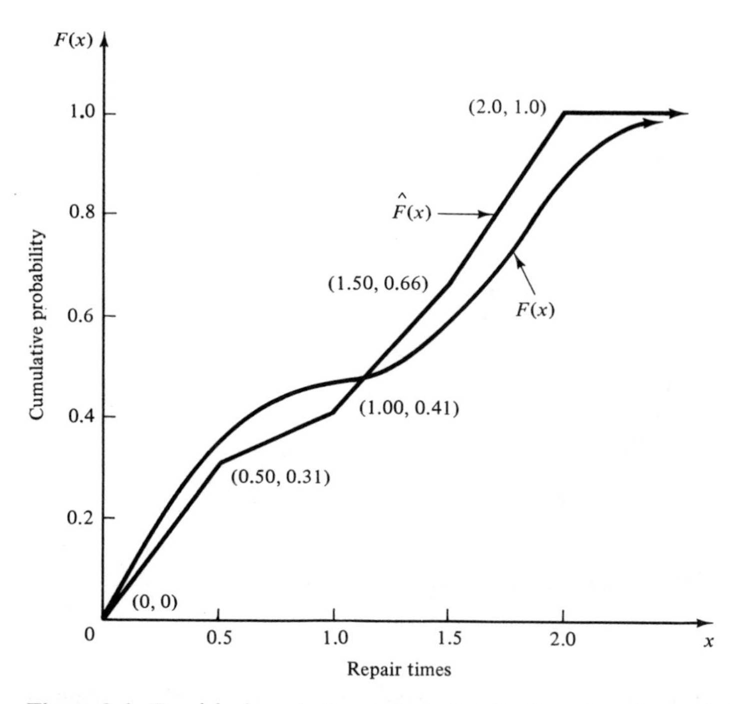

Figure 8.4. Empirical and theoretical distribution functions, for repair time data ($X \geq 0$).

8.4). The true shape of $F(x)$ is unknown and would always be unknown in practice, unless a virtually unlimited amount of data was available. The particular curve in Figure 8.4 illustrates one possible shape of this underlying distribution, and also that $\hat{F}(x)$ is an estimate of $F(x)$.

The empirical cdf $\hat{F}(x)$ is defined using the information in Table 8.2. Each interval defines two points on the graph, which are connected by a straight line. (This linear interpolation is not the only possibility but it is the most straightforward.) Note that four intervals result in five pairs of points to define four line segments. In Example 8.2, it can be seen that

$$\hat{F}(x) = 0 \qquad \text{for } x < 0$$

and

$$\hat{F}(x) = 1 \qquad \text{for } x > 2$$

but this fact is of no importance for generation purposes. It has been assumed that the random variable, X, for repair times satisfies $X \geq 0$ and that any value, no matter how small, is possible. This assumption leads to the point (0, 0) on the graph of Figure 8.4. On the other hand, suppose it is known that all repairs take at least 15 minutes, so that $X \geq 0.25$ hour always. Then the point (0, 0) should be replaced by (0.25, 0), as shown in Figure 8.5. Figure 8.5 will be used to illustrate the generation procedure.

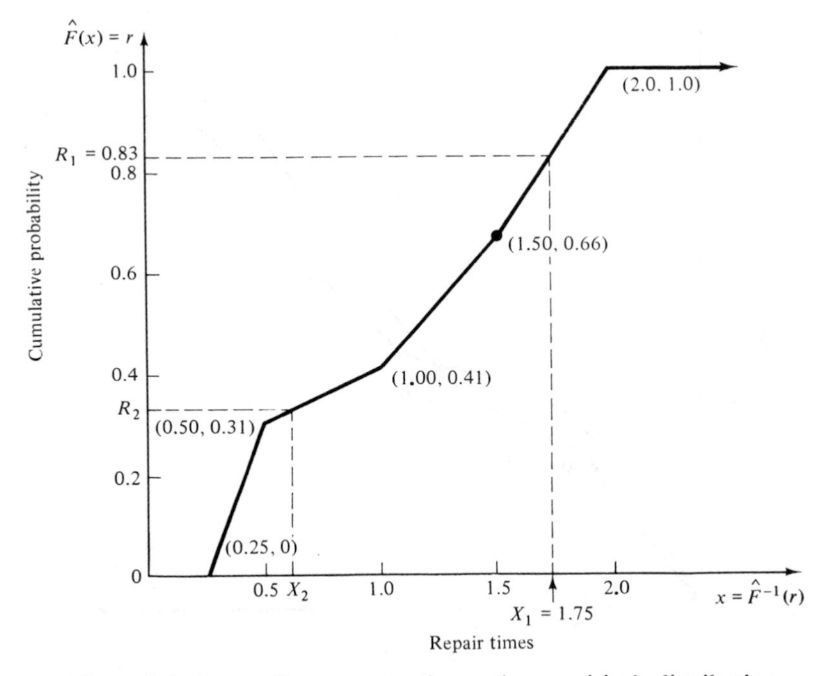

Figure 8.5. Generating variates from the empirical distribution function for repair time data ($X \geq .25$).

The inverse transform technique applies directly to generating repair time variates, X. Recalling the graphical interpretation of the technique, first generate a random number R_1, say $R_1 = 0.83$, and read X_1 off the graph of Figure 8.5. Symbolically, this is written as

$$X_1 = F^{-1}(R_1)$$

but algebraically, since R_1 is between 0.66 and 1.00, X_1 is computed by a linear interpolation between 1.5 and 2.0; that is,

$$X_1 = 1.5 + \left[\frac{R_1 - 0.66}{1.00 - 0.66}\right](2.0 - 1.5) = 1.75 \tag{8.9}$$

When $R_1 = 0.83$, note that $(R_1 - 0.66)/(1.00 - 0.66) = 0.5$, so that X_1 will be one-half of the distance between 1.5 and 2.0 since R_1 is one-half of the way between 0.66 and 1.00.

Note that for all R_1 in the interval (0.66, 1.00), the value $a_4 = (2.0 - 1.5)/(1.00 - 0.66) = 1.47$ will be needed to compute X_1. The value a_4 is the slope $\Delta x/\Delta r$ of the function $x = \hat{F}^{-1}(r)$, which is merely the reflection about the line $r = x$ of the function $r = \hat{F}(x)$ of Figure 8.5. The inverse function reverses the role of input and

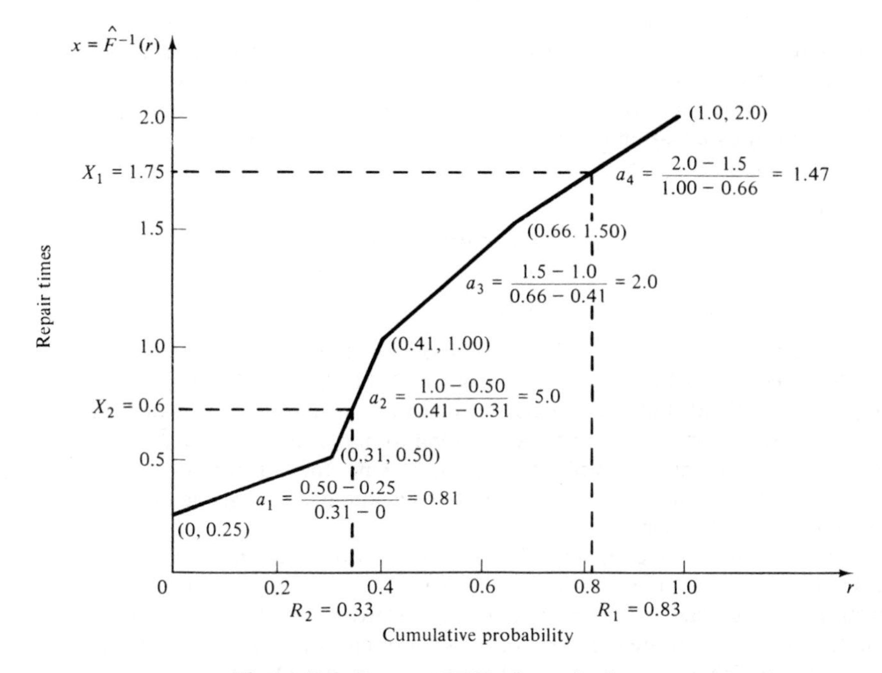

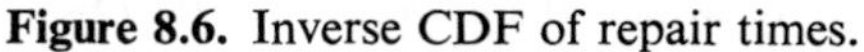

Figure 8.6. Inverse CDF of repair times.

output, as shown by Figure 8.6. The slopes of the four line segments are also given in Figure 8.6, and in Table 8.3, which can be used to generate variates, X, as follows:

Table 8.3. INTERVALS AND SLOPES FOR GENERATING REPAIR TIMES, X

i	*Input,* r_i	*Output,* x_i	*Slope,* a_i
1	0	0.25	0.81
2	0.31	0.5	5.0
3	0.41	1.0	2.0
4	0.66	1.5	1.47
5	1.00	2.0	—

Step 1. Generate R.

Step 2. Find the interval i in which R lies; that is, find i so that $r_i \leq R \leq r_{i+1}$.

Step 3. Compute X by

$$X = x_i + a_i(R - r_i) \tag{8.10}$$

For the repair-time data, the endpoints (r_i, x_i), $i = 1, \ldots, 5$, and slopes a_i, $i = 1, \ldots, 4$, are given in Table 8.3. If it is necessary to generate a large number of X's, it is advantageous to compute the slopes a_i ahead of time and to store them as in Table 8.3, for future use. Note that equation (8.9) is an application of the general interpolation formula given by equation (8.10). As another illustration, suppose that $R_2 = 0.33$. By Table 8.3, since $r_2 = 0.31 \leq R_2 = 0.33 < r_3 = 0.41$, R_2 lies in the interval $i = 2$, and, therefore,

$$\begin{aligned} X_2 &= x_2 + a_2(R_2 - r_2) \\ &= 0.5 + 5.0(0.33 - 0.31) \\ &= 0.6 \end{aligned}$$

The point $(R_2 = 0.33,\ X_2 = 0.6)$ is also shown in Figures 8.5 and 8.6.

Now reconsider Figure 8.4 and the data of Table 8.2. The data are restricted in the range $0 \leq X \leq 2.0$, but the underlying distribution may have a wider range. This provides one important reason for attempting to find a theoretical statistical distribution (such as the gamma or Weibull) for the data, since these distributions allow a wider range, namely $0 \leq X < \infty$. In general, it is recommended that the empirical cdf be used only as a last resort.

In addition, it is recommended that individual data points be collected, not just summary interval data as was done in Table 8.2. If data are summarized in terms of frequency in intervals, it is recommended that relatively short intervals be used, as this results in a more accurate portrayal of the underlying cdf. For example, for the repair-time data of Table 8.2, for which there were $n = 100$ observations, a much more accurate estimate could have been obtained by using 10 to 20 intervals, certainly not an excessive number, rather than the four fairly wide intervals actually used here for purposes of illustration.

Now consider an example for which all the raw data are available. The number of observations is kept small for purposes of illustration, but the technique can be applied to any number of observations.

EXAMPLE 8.3

Five observations of fire crew response times (in minutes) to incoming alarms have been collected to be used in a simulation investigating possible alternative staffing and crew scheduling policies. The data are

2.76 1.83 0.80 1.45 1.24

Before collecting more data, it is desired to develop a preliminary simulation model which uses a response-time distribution based on these five observations. Thus, a method for generating random variates from the response-time distribution is needed. Initially, it will be assumed that response times X have a range $0 \leq X \leq c$, where c is unknown, but will be estimated by $\hat{c} = \max\{X_i: i = 1, \ldots, n\} = 2.76$, where $\{X_i, i = 1, \ldots, n\}$ is the raw data and $n = 5$ is the number of observations.

Arrange the data from smallest to largest, as in Table 8.4, and assign a probability of $1/n = 1/5$ to each interval. The resulting empirical cdf, $\hat{F}(x)$, is illustrated in Figure 8.7, and the slopes, $\Delta x/\Delta r$, of the inverse cdf, $x = \hat{F}^{-1}(r)$, which are required to generate

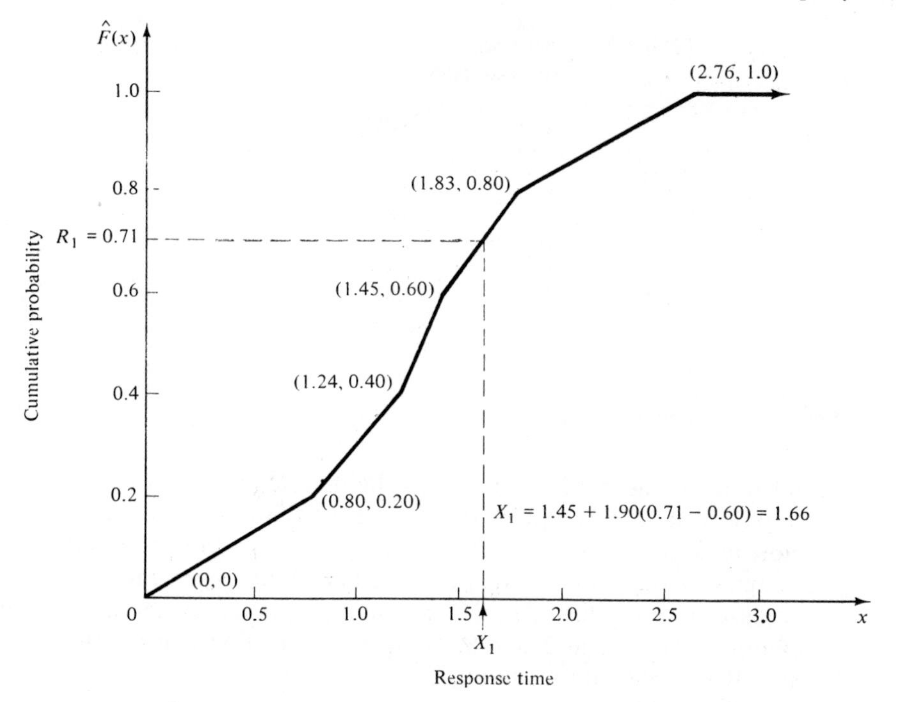

Figure 8.7. Empirical CDF of fire crew response times.

Table 8.4. SUMMARY OF FIRE CREW RESPONSE-TIME DATA

Interval (Minutes)	*Probability*	*Cumulative Probability*	*Slope, $\Delta x/\Delta r$*
$0.0 < x \leq 0.80$	0.2	0.2	4.00
$0.80 < x \leq 1.24$	0.2	0.4	2.20
$1.24 < x \leq 1.45$	0.2	0.6	1.05
$1.45 < x \leq 1.83$	0.2	0.8	1.90
$1.83 < x \leq 2.76$	0.2	1.0	4.65

response times, X, are given in Table 8.4. As an example, if a random number $R_1 = 0.71$ is generated, then R_1 is seen to lie in the fourth interval (between $r_4 = 0.60$ and $r_5 = 0.80$, so that by equation (8.10),

$$\begin{aligned} X_1 &= x_4 + a_4(R_1 - r_4) \\ &= 1.45 + 1.90(0.71 - 0.60) \\ &= 1.65 \end{aligned}$$

The reader is referred to Figure 8.7 for a graphical view of the generation procedure, and to Table 8.5, which summarizes the information in Table 8.4 in the form previously used for generation purposes.

Table 8.5. INVERSE EMPIRICAL CDF OF FIRE CREW RESPONSE TIMES

i	Input, r_i	Output, x_i	Slope, a_i
1	0	0	4.00
2	0.2	0.8	2.20
3	0.4	1.24	1.05
4	0.6	1.45	1.90
5	0.8	1.83	4.65
6	1.0	2.76	—

Several comments are in order:

1. When applying this last version of the inverse transform technique to empirical data, a computerized version of the procedure will become more inefficient as the number of observations, n, increases. A systematic computerized version is often called a table-lookup generation scheme, because given a value of R, the computer program must search an array of inputs such as in Table 8.5 to find the interval i in which R lies, namely the interval i satisfying

$$r_i \leq R < r_{i+1}$$

The more intervals there are, the longer on the average the search will take. The analyst should consider this trade-off between accuracy of the estimating cdf and computational efficiency when programming the procedure. If a large number of observations are available, the analyst may well decide to group the observations into from 20 to 50 intervals (say) and then use the procedure of Example 8.2.

2. In Example 8.3 it was assumed that response times X satisfied $0 \leq X \leq 2.76$. This assumption led to the inclusion of the points $(x_1, r_1) = (0, 0)$ and $(x_6, r_6) = (2.76, 1.00)$ in Figure 8.7 and Table 8.5. If it is known a priori that X falls in some other range, for example, if it is known that response times are always between 15 seconds and 3 minutes, that is,

$$0.25 \leq X \leq 3.0$$

then the points $(x_1, r_1) = (0.25, 0)$, $(x_6, r_6) = (2.76, 0.83)$, and $(x_7, r_7) = (3.0, 1.0)$ would be used in Figure 8.7 and Table 8.5 to estimate the empirical cdf of response times. Note that because of inclusion of the new point (3.0, 1.0) there are now six intervals instead of five and each interval is assigned probability $1/6 = 0.167$. Exercise 12 illustrates the use of these additional assumptions.

8.1.6. Table-lookup Procedures for approximations to Exponential and Normal Generation

In the discrete-event simulation language, GPSS V, there is no (direct) capability for computing logarithms, sines and cosines, or roots, and hence the exact generation procedures described for the exponential distribution in Section 8.1.1 and for the normal distribution in Section 8.2 cannot be applied. Standard piecewise linear approximations have been developed for use in GPSS to generate, at least approximately, both exponential and normal variates. Graphically, these approximations involve replacing the actual cdf, namely

$$F(x) = 1 - e^{-x}, \qquad x \geq 0 \tag{8.11}$$

for the exponential distribution with mean 1, and

$$\Phi(x) = \int_{-\infty}^{x} \sqrt{1/2\pi}\, e^{-t^2/2}\, dt \tag{8.12}$$

for the standard normal distribution with mean zero and variance 1, by a sequence of straight-line segments (i.e., a piecewise linear function), and then using the inverse transform technique on this piecewise linear approximation. As seen in Section 8.1.5, the inverse transform technique becomes a table-lookup procedure when the cdf is piecewise linear. Conceptually, the idea is similar to that illustrated in Figure 8.4, except that now the true cdf $F(x)$ is known [it is either equation (8.11) or (8.12)] and the endpoints of the straight-line segments are not estimated from data, but rather are calculated in such a way that the line segments will approximate the curve as closely as possible.

The standard approximations for generating exponential and normal random variates are given in Tables 8.6 and 8.7, respectively. To generate a value X from one of these distributions, first generate a random number R, then find the interval in which R lies, and finally compute X by the linear interpolation formula given by equation (8.10) with the appropriate value of r_i, x_i, and a_i taken from Table 8.6 or 8.7.

Example 8.4

Generate six values from an exponential distribution with mean 40 using the exact method of Equation (8.2a) and the approximate table-lookup procedure of Table 8.6. For the approximate method, first generate an exponential variate X with mean 1 and then use

$$Y = \beta X \tag{8.13}$$

to compute an exponential variate Y with the desired mean β. Equation (8.13) does not hold in general for other distributions. Multiplying a variate with mean 1 by a value β will yield a new variate Y with mean β, but usually the shape of the distribution will also change. Equation (8.13) can be used for the gamma family of distributions, which includes the exponential and Erlang distributions.

Table 8.6. TABLE FOR GENERATING AN EXPONENTIALLY DISTRIBUTED RANDOM VARIABLE (MEAN = 1)

Input, r_i	*Output*, x_i	*Slope*, a_i	*Input*, r_i	*Output*, x_i	*Slope*, a_i
0	0	1.04	0.90	2.30	11.0
0.1	0.104	1.18	0.92	2.52	14.5
0.2	0.222	1.33	0.94	2.81	18.0
0.3	0.355	1.54	0.95	2.99	21.0
0.4	0.509	1.81	0.96	3.20	30.0
0.5	0.690	2.25	0.97	3.50	40.0
0.6	0.915	2.85	0.98	3.90	70.0
0.7	1.20	3.60	0.99	4.60	140
0.75	1.38	4.40	0.995	5.30	300
0.80	1.60	5.75	0.998	6.20	800
0.84	1.83	7.25	0.999	7.0	3333
0.88	2.12	9.00	0.9997	8.0	—

Source: Geoffrey Gordon, *The Application of GPSS V to Discrete System Simulation*, © 1975, p. 177. Reprinted by permission of Prentice-Hall, Inc., Englewood Cliffs, N.J.

Table 8.7. TABLE FOR GENERATING A STANDARD NORMAL VARIABLE

Input, r_i	*Output*, x_i	*Slope*, a_i	*Input*, r_i	*Output*, x_i	*Slope*, a_i
0	−5.0	33,333	0.57926	0.2	2.63
0.00003	−4.0	756	0.65542	0.4	2.84
0.00135	−3.0	206	0.72575	0.6	3.21
0.00621	−2.5	30.2	0.78814	0.8	3.76
0.02275	−2.0	11.3	0.84134	1.0	4.59
0.06681	−1.5	6.22	0.88493	1.2	6.22
0.11507	−1.2	4.59	0.93319	1.5	11.3
0.15866	−1.0	3.76	0.97725	2.0	30.2
0.21186	−0.8	3.21	0.99379	2.5	206
0.27425	−0.6	2.84	0.99865	3.0	756
0.34458	−0.4	2.63	0.99997	4.0	33,333
0.42074	−0.2	2.52	1.0	5.0	—
0.50000	0	2.52			

Source: Geoffrey Gordon, *The Application of GPSS V to Discrete System Simulation*, © 1975, p. 181. Reprinted by permission of Prentice-Hall, Inc., Englewood Cliffs, N.J.

Suppose that $R_1 = 0.1636$; then by the exact method given by equation (8.2a),

$$\begin{aligned} Y_1 &= -\beta \ln (1 - R_1) \\ &= -40 \ln (1 - 0.1636) \\ &= 7.15 \end{aligned}$$

Using the approximate method, first note that $R_1 = 0.1636$ lies in interval $i = 2$, that is, $r_2 = 0.1 \leq R_1 = 0.1636 < r_3 = 0.2$, so that using Table 8.6,

$$\begin{aligned} X_1 &= x_2 + a_2(R_1 - r_2) \\ &= 0.104 + 1.18(0.1636 - 0.1) \\ &= 0.179 \end{aligned}$$

and using equation (8.13), with $\beta = 40$,

$$Y_1 = 40X_1 = 7.16$$

Five additional values plus the one here are exhibited in Table 8.8.

Table 8.8. GENERATING EXPONENTIAL VARIATES, BY EXACT AND APPROXIMATE METHODS

i	R_i	Y_i (*Exact*)	Y_i (*Approximate*)	*Percent Error*
1	0.1636	7.15	7.16	0.14
2	0.9040	93.74	93.76	0.02
3	0.1871	8.29	8.27	0.24
4	0.7824	61.00	60.90	0.16
5	0.5905	35.71	35.75	0.11
6	0.0500	2.05	2.08	1.38

The first five uniform variates were chosen from Table A.1, but the last value $R_6 = 0.05$, was chosen arbitrarily to show that the relative error in this particular piecewise linear approximation can be as high as 1.4%. Another criticism of the approximation is that the range of the generated variable is restricted to $0 \leq X \leq 8$, the maximum x_i value in Table 8.6, whereas the range of an exponential is all nonnegative values, $0 \leq X < \infty$. For an exponential random variable X with mean 1, the probability that X is greater than 8, namely

$$\Pr(X > 8) = e^{-8} = 0.00034$$

may seem quite small, and in fact such values will be generated by the exact method given by equation (8.2) only very rarely (about 34 such values in every 100,000 generated values), but in some situations if large numbers of variates will be generated and if a large value of X has a highly significant effect on the system, the limitations of the standard approximation in Table 8.6 become more important. Similar criticisms have been leveled at the approximation to the normal cdf in Table 8.7. In general, it is recommended that an exact method, such as one of those discussed in this chapter, be used whenever possible. (GPSS V is the only major discrete-event simulation language which relies exclusively on the table-lookup technique and numerical approximations. However, GPSS V does have the capability to call FORTRAN and PL/1 routines through use of the HELP blocks, and this capability can be exploited to use exact methods for generating variates from the standard statistical distributions.)

The use of Table 8.7 for generating approximate normal variates is left for Exercise 24. For further discussion of the table-lookup procedure in GPSS V, and in particular the method employed by GPSS to generate exponential and normal variates, Chapter 9 of Gordon [1975] is recommended.

8.1.7. Discrete Distributions

All discrete distributions can be generated using the inverse transform technique, either numerically through a table-lookup procedure, or in some cases algebraically with the final generation scheme in terms of a formula. Other techniques are sometimes used for certain distributions, such as the convolution technique for the binomial distribution. Some of these methods are discussed in later sections. This subsection gives examples covering both empirical distributions and two of the standard discrete distributions, the (discrete) uniform and the geometric.

Example 8.5 An Empirical Discrete Distribution

At the end of the day, the number of shipments on the loading docks of the IHW Company (whose main product is the famous incredibly huge widget) is either 0, 1, or 2, with observed relative frequency of occurrence of 0.50, 0.30, and 0.20, respectively. Internal consultants have been asked to develop a model to improve the efficiency of the loading and hauling operations, and as part of this model, they will need to be able to generate values, X, to represent the number of shipments on the loading docks at the end of each day. The consultants decide to model X as a discrete random variable with distribution as given in Table 8.9 and shown in Figure 8.8. The probability mass

Table 8.9. Distribution of Number of Shipments, X

x	$p(x)$	$F(x)$
0	0.50	0.50
1	0.30	0.80
2	0.20	1.00

function (pmf), $p(x)$, is given by

$$\begin{aligned} p(0) &= P(X = 0) = 0.50 \\ p(1) &= P(X = 1) = 0.30 \\ p(2) &= P(X = 2) = 0.20 \end{aligned}$$

and the cdf, $F(x) = P(X \leq x)$, is given by

$$F(x) = \begin{cases} 0 & x < 0 \\ 0.5 & 0 \leq x < 1 \\ 0.8 & 1 \leq x < 2 \\ 1.0 & 2 \leq x \end{cases}$$

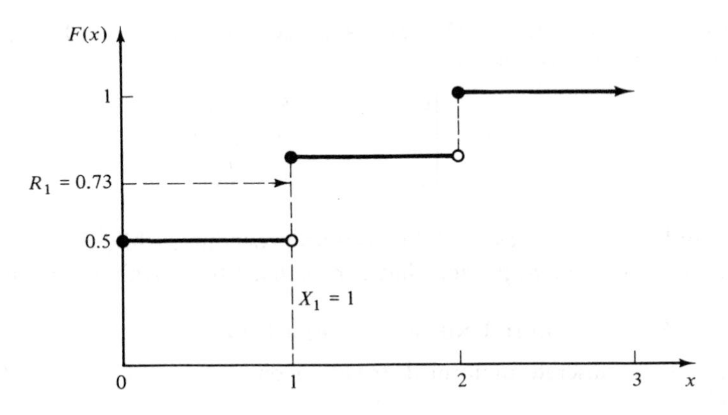

Figure 8.8. CDF of number of shipments, X.

Recall that the cdf of a discrete random variable always consists of horizontal line segments with jumps of size $p(x)$ at those points, x, which the random variable can assume. For example, in Figure 8.8, there is a jump of size $p(0) = 0.5$ at $x = 0$, of size $p(1) = 0.3$ at $x = 1$, and of size $p(2) = 0.2$ at $x = 2$.

For generating discrete random variables, the inverse transform technique becomes a table-lookup procedure, but unlike the case of continuous variables, interpolation is not required. To illustrate the procedure, suppose that $R_1 = 0.73$ is generated. Graphically, as illustrated in Figure 8.8, first locate $R_1 = 0.73$ on the vertical axis, next draw a horizontal line segment until it hits a "jump" in the cdf, and then drop a perpendicular to the horizontal axis to get the generated variate. Here $R_1 = 0.73$ is transformed to $X_1 = 1$. This procedure is analogous to the procedure used for empirical continuous distributions in Section 8.1.5 and illustrated in Figure 8.5, except that the final step of linear interpolation is eliminated.

The table-lookup procedure is facilitated by construction of a table such as Table 8.10. When $R_1 = 0.73$ is generated, first find the interval in which R_1 lies. In general, for $R = R_1$, if

$$F(x_{i-1}) = r_{i-1} < R \leq r_i = F(x_i) \tag{8.14}$$

then set $X_1 = x_i$. Here $r_0 = 0$, $x_0 = -\infty$, while $x_1, x_2, \ldots, x_n$ are the possible values of the random variable, and $r_k = p(x_1) + \cdots + p(x_k)$, $k = 1, 2, \ldots, n$. For this example, $n = 3$, $x_1 = 0$, $x_2 = 1$, $x_3 = 2$, and hence $r_1 = 0.5$, $r_2 = 0.8$, and $r_3 = 1.0$. (Note that $r_n = 1.0$ in all cases.)

Table 8.10. TABLE FOR GENERATING THE DISCRETE VARIATE X

i	*Input,* r_i	*Output,* x_i
1	0.50	0
2	0.80	1
3	1.00	2

Since $r_1 = 0.5 < R_1 = 0.73 \leq r_2 = 0.8$, set $X_1 = x_2 = 1$. The generation scheme is summarized as follows:

$$X = \begin{cases} 0, & R \leq 0.5 \\ 1, & 0.5 < R \leq 0.8 \\ 2, & 0.8 < R \leq 1.0 \end{cases}$$

Example 8.5 illustrates the table-lookup procedure, while the next example illustrates an algebraic approach that can be used for certain distributions.

EXAMPLE 8.6 A DISCRETE UNIFORM DISTRIBUTION

Consider the discrete uniform distribution on $\{1, 2, \ldots, k\}$ with pmf and cdf given by

$$p(x) = \frac{1}{k}, \qquad x = 1, 2, \ldots, k$$

and

$$F(x) = \begin{cases} 0, & x < 1 \\ \frac{1}{k}, & 1 \leq x < 2 \\ \frac{2}{k}, & 2 \leq x < 3 \\ \vdots & \vdots \\ \frac{k-1}{k}, & k-1 \leq x < k \\ 1 & k \leq x \end{cases}$$

Let $x_i = i$ and $r_i = p(1) + \cdots + p(x_i) = F(x_i) = i/k$ for $i = 1, 2, \ldots, k$. Then by using inequality (8.14) it can be seen that if the generated random number R satisfies

$$r_{i-1} = \frac{i-1}{k} < R \leq r_i = \frac{i}{k} \tag{8.15}$$

then X is generated by setting $X = i$. Now, inequality (8.15) can be solved for i:

$$i - 1 < Rk \leq i$$
$$Rk \leq i < Rk + 1 \tag{8.16}$$

Let $\lceil y \rceil$ denote the smallest integer $\geq y$. For example, $\lceil 7.82 \rceil = 8$, $\lceil 5.13 \rceil = 6$, and $\lceil -1.32 \rceil = -1$. For $y \geq 0$, $\lceil y \rceil$ is a function that rounds up. This notation and inequality (8.16) yield a formula for generating X, namely

$$X = \lceil Rk \rceil \tag{8.17}$$

For example, consider generating a random variate X, uniformly distributed on $\{1, 2, \ldots, 10\}$. The variate, X, might represent the number of pallets to be loaded onto a truck. Using Table A.1 as a source of random numbers, R, and equation (8.17) with $k = 10$ yields

$$\begin{aligned}
R_1 &= 0.78 & X_1 &= \lceil 7.8 \rceil = 8 \\
R_2 &= 0.03 & X_2 &= \lceil 0.3 \rceil = 1 \\
R_3 &= 0.23 & X_3 &= \lceil 2.3 \rceil = 3 \\
R_4 &= 0.97 & X_4 &= \lceil 9.7 \rceil = 10
\end{aligned}$$

The procedure discussed here can be modified to generate a discrete uniform random variate with any range consisting of consecutive integers. Exercise 13 asks the student to devise a procedure for one such case.

EXAMPLE 8.7

Consider the discrete distribution with pmf given by

$$p(x) = \frac{2x}{k(k+1)}, \qquad x = 1, 2, \ldots, k$$

(This example is taken from Schmidt and Taylor [1970].) For integer values of x in the range $\{1, 2, \ldots, k\}$, the cdf is given by

$$\begin{aligned}
F(x) &= \sum_{i=1}^{x} \frac{2i}{k(k+1)} \\
&= \frac{2}{k(k+1)} \sum_{i=1}^{x} i \\
&= \frac{2}{k(k+1)} \frac{x(x+1)}{2} \\
&= \frac{x(x+1)}{k(k+1)}
\end{aligned}$$

Generate R and use inequality (8.14) to conclude that $X = x$ whenever

$$F(x-1) = \frac{(x-1)x}{k(k+1)} < R \leq \frac{x(x+1)}{k(k+1)} = F(x)$$

or, whenever

$$(x-1)x < k(k+1)R \leq x(x+1)$$

To solve this inequality for x in terms of R, first find a value of x that satisfies

$$(x-1)x = k(k+1)R$$

or

$$x^2 - x - k(k+1)R = 0$$

Then by rounding up, the solution is $X = \lceil x - 1 \rceil$. By the quadratic formula, namely

$$x = \frac{-b \pm \sqrt{b^2 - 4ac}}{2a}$$

with $a = 1$, $b = -1$, and $c = -k(k+1)R$, the solution to the quadratic equation is

$$x = \frac{1 \pm \sqrt{1 + 4k(k+1)R}}{2} \tag{8.18}$$

The positive root in equation (8.18) is the correct one to use (why?), so X is generated by

$$X = \left\lceil \frac{1 + \sqrt{1 + 4k(k+1)R}}{2} - 1 \right\rceil \tag{8.19}$$

Exercise 14 asks the student to generate a few values from this distribution.

EXAMPLE 8.8 THE GEOMETRIC DISTRIBUTION

Consider the geometric distribution with pmf

$$p(x) = p(1-p)^x, \qquad x = 0, 1, 2, \ldots$$

where $0 < p < 1$. Its cdf is given by

$$\begin{aligned} F(x) &= \sum_{j=0}^{x} p(1-p)^j \\ &= \frac{p\{1-(1-p)^{x+1}\}}{1-(1-p)} \\ &= 1-(1-p)^{x+1} \end{aligned}$$

for $x = 0, 1, 2, \ldots$. Using the inverse transform technique [i.e., inequality (8.14)], recall that a geometric random variable X will assume the value x whenever

$$F(x-1) = 1-(1-p)^x < R \leq 1-(1-p)^{x+1} = F(x) \tag{8.20}$$

where R is a generated random number assumed $0 < R < 1$. Solving inequality (8.20) for x proceeds as follows:

$$\begin{aligned} (1-p)^{x+1} &\leq 1-R < (1-p)^x \\ (x+1)\,\ell\text{n}\,(1-p) &\leq \ell\text{n}\,(1-R) < x\,\ell\text{n}\,(1-p) \end{aligned}$$

But $1-p < 1$ implies that $\ell\text{n}\,(1-p) < 0$, so that

$$\frac{\ell\text{n}\,(1-R)}{\ell\text{n}\,(1-p)} - 1 \leq x < \frac{\ell\text{n}\,(1-R)}{\ell\text{n}\,(1-p)} \tag{8.21}$$

Thus, $X = x$ for that integer value of x satisfying inequality (8.21), or, in brief, using the round-up function $\lceil \cdot \rceil$,

$$X = \left\lceil \frac{\ell\text{n}\,(1-R)}{\ell\text{n}\,(1-p)} - 1 \right\rceil \tag{8.22}$$

Since p is a fixed parameter, let $\beta = -1/\ell\text{n}\,(1-p)$. Then $\beta > 0$ and by equation (8.22), $X = \lceil -\beta\,\ell\text{n}\,(1-R) - 1 \rceil$. By equation (8.1), $-\beta\,\ell\text{n}\,(1-R)$ is an exponentially distributed random variable with mean β, so that one way of generating a geometric variate with parameter p is to generate (by any method) an exponential variate with parameter $\beta^{-1} = -\ell\text{n}\,(1-p)$, subtract one, and round up.

Occasionally, a geometric variate X is needed which can assume values $\{q, q+1,$

$q + 2, \ldots\}$ with pmf $p(x) = p(1 - p)^{x-q}$ $(x = q, q + 1, \ldots)$. Such a variate, X, can be generated, using equation (8.22), by

$$X = q + \left\lceil \frac{\ell n\,(1 - R)}{\ell n\,(1 - p)} - 1 \right\rceil \tag{8.23}$$

One of the most common cases is $q = 1$.

EXAMPLE 8.9

Generate three values from a geometric distribution on the range $\{X \geq 1\}$ with mean 2. Such a geometric distribution has pmf $p(x) = p(1 - p)^{x-1}$ $(x = 1, 2, \ldots)$ with mean $1/p = 2$, or $p = 1/2$. Thus, X can be generated by equation (8.23) with $q = 1$, $p = 1/2$, and $1/\ell n\,(1 - p) = -1.443$. Using Table A.1, $R_1 = 0.932$, $R_2 = 0.105$, and $R_3 = 0.687$, which yields

$$\begin{aligned} X_1 &= 1 + \lceil -1.443\,\ell n\,(1 - 0.932) - 1 \rceil \\ &= 1 + \lceil 3.878 - 1 \rceil = 4 \\ X_2 &= 1 + \lceil -1.443\,\ell n\,(1 - 0.105) - 1 \rceil = 1 \\ X_3 &= 1 + \lceil -1.443\,\ell n\,(1 - 0.687) - 1 \rceil = 2 \end{aligned}$$

Exercise 15 deals with an application of the geometric distribution.

8.2. Direct Transformation for the Normal Distribution

Many methods have been developed for generating normally distributed random variates. The inverse transform technique cannot be applied, however, because the inverse cdf cannot be computed analytically. (The inverse transform technique was applied in Section 8.1.6 to a piecewise linear approximation to the normal cdf.) The standard normal cdf is given by

$$\Phi(x) = \int_{-\infty}^{x} \frac{1}{\sqrt{2\pi}} e^{-t^2/2}\, dt, \qquad -\infty < x < \infty$$

To use the inverse transform technique, it is necessary to be able to solve (in closed form) $\Phi(X) = R$ for X in terms of R (try it—it is impossible!). Other techniques that have been used include the convolution technique (Section 8.3) and acceptance–rejection methods (Section 8.4). This section describes an intuitively appealing direct transformation that produces an independent pair of standard normal variates with mean zero and variance 1. The method is due to Box and Muller [1958]. Although not as efficient as many more modern techniques, it is easy to program in a scientific language such as FORTRAN.

Consider two standard normal random variables, Z_1 and Z_2, plotted as a point in the plane as shown in Figure 8.9 and represented in polar coordinates

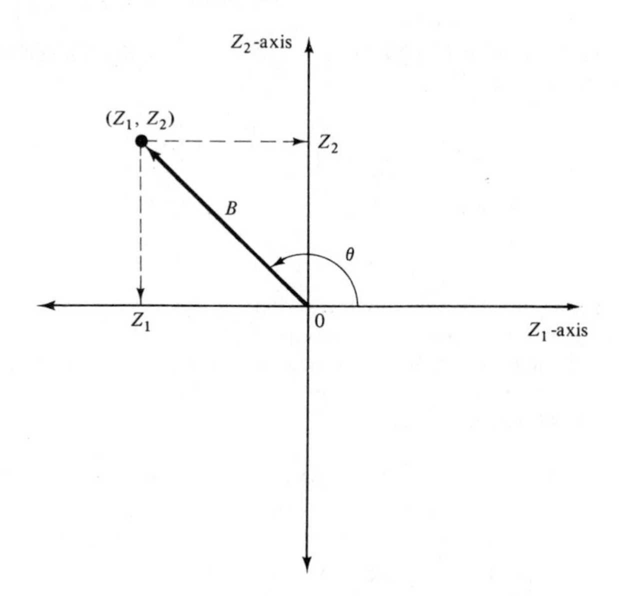

Figure 8.9. Polar representation of a pair of standard normal variables.

as

$$Z_1 = B \cos \theta$$

and

$$Z_2 = B \sin \theta \tag{8.24}$$

It is known that $B^2 = Z_1^2 + Z_2^2$ has the chi-square distribution with 2 degrees of freedom, which is equivalent to an exponential distribution with mean 2. Thus, the radius, B, can be generated by use of equation (8.2b):

$$B = (-2 \ell\text{n } R)^{1/2} \tag{8.25}$$

By the symmetry of the normal distribution, it seems reasonable to suppose, and indeed it is the case, that the angle θ is uniformly distributed between 0 and 2π radians. In addition, the radius, B, and the angle, θ, are mutually independent. Combining Equations (8.24) and (8.25) gives a direct method for generating two independent standard normal variates, Z_1 and Z_2, from two independent random numbers R_1 and R_2:

$$\begin{aligned} Z_1 &= (-2 \ell\text{n } R_1)^{1/2} \cos (2\pi R_2) \\ Z_2 &= (-2 \ell\text{n } R_1)^{1/2} \sin (2\pi R_2) \end{aligned} \tag{8.26}$$

An example of the use of equations (8.26) is given in Section 8.3.2.

8.3. Convolution Method

The probability distribution of a sum of two or more independent random variables is called a convolution of the distributions of the original variables. The convolution method thus refers to adding together two or more random variables to obtain a new random variable with the desired distribution. This technique can be applied to obtain Erlang variates, approximately normally distributed variates, and binomial variates. What is important is not the cdf of the desired random variable, but rather its relation to other more easily generated variates.

8.3.1. Erlang Distribution

As discussed in Section 4.4, an Erlang random variable X with parameters (K, θ) can be shown to be the sum of K independent exponential random variables, X_i $(i = 1, \ldots, K)$, each having mean $1/K\theta$; that is,

$$X = \sum_{i=1}^{K} X_i$$

Since each X_i can be generated by equation (8.2b) with $1/\lambda = 1/K\theta$, an Erlang variate can be generated by

$$\begin{aligned} X &= \sum_{i=1}^{K} \frac{-1}{K\theta} \ln R_i \\ &= \frac{-1}{K\theta} \ln \left(\prod_{i=1}^{K} R_i \right) \end{aligned} \tag{8.27}$$

In equation (8.27), Π stands for product. It is more efficient computationally to multiply all the random numbers first and then to compute only one logarithm.

Example 8.10

Trucks arrive at a large warehouse in a completely random fashion which is modeled as a Poisson process with arrival rate $\lambda = 10$ trucks per hour. The guard at the entrance sends trucks alternately to the north and south docks. An analyst has developed a model to study the loading/unloading process at the south docks, and needs a model of the arrival process at the south docks alone. An interarrival time X between successive truck arrivals at the south docks is equal to the sum of two interarrival times at the entrance and thus it is the sum of two exponential random variables each with mean 0.1 hour, or 6 minutes. Thus, X has the Erlang distribution with $K = 2$ and mean $1/\theta = 2/\lambda = 0.2$ hour. To generate the variate X, first obtain $K = 2$ random numbers from Table A.1, say $R_1 = 0.937$ and $R_2 = 0.217$. Then by equation (8.27),

$$\begin{aligned} X &= -0.1 \ln [0.937(0.217)] \\ &= 0.159 \text{ hour} = 9.56 \text{ minutes} \end{aligned}$$

In general, equation (8.27) implies that K uniform random numbers are needed for each Erlang variate generated. If K is large, it is more efficient to generate Erlang variates by other techniques, such as one of the many acceptance–rejection techniques for the gamma distribution given by Fishman [1978].

8.3.2. Generating Approximate Normal Variates

The central limit theorem asserts that the sum of n independent and identically distributed random variables, $X_1, X_2, \ldots, X_n$, each with mean μ_X and finite variance σ_X^2, is approximately normally distributed with mean $n\mu_X$ and variance $n\sigma_X^2$. Applying this to uniform random variables on (0, 1) which have mean $\mu = 0.5$ and variance $\sigma^2 = 1/12$, it follows that

$$Z = \frac{\sum_{i=1}^{n} R_i - 0.5n}{(n/12)^{1/2}} \tag{8.28}$$

is approximately normally distributed with mean zero and variance 1. The approximation becomes better the larger the value of n, but many textbook authors assert that $n = 12$ is sufficiently large for an adequate approximation to normality, and moreover, the use of $n = 12$ is the most efficient for a computerized routine because a square root and a division operation are avoided, as shown by setting $n = 12$ into equation (8.28) to obtain the generation scheme:

$$Z = \sum_{i=1}^{12} R_i - 6 \tag{8.29}$$

for an approximately normal random variable with mean zero and variance 1. If it is desired to generate a normal variate Y with mean μ_Y and variance σ_Y^2, first generate Z by equation (8.29) and then use the transformation

$$Y = \mu_Y + \sigma_Y Z \tag{8.30}$$

Contrast equation (8.30), for converting a standard normal variate to the desired normal variate, with equation (8.13) for converting a standard exponential variate to the desired exponential variate. In the first case the standard variate is multiplied by the standard deviation, σ, whereas in the latter case, it is multiplied by the mean.

Example 8.11

Service times at a cashier's window are normally distributed with mean $\mu = 7.3$ minutes and variance $\sigma^2 = 11.7$ minutes². To generate a typical service time, first obtain 12 random numbers from Table A.1:

0.1758	0.1489	0.2774	0.6033	0.9813	0.1052
0.1816	0.7484	0.1699	0.7350	0.6430	0.8803

Then use equations (8.29) and (8.30) to obtain

$$Y = 7.3 + \sqrt{11.7}\left(\sum_{i=1}^{12} R_i - 6\right) = 6.10$$

Many authors of simulation texts recommend this technique for generating approximate normal random variates, but an exact technique, such as the one in Section 8.2, is always preferable to an approximate technique. Many exact techniques are known, and some are both easy to use and efficient in a computerized routine. (The interested reader is referred to Fishman [1978].) To illustrate one exact generation scheme, consider equation (8.26) with $R_1 = 0.1758$ and $R_2 = 0.1489$. Two normal random variates are generated as follows:

$$Z_1 = [-2 \ln (0.1758)]^{1/2} \cos 2\pi(0.1489) = 1.11$$
$$Z_2 = [-2 \ln (0.1758)]^{1/2} \sin 2\pi(0.1489) = 1.50$$

This technique requires one-twelfth the random numbers required in the approximate technique; however, sine, cosine, and logarithm calculations are relatively inefficient on a computer. More efficient techniques are discussed by Fishman [1978] and Schmeiser [1980].

8.4. Acceptance-Rejection Technique

Suppose that an analyst needed to devise a method for generating random variates, X, uniformly distributed between 1/4 and 1. One way to proceed would be to follow these steps:

Step 1. Generate a random number R.

Step 2a. If $R \geq 1/4$, accept $X = R$, then go to step 3.

Step 2b. If $R < 1/4$, reject R, and return to step 1.

Step 3. If another uniform random variate on [1/4, 1] is needed, repeat the procedure beginning at step 1. If not, stop.

Each time step 1 is executed, a new random number R must be generated. Step 2a is an "acceptance" and step 2b is a "rejection" in this acceptance–rejection technique. To summarize the technique, random variates (R) with some distribution (here uniform on [0, 1]) are generated until some condition ($R > 1/4$) is satisfied. When the condition is finally satisfied, the desired random variate, X

(here uniform on [1/4, 1]), can be computed ($X = R$). This procedure can be shown to be correct by recognizing that the accepted values of R are conditioned values; that is, R itself does not have the desired distribution, but R conditioned on the event $\{R \geq 1/4\}$ does have the desired distribution. To show this, take $1/4 \leq a < b \leq 1$; then

$$P(a < R \leq b \mid 1/4 \leq R \leq 1) = \frac{P(a < R \leq b)}{P(1/4 \leq R \leq 1)} = \frac{b - a}{3/4} \tag{8.31}$$

which is the correct probability for a uniform distribution on [1/4, 1]. Equation (8.31) says that the probability distribution of R, given that R is between 1/4 and 1 (all other values of R are thrown out), is the desired distribution. Therefore, if $1/4 \leq R \leq 1$, set $X = R$.

The efficiency of an acceptance–rejection technique depends heavily on being able to minimize the number of rejections. In this example, the probability of a rejection is $P(R < 1/4) = 1/4$, so that the number of rejections is a geometrically distributed random variable with probability of "success" being $p = 3/4$ and mean number of rejections $(1/p - 1) = 4/3 - 1 = 1/3$. (Example 8.8 discussed the geometric distribution.) The mean number of random numbers R required to generate one variate X is one more than the number of rejections; hence, it is $4/3 = 1.33$. In other words, to generate 1000 values of X would require approximately 1333 random numbers R.

In the present situation an alternative procedure exists for generating a uniform variate on [1/4, 1], namely equation (8.4), which reduces to $X = 1/4 + (3/4)R$. Whether the acceptance–rejection technique or an alternative procedure such as the inverse-transform technique [equation (8.4)] is the more efficient depends on several considerations. The computer being used, the skills of the programmer and the relative efficiency of generating the additional (rejected) random numbers needed by acceptance–rejection should be compared to the computations required by the alternative procedure. In practice, concern with generation efficiency is left to specialists who conduct extensive tests comparing alternative methods (i.e., until a simulation model begins to require excessive computer runtime due to the generator being used).

For the uniform distribution on [1/4, 1], the inverse transform technique of equation (8.4) is undoubtedly much easier to apply and probably more efficient than the acceptance–rejection technique. The main purpose of this example was to explain and motivate the basic concept of the acceptance–rejection technique. However, for some important distributions such as the gamma, the inverse cdf does not exist in closed form and therefore the inverse transform technique is not applicable. For other important distributions such as the exponential and normal, acceptance–rejection and other more advanced techniques lead to much more efficient generation schemes. These more advanced techniques are summarized by Fishman [1978].

In the following subsections, the acceptance–rejection technique is illustrated for the generation of random variates for the Poisson and gamma distributions.

8.4.1. Poisson Distribution

A Poisson random variable, N, with mean $\alpha > 0$, has pmf

$$p(n) = P(N = n) = \frac{e^{-\alpha}\alpha^n}{n!}, \qquad n = 0, 1, 2, \ldots$$

but more important, N can be interpreted as the number of arrivals from a Poisson arrival process in one unit of time. Recall from Section 4.5 that the interarrival times, $A_1, A_2, \ldots$ of successive customers are exponentially distributed with rate α (i.e., α is the mean number of arrivals per unit time); in addition, an exponential variate can be generated by Equation (8.2b). Thus there is a relationship between the (discrete) Poisson distribution and the (continuous) exponential distribution, namely:

$$N = n \tag{8.32a}$$

if and only if

$$A_1 + A_2 + \cdots + A_n \leq 1 < A_1 + \cdots + A_n + A_{n+1} \tag{8.32b}$$

Equation (8.32a), $N = n$, says there were exactly n arrivals during one unit of time; but relation (8.32b) says that the nth arrival occurred before time 1 while the $(n + 1)$st arrival occurred after time 1. Clearly, these two statements are equivalent. Proceed now by generating exponential interarrival times until some arrival, say $n + 1$, occurs after time 1; then set $N = n$.

For efficient generation purposes, relation (8.32b) is usually simplified by first using equation (8.2b), $A_i = (-1/\alpha)\, \ell\text{n}\, R_i$, to obtain

$$\sum_{i=1}^{n} \frac{-1}{\alpha}\, \ell\text{n}\, R_i \leq 1 < \sum_{i=1}^{n+1} \frac{-1}{\alpha}\, \ell\text{n}\, R_i$$

Next multiply through by $-\alpha$, which reverses the sign of the inequality, and use the fact that a sum of logarithms is the logarithm of a product, to get

$$\ell\text{n} \prod_{1}^{n} R_i = \sum_{i=1}^{n} \ell\text{n}\, R_i \geq -\alpha > \sum_{i=1}^{n+1} \ell\text{n}\, R_i = \ell\text{n} \prod_{1}^{n+1} R_i$$

Finally, use the relation $e^{\ell\text{n} x} = x$ for any number x to obtain

$$\prod_{1}^{n} R_i \geq e^{-\alpha} > \prod_{1}^{n+1} R_i \tag{8.33}$$

which is equivalent to relation (8.32b). The procedure for generating a Poisson random variate, N, is given by the following steps:

Step 1. Set $n = 0, P = 1$.

Step 2. Generate a random number R_{n+1} and replace P by $P \cdot R_{n+1}$.

Step 3. If $P < e^{-\alpha}$, then accept $N = n$. Otherwise, reject the current n, increase n by one, and return to step 2.

Note that upon completion of step 2, P is equal to the rightmost expression in relation (8.33). The basic idea of a rejection technique is again exhibited; if $P \geq e^{-\alpha}$ in step 3, then n is rejected and the generation process must proceed through at least one more trial.

How many random numbers will be required, on the average, to generate one Poisson variate, N? If $N = n$, then $n + 1$ random numbers are required, so the average number is given by

$$E(N + 1) = \alpha + 1$$

which is quite large if the mean, α, of the Poisson distribution is large.

EXAMPLE 8.12

Generate three Poisson variates with mean $\alpha = 0.2$. First compute $e^{-\alpha} = e^{-0.2} = 0.8187$. Next get a sequence of random numbers R from Table A.1 and follow steps 1 to 3 above:

Step 1. Set $n = 0, P = 1$.

Step 2. $R_1 = 0.4357, P = 1 \cdot R_1 = 0.4357$.

Step 3. Since $P = 0.4357 < e^{-\alpha} = 0.8187$, accept $N = 0$.

Step 1-3. ($R_1 = 0.4146$ leads to $N = 0$.)

Step 1. Set $n = 0, P = 1$.

Step 2. $R_1 = 0.8353, P = 1 \cdot R_1 = 0.8353$.

Step 3. Since $P \geq e^{-\alpha}$, reject $n = 0$ and return to step 2 with $n = 1$.

Step 2. $R_2 = 0.9952$, $P = R_1 \cdot R_2 = 0.8313$.

Step 3. Since $P \geq e^{-\alpha}$, reject $n = 1$ and return to step 2 with $n = 2$.

Step 2. $R_3 = 0.8004$, $P = R_1 R_2 R_3 = 0.6654$.

Step. 3. Since $P < e^{-\alpha}$, accept $N = 2$.

The calculations required for the generation of these three Poisson random variates are summarized as follows:

n	R_{n+1}	P	*Accept/Reject*	*Result*
0	0.4357	0.4357	$P < e^{-\alpha}$ (accept)	$N = 0$
0	0.4146	0.4146	$P < e^{-\alpha}$ (accept)	$N = 0$
0	0.8353	0.8353	$P \geq e^{-\alpha}$ (reject)	
1	0.9952	0.8313	$P \geq e^{-\alpha}$ (reject)	
2	0.8004	0.6654	$P < e^{-\alpha}$ (accept)	$N = 2$

It took five random numbers, R, to generate three Poisson variates here ($N = 0$, $N = 0$, and $N = 2$), but in the long run, to generate, say, 1000 Poisson variates with mean $\alpha = 0.2$, it would require approximately $1000(\alpha + 1)$ or 1200 random numbers.

EXAMPLE 8.13

Buses arrive at the bus stop at Peachtree and North Avenue according to a Poisson process with a mean of one bus per 15 minutes. Generate a random variate, N, which represents the number of arriving buses during a 1-hour time slot. Now, N is Poisson distributed with a mean of four buses per hour. First compute $e^{-\alpha} = e^{-4} = 0.0183$. Using the same sequence of random numbers as was used in Example 8.11 yields the following summarized results:

n	R_{n+1}	P	*Accept/Reject*	*Result*
0	0.4357	0.4357	$P \geq e^{-\alpha}$ (reject)	
1	0.4146	0.1806	$P \geq e^{-\alpha}$ (reject)	
2	0.8353	0.1508	$P \geq e^{-\alpha}$ (reject)	
3	0.9952	0.1502	$P \geq e^{-\alpha}$ (reject)	
4	0.8004	0.1202	$P \geq e^{-\alpha}$ (reject)	
5	0.7945	0.0955	$P \geq e^{-\alpha}$ (reject)	
6	0.1530	0.0146	$P < e^{-\alpha}$ (accept)	$N = 6$

It is immediately seen that a larger value of α (here $\alpha = 4$) usually requires more random numbers; if 1000 Poisson variates were desired, approximately $1000(\alpha + 1) = 5000$ random numbers would be required.

When α is large, say $\alpha \geq 15$, the rejection technique outlined here becomes quite expensive, but fortunately an approximate technique based on the normal distribution works quite well. When the mean, α, is large,

$$Z = \frac{N - \alpha}{\sqrt{\alpha}}$$

is approximately normally distributed with mean zero and variance 1, which suggests an approximate technique. First generate a standard normal variate Z, by equation (8.26) or (8.28), then generate the desired Poisson variate, N, by

$$N = \lceil \alpha + \sqrt{\alpha}\, Z - 0.5 \rceil \qquad (8.34)$$

where $\lceil \cdot \rceil$ is the round up function described in Section 8.1.7. (If $\alpha + \sqrt{\alpha}\, Z - 0.5 < 0$, then set $N = 0$.) The "0.5" used in the formula makes the round up function become a "round to the nearest integer" function. Equation (8.34) is not an acceptance–rejection technique, but used as an alternative to the acceptance–rejection method, it provides a fairly efficient and accurate method for generating Poisson variates with a large mean.

8.4.2. Gamma Distribution

Several acceptance–rejection techniques for generating gamma random variates have been developed [Fishman, 1978]. One of the most efficient is due to Cheng [1977]; the mean number of trials is between 1.13 and 1.47 for any value of the shape parameter $\beta \geq 1$.

If the shape parameter β is an integer, say $\beta = k$, one possibility is to use the convolution technique in Section 8.3.1 since the Erlang distribution is a special case of the more general gamma distribution. On the other hand, the acceptance-rejection technique described here would be a highly efficient method for the Erlang distribution especially if $\beta = k$ were large. The routine generates gamma random variates with scale parameter θ and shape parameter β, that is, with mean $1/\theta$ and variance $1/\beta\theta^2$. The steps are as follows:

Step 1. Compute $a = (2\beta - 1)^{1/2}$, $b = 2\beta - \ln 4 + 1/a$.

Step 2. Generate R_1 and R_2.

Step 3. Compute $X = \beta[R_1/(1 - R_1)]^a$.

Step 4a. If $X > b - \ln(R_1^2 R_2)$, reject X and return to step 2.

Step 4b. If $X \leq b - \ln(R_1^2 R_2)$, use X as the desired variate.
The generated variates from step 4b will have mean and variance both equal to β. If it is desired to have mean $1/\theta$ and variance $1/\beta\theta^2$ as in Section 4.4, then include

Step 5. Replace X by $X/\beta\theta$.

The basic idea of all acceptance–rejection methods is again illustrated here, but the proof of this example is beyond the scope of this book. In step 3, $X = \beta[R_1/(1 - R_1)]^a$ is not gamma distributed, but rejection of certain values of X in step 4a guarantees that the accepted values in step 4b do have the gamma distribution.

EXAMPLE 8.14

Downtimes for a high-production candy-making machine have been found to be gamma distributed with mean 2.2 minutes and variance 2.10 minutes^2. Thus, $1/\theta = 2.2$ and $1/\beta\theta^2 = 2.10$, which implies that $\beta = 2.30$ and $\theta = 0.4545$.

Step 1. $a = 1.90$, $b = 3.74$.

Step 2. Generate $R_1 = 0.832$, $R_2 = 0.021$.

Step 3. Compute $X = 2.3(0.832/0.168)^{1.9} = 48.1$.

Step 4. $X = 48.1 > 3.74 - \ell n\,[(0.832)^2 0.021] = 7.97$, so reject X and return to step 2.

Step 2. Generate $R_1 = 0.434$, $R_2 = 0.716$.

Step 3. Compute $X = 2.3(0.434/0.566)^{1.9} = 1.389$.

Step 4. Since $X = 1.389 \leq 3.74 - \ell n\,[(0.434)^2 0.716] = 5.74$, accept X.

Step 5. Divide X by $\beta\theta = 1.045$ to get $X = 1.329$.

This example took two trials (i.e., one rejection) to generate an acceptable gamma-distributed random variate, but on the average to generate, say, 1000 gamma variates, the method will require between 1130 and 1470 trials, or equivalently, between 2260 and 2940 random numbers. The method is somewhat cumbersome for hand calculations, but is easy to program on the computer and presently is one of the most efficient gamma generators known.

8.5. Summary

The basic principles of random variate generation using the inverse transform technique, the convolution method, and acceptance–rejection techniques have been introduced and illustrated by examples. Methods for generating most of the important continuous and discrete distributions, as well as empirical distributions, have been given. For a state-of-the-art treatment, the reader is referred to Fishman [1978] or Schmeiser [1981].

REFERENCES

BOX, G. E. P., AND M. F. MULLER [1958], "A Note on the Generation of Random Normal Deviates," *Annals of Mathematical Statistics*, Vol. 29, pp. 610–11.

CHENG, R. C. H. [1977], "The Generation of Gamma Variables," *Applied Statistician*, Vol. 26, No. 1, pp. 71–75.

FISHMAN, GEORGE S. [1978], *Principles of Discrete Event Simulation*, Wiley, New York.

GORDON, GEOFFREY [1975], *The Application of GPSS V to Discrete System Simulation*, Prentice-Hall, Englewood Cliffs, N.J.

SCHMEISER, BRUCE W. [1980], "Random Variate Generation: A Survey," in *Simulation with Discrete Models: A State of the Art View*, T. I. Ören, C. M. Shub, and P. F. Roth, eds.

SCHMIDT, J. W., AND R. E. TAYLOR [1970], *Simulation and Analysis of Industrial Systems*, Irwin, Homewood, Ill.

EXERCISES

1. Develop a random variate generator for a random variable X with the pdf

$$f(x) = \begin{cases} e^{2x}, & -\infty < x \leq 0 \\ e^{-2x}, & 0 < x < \infty \end{cases}$$

2. Develop a generation scheme for the triangular distribution with pdf

$$f(x) = \begin{cases} \frac{1}{2}(x-2), & 2 \leq x \leq 3 \\ \dfrac{1}{2}\left(2 - \dfrac{x}{3}\right), & 3 < x \leq 6 \\ 0, & \text{otherwise} \end{cases}$$

   Generate 10 values of the random variate, compute the sample mean, and compare it to the true mean of the distribution.

3. Develop a generator for a triangular distribution with range (1, 10) and mode at $x = 4$.
4. Develop a generator for a triangular distribution with range (1, 10) and a mean of 4.
5. Given the following cdf for a continuous variable with range -3 to 4, develop a generator for the variable.

$$F(x) = \begin{cases} 0, & x \leq -3 \\ \dfrac{1}{2} + \dfrac{x}{6}, & -3 < x \leq 0 \\ \dfrac{1}{2} + \dfrac{x^2}{32}, & 0 < x \leq 4 \\ 1, & x > 4 \end{cases}$$

6. Given the cdf $F(x) = x^4/16$ on $0 \leq x \leq 2$, develop a generator for this distribution.

7. Given the pdf $f(x) = x^2/9$ on $0 \leq x \leq 3$, develop a generator for this distribution.

8. Develop a generator for a random variable whose pdf is

$$f(x) = \begin{cases} \dfrac{1}{3}, & 0 \leq x \leq 2 \\ \dfrac{1}{24}, & 2 < x \leq 10 \\ 0, & \text{otherwise} \end{cases}$$

9. The cdf of a discrete random variable X is given by

$$F(x) = \frac{x(x+1)(2x+1)}{n(n+1)(2n+1)}, \qquad x = 1, 2, \ldots, n$$

When $n = 4$, generate three values of X using $R_1 = 0.83$, $R_2 = 0.24$, and $R_3 = 0.57$.

10. Times to failure for an automated production process have been found to be randomly distributed with a Weibull distribution with parameters $\beta = 2$ and $\alpha = 10$. Derive equation (8.5) and then use it to generate five values from this Weibull distribution, using five random numbers taken from Table A.1.

11. Data have been collected on service times at a drive-in bank window at the Shady Lane National Bank. This data are summarized into intervals as follows:

Interval (Seconds)	*Frequency*
15–30	10
30–45	20
45–60	25
60–90	35
90–120	30
120–180	20
180–300	10

Set up a table like Table 8.3 for generating service times by the table-lookup method and generate five values of service time using four-digit random numbers.

12. In Example 8.3, assume that fire crew response times satisfy $0.25 \leq x \leq 3$. Modify Table 8.5 to accommodate this assumption. Then generate five values of response time using four-digit uniform random numbers from Table A.1.

13. For a preliminary version of a simulation model, the number of pallets, X, to be loaded onto a truck at a loading dock was assumed to be uniformly distributed between 8 and 24. Devise a method for generating X assuming that the loads on successive trucks are independent. Use the technique of Example 8.6 for discrete uniform distributions. Finally, generate loads for 10 successive trucks by using four-digit random numbers.

14. After collecting more data, it was found that the distribution of Example 8.7 was a better approximation to the number of pallets loaded than was the uniform distribution, as was assumed in Exercise 13. Using equation (8.19) generate loads for 10 successive trucks using the same random numbers as were used in Exercise 13. Compare the results to the results of Exercise 13.

15. The weekly demand, X, for a slow-moving item has been found to be well approximated by a geometric distribution on the range $\{0, 1, 2, \ldots\}$ with mean weekly demand of 2.5 items. Generate 10 values of X, demand per week, using random numbers from Table A.1. (*Hint:* For a geometric distribution on the range $\{q, q+1, \ldots\}$ with parameter p, the mean is $1/p + q - 1$.)

16. In Exercise 15, suppose that the demand has been found to have a Poisson distribution with mean 2.5 items per week. Generate 10 values of X, demand per week, using random numbers from Table A.1. Discuss the differences between the geometric and the Poisson distributions.

17. Lead times have been found to be exponentially distributed with mean 3.7 days. Generate five random lead times from this distribution.

18. Regular maintenance of a production routine has been found to vary and has been modeled as a normally distributed random variable with mean 33 minutes and variance 4 minutes^2. Generate five random maintenance times, with the given distribution, by one of the methods of this chapter.

19. A machine is taken out of production if it fails, or after 5 hours, whichever comes first. By running similar machines until failure, it has been found that time to failure, X, has the Weibull distribution with $\alpha = 8$, $\beta = 0.75$, and $\nu = 0$ (refer to Sections 4.4 and 8.1.3). Thus, the time until the machine is taken out of production can be represented as $Y = \min(X, 5)$. Develop a step-by-step procedure for generating Y.

20. The time until a component is taken out of service is uniformly distributed on 0 to 8 hours. Two such independent components are put in series, and the whole system goes down when one of the components goes down. If $X_i (i = 1, 2)$ represents the component runtimes, then $Y = \min(X_1, X_2)$ represents the system lifetime. Devise two distinct ways to generate Y. [*Hint:* One way is relatively straightforward. For a second method, first compute the cdf of Y: $F_Y(y) = P(Y \leq y) = 1 - P(Y > y)$, for $0 \leq y \leq 8$. Use the equivalence $\{Y > y\} = \{X_1 > y \text{ and } X_2 > y\}$ and the independence of X_1 and X_2. After finding $F_Y(y)$, proceed with the inverse transform technique.]

21. In Exercise 20, component lifetimes are exponentially distributed, one with mean 2 hours and the other with mean 6 hours. Rework Exercise 20 under this new assumption. Discuss the relative efficiency of the two generation schemes devised.

22. Develop a technique for generating a binomial random variable, X, using the convolution technique. [*Hint:* X can be represented as the number of successes in n independent Bernoulli trials, each success having probability p. Thus, $X = \sum_{i=1}^{n} X_i$, where $P(X_i = 1) = p$ and $P(X_i = 0) = 1 - p$.]

23. Develop an acceptance–rejection technique for generating a geometric random variable, X, with parameter p on the range $\{0, 1, 2, \ldots\}$. (*Hint:* X can be thought of as the number of trials before the first success occurs in a sequence of independent Bernoulli trials.)

24. Write a computer routine in FORTRAN or BASIC to generate standard normal variates by the exact method discussed in this chapter. Use it to generate 1000 values. Compare the true probability, $\Phi(z)$, that a value lies in $(-\infty, z)$ to the actual observed relative frequency that values were $\leq z$, for $z = -4, -3, -2, -1, 0, 1, 2, 3$, and 4. Repeat for each of the two approximate methods. Compare the three methods.

25. Write a computer routine in FORTRAN or BASIC to generate gamma variates with shape parameter β and scale parameter θ. Generate 1000 values with $\beta = 2.5$ and $\theta = 0.2$ and compare the true mean, $1/\theta = 5$, to the sample mean.

26. Write a computer routine in FORTRAN or BASIC to generate 200 values from one of the variates in Exercises 1 to 23. Make a histogram of the 200 values and compare it to the theoretical density function (or probability mass function for discrete random variables).

part four

ANALYSIS OF SIMULATION DATA

INPUT DATA ANALYSIS

Input data provide the driving force for a simulation model. In the simulation of a queueing system, typical input data might be the distributions of time between arrivals and service times. For an inventory system simulation, input data include the distributions of demand and lead time. For the simulation of a reliability system, an example is the distribution of time to failure of a component.

The examples and exercises in Chapters 2 and 3 specify the distributions. However, determining appropriate distributions for input data is a major task in simulation from the standpoint of time and resource requirements. Additionally, regardless of the sophistication level of the analyst, faulty assumptions on inputs will lead to outputs whose interpretation may give rise to misleading recommendations.

There are four steps in the development of a valid model of input data. The first step, the collection of raw data, is often a major time consumer in a real-world simulation. The second step, the identification of the underlying statistical distribution, begins by developing a frequency distribution, or histogram, of the data. From the frequency distribution, a distributional assumption is made. Fortunately, as described in Chapter 4, several well-known distributions occur rather frequently in practice. In the third step, estimates are made of the parameters that characterize the distribution.

Finally, in the fourth step, the distributional assumption and the associated parameter estimates are tested for goodness of fit. The chi-square and the Kolmogorov–Smirnov tests are frequently used. If the null hypothesis that the

9

data follow the distributional assumption fails, then the analyst returns to the second step, makes a different distributional assumption, and repeats the procedure. If several iterations of this procedure fail to yield a fit between an assumed distributional form and the collected data, the empirical form of the distribution may be used.

Each of these steps is discussed in this chapter. Additionally, there is a discussion of the treatment of data when there is a relationship between two or more variables of interest.

9.1. Data Collection

Problems are found at the end of each chapter, as exercises for the reader, in mathematics, physics, chemistry, and other technical subject texts. Years and years of working these problems may give the student the impression that data are readily available. Nothing could be further from the truth. Data collection is one of the biggest tasks in solving a real problem. It is one of the most important and difficult problems in simulation.

"GIGO," or "garbage in–garbage out" is a basic concept in the computer science world and it applies equally in the area of discrete system simulation. Many are fooled by a pile of computer printouts, as if this sheaf of papers is the absolute truth. Even if the model structure is valid, if the input data are inaccurately collected, inappropriately analyzed, or not representative of the environment, the simulation output data will be misleading and possibly damaging or costly when used for policy or decision making.

EXAMPLE 9.1 (THE LAUNDROMAT)

As budding simulation students, both authors had assignments to simulate the operation of an ongoing system. One of these systems which seemed to be a rather simple operation was a self-service laundromat with 10 washing machines and six dryers.

However, the data collection aspect of the problem rapidly became rather enormous. The interarrival time distribution was not homogeneous; that is, the distribution changed by time of day and by day of the week. The laundromat was open 7 days a week for 16 hours per day, or 112 hours per week. It would have been impossible to cover the operation of the laundromat with the limited resources available (two students who were also taking four other courses) and with a tight time constraint (the simulation was to be completed in a 4-week period). Additionally, the distribution of time between arrivals during one week may not have been followed during the next week. As a compromise, a sample of times was selected, and the interarrival-time distributions were determined and classified according to arrival rate (perhaps inappropriately) as "high," "medium," and "low."

Service time distributions also presented a difficult problem from many perspectives. The proportion of customers demanding the various service combinations had to be observed and recorded. The simplest case was the customer desiring one washer followed by one dryer. However, a customer might choose two washing machines followed by one dryer, one dryer only, and so on. Since the customers used numbered machines, it was possible to follow them using that reference, rather than remembering them by personal characteristics. Because of the dependence between washer demand and dryer demand for an individual customer, it would have been inappropriate to treat the service times for washers and dryers separately as independent variables.

Some customers waited patiently for their clothes to complete the washing or drying cycle, and then they removed their clothes promptly. Others left the premises and returned after their clothes had finished their cycle on the machine being used. In a very busy period, the manager would remove a customer's clothes after the cycle and set them aside in a basket. It was decided that service termination would be measured as the point in time when the machine was emptied of its contents.

Also, machines would break down from time to time. The length of the breakdown varied from a few moments, when the manager repaired the machine, to several days (a breakdown on Friday night, requiring a part not in the laundromat storeroom, would not be fixed until the following Monday). The short-term repair times were recorded by the student team. The long-term repair completion times were estimated by the manager. Breakdowns then became part of the simulation.

Many lessons can be learned from an actual experience at data collection. The first five exercises at the end of this chapter suggest some situations in which the student can gain such experience.

The following suggestions may enhance and facilitate the conduct of the data collection exercise, although they are not all inclusive.

1. A useful expenditure of time is in planning. This could begin by preobserving the situation. Try to collect data while preobserving. Devise forms for this purpose. It is very likely that these forms will have to be modified several times before the actual data collection begins. Watch for unusual circumstances and consider how they will be handled.

2. Try to analyze the data as it is being collected. Determine if the data being collected are adequate to provide the distributions needed as input in the simulation. Determine if any data being collected are useless to the simulation. There is no need to collect superfluous data.

3. Try to combine homogeneous data sets. Check data for homogeneity in successive time periods and during the same time period on successive days. For example, check for homogeneity of data from 2:00 P.M. to 3:00 P.M. and 3:00 P.M. to 4:00 P.M., and check to see if the data are homogeneous for 2:00 P.M. to 3:00 P.M. on Thursday and Friday. When checking for homogeneity, an initial test is to see if the means of the distributions (the average interarrival times, for example) are the same. The two sample t test can be used for this purpose. A more thorough analysis would require a determination of the equivalence of the distributions using a two-sample Smirnov test [Connover, 1980].

4. To determine whether there is a relationship between two variables, build a scatter diagram. Sometimes, an eyeball scan of the scatter diagram will indicate if there is a relationship between two variables of interest. Section 9.5 describes both regression analysis, a statistical technique to determine the relationship, and an associated test which determines the significance of the relationship.

5. Consider the possibility that a sequence of observations which appear to be independent may possess autocorrelation. Autocorrelation may exist in successive time periods or for successive customers. For example, the service time for the ith customer may be related to the service time for the $(i + n)$th customer. A brief introduction to autocorrelation was provided in Section 7.4.3.

Again, these are just a few suggestions. Data collection and analysis must be approached with great care.

9.2. Identifying the Distribution

9.2.1. HISTOGRAMS

A frequency distribution or histogram is useful in identifying the shape of a distribution. A histogram is constructed as follows:

1. Divide the range of the data into intervals (intervals are usually of equal width; however, unequal widths may be used if the heights of the frequencies are adjusted).
2. Label the horizontal axis to conform to the intervals selected.
3. Determine the frequency of occurrences within each interval.

4. Label the vertical axis so that the total occurrences can be plotted for each interval.
5. Plot the frequencies on the vertical axis.

The number of class intervals depends on the number of observations and the amount of scatter or dispersion in the data. Hines and Montgomery [1980] state that choosing the number of class intervals approximately equal to the square root of the sample size often works well in practice. They also state that judgment can be used in order to give a reasonable display. Nelson [1979] says that judgment usually works better than rules of thumb. If the intervals are too wide, the histogram will be coarse, or blocky, and its shape and other details will not show well. If the intervals are too narrow, the histogram will be ragged and will not smooth the data. Examples of a coarse, ragged, and appropriate histogram using the same data are shown in Figure 9.1.

The histogram for continuous data corresponds to the probability density function of a theoretical distribution. If continuous, a line drawn through the center point of each class interval frequency should result in a shape like that of a pdf.

Histograms for discrete data, where there are a large number of data points, should have a cell for each value in the range of the data. However, if there are few data points, it may be necessary to combine adjacent cells to eliminate the ragged appearance of the histogram. If the histogram is associated with discrete data, it should look like a probability mass function.

EXAMPLE 9.2 (DISCRETE DATA)

The number of vehicles arriving at the northwest corner of an intersection in a 5-minute period between 7:00 A.M. and 7:05 A.M. was monitored for five workdays over a 20-week period. Table 9.1 shows the resulting data. The first entry in the table indicates that there were 12 5-minute periods during which zero vehicles arrived, 10 periods during which one vehicle arrived, and so on.

Table 9.1. NUMBER OF ARRIVALS IN A 5-MINUTE PERIOD

Arrivals per Period	*Frequency*	*Arrivals per Period*	*Frequency*
0	12	6	7
1	10	7	5
2	19	8	5
3	17	9	3
4	10	10	3
5	8	11	1

Since the number of automobiles is a discrete variable, and since there are ample data, the histogram can have a cell for each possible value in the range of the data. The resulting histogram is shown in Figure 9.2.

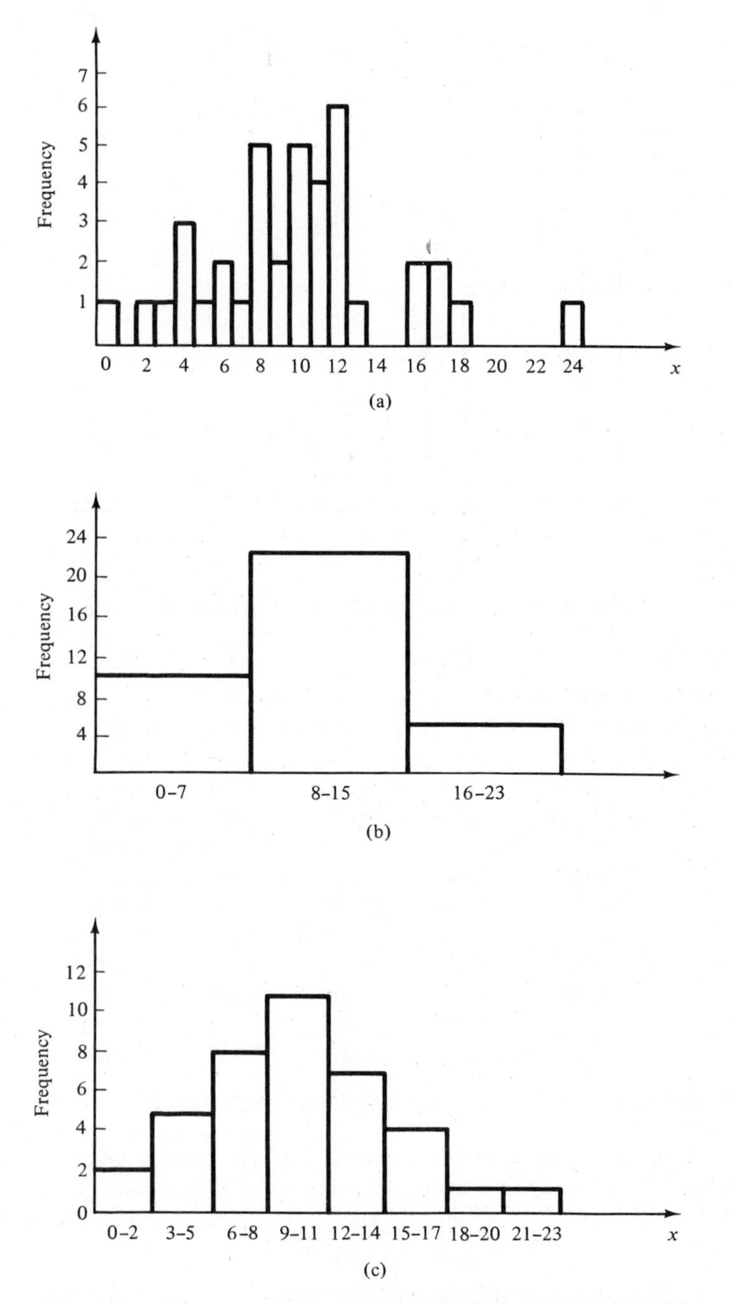

Figure 9.1. Coarse, ragged, and appropriate histograms: (a) Original data—too ragged; (b) combining adjacent cells—too coarse; (c) combining adjacent cells—appropriate.

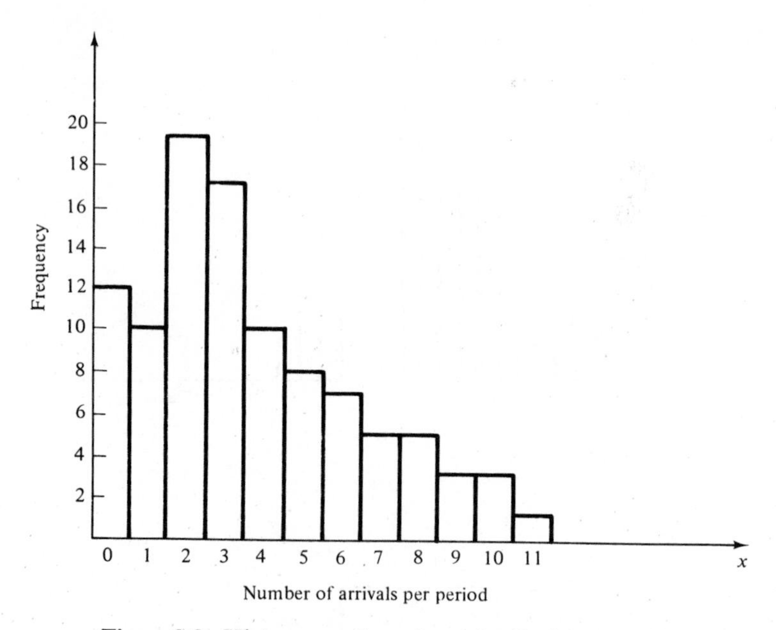

Figure 9.2. Histogram of number of arrivals per period.

EXAMPLE 9.3 (CONTINUOUS DATA)

Life tests were performed on a random sample of 50 PDP-11 electronic chips at 1.5 times the nominal voltage, and their lifetime (or time to failure) in days was recorded:

79.919	3.081	0.062	1.961	5.845
3.027	6.505	0.021	0.013	0.123
6.769	59.899	1.192	34.760	5.009
18.387	0.141	43.565	24.420	0.433
144.695	2.663	17.967	0.091	9.003
0.941	0.878	3.371	2.157	7.579
0.624	5.380	3.148	7.078	23.960
0.590	1.928	0.300	0.002	0.543
7.004	31.764	1.005	1.147	0.219
3.217	14.382	1.008	2.336	4.562

Lifetime, usually considered a continuous variable, is recorded here to three-decimal-place accuracy. The histogram is prepared by placing the data in class intervals. The range of the data is rather extensive, from 0.002 day to 144.695 days. However, most of the values (30 of 50) are in the zero- to 5-day range. Using intervals of width three results in Table 9.2. The data of Table 9.2 are then used to prepare the histogram shown in Figure 9.3.

9.2.2. THE DISTRIBUTIONAL ASSUMPTION

In Chapter 4, the distributions that often arise in simulation were described. Additionally, the shapes of these distributions were displayed. The purpose of preparing a histogram is to infer a known pdf or pmf. A distributional assump-

Table 9.2. ELECTRONIC CHIP DATA

Chip Life (Days)	*Frequency*
$0 \leq x_j < 3$	23
$3 \leq x_j < 6$	10
$6 \leq x_j < 9$	5
$9 \leq x_j < 12$	1
$12 \leq x_j < 15$	1
$15 \leq x_j < 18$	2
$18 \leq x_j < 21$	0
$21 \leq x_j < 24$	1
$24 \leq x_j < 27$	1
$27 \leq x_j < 30$	0
$30 \leq x_j < 33$	1
$33 \leq x_j < 36$	1
⋮	⋮
$42 \leq x_j < 45$	1
⋮	⋮
$57 \leq x_j < 60$	1
⋮	⋮
$78 \leq x_j < 81$	1
⋮	⋮
$143 \leq x_j < 147$	1

tion is made on the basis of what might arise in the context being investigated along with the shape of the histogram. Thus, if interarrival time data have been collected, and the histogram has a shape similar to the pdf in Figure 4.9, the assumption of an exponential distribution would be warranted. Similarly, if measurements of the weights of pallets of freight are being made, and the histogram appears symmetric about the mean with a shape like that shown in Figure 4.12, the assumption of a normal distribution would be warranted.

The exponential, normal, and Poisson distributions are frequently encountered, and are not difficult to analyze from a computational standpoint. Although more difficult to analyze, the gamma and Weibull distribution provide a wide array of shapes, and should not be overlooked when modeling an underlying probabilistic process. Perhaps an exponential distribution was assumed, but it was found not to fit the data. The next step would be to examine where the lack of fit occurred. If the lack of fit was in one of the tails of the distribution, perhaps a gamma or Weibull distribution would more adequately fit the data.

The student is encouraged to complete Exercises 6 through 11 to learn more

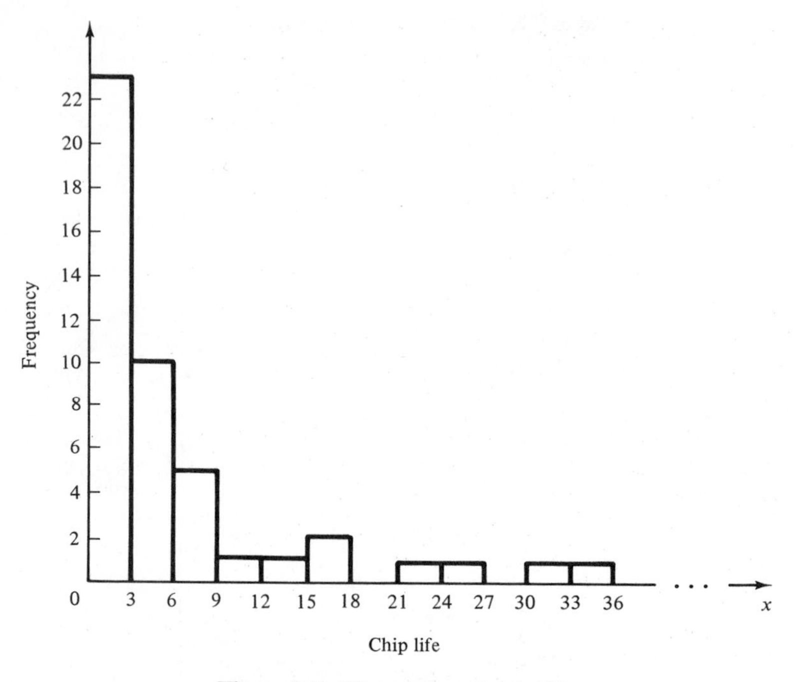

Figure 9.3. Histogram of chip life.

about the shape of the distributions mentioned in this section. Examining the variations in shape as the parameters change will be very instructive.

9.2.3. Probability Plotting

The construction of histograms, as discussed in Section 9.2.1, and the recognition of a distributional shape, as discussed in Section 9.2.2, are necessary ingredients in determining the distribution of input data. Probability plotting is an informal means of evaluating a distributional assumption. When there are a small number of data points, say 30 or fewer, a histogram would be rather ragged, but a probability plot would provide a good lead in determining the underlying distribution. In fact, it is rather cumbersome to use probability plots when sample sizes are large.

Monographs by Nelson [1979] and Shapiro [1980], both sponsored by the American Society for Quality Control, describe probability plotting in considerable detail. The basic idea behind the method is that a special transformation has been applied to the vertical scale of a graph of the assumed cumulative distribution function. This transforms all cumulative distribution functions of the assumed type into straight lines.

If the assumed distribution is correct, the sample order statistics will be nearly linear when plotted on probability paper. There will usually be random deviations from linearity, but the larger the sample size, the greater is the tendency toward linearity if the distributional assumption is correct. If the distributional assumption is incorrect, the plotted points will be nonlinear in a systematic manner.

A wide assortment of probability papers are available from TEAM at Box 25, Tamworth, NJ 03886. Codex and Keuffel & Esser also carry some papers. The available types of probability paper include those for the normal, exponential, Weibull, and chi-square distributions. Probability paper for the chi-square distribution can be used to plot the gamma distribution.

Shapiro [1980] gives the basic steps in the preparation of probability plots:

1. Make a distributional assumption and obtain the relevant probability paper.
2. Let $\{x_i, i = 1, 2, \ldots, n\}$ be the sample data, with n observations. Order the observations from the smallest to the largest. Denote these as $\{y_j, j = 1, 2, \ldots, n\}$, where $y_1 \leq y_2 \leq \cdots \leq y_n$. Let j denote the ranking or order number. Therefore, $j = 1$ for the smallest and $j = n$ for the largest.
3. Plot the y_j's against the quantity $P_j = 100(j - \frac{1}{2})/n$ on probability paper of the correct type. (To simplify this computation, determine $P_1 = 50/n$ and $P_{j+1} = P_j + 100/n, j = 1, 2, \ldots, n - 1$.)
4. If the assumed distribution is appropriate, the plotted points will appear as a straight line. If the assumed distribution is inappropriate, the points will deviate from a straight line, usually in a systematic manner. The decision of whether or not to reject some hypothesized model is subjective.

EXAMPLE 9.4 (NORMAL PROBABILITY PLOTTING)

A robot is used to install the doors on automobiles along an assembly line. It was thought that the installation times followed a normal distribution. The robot is capable of accurately measuring installation times. A sample of 20 installation times was automatically taken by the robot with the following results, where the values are in seconds:

99.79	99.56	100.17	100.33
100.26	100.41	99.98	99.83
100.23	100.27	100.02	100.47
99.55	99.62	99.65	99.82
99.96	99.90	100.06	99.85

The observations are now ranked from smallest to largest as follows:

Rank	*Value*	*Rank*	*Value*	*Rank*	*Value*	*Rank*	*Value*
1	99.55	6	99.82	11	99.98	16	100.26
2	99.56	7	99.83	12	100.02	17	100.27
3	99.62	8	99.85	13	100.06	18	100.33
4	99.65	9	99.90	14	100.17	19	100.41
5	99.79	10	99.96	15	100.23	20	100.47

The ordered observations are then plotted against P_j. The special-purpose graph paper used for this example has a prelabeled horizontal axis for plotting the P_j values.

The y_j's are plotted on the vertical axis after a scale is drawn. The scale should be drawn to accommodate the smallest and largest values. The smallest sample value, $y_1 = 99.55$, is plotted against $P_1 = 50/20 = 2.5$, the second smallest, $y_2 = 99.56$, is plotted against $P_2 = 2.5 + 100/20 = 7.5$, and so on until, finally, the highest value, $y_{20} = 100.57$, is plotted against $P_{20} = 97.5$. The plotted values are shown in Figure 9.4.

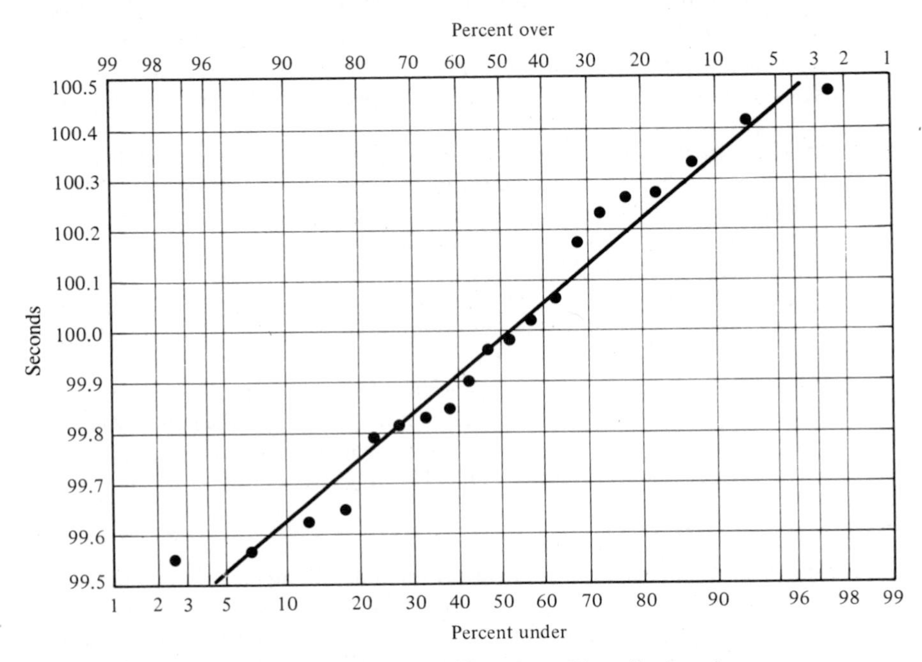

Figure 9.4. Normal probability plot of installation times.

A straight line, using the "eyeball" method, has been drawn through the plotted points. In the evaluation of the linearity of the plot, the following should be considered:

1. The observed values are random and will never fall exactly on a straight line.
2. The ordered values are not independent since they have been ranked. Hence, if one point is above the straight line, it is likely that the next point will also lie above the line. It is unlikely that the points will be scattered about the line.
3. The variances of the extremes (largest and smallest values) are much higher than the variances in the middle of the plot. Greater discrepancies can be accepted at the extremes. The linearity of the points in the middle of the plot is more important than the linearity at the extremes.

The plotted points in Figure 9.4 fall close to a straight line. Informally, the hypothesis of normality is not rejected.

In this example there were only 20 data points. An advantage of using probability paper is that small sample sizes can be used. A large sample size (say $n > 50$) would be tedious to plot manually. When a large sample is available, it is not necessary to plot every point. Thus, if 200 sample values are available, a random subsample of 40 points could be selected, ordered, and plotted to test the distributional assumption. More formal and objective tests that use all the data are presented in Section 9.4.

9.3. Parameter Estimation

After a distributional assumption has been made, the next step is to estimate the parameters of the distribution. When using probability paper, rough estimates of the parameter values can be made as described in Hines and Montgomery [1980]. However, better estimates are available and can be obtained in the manner described in this section.

9.3.1. Preliminary Statistics: Sample Mean and Sample Variance

In a number of instances the sample mean, or the sample mean and sample variance, are used to estimate the parameters of a hypothesized distribution. In the following paragraphs, three sets of equations are given for computing the sample mean and sample variance. Equations (9.1) and (9.2) can be used when discrete or continuous raw data are available. Equations (9.3) and (9.4) are used when the data are discrete and have been grouped in a frequency distribution. Equations (9.5) and (9.6) are used when the data are discrete or continuous and have been placed in class intervals. Equations (9.5) and (9.6) are approximations and should be used only when the raw data are unavailable.

If the observations in a sample of size n are $X_1, X_2, \ldots, X_n$, the sample mean ($\bar{X}$) is defined by

$$\bar{X} = \frac{\sum_{i=1}^{n} X_i}{n} \tag{9.1}$$

and the sample variance, S^2, is defined by

$$S^2 = \frac{\sum_{i=1}^{n} X_i^2 - n\bar{X}^2}{n-1} \tag{9.2}$$

If the data are discrete and grouped in a frequency distribution, equations (9.1) and (9.2) can be modified to provide for much greater computational efficiency. The sample mean can be computed by

$$\bar{X} = \frac{\sum_{j=1}^{k} f_j X_j}{n} \tag{9.3}$$

and the sample variance by

$$S^2 = \frac{\sum_{j=1}^{k} f_j X_j^2 - n\bar{X}^2}{n-1} \tag{9.4}$$

where k is the number of distinct values of X and f_j is the observed frequency of the value X_j of X.

EXAMPLE 9.5 (GROUPED DATA)

The data in Table 9.1 can be analyzed to obtain $n = 100, f_1 = 12, X_1 = 0, f_2 = 10, X_2 = 1, \ldots, \sum_{j=1}^{k} f_j X_j = 364$, and $\sum_{j=1}^{k} f_j X_j^2 = 2080$. From equation (9.3),

$$\bar{X} = \frac{364}{100} = 3.64$$

and from equation (9.4),

$$S^2 = \frac{2080 - 100(3.64)^2}{99} = 7.63$$

The sample standard deviation, S, is just the square root of the sample variance. In this case $S = \sqrt{7.63} = 2.76$. Equations (9.1) and (9.2) would have yielded exactly the same results for $\bar{X}$ and S^2.

It is preferable to use the raw data, if possible, when the values are continuous. However, the data may have been received after they have been placed in class intervals. Then, it is no longer possible to obtain the exact sample mean and variance. In such cases, the sample mean and sample variance are approximated from the following equations:

$$\bar{X} \doteq \frac{\sum_{j=1}^{c} f_j m_j}{n} \tag{9.5}$$

and

$$S^2 \doteq \frac{\sum_{j=1}^{c} f_j m_j^2 - n\bar{X}^2}{n-1} \tag{9.6}$$

where f_j is the observed frequency in the jth class interval, m_j the midpoint of the jth interval, and c the number of class intervals.

EXAMPLE 9.6 (CONTINUOUS DATA IN CLASS INTERVALS)

Assume that the raw data on chip life shown in Example 9.3 were either discarded or lost. However, the data shown in Table 9.2 are still available. To determine approximate values of $\bar{X}$ and S^2, equations (9.5) and (9.6) are used. The following values are determined: $f_1 = 23$, $m_1 = 1.5$, $f_2 = 10$, $m_2 = 4.5, \ldots, \sum_{j=1}^{49} f_j m_j = 614$ and $\sum_{j=1}^{49} f_j m_j^2 = 37{,}226.5$. With $n = 50$, $\bar{X}$ is approximated from equation (9.5) as

$$\bar{X} \doteq \frac{614}{50} = 12.28$$

Then, S^2 is approximated from equation (9.6) as

$$S^2 \doteq \frac{37{,}226.5 - 50(12.28)^2}{49} = 605.849$$

and

$$S \doteq 24.614$$

Applying equations (9.1) and (9.2) to the original data in Example 9.3 results in $\bar{X} = 11.894$ and $S = 24.953$. Thus, when the raw data are either discarded or lost, some inaccuracies may result.

9.3.2. Suggested Estimators

Numerical estimates of the distribution parameters are needed to reduce the distributional assumption to a specific distribution and to test the resulting hypothesis. Table 9.3 contains suggested estimators for distributions often used

Table 9.3. SUGGESTED ESTIMATORS FOR DISTRIBUTIONS OFTEN USED IN SIMULATION

Distribution	*Parameter(s)*	*Suggested Estimator(s)*
Poisson	α	$\hat{\alpha} = \bar{X}$
Exponential	λ	$\hat{\lambda} = \dfrac{1}{\bar{X}}$
Gamma	β, θ	$\hat{\beta}$ (see Table A.8) $\hat{\theta} = \dfrac{1}{\bar{X}}$
Uniform on (0, b)	b	$\hat{b} = \dfrac{n+1}{n}[\max(X)]$ (unbiased)
Normal	μ, σ^2	$\hat{\mu} = \bar{X}$ $\hat{\sigma}^2 = S^2$ (unbiased)
Weibull with $\nu = 0$	α, β	$\hat{\beta}_0 = \dfrac{\bar{X}}{S}$ $\hat{\beta}_j = \hat{\beta}_{j-1} - \dfrac{f(\hat{\beta}_{j-1})}{f'(\hat{\beta}_{j-1})}$ See equations (9.12) and (9.15) for $f(\hat{\beta})$ and $f'(\hat{\beta})$ Iterate until convergence $\hat{\alpha} = \left(\dfrac{1}{n}\sum_{i=1}^{n} X_i^{\hat{\beta}}\right)^{1/\hat{\beta}}$

in simulation, all of which were described in Chapter 4. Except for an adjustment to remove bias in the estimate of σ^2 of the normal distribution and in the estimate of b of the uniform distribution, these estimators are the maximum likelihood estimators based on the raw data. (If the data are in class intervals, these estimators must be modified.) The reader is referred to Fishman [1973] for more advanced treatment, including the maximum likelihood equations and sampling properties of each estimator. Examples of the use of the estimators are given in the following paragraphs. The reader should keep in mind that a parameter is an unknown constant, but the estimator is a statistic or random variable because it depends on the sample values. To distinguish the two clearly, if, say, a parameter is denoted by α, the estimator will be denoted by $\hat{\alpha}$.

EXAMPLE 9.7 (POISSON DISTRIBUTION)

Assume that the arrival data in Table 9.1 require analysis. By comparison to Figure 4.7, an examination of Figure 9.1 suggests a Poisson distributional assumption with unknown parameter α. From Table 9.3, the estimator of α is $\bar{X}$, which was determined in Example 9.5. Thus, $\hat{\alpha} = 3.64$. It will be recalled that the true mean and variance are equal for the Poisson distribution. In Example 9.5, the sample variance was estimated by $S^2 = 7.63$. However, it should never be expected that the sample mean and the sample variance will be equal, since both are random variables.

EXAMPLE 9.8 (UNIFORM DISTRIBUTION)

A spy is trying to determine the number of tanks in the enemy army. The enemy labels every tank with a number. The lowest number is 100, and the numbers go up in sequence from 100 to some unknown maximum $100 + b$. The spy watches tanks cross a road junction for a day and observes the following numbers:

1783, 1522, 920, 587, 3653, 146,
2937, 1492, 736, 372, 3104, 3535

Since the values are all greater than or equal to 100, each observation should be adjusted by subtracting 100 so that they will be in the range $[0, b]$. Although the data are discrete, it is assumed that the adjusted observations constitute a random sample from the continuous uniform distribution on $[0, b]$. The expected number of tanks in the enemy army can then be approximated by the parameter $\hat{b}$ of the uniform distribution. This approximation is quite acceptable since the observed values have a wide range. The estimated number of tanks in the enemy army is given by

$$\hat{b} = \frac{13}{12}(3653 - 100) = 3849$$

EXAMPLE 9.9 (NORMAL DISTRIBUTION)

The parameters of the normal distribution, μ and σ^2, are estimated by $\bar{X}$ and S^2, as shown in Table 9.3. The probability plot in Example 9.4 leads to a distributional assumption that the installation times are normal. Using equations (9.1) and (9.2), the data in Example 9.4 yield $\hat{\mu} = \bar{X} = 100.0065$ and $\hat{\sigma}^2 = S^2 = 0.0698$ second^2.

EXAMPLE 9.10 (GAMMA DISTRIBUTION)

The estimator, $\hat{\beta}$, for the gamma distribution is determined by the use of Table A.8 from Choi and Wette [1969]. Table A.8 requires the computation of the quantity $1/M$, where

$$M = \ln \bar{X} - \frac{1}{n} \sum_{i=1}^{n} \ln X_i \qquad (9.7)$$

Also, it can be seen in Table 9.3 that $\hat{\theta}$ is given by

$$\hat{\theta} = \frac{1}{\bar{X}} \tag{9.8}$$

In Chapter 4 it was stated that lead time is often gamma distributed. Suppose that the lead times (in days) associated with 20 orders have been accurately measured as follows:

Order	*Lead Time (Days)*	*Order*	*Lead Time (Days)*
1	70.292	11	30.215
2	10.107	12	17.137
3	48.386	13	44.024
4	20.480	14	10.552
5	13.053	15	37.298
6	25.292	16	16.314
7	14.713	17	28.073
8	39.166	18	39.019
9	17.421	19	32.330
10	13.905	20	36.547

To determine $\hat{\beta}$ and $\hat{\theta}$, it is first necessary to determine M using equation (9.7). Here, $\bar{X}$ is determined from equation (9.1) to be

$$\bar{X} = \frac{564.32}{20} = 28.22$$

Then,

$$\ell n\, \bar{X} = 3.34$$

Next,

$$\sum_{i=1}^{20} \ell n\, X_i = 63.99$$

Then,

$$M = 3.34 - \frac{63.99}{20} = 0.14$$

and

$$1/M = 7.14$$

By interpolation in Table A.8, $\hat{\beta} = 3.728$. Finally, equation (9.8) results in

$$\hat{\theta} = \frac{1}{28.22} = 0.035$$

Example 9.11 (Exponential Distribution)

Assuming that the data in Example 9.3 come from an exponential distribution, the parameter estimate, $\hat{\lambda}$, can be determined. In Table 9.3, $\hat{\lambda}$ is obtained using $\bar{X}$ as follows:

$$\hat{\lambda} = \frac{1}{\bar{X}} = \frac{1}{11.894} = 0.084 \text{ per day}$$

EXAMPLE 9.12 (WEIBULL DISTRIBUTION)

Suppose that a random sample of size n, $X_1, X_2, \ldots, X_n$, has been taken, and the observations are assumed to come from a Weibull distribution. The likelihood function derived using the pdf given by equation (4.45b) can be shown to be

$$L(\alpha, \beta) = \frac{\beta^n}{\alpha^{\beta n}}\left[\prod_{i=1}^{n} X_i^{(\beta-1)}\right] \exp\left[-\sum_{i=1}^{n}\left(\frac{X_i}{\alpha}\right)^{\beta}\right] \tag{9.9}$$

The maximum likelihood estimates are those values of α and β that maximize $L(\alpha, \beta)$, or equivalently maximize $\ell n\, L(\alpha, \beta)$, denoted by $l(\alpha, \beta)$. The maximum value of $l(\alpha, \beta)$ is obtained by taking the partial derivatives $\partial l(\alpha, \beta)/\partial\alpha$ and $\partial l(\alpha, \beta)/\partial\beta$, setting each to zero, and solving the resulting equations, which after substitution become

$$f(\beta) = 0 \tag{9.10}$$

and

$$\alpha = \left(\frac{1}{n}\sum_{i=1}^{n} X_i^{\beta}\right)^{1/\beta} \tag{9.11}$$

where

$$f(\beta) = \frac{n}{\beta} + \sum_{i=1}^{n} \ell n\, X_i - \frac{n \sum_{i=1}^{n} X_i^{\beta}\, \ell n\, X_i}{\sum_{i=1}^{n} X_i^{\beta}} \tag{9.12}$$

The maximum likelihood estimates, $\hat{\alpha}$ and $\hat{\beta}$, are the solutions of equations (9.10) and (9.11). First $\hat{\beta}$ is determined through the iterative procedure explained below. Then $\hat{\alpha}$ is determined using equation (9.11) with $\beta = \hat{\beta}$.

Since equation (9.10) is nonlinear, it is necessary to use a numerical analytic technique to solve it. In Table 9.3 an iterative method for determining $\hat{\beta}$ is given as

$$\hat{\beta}_j = \hat{\beta}_{j-1} - \frac{f(\hat{\beta}_{j-1})}{f'(\hat{\beta}_{j-1})} \tag{9.13}$$

Equation (9.13) employs Newton's method in reaching $\hat{\beta}$, where $\hat{\beta}_j$ is the jth iteration beginning with an initial estimate for $\hat{\beta}_0$, given in Table 9.3, as follows:

$$\hat{\beta}_0 = \frac{\bar{X}}{S} \tag{9.14}$$

If the initial estimate, $\hat{\beta}_0$, is sufficiently close to the solution $\hat{\beta}$, then as $j \longrightarrow \infty$, $\hat{\beta}_j$ approaches $\hat{\beta}$. When using Newton's method, $\hat{\beta}$ is approached through increments of size $f(\hat{\beta}_{j-1})/f'(\hat{\beta}_{j-1})$. Equation (9.12) is used to compute $f(\hat{\beta}_{j-1})$ and equation (9.15) is used to compute $f'(\hat{\beta}_{j-1})$ as follows:

$$f'(\beta) = \frac{-n}{\beta^2} - \frac{n \sum_{i=1}^{n} X_i^{\beta}(\ell n\, X_i)^2}{\sum_{i=1}^{n} X_i^{\beta}} + \frac{n\left(\sum_{i=1}^{n} X_i^{\beta}\, \ell n\, X_i\right)^2}{\left(\sum_{i=1}^{n} X_i^{\beta}\right)^2} \tag{9.15}$$

Equation (9.15) can be derived from equation (9.12) by differentiating $f(\beta)$ with respect to β. The iterative process continues until $f(\hat{\beta}_j) \doteq 0$, for example, until $|f(\hat{\beta}_j)| \leq 0.001$.

Consider the data given in Example 9.3. These data concern the failure of electronic components and may come from an exponential distribution. In Example 9.11, the parameter $\hat{\lambda}$ was estimated on the hypothesis that the data were from an exponential distribution. If the hypothesis that the data came from an exponential distribution is rejected, an alternative hypothesis is that the data come from a Weibull distribution. The Weibull distribution is suspected since the data pertain to electronic component failures which occur suddenly.

Equation (9.14) is used to determine $\hat{\beta}_0$. For the data in Example 9.3, $n = 50$, $\bar{X} = 11.894$, $\bar{X}^2 = 141.467$, and $\sum_{i=1}^{50} X_i^2 = 37{,}575.850$, so that S^2 is found by equation (9.2) to be

$$S^2 = \frac{37{,}578.850 - 50(141.467)}{49} = 622.650$$

and $S = 24.953$. Thus,

$$\hat{\beta}_0 = \frac{11.894}{24.953} = 0.477$$

To compute $\hat{\beta}_1$ using equation (9.13) requires the determination of $f(\hat{\beta}_0)$ and $f'(\hat{\beta}_0)$ using equations (9.12) and (9.15). The following additional values are needed: $\sum_{i=1}^{50} X_i^{\hat{\beta}_0} = 115.125$, $\sum_{i=1}^{50} \ell n\, X_i = 38.294$, $\sum_{i=1}^{50} X_i^{\hat{\beta}_0} \ell n\, X_i = 292.629$, and $\sum_{i=1}^{50} X_i^{\hat{\beta}_0} (\ell n\, X_i)^2 = 1057.781$. Thus,

$$f(\hat{\beta}_0) = \frac{50}{0.477} + 38.294 - \frac{50(292.629)}{115.125} = 16.024$$

and

$$f'(\hat{\beta}_0) = \frac{-50}{(0.477)^2} - \frac{50(1057.781)}{115.125} + \frac{50(292.629)^2}{(115.125)^2} = -356.110$$

Then, by equation (9.13),

$$\hat{\beta}_1 = 0.477 - \frac{16.024}{-356.110} = 0.522$$

After four iterations, $|f(\hat{\beta}_3)| \leq 0.001$, at which point $\hat{\beta} \doteq \hat{\beta}_4 = 0.525$ is the approximate solution to equation (9.10). Table 9.4 contains the values needed to complete each iteration.

Table 9.4. ITERATIVE ESTIMATION OF PARAMETERS OF THE WEIBULL DISTRIBUTION

j	$\hat{\beta}_j$	$\sum_{i=1}^{50} X_i^{\hat{\beta}_j}$	$\sum_{i=1}^{50} X_i^{\hat{\beta}_j} \ell n\, X_i$	$\sum_{i=1}^{50} X_i^{\hat{\beta}_j}(\ell n\, X_i)^2$	$f(\hat{\beta}_j)$	$f'(\hat{\beta}_j)$	$\hat{\beta}_{j+1}$
0	0.477	115.125	292.629	1057.781	16.024	−356.110	0.522
1	0.522	129.489	344.713	1254.111	1.008	−313.540	0.525
2	0.525	130.603	348.769	1269.547	0.004	−310.853	0.525
3	0.525	130.608	348.786	1269.614	0.000	−310.841	0.525

Now, $\hat{\alpha}$ can be determined using equation (9.11) with $\beta = \hat{\beta} = 0.525$ as follows:

$$\hat{\alpha} = \left[\frac{130.608}{50}\right]^{1/0.525} = 6.227$$

If $\hat{\beta}_0$ is sufficiently close to $\hat{\beta}$, the procedure converges quickly, usually in four to five iterations. However, if the procedure appears to be diverging, try other initial guesses for $\hat{\beta}_0$, for example, one-half the initial estimate or twice the initial estimate.

9.4. Goodness-of-Fit Tests

Hypothesis testing was discussed in Section 7.4 with respect to testing random numbers. In Section 7.4.1 the Kolmogorov–Smirnov test and the chi-square test were introduced. These two tests are applied in this section to hypotheses about distributional forms of input data.

9.4.1. Chi-Square Test

One procedure for testing the hypothesis that a random sample of size n of the random variable X follows a specific distributional form is the chi-square "goodness-of-fit" test. The test is valid for large sample sizes, for both discrete and continuous distributional assumptions, when parameters are estimated by maximum likelihood. The test procedure begins by arranging the n observations into a set of k class intervals or cells. The test statistic is given by

$$\chi_0^2 = \sum_{i=1}^{k} \frac{(O_i - E_i)^2}{E_i} \tag{9.16}$$

where O_i is the observed frequency in the ith class interval and E_i is the expected frequency in that class interval. The expected frequency for each class interval is computed as $E_i = np_i$, where p_i is the theoretical, hypothesized probability associated with the ith class interval.

It can be shown that χ_0^2 approximately follows the chi-square distribution with $k - s - 1$ degrees of freedom, where s represents the number of parameters of the hypothesized distribution estimated by the sample statistics. The hypotheses are:

H_0: the random variable, X, conforms to the distributional assumption with the parameter(s) given by the parameter estimate(s)

H_1: the random variable X does not conform

The critical value $\chi^2_{\alpha, k-s-1}$ is found in Table A.5. The null hypothesis, H_0, is rejected if $\chi_0^2 > \chi^2_{\alpha, k-s-1}$.

In applying the test, if expected frequencies are too small, χ_0^2 will not only reflect the departure of the observed from the expected frequency, but the smallness of the expected frequency as well. Although there is no general agreement regarding the minimum size of E_i, values of 3, 4, and 5 have been widely used. In Section 7.4.1, when the chi-square test was discussed, the minimum expected frequency of five was suggested. If an E_i value is too small, it can be combined with expected frequencies in adjacent class intervals. The corresponding O_i values should also be combined and k should be reduced by one for each cell that is combined.

If the distribution being tested is discrete, each value of the random variable should be a class interval, unless it is necessary to combine adjacent class intervals. For the discrete case, if combining adjacent cells is not required,

$$p_i = p(x_i) = P(X = x_i)$$

Otherwise, p_i is determined by summing the probabilities of appropriate adjacent cells.

If the distribution being tested is continuous, the class intervals are given by $[a_{i-1}, a_i)$, where a_{i-1} and a_i are the endpoints of the ith class interval. For the continuous case with assumed pdf $f(x)$, or assumed cdf $F(x)$, p_i can be computed by either expression:

$$p_i = \int_{a_{i-1}}^{a_i} f(x)\,dx = F(a_i) - F(a_{i-1})$$

For the discrete case, the number of class intervals is determined by the number of cells resulting after combining adjacent cells as necessary. However, for the continuous case the number of class intervals must be specified. Although there are no general rules to be followed, the recommendations in Table 9.5 are made to aid in determining the number of class intervals for continuous data.

Table 9.5. RECOMMENDATIONS FOR NUMBER OF CLASS INTERVALS FOR CONTINUOUS DATA

Sample Size, n	*Number of Class Intervals,* k
20	Do not use the chi-square test
50	5 to 10
100	10 to 20
>100	$\sqrt{n}$ to $n/5$

EXAMPLE 9.13 (CHI-SQUARE TEST APPLIED TO POISSON ASSUMPTION)

In Example 9.7, the vehicle arrival data presented in Example 9.2 were analyzed. Since the histogram of the data, shown in Figure 9.2, appeared to follow a Poisson distribution, the parameter, $\hat{\alpha} = 3.64$, was determined. Thus, the following hypotheses are formed:

H_0: the random variable is Poisson distributed
H_1: the random variable is not Poisson distributed

The pmf for the Poisson distribution was given in equation (4.15) as follows:

$$p(x) = \begin{cases} \dfrac{e^{-\alpha}\alpha^x}{x!} & x = 0, 1, 2, \ldots \\ 0, & \text{otherwise} \end{cases} \tag{9.17}$$

For $\alpha = 3.64$, the probabilities associated with various values of x are obtained using equation (9.17) with the following results:

$$\begin{aligned} p(0) &= 0.026 & p(6) &= 0.085 \\ p(1) &= 0.096 & p(7) &= 0.044 \\ p(2) &= 0.174 & p(8) &= 0.020 \\ p(3) &= 0.211 & p(9) &= 0.008 \\ p(4) &= 0.192 & p(10) &= 0.003 \\ p(5) &= 0.140 & p(11) &= 0.001 \end{aligned}$$

With this information, Table 9.6 is constructed. The value of E_1 is given by $np_1 = 100(0.026) = 2.6$. In a similar manner, the remaining E_i values are determined. Since $E_1 = 2.6 < 5$, E_1 and E_2 are combined. In that case O_1 and O_2 are also combined and k is reduced by one. The last five class intervals are also combined for the same reason and k is further reduced by four.

Table 9.6. CHI-SQUARE GOODNESS-OF-FIT TEST FOR EXAMPLE 9.13

x_i	Observed Frequency, O_i		Expected Frequency, E_i		$\frac{(O_i - E_i)^2}{E_i}$
0	12	22	2.6	12.2	7.87
1	10		9.6		
2	19		17.4		0.15
3	17		21.1		0.80
4	10		19.2		4.41
5	8		14.0		2.57
6	7		8.5		0.26
7	5		4.4		
8	5		2.0		
9	3	17	0.8	7.6	11.62
10	3		0.3		
11	1		0.1		
	100		100.0		27.68

The calculated χ_0^2 is 27.68. The degrees of freedom for the tabulated value of χ^2 is $k - s - 1 = 7 - 1 - 1 = 5$. Here, $s = 1$, since one parameter was estimated from the data. At $\alpha = 0.05$, the critical value $\chi^2_{0.05,5}$ is 11.1. Thus, H_0 would be rejected at level of significance 0.05. The analyst must now search for a better-fitting model or use the empirical distribution of the data.

9.4.2. Chi-Square Test with Equal Probabilities

If a continuous distributional assumption is being tested, class intervals that are equal in probability rather than equal in width of interval should be used. This has been recommended by a number of authors [Mann and Wald,

1942; Gumbel, 1943; Kendall and Stuart, 1979]. It should be noted that the procedure is not applicable to data collected in class intervals, where the raw data have been discarded or lost.

Unfortunately, there is as yet no method for determining the probability associated with each interval that maximizes the power for a test of a given size. (The power of a test is defined as the probability of rejecting a false hypothesis.) However, if using equal probabilities, $p_i = 1/k$. Since

$$E_i = np_i \geq 5$$

substituting for p_i yields

$$\frac{n}{k} \geq 5$$

and solving for k yields

$$k \leq \frac{n}{5} \tag{9.18}$$

Equation (9.18) was used in determining the recommended maximum number of class intervals in Table 9.5.

If the assumed distribution is normal, exponential or Weibull, the methodology described in this section is straightforward. Examples 9.14 and 9.15 indicate how the procedure is accomplished for the exponential and Weibull distributions. Example 9.16 shows how the E_i values are determined for the normal distribution. If the assumed distribution is gamma (but not Erlang), the computation of endpoints for class intervals is complex. A mathematical/statistical package, such as IMSL, is available for many digital computers and can be used to evaluate the cumulative gamma distribution (known as the incomplete gamma function). The incomplete gamma function was first tabulated by Pearson [1922] for various values of the shape parameter, β. Recall from Section 4.4 that tables for the Poisson variate can be used to evaluate the cumulative distribution function for the Erlang distribution.

EXAMPLE 9.14 (CHI-SQUARE TEST FOR EXPONENTIAL DISTRIBUTION)

In Example 9.11, the failure data presented in Example 9.3 were analyzed. Since the histogram of the data, shown in Figure 9.3, appeared to follow an exponential distribution, the parameter $\hat{\lambda} = 1/\bar{X} = 0.084$ was determined. Thus, the following hypotheses are formed:

H_0: the random variable is exponentially distributed
H_1: the random variable is not exponentially distributed

In order to perform the chi-square test with intervals of equal probability, the endpoints of the class intervals must be determined. Equation (9.18) indicates that the number of intervals should be less than or equal to $n/5$. Here, $n = 50$, so that $k \leq 10$. In Table 9.5, it is recommended that 7 to 10 class intervals be used. Let $k = 8$, then each interval will have probability $p = 0.125$. The endpoints for each interval are computed from the cdf for the exponential distribution, given in equation (4.24), as follows:

$$F(a_i) = 1 - e^{-\lambda a_i} \tag{9.19}$$

where a_i represents the endpoint of the ith interval, $i = 1, 2, \ldots, k$. Since $F(a_i)$ is the cumulative area from zero to a_i, $F(a_i) = ip$, so equation (9.19) can be written as

$$ip = 1 - e^{-\lambda a_i}$$

or

$$e^{-\lambda a_i} = 1 - ip$$

Taking the logarithm of both sides and solving for a_i gives a general result for the endpoints of k equiprobable intervals for the exponential distribution, namely

$$a_i = -\frac{1}{\lambda} \ell\text{n}\,(1 - ip), \qquad i = 0, 1, \ldots, k \tag{9.20}$$

Regardless of the value of λ, equation (9.20) will always result in $a_0 = 0$ and $a_k = \infty$.

With $\hat{\lambda} = 0.084$ and $k = 8$, a_1 is determined from equation (9.20) as

$$a_1 = -\frac{1}{0.084} \ell\text{n}\,(1 - 0.125) = 1.590$$

Continued application of equation (9.20) for $i = 2, 3, \ldots, 7$ results in $a_2, \ldots, a_7$ as 3.425, 5.595, 8.252, 11.677, 16.503, and 24.755. Since $k = 8$, $a_8 = \infty$. The first interval is [0, 1.590), the second interval is [1.590, 3.425), and so on. The expectation is that 0.125 of the observations will fall in each interval. The observations, expectations, and the contributions to the calculated value of χ_0^2 are shown in Table 9.7. The calculated

Table 9.7. CHI-SQUARE GOODNESS-OF-FIT TEST FOR EXAMPLE 9.14

Class Interval	*Observed Frequency,* O_i	*Expected Frequency,* E_i	$\frac{(O_i - E_i)^2}{E_i}$
[0, 1.590)	19	6.25	26.01
[1.590, 3.425)	10	6.25	2.25
[3.425, 5.595)	3	6.25	0.81
[5.595, 8.252)	6	6.25	0.01
[8.252, 11.677)	1	6.25	4.41
[11.677, 16.503)	1	6.25	4.41
[16.503, 24.755)	4	6.25	0.81
[24.755, ∞)	6	6.25	0.01
	50	50	39.6

value of χ_0^2 is 39.6. The degrees of freedom are given by $k - s - 1 = 8 - 1 - 1 = 6$. At $\alpha = 0.05$, the tabulated value of $\chi^2_{0.05,6}$ is 12.6. Since $\chi^2_{0.05,6} \leq \chi_0^2$, the null hypothesis is rejected. (The value of $\chi^2_{0.01,6}$ is 16.8, so the null hypothesis would also be rejected at level of significance $\alpha = 0.01$.)

EXAMPLE 9.15 (CHI-SQUARE TEST FOR WEIBULL DISTRIBUTION)

In Example 9.12, the parameters $\hat{\beta}$ and $\hat{\theta}$ of the Weibull distribution were computed for the sample of chip failure data which was rejected as exponentially distributed in Example 9.14. In Example 9.12, it was stated that if the data are rejected as being exponentially distributed, they should be tested for a Weibull fit.

The hypotheses to be tested are as follows:

H_0: the random variable, X, is Weibull distributed
H_1: the random variable, X, is not Weibull distributed

To test the data as Weibull distributed requires essentially the same steps as in Example 9.14, the first step being the determination of endpoints. Again, let $k = 8$ so that each interval will have probability $p = 0.125$. The cdf for the Weibull is given in equation (4.48) as

$$F(x) = 1 - \exp\left[-\left(\frac{x-\nu}{\alpha}\right)^{\beta}\right], \qquad x \geq \nu \tag{9.21}$$

In Example 9.12, it was assumed that $\nu = 0$. Letting the endpoints of the class intervals be denoted by a_i, we have

$$F(a_i) = 1 - \exp\left[-\left(\frac{a_i}{\alpha}\right)^{\beta}\right] \tag{9.22}$$

Proceeding as in the derivation of equation (9.20), using

$$F(a_i) = ip$$

equation (9.22) can be written as

$$ip = 1 - \exp\left[-\left(\frac{a_i}{\alpha}\right)^{\beta}\right]$$

or

$$\exp\left[-\left(\frac{a_i}{\alpha}\right)^{\beta}\right] = 1 - ip$$

Taking the logarithm of both sides yields

$$\left(\frac{a_i}{\alpha}\right)^{\beta} = -\ell n\,(1 - ip)$$

Then, solving for a_i yields

$$a_i = \alpha[-\ell n\,(1 - ip)]^{1/\beta} \tag{9.23}$$

Regardless of the value of α and β, equation (9.23) will always result in $a_0 = 0$ and $a_k = \infty$.

In Example 9.12, $\hat{\alpha} = 6.23$ and $\hat{\beta} = 0.525$. Using these estimates in equation (9.23) together with $k = 8$ and $p = 0.125$, the value of a_1 is determined by

$$a_1 = 6.23[-\ell n\,(1 - 0.125)]^{1/0.525} = 0.134$$

Continued application of equation (9.23) for $i = 2, 3, \ldots, 7$ results in $a_2, \ldots, a_7$ as 0.578, 1.471, 3.083, 5.970, 11.537, and 24.968. Since $k = 8$, $a_8 = \infty$. The first interval is [0, 0.134), the second interval is given by [0.134, 0.578), and so on. The expectation is that 0.125 of the observations will fall in each interval. The observations, expectations, and contributions to the calculated value of χ_0^2 are shown in Table 9.8. The calculated value of χ_0^2 is 1.20. The degrees of freedom are given by $k - s - 1 = 8 - 2 - 1 = 5$, where $s = 2$, since two parameters β and θ, were estimated. At a level of significance $\alpha = 0.05$, the tabulated value of $\chi^2_{0.05,5}$ is 11.1. Since $\chi^2_{0.05,5} > \chi_0^2$, H_0 is not rejected. This test provides no reason to doubt the Weibull assumption.

Table 9.8. CHI-SQUARE GOODNESS-OF-FIT TEST FOR EXAMPLE 9.15

Class Interval	*Observed Frequency,* O_i	*Expected Frequency,* E_i	$\frac{(O_i - E_i)^2}{E_i}$
[0, 0.134)	6	6.25	0.01
[0.134, 0.578)	5	6.25	0.25
[0.578, 1.471)	8	6.25	0.49
[1.471, 3.083)	7	6.25	0.09
[3.083, 5.970)	7	6.25	0.09
[5.970, 11.537)	6	6.25	0.01
[11.537, 24.968)	5	6.25	0.25
[24.968, ∞)	6	6.25	0.01
	50	50	1.20

EXAMPLE 9.16 (COMPUTING INTERVALS FOR THE NORMAL DISTRIBUTION)

The cdf of the normal distribution was given by equation (4.39) as

$$F(x) = \Phi\left(\frac{x - \mu}{\sigma}\right), \qquad -\infty \leq x \leq \infty \tag{9.24}$$

where $\Phi(\cdot)$ is the cdf of a standard normal distribution with mean zero and variance 1. If $x = a_i$, where a_i are the interval endpoints, equation (9.24), can be written as

$$F(a_i) = \Phi\left(\frac{a_i - \mu}{\sigma}\right)$$

where $i = 1, 2, \ldots, k$. The first interval will always be $[-\infty, a_1)$ and the last interval will always be $[a_{k-1}, \infty]$. For $k = 8$, the probability of each interval is $p = 0.125$, so $F(a_1) = 0.125$. Thus, using Table A.3 and $F(a_1) = \Phi[(a_1 - \mu)/\sigma] = \Phi(-1.152)$, or $a_1 = \mu - 1.152\sigma$.

Repeating this process leads to a set of intervals as follows:

$$\begin{array}{l}
(-\infty, \mu - 1.152\sigma) \\
[\mu - 1.152\sigma, \mu - 0.674\sigma) \\
[\mu - 0.674\sigma, \mu - 0.319\sigma) \\
[\mu - 0.319\sigma, \mu) \\
[\mu, \mu + 0.319\sigma) \\
[\mu + 0.319\sigma, \mu + 0.674\sigma) \\
[\mu + 0.674\sigma, \mu + 1.152\sigma) \\
[\mu + 1.152\sigma, \infty)
\end{array}$$

Testing random samples of data for normality are given as exercises for the reader.

9.4.3. Kolmogorov–Smirnov Test for Goodness of Fit

The Kolmogorov–Smirnov test was presented in Section 7.4.1 to test for the uniformity of numbers and again in Section 7.4.4 to perform the gap test. Both of these uses fall into the category of testing for goodness of fit. Any continuous distributional assumption can be tested for goodness of fit using the method of Section 7.4.1, while discrete distributional assumptions can be tested using the method of Section 7.4.4.

The Kolmogorov–Smirnov test is particularly useful when sample sizes are small and when no parameters have been estimated from the data. When parameter estimates have been made, the critical values in Table A.7 are biased; they are too conservative. In this context "conservative" means that the critical values will be too large, resulting in smaller Type I (α) errors than those specified. The exact value of α can be determined in some instances as discussed at the end of this section.

The chi-square goodness-of-fit test can accommodate the estimation of parameters from the data with a resultant decrease in the degrees of freedom (one for each parameter estimated). The chi-square test requires that the data be placed in class intervals, and in the case of a continuous distributional assumption, this grouping is arbitrary. Changing the number of classes and the interval width affects the value of the calculated, and tabulated chi-square. A hypothesis may be accepted when the data are grouped one way, but rejected when grouped another way. Also, the distribution of the chi-square test statistic is known only approximately, and the power of the test is sometimes rather low. As a result of these considerations, goodness-of-fit tests, other than the chi-square, are desired.

The Kolmogorov–Smirnov test does not take any special tables when an exponential distribution is assumed. The following example indicates how the test is applied in this instance. (Note that it is not necessary to estimate the parameter of the distribution in this example, permitting the use of Table A.7.)

Example 9.17 (Kolmogorov–Smirnov Test for Exponential Distribution)

Suppose that 50 interarrival times (in minutes) are collected over the following 100-minute interval (arranged in order of occurrence):

0.44, 0.53, 2.04, 2.74, 2.00, 0.30, 2.54, 0.52, 2.02, 1.89, 1.53, 0.21,
2.80, 0.04, 1.35, 8.32, 2.34, 1.95, 0.10, 1.42, 0.46, 0.07, 1.09, 0.76,
5.55, 3.93, 1.07, 2.26, 2.88, 0.67, 1.12, 0.26, 4.57, 5.37, 0.12, 3.19,
1.63, 1.46, 1.08, 2.06, 0.85, 0.83, 2.44, 2.11, 3.15, 2.90, 6.58, 0.64

The null hypothesis and its alternate are formed as follows:

H_0: the interarrival times are exponentially distributed
H_1: the interarrival times are not exponentially distributed

The data were collected over the interval 0 to $T = 100$ minutes. It can be shown that if the underlying distribution of interarrival times $\{T_1, T_2, \ldots\}$ is exponential, the arrival times are uniformly distributed on the interval $(0, T)$. The arrival times $T_1, T_1 + T_2, T_1 + T_2 + T_3, \ldots, T_1 + \ldots + T_{50}$ are obtained by adding interarrival times. The arrival times are then normalized to a (0, 1) interval so that the Kolmogorov–Smirnov test, as presented in Section 7.4.1, can be applied. On a (0, 1) interval, the points will be $[T_1/T, (T_1 + T_2)/T, \ldots, (T_1 + \ldots + T_{50})/T]$. The resulting 50 data points are as follows:

0.0044, 0.0097, 0.0301, 0.0575, 0.0775, 0.0805, 0.1147, 0.1111, 0.1313, 0.1502,
0.1655, 0.1676, 0.1956, 0.1960, 0.2095, 0.2927, 0.3161, 0.3356, 0.3366, 0.3508,
0.3553, 0.3561, 0.3670, 0.3746, 0.4300, 0.4694, 0.4796, 0.5027, 0.5315, 0.5382,
0.5494, 0.5520, 0.5977, 0.6514, 0.6526, 0.6845, 0.7008, 0.7154, 0.7262, 0.7468,
0.7553, 0.7636, 0.7880, 0.7982, 0.8206, 0.8417, 0.8732, 0.9022, 0.9680, 0.9744

Following the procedure in Example 7.13 yields a D^+ of 0.1054 and a D^- of 0.0080. Therefore, the Kolmogorov–Smirnov statistic is $D = \max(0.1054, 0.0080) = 0.1054$. The critical value of D obtained from Table A.7 for a level of significance of $\alpha = .05$ and $n = 50$ is $D_{0.05} = 1.36/\sqrt{n} = 0.1923$. Since $D = 0.1054$, the hypothesis that the interarrival times are exponentially distributed cannot be rejected.

The Kolmogorov–Smirnov test has been modified so that it can be used in several situations where the parameters are estimated from the data. The computation of the test statistic is the same, but different tables of critical values are used. Different tables of critical values are required for different distributional assumptions. Lilliefors [1967] developed a test for normality. The null hypothesis states that the population is one of the family of normal distributions without specifying the parameters of the distribution. The interested reader may wish to study Lilliefors' original work, as he describes how simulation was used to develop the critical values.

Lilliefors [1969] also modified the critical values of the Kolmogorov–Smirnov test for the exponential distribution. Lilliefors again used random sampling to obtain approximate critical values, but Durbin [1975] subsequently obtained the exact distribution.

Connover [1980] gives examples of Kolmogorov–Smirnov tests for the normal and exponential distributions. He also refers to several other Kolmogorov–Smirnov-type tests which may be of interest to the reader.

9.5. Bivariate Data

In Sections 9.1 to 9.4, the random variables presented were considered to be independent of any other variables within the context of the problem. However, variables may be related, and if the variables appear in a simulation model, either as inputs, outputs, or both inputs and outputs, the relationship should be determined and taken into consideration. Regression analysis is a statistical technique to determine the relationship between the variables.

EXAMPLE 9.18

Suppose that lead time and demand for industrial robots are related. An increase in demand results in an increase in lead time since the final assembly of the robots must be made according to the specifications of the purchaser. Rather than treat lead time and demand as independent random variables in a simulation, regression analysis could be used to determine the relationship between the two variables. Then, when a pair of values is needed, a random demand could be generated and the associated expected lead time computed from the regression equation.

In general, suppose that there is a single dependent variable or response, y, that is related to k independent variables, say $x_1, x_2, \ldots, x_k$. There are two cases to be considered. In the first case, the dependent variable, y, is a random variable, while the independent variables $x_1, x_2, \ldots, x_k$ are measured with negligible error. The $\{x_i\}$ are called mathematical variables and are usually subject to control. In the second case, $y, x_1, x_2, \ldots, x_k$ are jointly distributed random variables. Regardless, whether the first or second case is being considered, the analysis presented in this section is appropriate.

The relationship between the variables, y and x, is characterized by a mathematical model called a regression equation. In some cases the analyst will know the exact form of the true functional relationship between y and $x_1, x_2, \ldots, x_k$, say $y = f(x_1, x_2, \ldots, x_k)$. However, in most cases, the true functional relationship is unknown, and the analyst must select an appropriate function to approximate f. The regression equation contains unknown parameters whose values are estimated from a set of data (y, x).

9.5.1. SIMPLE LINEAR REGRESSION

Suppose that it is desired to estimate the relationship between a single independent variable x and a dependent variable y. Suppose that the true relationship between y and x is a linear relationship, where the observation, y, is a random variable and x is a mathematical variable. The expected value of y for a given value of x is assumed to be

$$E(y|x) = \beta_0 + \beta_1 x \tag{9.25}$$

where β_0 = intercept on the y axis; an unknown constant
β_1 = slope, or change in y for a unit change in x; an unknown constant
It is assumed that each observation of y can be described by the model

$$y = \beta_0 + \beta_1 x + \epsilon \tag{9.26}$$

where ϵ is a random error with mean zero and constant variance σ^2. The regression model given by equation (9.26) involves a single variable x and is commonly called a simple linear regression model.

Suppose that there are n pairs of observations (y_1, x_1), (y_2, x_2), . . . , (y_n, x_n). These observations may be used to estimate β_0 and β_1 in equation (9.26). The method of least squares is commonly used to form the estimates. In the method of least squares, β_0 and β_1 are estimated such that the sum of the squares of the deviations between the observations and the regression line is minimized. The individual observations in equation (9.26) may be written as

$$y_i = \beta_0 + \beta_1 x_i + \epsilon_i, \qquad i = 1, 2, \ldots, n \tag{9.27}$$

where $\epsilon_1, \epsilon_2, \ldots$ are assumed to be uncorrelated random variables.

Each ϵ_i in equation (9.27) is given by

$$\epsilon_i = y_i - \beta_0 - \beta_1 x_i \tag{9.28}$$

and represents the difference between the observed response, y_i, and the expected response, $\beta_0 + \beta_1 x_i$, predicted by the model in equation (9.25). Figure 9.5 shows how ϵ_i is related to x_i, y_i, and $E(y_i | x_i)$.

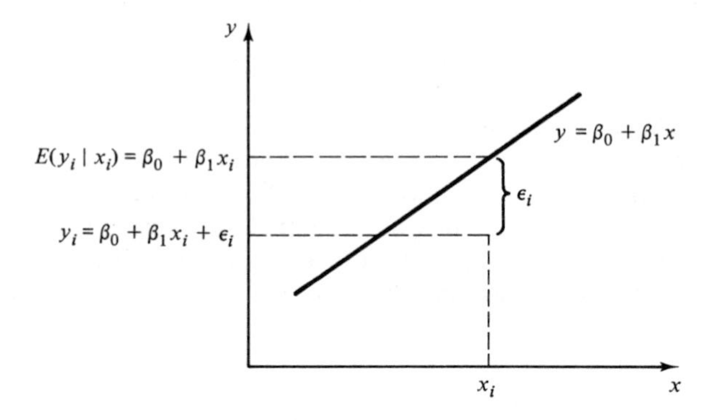

Figure 9.5. Relationship of ϵ_i to x_i, y_i, and $E(y_i | x_i)$.

The sum of squares of the deviations given in equation (9.28) is given by

$$L = \sum_{i=1}^{n} \epsilon_i^2 = \sum_{i=1}^{n} (y_i - \beta_0 - \beta_1 x_i)^2 \tag{9.29}$$

where L is called the least-squares function. It is convenient to rewrite y_i as follows:

$$y_i = \beta_0' + \beta_1(x_i - \bar{x}) + \epsilon_i \tag{9.30}$$

where $\beta_0' = \beta_0 + \beta_1 \bar{x}$ and $\bar{x} = \sum_{i=1}^{n} x_i/n$. Equation (9.30) is often called the transformed linear regression model. Using equation (9.30), equation (9.29) becomes

$$L = \sum_{i=1}^{n} [y_i - \beta_0' - \beta_1(x_i - \bar{x})]^2$$

To minimize L, find $\partial L/\partial \beta_0'$ and $\partial L/\partial \beta_1$, set each to zero, and solve for $\hat{\beta}_0'$ and $\hat{\beta}_1$. Taking the partial derivatives and setting each to zero yields

$$n\hat{\beta}_0' = \sum_{i=1}^{n} y_i$$
$$\hat{\beta}_1 \sum_{i=1}^{n} (x_i - \bar{x})^2 = \sum_{i=1}^{n} y_i(x_i - \bar{x}) \tag{9.31}$$

Equations (9.31) are often called the "normal equations," which have the solutions

$$\hat{\beta}_0' = \bar{y} = \sum_{i=1}^{n} \frac{y_i}{n} \tag{9.32}$$

and

$$\hat{\beta}_1 = \frac{\sum_{i=1}^{n} y_i(x_i - \bar{x})}{\sum_{i=1}^{n} (x_i - \bar{x})^2} \tag{9.33}$$

The numerator in equation (9.33) is rewritten for computational purposes as

$$S_{xy} = \sum_{i=1}^{n} y_i(x_i - \bar{x}) = \sum_{i=1}^{n} x_i y_i - \frac{\left(\sum_{i=1}^{n} x_i\right)\left(\sum_{i=1}^{n} y_i\right)}{n} \tag{9.34}$$

where S_{xy} denotes the corrected sum of cross products of x and y. The denominator of Equation (9.33) is rewritten for computational purposes as

$$S_{xx} = \sum_{i=1}^{n} (x_i - \bar{x})^2 = \sum_{i=1}^{n} x_i^2 - \frac{\left(\sum_{i=1}^{n} x_i\right)^2}{n} \tag{9.35}$$

where S_{xx} denotes the corrected sum of squares of x.

The value of $\hat{\beta}_0$ can be retrieved easily as

$$\hat{\beta}_0 = \hat{\beta}_0' - \hat{\beta}_1 \bar{x} \tag{9.36}$$

EXAMPLE 9.19 (CALCULATING $\hat{\beta}_0$ AND $\hat{\beta}_1$)

A catalog sales firm processes orders at a central warehouse. The orders are assembled in baskets that pass through the warehouse on a roller conveyor. The baskets are taken to the packaging area in racks that hold 10 baskets. The packers can easily see how many orders are in the queue.

A simulation of the packaging area is to be conducted. However, there seems to be a relationship between the length of the queue and the packing rate. If there is a relationship, the rate of service will change during the simulation as a function of the queue length.

A study of the operation was conducted with the results shown in Table 9.9. The observers noted the instantaneous queue length, x, at 15 predetermined points in time and then counted the number of packages, y, that were placed on the conveyor headed for the mail room in the subsequent 10-minute period. The graphical relationship between the packing rate and queue length is shown in Figure 9.6. Such a display is

Table 9.9. HISTORICAL QUEUE LENGTHS AND PACKING RATES

Observation, i	*Queue Length,* x_i	*Packing Rate,* y_i *(Completions in 10 Minutes)*
1	20	20
2	30	24
3	30	29
4	50	34
5	30	27
6	40	33
7	40	31
8	60	39
9	30	23
10	20	18
11	40	34
12	40	32
13	50	36
14	20	21
15	40	30

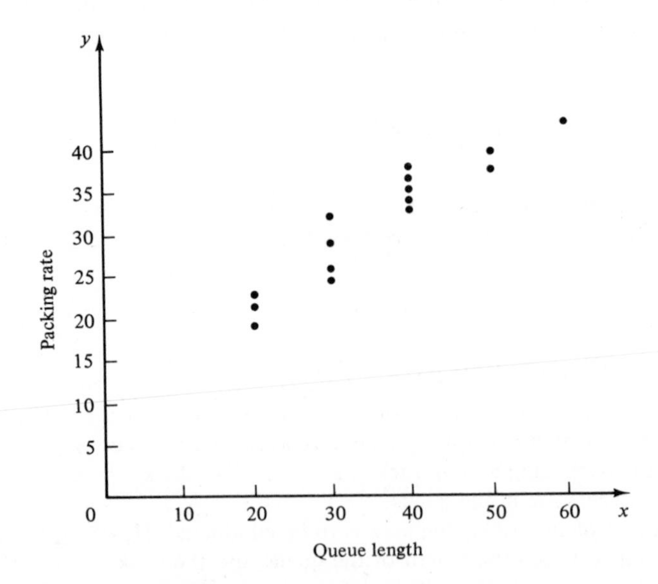

Figure 9.6. Relationship between packing rate and queue length.

called a scatter diagram. Examination of this scatter diagram indicates that there is a strong relationship between packing rate and queue length. The tentative assumption of the linear model given by Equation (9.26) appears to be reasonable.

With the packing rates as the y_i values (the dependent variables) and the queue lengths as the x_i values (the independent variables), $\hat{\beta}_0$ and $\hat{\beta}_1$ can be determined by the following computations: $n = 15$, $\sum_{i=1}^{15} x_i = 540$, $\sum_{i=1}^{15} y_i = 431$, $\sum_{i=1}^{15} x_i^2 = 21{,}400$, and $\sum_{i=1}^{15} x_i y_i = 16{,}510$. Now, $\bar{x} = 540/15 = 36$.

From equation (9.34) S_{xy} is calculated as

$$S_{xy} = 16{,}510 - \frac{(540)(431)}{15} = 994$$

From equation (9.35), S_{xx} is calculated as

$$S_{xx} = 21{,}400 - \frac{(540)^2}{15} = 1960$$

Then, $\hat{\beta}_1$ is calculated from equation (9.33) as

$$\hat{\beta}_1 = \frac{S_{xy}}{S_{xx}} = \frac{994}{1960} = 0.507$$

As shown in equation (9.32), $\hat{\beta}'_0$ is just $\bar{y}$, or

$$\hat{\beta}'_0 = \frac{431}{15} = 28.73$$

To express the model in the original terms, compute $\hat{\beta}_0$ from equation (9.36) as

$$\hat{\beta}_0 = 28.73 - 0.507(36) = 10.48$$

Then an estimate of the mean of y given x, $E(y|x)$, is given by

$$\hat{y} = \hat{\beta}_0 + \hat{\beta}_1 x = 10.48 + 0.507x \tag{9.37}$$

Regression analysis is widely used and frequently misused. Several of the common abuses are briefly mentioned. Relationships derived in the manner of equation (9.37) are valid for values of the independent variable within the range of the original data. The linear relationship that has been tentatively assumed may not be valid outside the original range. Therefore, equation (9.37) can only be considered as valid for $20 \leq x \leq 60$. Regression models are not advised for extrapolation purposes.

Care should be taken in selecting variables that have a plausible causal relationship with each other. It is quite possible to develop statistical relationships that are unrelated in a practical sense. For example, an attempt might be made to relate monthly output of a steel mill to the weight of computer printouts appearing on a manager's desk during the month. A straight line may appear to provide a good model for the data, but the relationship between the two variables is tenuous. A strong observed relationship does not imply that a causal relationship exists between the variables. Causality can be inferred only when analysis uncovers some plausible reasons for its existence. In Example 9.19 it has been observed that workers do work faster as the queue length increases, so that a relationship of the form of equation (9.37) is at least plausible.

9.5.2 Testing for Significance of Regression

In Section 9.5.1 it was assumed that a linear relationship existed between y and x. In Example 9.19 a scatter diagram, shown in Figure 9.6, relating queue length and packing rate was prepared to determine whether a linear model was a reasonable tentative assumption prior to the calculation of $\hat{\beta}_0$ and $\hat{\beta}_1$. However, the adequacy of the simple linear relationship should be tested prior to using the model for predicting the response, y_i, given an input, x_i. There are several tests which can be conducted to aid in determining model adequacy. Testing whether the order of the model tentatively assumed is correct, commonly called the "lack-of-fit test," is suggested. The procedure is explained by Hines and Montgomery [1980].

Testing for the significance of regression provides another means for assessing the adequacy of the model. The hypothesis test described below requires the additional assumption that the error component ϵ_i is normally distributed. Thus, the complete assumptions are that the errors are NID $(0, \sigma^2)$, that is, normally and independently distributed with mean zero and constant variance σ^2. The adequacy of the assumptions can and should be checked by residual analysis, discussed by Draper and Smith [1966] and by Hines and Montgomery [1980].

Testing for significance of regression is one of many hypothesis tests that can be developed from the variance properties of $\hat{\beta}_0$ and $\hat{\beta}_1$. The interested reader is referred to Draper and Smith [1966] and Montgomery and Peck [1982] for an intensive discussion of hypothesis testing in regression. Just the highlights of testing for significance of regression are given in this section.

Suppose that the alternative hypotheses are that

$$H_0\colon\ \beta_1 = 0$$
$$H_1\colon\ \beta_1 \neq 0$$

Failure to reject H_0 indicates that there is no linear relationship between x and y. This situation is illustrated in Figure 9.7. Note that two possibilities exist. In Figure 9.7(a), the implication is that x is of little value in explaining the variability in y, and that $\hat{y} = \bar{y}$ is the best estimator. In Figure 9.7(b), the implication is that the true relationship is not linear.

Alternatively, if H_0 is rejected, the implication is that x is of value in explaining the variability in y. This situation is illustrated in Figure 9.8. Here, also, two possibilities exist. In Figure 9.8(a), the straight-line model is adequate. However, in Figure 9.8(b), even though there is a linear effect of x, a model with higher-order terms (such as $x^2, x^3, \ldots$) is necessary. Thus, even though there

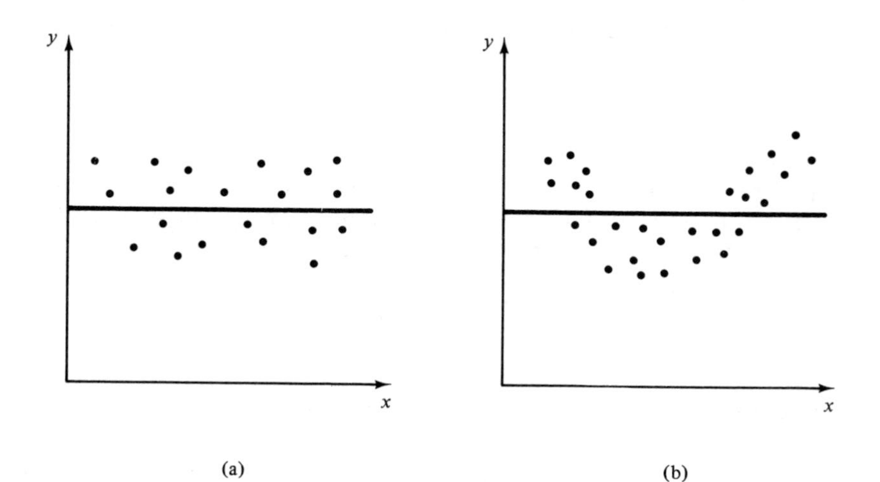

Figure 9.7. Failure to reject $H_0: \beta_1 = 0$.

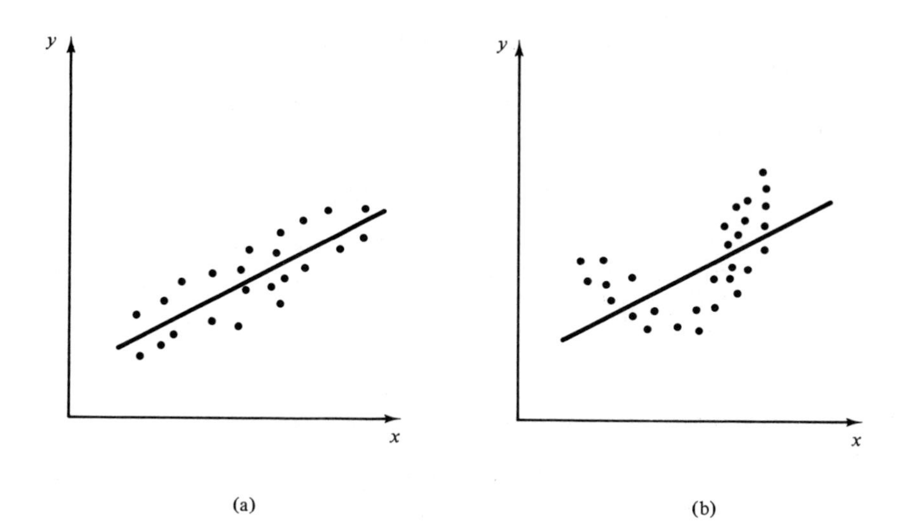

Figure 9.8. $H_0: \beta_1 = 0$ is rejected.

may be significance of regression, testing of the residuals and testing for lack of fit are needed to assure the adequacy of the model.

The appropriate test statistic for significance of regression is given by

$$t_0 = \frac{\hat{\beta}_1}{\sqrt{MS_E/S_{xx}}} \tag{9.38}$$

where MS_E is the mean squared error. The error is the difference between the observed value, y_i, and the predicted value, $\hat{y}_i$, at x_i, or $e_i = y_i - \hat{y}_i$. The squared error is given by $\sum_{i=1}^{n} e_i^2$, and the mean squared error, given by

$$\text{MS}_\text{E} = \sum_{i=1}^{n} \frac{e_i^2}{n-2} \tag{9.39}$$

is an unbiased estimator of $\sigma^2 = V(\epsilon_i)$. The direct method can be used to calculate $\sum_{i=1}^{n} e_i^2$; that is, calculate each $\hat{y}_i$, compute e_i, compute e_i^2, and sum all the e_i^2 values, $i = 1, 2, \ldots, n$. However, it can be shown that

$$\sum_{i=1}^{n} e_i^2 = S_{yy} - \hat{\beta}_1 S_{xy} \tag{9.40}$$

where S_{yy}, the corrected sum of squares of y, is given by

$$S_{yy} = \sum_{i=1}^{n} y_i^2 - \frac{\left(\sum_{i=1}^{n} y_i\right)^2}{n} \tag{9.41}$$

and S_{xy} is given by Equation (9.34). Equation (9.40) may be easier to use than the direct method.

The statistic defined by equation (9.38) has the t distribution with $n-2$ degrees of freedom. The null hypothesis H_0 is rejected if $|t_0| > t_{\alpha/2,n-2}$.

EXAMPLE 9.20 (TESTING FOR SIGNIFICANCE OF REGRESSION)

Given the results in Example 9.19, the test for the significance of regression is conducted. One more computation is needed prior to conducting the test. That is, $\sum_{i=1}^{n} y_i^2 = 12{,}943$. Using equation (9.41) yields

$$S_{yy} = 12{,}943 - \frac{(431)^2}{15} = 558.93$$

Then $\sum_{i=1}^{15} e_i^2$ is computed according to equation (9.40) as

$$\sum_{i=1}^{15} e_i^2 = 558.93 - 0.507(994) = 54.97$$

Now, the value of MS_E is calculated from equation (9.39) as

$$\text{MS}_\text{E} = \frac{54.97}{13} = 4.22$$

The value of t_0 can be calculated using equation (9.38) as

$$t_0 = \frac{0.507}{\sqrt{4.22/1960}} = 11.02$$

Since $t_{0.025,13} = 2.16$ from Table A.4, reject the hypothesis that $\beta_1 = 0$. Thus there is significant evidence that x and y are related.

9.5.3. Use of the Regression Equation in Simulation

Assuming that model assumptions have been tested by a lack-of-fit test and residual analysis, as discussed earlier, and the regression is significant, as tested in Section 9.5.2, a relationship such as that given by equation (9.37) can be used in a simulation. Recall that $\hat{y}$ is an estimate of the mean of y for given x, $E(y|x)$. The simulation requires a value of y, say y', given by

$$y' = \hat{y} + \epsilon \tag{9.42}$$

where ϵ is normally distributed with mean zero and variance σ^2. This variance can be estimated by the mean square error, MS_E, given by equation (9.39).

In Example 9.20, σ^2 was estimated by $MS_E = 4.22$. In the simulation, if the instantaneous queue length is observed to be x, then a $N(0, \sigma^2)$ random variate with $\sigma^2 = 4.22$ can be generated to represent ϵ, and then x and ϵ used in equations (9.37) and (9.42) to calculate y', the service rate for the subsequent 10-minute period.

In simulation, there are two situations in which regression might be used. In the first situation, the input variable, x, is observed and the random output is to be generated. In that case a normally distributed ϵ is generated as described in the preceding paragraph. In the second situation, the input variable, x, is randomly generated in addition to a $N(0, \sigma^2)$ random variate, ϵ, and together, these randomly generated variates are used to generate a random output, y. Example 9.18, which related lead time and demand for industrial robots, would require this second procedure in an application. Similarly, Exercise 23, which relates milling time and planing time in a job shop, would be another instance of the second situation in an application to simulation.

9.5.4. Multivariate Linear Regression

If the simple linear regression model of Section 9.5.1 is inadequate, several other possibilities exist. There may be several independent variables, so that the relationship is of the form

$$y = \beta_0 + \beta_1 x_1 + \beta_2 x_2 + \cdots + \beta_m x_m + \epsilon \tag{9.43}$$

Note that this model is still linear, but has more than one independent variable. Regression models having the form shown in equation (9.43) are called multivariate linear regression models.

Another possibility is that the model is of a quadratic form such as

$$y = \beta_0 + \beta_1 x + \beta_2 x^2 + \epsilon \tag{9.44}$$

Equation (9.44) is also a linear model which may be transformed to the form of equation (9.43) by letting $x_1 = x$ and $x_2 = x^2$.

Yet another possibility is a model of the form such as

$$y = \beta_0 + \beta_1 x_1 + \beta_2 x_2 + \beta_3 x_1 x_2 + \epsilon$$

which is also a linear model. The analysis of these three models with the forms shown above, and related models, can be found in Draper and Smith [1966], Hines and Montgomery [1980], Montgomery and Peck [1982], and other applied statistics texts.

9.6. Summary

Input data collection and analysis require major time and resource commitments in a discrete-event simulation project. However, regardless of the validity or sophistication of the simulation model, unreliable inputs are going to lead to outputs whose subsequent interpretation will probably result in faulty recommendations.

This chapter discussed four steps in the development of models of input data—collecting the raw data, identifying the underlying statistical distribution, estimating the parameters, and testing for goodness of fit.

Some suggestions were given for facilitating the data collection step. However, experience, such as that obtained by completing any of Exercises 1 through 5, will increase awareness of the difficulty of problems that may arise in data collection and the need for planning.

Once the data have been collected, a statistical model should be hypothesized. Constructing a histogram is very useful at this point if sufficient data are available. Based on the underlying process and the shape of the histogram, a distribution can usually be selected for further investigation.

The investigation proceeds with the estimation of parameters for the hypothesized distribution. Suggested estimators were given for distributions often used in simulation. In a number of instances, these are functions of the sample mean and sample variance.

The last step in the process is the testing of the distributional hypothesis. The Kolmogorov–Smirnov and chi-square goodness-of-fit tests can be applied to many distributional assumptions. When a distributional assumption is rejected, another distribution is tried. When all else fails, the empirical distribution may be used in the model.

REFERENCES

Choi, S. C., and R. Wette [1969], "Maximum Likelihood Estimation of the Parameters of the Gamma Distribution and Their Bias," *Technometrics*, Vol. 11, No. 4, pp. 683–890.

CONNOVER, W. J. [1980], *Practical Nonparametric Statistics*, 2nd ed., Wiley, New York.

DRAPER, N. R., AND H. SMITH [1966], *Applied Regression Analysis*, Wiley, New York.

DURBIN, J. [1975], "Kolmogorov–Smirnov Tests When Parameters Are Estimated with Applications to Tests of Exponentiality and Tests on Spacings," *Biometrika*, Vol. 65, pp. 5–22.

FISHMAN, G. S. [1973], *Concepts and Methods in Discrete Event Digital Simulation*, Wiley, New York.

GUMBEL, E. J. [1943], "On the Reliability of the Classical Chi-Square Test," *Annals of Mathematical Statistics*, Vol. 14, pp. 253ff.

HINES, W. W., AND D. C. MONTGOMERY [1980], *Probability and Statistics in Engineering and Management Science*, 2nd ed., Wiley, New York.

KENDALL, M., AND A. STUART [1979], *The Advanced Theory of Statistics*, 4th ed., Vol. 2, Macmillan, New York.

LILLIEFORS, H. W. [1967], "On the Kolmogorov–Smirnov Test for Normality with Mean and Variance Unknown," *Journal of the American Statistical Association*, Vol. 62, pp. 339–402.

LILLIEFORS, H. W. [1969], "On the Kolmogorov–Smirnov Test for the Exponential Distribution with Mean Unknown," *Journal of the American Statistical Association*, Vol. 64, pp. 387–89.

MANN, H. B., AND A. WALD [1942], "On the Choice of the Number of Intervals in the Application of the Chi-Square Test," *Annals of Mathematical Statistics*, Vol. 18, pp. 50ff.

MONTGOMERY, D. C., AND E. A. PECK [1982], *Introduction to Linear Regression Analysis*, Wiley, New York.

NELSON, W. [1979], "How to Analyze Data with Simple Plots," American Society for Quality Control, Milwaukee, Wis.

PEARSON, K. [1922], "Tables of the Incomplete Gamma Function," *Biometrika*, London.

SHAPIRO, S. S. [1980], "How to Test Normality and Other Distributional Assumptions," American Society for Quality Control, Milwaukee, Wis.

THOMON, D. R., L. J. BAIN, AND C. E. ANTLE [1969], "Inferences on the Parameters of the Weibull Distribution," *Technometrica*, Vol. 11, No. 3, pp. 445–60.

EXERCISES

1. Go to a small appliance repair shop and determine the interarrival and service-time distributions. If there are several workers, how do the service-time distributions compare to each other? Do service-time distributions need to be constructed for each type of appliance? (Make sure that the management gives permission to perform this study.)

2. Go to a cafeteria and collect data on the distributions of interarrival and service times. The distribution of interarrival times is probably different for each of the three daily meals and may also vary during the meal; that is, the interarrival time distribution for 11:00 A.M. to 12:00 noon may be different than from 12:00 noon to 1:00 P.M. Define service time as the time from when the customer reaches the point at which the first selection could be made until exiting from the cafeteria line. (Any reasonable modification of this definition is acceptable.) The service-time distribution probably changes for each meal. Can times of the day or days of the week for either distribution be grouped due to homogeneity of the data? (Make sure that the management gives permission to perform this study.)

3. Go to a major traffic intersection and determine the interarrival time distributions from each direction. Some arrivals want to go straight, some turn left, some turn right. The interarrival time distribution varies during the day and by day of the week. Every now and then an accident occurs.
4. Go to a grocery store and determine the interarrival and service distributions at the checkout counters. These distributions may vary by time of day and by day of the week. Record, also, the number of service channels available at all times. (Make sure that the management gives permission to perform this study.)
5. Go to a laundromat and "relive" the author's data collection experience as discussed in Example 9.1. (Make sure that the management gives permission to perform this study.)
6. Prepare four theoretical normal density functions, all on the same figure, each distribution having mean zero, but let the standard deviations be 1/4, 1/2, 1, and 2.
7. On one figure, draw the pdf of the Erlang distributions where $\theta = 1/2$ and $k = 1$, 2, 4, and 8.
8. On one figure, draw the pdf of the Erlang distributions where $\theta = 2$ and $k = 1$, 2, 4, and 8.
9. Draw the pmf of the Poisson distribution that results when the parameter α is equal to the following:
   (a) $\alpha = 1/2$
   (b) $\alpha = 1$
   (c) $\alpha = 2$
   (d) $\alpha = 4$
10. On one figure draw the two exponential pdf's that result when the parameter, λ, equals 0.6 and 1.2.
11. On one figure draw the three Weibull pdf's which result when $\nu = 0$, $\alpha = 1/2$, and $\beta = 1$, 2, and 4.
12. The following data are randomly generated from a gamma distribution:

1.691	1.437	8.221	5.976
1.116	4.435	2.345	1.782
3.810	4.589	5.313	10.90
2.649	2.432	1.581	2.432
1.843	2.466	2.833	2.361

    Determine the maximum likelihood estimators $\hat{\beta}$ and $\hat{\theta}$.
13. The following data are randomly generated from a Weibull distribution where $\nu = 0$.

7.936	5.224	3.937	6.513
4.599	7.563	7.172	5.132
5.259	2.759	4.278	2.696
6.212	2.407	1.857	5.002
4.612	2.003	6.908	3.326

Determine the maximum likelihood estimators $\hat{\alpha}$ and $\hat{\beta}$. (The exercise requires a programmable calculator, a computer or a lot of patience.)

14. The highway between Atlanta, Georgia, and Athens, Georgia, has a high incidence of accidents along its 100 kilometers. Public safety officers say that the occurrence of accidents along the highway is randomly (uniformly) distributed, but the news media say otherwise. The Georgia Department of Public Safety published records for the month of September. These records indicated the point at which 30 accidents involving an injury or death occurred, as follows (the data points represent the distance from the city limits of Atlanta):

88.3	40.7	36.3	27.3	36.8
91.7	67.3	7.0	45.2	23.3
98.8	90.1	17.2	23.7	97.4
32.4	87.8	69.8	62.6	99.7
20.6	73.1	21.6	6.0	45.3
76.6	73.2	27.3	87.6	87.2

Use the Kolmogorov–Smirnov test to determine whether the distribution of location of accidents is uniformly distributed for the month of September.

15. Show that the Kolmogorov–Smirnov test statistic for Example 9.17 is $D = 0.1054$.

16. Records pertaining to the monthly number of job related injuries at an underground coal mine were being studied by a federal agency. The values for the past 100 months were as follows:

Injuries per Month	*Frequency of Occurrence*
0	35
1	40
2	13
3	6
4	4
5	1
6	1

(a) Apply the chi-square test to these data to test the hypothesis that the underlying distribution is Poisson. Use a level of significance of $\alpha = .05$.

(b) Apply the chi-square test to these data to test the hypothesis that the distribution is Poisson with mean 1.0. Again let $\alpha = 0.05$.

(c) What are the differences in parts (a) and (b), and when might each case arise?

17. The time required for 50 different employees to compute and record the number of hours worked during the week was measured with the following results in minutes:

Employee	*Time (Minutes)*	*Employee*	*Time (Minutes)*
1	1.88	26	0.04
2	0.54	27	1.49
3	1.90	28	0.66
4	0.15	29	2.03
5	0.02	30	1.00
6	2.81	31	0.39
7	1.50	32	0.34
8	0.53	33	0.01
9	2.62	34	0.10
10	2.67	35	1.10
11	3.53	36	0.24
12	0.53	37	0.26
13	1.80	38	0.45
14	0.79	39	0.17
15	0.21	40	4.29
16	0.80	41	0.80
17	0.26	42	5.50
18	0.63	43	4.91
19	0.36	44	0.35
20	2.03	45	0.36
21	1.42	46	0.90
22	1.28	47	1.03
23	0.82	48	1.73
24	2.16	49	0.38
25	0.05	50	0.48

Use the chi-square test (as in Example 9.14) to test the hypothesis that these service times are exponentially distributed. Let the number of class intervals be $k = 6$. Use a level of significance of $\alpha = 0.05$.

18. Studentwiser Beer Company is trying to determine the distribution of the breaking strength of their glass bottles. Fifty bottles are selected at random and tested for breaking strength, with the following results (in pounds per square inch):

218.95	232.75	212.80	231.10	215.95
237.55	235.45	228.25	218.65	212.80
230.35	228.55	216.10	229.75	229.00
199.75	225.10	208.15	213.85	205.45
219.40	208.15	198.40	238.60	219.55
243.10	198.85	224.95	212.20	222.90
218.80	203.35	223.45	213.40	206.05
229.30	239.20	201.25	216.85	207.25
204.85	219.85	226.15	230.35	211.45
227.95	229.30	225.25	201.25	216.10

(a) Use the chi-square test with equiprobable intervals to test these breaking strengths for normality at a level of significance of $\alpha = 0.05$. (*Hint:* $\bar{X} = 219.4$ and $S^2 = 137.3$.)

(b) The manufacturer of the bottles says the breaking strengths are normally distributed with $\mu = 200$ and $\sigma = 15$. Test the manufacturer's claim at a level of significance of $\alpha = 0.05$.

19. The Crosstowner is a bus that cuts a diagonal path from northeast Atlanta to southwest Atlanta. The time required to complete the route is maintained by the bus operator. The bus runs Monday through Friday. The times of the last fifty 8:00 A.M. runs, in minutes, are as follows:

92.3	92.8	106.8	108.9	106.6
115.2	94.8	106.4	110.0	90.9
104.6	72.0	86.0	102.4	99.8
87.5	111.4	105.9	90.7	99.2
97.8	88.3	97.5	97.4	93.7
99.7	122.7	100.2	106.5	105.5
80.7	107.9	103.2	116.4	101.7
84.8	101.9	99.1	102.2	102.5
111.7	101.5	95.1	92.8	88.5
74.4	98.9	111.9	96.5	95.9

How are these run times distributed? Develop and test a suitable model.

20. The time required for the transmission of a message (in minutes) is sampled electronically at a communications center. The last 50 values in the sample are as follows:

7.936	4.612	2.407	4.278	5.132
4.599	5.224	2.003	1.857	2.696
5.259	7.563	3.937	6.908	5.002
6.212	2.759	7.172	6.513	3.326
8.761	4.502	6.188	2.566	5.515
3.785	3.742	4.682	4.346	5.359
3.535	5.061	4.629	5.298	6.492
3.502	4.266	3.129	1.298	3.454
5.289	6.805	3.827	3.912	2.969
4.646	5.963	3.829	4.404	4.924

How are the transmission times distributed? Develop and test an appropriate model.

21. The time (in minutes) between requests for the hookup of electric service was accurately maintained at the Gotwatts Flash and Flicker Company with the following results for the last 50 requests:

0.661	4.910	8.989	12.801	20.249
5.124	15.033	58.091	1.543	3.624
13.509	5.745	0.651	0.965	62.146
15.512	2.758	17.602	6.675	11.209
2.731	6.892	16.713	5.692	6.636
2.420	2.984	10.613	3.827	10.244
6.255	27.969	12.107	4.636	7.093
6.892	13.243	12.711	3.411	7.897
12.413	2.169	0.921	1.900	0.315
4.370	0.377	9.063	1.875	0.790

How are the times between requests for service distributed? Develop and test a suitable model.

22. Daily demands for transmission overhaul kits for the D-3 dragline were maintained by Earth Moving Tractor Company with the following results:

0	2	0	0	0
1	0	1	1	1
0	1	0	0	0
2	0	1	0	1
0	1	0	0	2
1	0	1	0	0
0	0	0	0	0
1	0	1	0	1
0	0	3	0	1
1	0	0	0	0

How are the daily demands distributed? Develop and test an appropriate model.

23. A simulation is to be conducted of a job shop that performs two operations, milling and planing, in that order. It would be possible to collect data about processing times for each operation, then generate random occurrences from each distribution. However, the shop manager says that the times might be related; large milling jobs take lots of planing. Data are collected for the next 25 orders with the following results in minutes:

Order	*Milling Time (Minutes)*	*Planing Time (Minutes)*	*Order*	*Milling Time (Minutes)*	*Planing Time (Minutes)*
1	12.3	10.6	14	24.6	16.6
2	20.4	13.9	15	28.5	21.2
3	18.9	14.1	16	11.3	9.9
4	16.5	10.1	17	13.3	10.7
5	8.3	8.4	18	21.0	14.0
6	6.5	8.1	19	19.5	13.0
7	25.2	16.9	20	15.0	11.5
8	17.7	13.7	21	12.6	9.9
9	10.6	10.2	22	14.3	13.2
10	13.7	12.1	23	17.0	12.5
11	26.2	16.0	24	21.2	14.2
12	30.4	18.9	25	28.4	19.1
13	9.9	7.7			

(a) Plot milling time on the horizontal axis and planing time on the vertical axis. Does a linear model seem appropriate for these data?

(b) With milling time as the independent variable and planing time as the dependent variable, estimate β_0 and β_1 using the model given by equation (9.26).

(c) Test $\hat{\beta}_1$ for significance of regression.

24. Write a computer program in FORTRAN, BASIC, or some other high-level language to compute the maximum likelihood estimates $(\hat{\alpha}, \hat{\beta})$ of the Weibull distribution. Inputs to the program should include the sample size, n; the observations, $x_1, x_2, \ldots, x_n$; a stopping criterion, ϵ (stop when $|f(\hat{\beta}_j)| \leq \epsilon$); and a print option, OPT (usually set $= 0$). Output would be the estimates $\hat{\alpha}$ and $\hat{\beta}$. If OPT $= 1$, additional output would be printed as in Table 9.4 showing convergence. Make the program as "user friendly" as possible.

25. Examine a computer program library to which you have accessibility. Obtain documentation on programs that would be useful in solving exercises 7 through 23. Use the library programs as an aid in solving selected problems.

VERIFICATION AND VALIDATION OF SIMULATION MODELS

One of the most important and difficult tasks facing a model developer is the verification and validation of the simulation model. The users of a model—the engineers and analysts who use the model outputs to aid in making design recommendations, and the managers who make decisions based on these recommendations—justifiably look upon a model with some degree of skepticism about its validity. To reduce this skepticism and to increase the model's credibility, it is the job of the model developer to work closely with the end users throughout the period of development and validation. The goal of the validation process is twofold: (1) to produce a model that represents true system behavior closely enough for the model to be used as a substitute for the actual system for the purpose of experimenting with the system; (2) to increase to an acceptable level the credibility of the model, so that the model will be used by managers and other decision makers.

Validation should not be seen as an isolated set of procedures that follows model development, but rather as an integral part of model development. Conceptually, however, the verification and validation process may be broken into two major components:

1. Verification refers to the comparison of the conceptual model to the computer code that implements that conception. It asks the questions: Is the model implemented correctly in computer code? Are the input

10

parameters and logical structure of the model correctly represented in the code?

2. Validation refers to the act of determining that a model is an accurate representation of the real system. Validation is usually achieved through the calibration of the model, an iterative process of comparing the model to actual system behavior and using the discrepancies between the two, and the insights gained, to improve the model. This process is repeated until model accuracy is judged to be acceptable.

This chapter describes methods that have been recommended and used in the verification and validation process. Most of the methods are informal subjective comparisons, while a few are formal statistical procedures. The use of the latter procedures involves issues related to output analysis, the subject of Chapters 11 and 12. Output analysis refers to analysis of the data produced by a simulation and drawing inferences from these data about the behavior of the real system. To summarize their relationship, validation is the process by which model users gain confidence that output analysis is making valid inferences about the real system under study.

Many reports and papers have been written on validation in general, and on the validation effort for a specific model. For discussions of the main issues in model validation, the reader is referred to the papers by Gass [1977], Law

[1979a], Naylor and Finger [1967], Ören [1981], Sargent [1980], Shannon [1975], and Van Horn [1969, 1971]. For statistical techniques relevant to different aspects of validation, the reader can obtain the foregoing articles plus those by Balci and Sargent [1981], Law [1979b], and Schruben [1980]. For case studies in which validation is emphasized, the reader is referred to Carson et al. [1981a, b], Gafarian and Walsh [1970], and Shechter and Lucas [1980]. A bibliography of work on validation has been published by Balci and Sargent [1980].

10.1 Model Building, Verification, and Validation

The first step in model building consists of observing the real system and the interactions among its various components, and collecting data on its behavior. But observation alone seldom yields sufficient understanding of system behavior. Persons familiar with the system, or any subsystem, should be questioned to take advantage of their special knowledge. Operators, technicians, repair and maintenance personnel, engineers, supervisors, and managers understand certain aspects of the system which may be unfamiliar to others. As model development proceeds, new questions may arise and the model developers will return to this step of learning true system structure and behavior.

The second step in model building is the construction of a conceptual

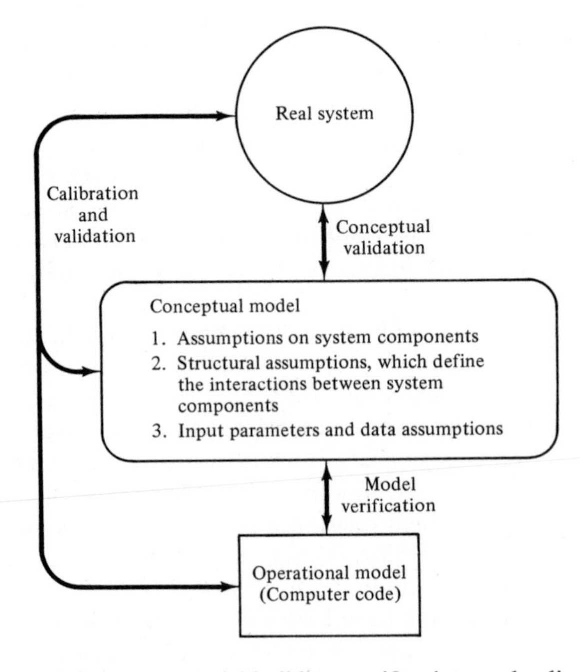

Figure 10.1. Model building, verification, and validation.

model—a collection of assumptions on the components and the structure of the system, plus hypotheses on the values of model input parameters. As illustrated by Figure 10.1, conceptual validation is the comparison of the real system to the conceptual model.

The third step is the coding of an operational model—the computer code that implements the conceptual model. In actuality, model building is not a linear process with three steps; instead, the model builder will return to each of these steps many times while building, verifying, and validating the model. Figure 10.1 depicts the ongoing model building process in which the need for verification and validation cause continual comparison of the real system to the conceptual model and to the operational model, and repeated modification of the model to improve its accuracy.

10.2 Verification of Simulation Models

The purpose of model verification is to assure that the conceptual model is reflected accurately in the computer code. The conceptual model quite often involves some degree of abstraction about system operations, and/or some amount of simplification of actual operations. Verification asks the question: Is the conceptual model (assumptions on system components and system structure, parameter values, abstractions and simplifications) accurately represented by the operational model (i.e., by the computer code)?

Many commonsense suggestions can be given for use in the verification process.

1. Have the code checked by someone other than the programmer.
2. Make a flow diagram which includes each logically possible action a system can take when an event occurs, and follow the model logic for each action for each event type. (An example of a logic flow diagram is given in Figures 2.2 and 2.3 for the model of a single-server queue.)
3. Closely examine the model output for reasonableness under a variety of settings of the input parameters. Have the code print out a wide variety of output statistics.
4. Have the computerized model print the input parameters at the end of the simulation, to be sure that these parameter values have not been changed inadvertently.
5. Make the computer code as self-documenting as possible. Give a precise definition of every variable used, and a general description of the purpose of each major section of code.

These suggestions are basically the same ones any programmer would follow when debugging a computer program.

Among these commonsense suggestions, one that is most easily implemented, but quite often overlooked, especially by students who are learning simulation, is a close and thorough examination of model output for reasonableness (suggestion 3). For example, consider a model of a complex network of queues consisting of many service centers in series and parallel configurations. Suppose that the modeler is interested mainly in the response time, defined as the time required for a customer to pass through a designated part of the network. During the verification (and calibration) phase of model development, it is recommended that the program collect and print out many statistics in addition to response times, such as utilizations of servers and time-average number of customers in various subsystems. Examination of the utilization of a server, for example, may reveal that it is unreasonably low (or high), a possible error that may be caused by wrong specification of mean service time, or a mistake in model logic that sends too few (or too many) customers to this particular server, or any number of other possible parameter misspecifications or errors in logic.

In a language such as GPSS or SLAM, which automatically collects many standard statistics (average queue lengths, average waiting times, etc.), it takes little or no extra programming effort to display almost all statistics of interest. When simulating with FORTRAN, GASP, or SIMSCRIPT, extra coding may be required to collect and print the additional statistics. (The effort required can be considerably greater in a general-purpose language such as FORTRAN, which does not have statistics-gathering capabilities to aid the programmer.)

Two sets of satistics that can give a quick indication of model reasonableness are "current contents" and "total count." These statistics apply to any system having items of some kind flowing through it, whether these items be called customers, transactions, inventory, or vehicles. Current contents refers to the number of items in each component of the system at a given time. Total count refers to the total number of items that have entered each component of the system by a given time. In GPSS, these statistics are automatically kept and can be printed at any point in simulation time. In other languages, simple counters may have to be added to the code and printed at appropriate times. If the current contents in some portion of the system is high, this indicates that a large number of entities are delayed. If the output is printed for successively longer simulation run times and this current contents tends to grow in a more or less linear fashion, it is highly likely that a queue is unstable, that is, its theoretical utilization is larger than 1 and the server(s) will fall further behind as time goes on. This indicates that possibly the number of servers is too small, or that a service time is misspecified. (Unstable queues were discussed in Chapter 5.) On the other hand, if the total count for some subsystem is zero, this indicates that no items entered that subsystem, again a highly suspect occurrence. Careful evaluation of these statistics for various run lengths can aid in the detection of mistakes in model logic and data misspecifications. Checking for output reasonableness will usually fail to detect the more subtle errors, but it is one of the quickest ways to discover gross errors. To aid this error detection, it is best if the model developer forecasts a reasonable range for the value of selected output statistics before making a run of the model.

Such a forecast reduces the possibility of rationalizing a discrepancy and failing to investigate the cause of unusual output.

For certain models, it is possible to consider more than whether a particular statistic is reasonable. It is possible to compute certain long-run measures of performance. For example, as was seen in Chapter 5, the analyst can compute the long-run server utilization for a large number of queueing systems without any special assumptions regarding interarrival or service-time distributions. Typically, the only information needed is the network configuration plus arrival and service rates. Any measure of performance that can be computed analytically and then compared to its simulated counterpart provides another valuable tool for verification. Presumably, the objective of the simulation is to estimate some measure of performance, such as mean response time, which cannot be computed analytically. But as was illustrated by the formulas in Chapter 5 for a number of special queues ($M/M/1$, $M/G/1$, etc.), all the measures of performance in a queueing system are interrelated. Thus, if a simulation model is predicting one such measure (such as utilization) correctly, then confidence in the model's predictive ability for other related measures (such as response time) is increased (even though the exact relation between the two measures is, of course, unknown in general and varies from model to model). Conversely, if a model incorrectly predicts utilization, its prediction of other quantities, such as mean response time, is highly suspect.

Another important way to aid the verification process is the oft-neglected documentation phase of programming. If a model builder writes brief comments in the code itself, plus definitions of all variables and parameters, and descriptions of each major section of code, it becomes much simpler for someone else, or the model builder at a later date, to verify the model logic. Documentation is also important when initially coding the model as a means of clarifying the logic of a model and verifying its completeness.

A more sophisticated technique is the use of a trace. In general, a trace is a detailed computer printout which gives the value of every variable (in a specified set of variables) in a computer program, every time that one of these variables changes in value. A trace designed specifically for use in a simulation program would give the value of selected variables each time the simulation clock was incremented (i.e., each time an event occurred). Thus, a simulation trace is nothing more than a detailed printout of the state of the simulation model as it changes over time.

EXAMPLE 10.1

When verifying the computer implementation (in FORTRAN, SIMSCRIPT, or GASP) of the single-server queue model of Example 2.1, an analyst made a run over 16 units of time and observed that the time-average length of the waiting line was $\hat{L}_Q =$ 0.4375 customer, which is certainly not unreasonable for a short run of only 16 time units. Nevertheless, the analyst decided that a more detailed verification would be of value.

The trace in Figure 10.2 gives the hypothetical printout from simulation time CLOCK = 0 to CLOCK = 16 for the simple single-server queue of Example 2.1. This example illustrates how an error can be found with a trace, when no error was apparent from the examination of the summary output statistics (such as $\hat{L}_Q$). Note that at simulation time CLOCK = 3, the number of customers in the system is NCUST = 1, but the server is idle (STATUS = 0). The source of this error could be incorrect logic, or simply not setting the attribute STATUS to a value of 1 (when coding in FORTRAN or SIMSCRIPT).

Definitions of Variables:

CLOCK = Simulation clock
EVTYP = Event type (start, arrival, departure, or stop)
NCUST = Number of customers in system at time 'CLOCK'
STATUS = Status of server (1–busy, 0–idle)

State of System Just After the Named Event Occurs:

CLOCK = 0	EVTYP = 'Start'	NCUST = 0	STATUS = 0
CLOCK = 3	EVTYP = 'Arrival'	NCUST = 1	STATUS = 0
CLOCK = 5	EVTYP = 'Depart'	NCUST = 0	STATUS = 0
CLOCK = 11	EVTYP = 'Arrival'	NCUST = 1	STATUS = 0
CLOCK = 12	EVTYP = 'Arrival'	NCUST = 2	STATUS = 1
CLOCK = 16	EVTYP = 'Depart'	NCUST = 1	STATUS = 1

·
·
·

Figure 10.2. Simulation trace of Example 2.1.

In any case the error must be found and corrected. Note that the less sophisticated practice of examining the summary measures, or output, did not detect the error. By using equation (5.4), the reader can verify that $\hat{L}_Q$ was computed correctly from the data ($\hat{L}_Q$ is the time-average value of 'NCUST minus STATUS'):

$$\hat{L}_Q = \frac{(0-0)3 + (1-0)2 + (0-0)6 + (1-0)1 + (2-1)4}{3+2+6+1+4}$$

$$= \frac{7}{16} = 0.4375$$

as previously mentioned. Thus, the output measure, $\hat{L}_Q$, had a reasonable value and was computed correctly from the data, but its value was indeed wrong because the attribute STATUS was not assuming correct values. As seen from Figure 10.2, a trace yields information on the actual history of the model which is more detailed and informative than the summary measures alone.

Most modern simulation languages have a built-in capability to conduct a trace at the request of the programmer, without the programmer having to do any extensive programming. The most widely used simulation languages—GPSS, SIMSCRIPT, GASP, SLAM—do indeed provide a built-in trace capability. In addition, most versions of FORTRAN have a tracing capability which can be of value to the simulator.

As can be easily imagined, a trace over a large span of simulation time can quickly produce an extremely large amount of computer printout, which would be extremely cumbersome to check in detail for correctness. The purpose of the trace is to verify the correctness of the computer program by making detailed paper-and-pencil calculations. To make this practical, a simulation with a trace is usually restricted to a very short period of time. It is desirable, of course, to ensure that each type of event (such as "ARRIVAL") occurs at least once, so that its consequences and effect on the model can be checked for accuracy. If an event is especially rare in occurrence, it may be necessary to use artificial data to force it to occur during a simulation of short duration. This is legitimate, as the purpose is to verify that the effect on the system of the rare event is as anticipated.

Of the three classes of techniques—the commonsense techniques, thorough documentation, and traces—it is recommended that the first two always be carried out. Close examination of model output for reasonableness is especially valuable and informative. A trace is more time consuming and expensive than the other techniques, and usually is used only as a last resort to detect the cause of elusive errors; however, it does provide a valuable technique for increasing confidence in the model.

10.3 Calibration and Validation of Models

Verification and validation, although conceptually distinct, usually are conducted simultaneously by the modeler. Validation is the overall process of comparing the model and its behavior to the real system and its behavior. Calibration is the iterative process of comparing the model to the real system, making adjustments (or even major changes) to the model, comparing the revised model to reality, making additional adjustments, comparing again, and so on. Figure 10.3 shows the relationship of model calibration to the overall validation process. The comparison of the model to reality is carried out by a variety of tests—some subjective and others objective. Subjective tests usually involve people who are knowledgeable about one or more aspects of the system making judgments about the model and its output. Objective tests always require data on the system's behavior plus the corresponding data produced by the model. Then one or more statistical tests are performed to compare some aspect of the system data set to the same aspect of the model data set. This iterative process of comparing model and system, and revising both the conceptual and operational models to accommodate any perceived model deficiencies, is continued until the model is judged to be sufficiently accurate.

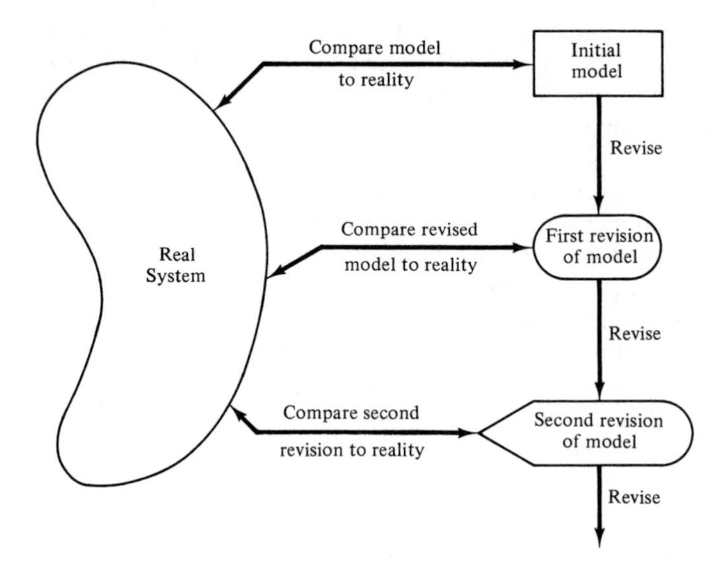

Figure 10.3. Iterative process of calibrating a model.

A possible criticism of the calibration phase, were it to stop at this point, is that the model has been validated only for the one data set used; that is, the model has been "fit" to one data set. One way to alleviate this criticism is to collect a new set of system data (or to reserve a portion of the original system data) to be used at this final stage of validation. That is, after the model has been calibrated using the original system data set, a "final" validation is conducted using the second system data set. If unacceptable discrepancies between the model and the real system are discovered in the "final" validation effort, the modeler must return to the calibration phase and modify the model until it becomes acceptable.

Validation is not an either/or proposition—no model is ever totally representative of the system under study. In addition, each revision of the model, as pictured in Figure 10.3, involves some cost, time, and effort. The modeler must weigh the possible, but not guaranteed, increase in model accuracy versus the cost of increased validation effort. Usually, the modeler (and model users) have some maximum discrepancy between model predictions and system behavior that would be acceptable. If this level of accuracy cannot be obtained within the budget constraints, either expectations of model accuracy must be lowered, or the model must be abandoned.

As an aid in the validation process, Naylor and Finger [1967] formulated a three-step approach which has been widely followed:

1. Build a model that has high face validity.
2. Validate model assumptions.

3. Compare the model input–output transformations to corresponding input–output transformations for the real system.

The next five subsections investigate these three steps in detail.

10.3.1. Face Validity

The first goal of the simulation modeler is to construct a model that appears reasonable on its face to model users and others who are knowledgeable about the real system being simulated. The potential users of a model should be involved in model construction from its conceptualization to its implementation to ensure that a high degree of realism is built into the model through reasonable assumptions regarding system structure, and reliable data. Potential users and knowledgeable persons can also evaluate model output for reasonableness, and can aid in identifying model deficiencies. Thus, the users can be involved in the calibration process as the model is iteratively improved based on the insights gained from the initial model deficiencies. Another advantage of user involvement is the increase in the model's perceived validity, or credibility, without which a manager would not be willing to trust simulation results as a basis for decision making.

Sensitivity analysis can also be used to check a model's face validity. The model user is asked if the model behaves in the expected way when one or more input variables is changed. For example, in most queueing systems, if the arrival rate of customers (or demands for service) were to increase, it would be expected that utilizations of servers, lengths of lines, and delays would tend to increase (although by how much might well be unknown). Based on experience and observations on the real system (or similar related systems), the model user and model builder would probably have some notion at least of the direction of change in model output when an input variable is increased or decreased. For most large-scale simulation models, there are many input variables and thus many possible sensitivity tests. The model builder must attempt to choose the most critical input variables for testing if it is too expensive or time consuming to vary all input variables. If real system data are available for at least two settings of the input parameters, objective scientific sensitivity tests can be conducted using appropriate statistical techniques.

10.3.2. Validation of Model Assumptions

Model assumptions fall into two general classes: structural assumptions and data assumptions. Structural assumptions involve questions of how the system operates and usually involve simplifications and abstractions of reality. For example, consider the customer queueing and service facility in a bank. Customers may form one line, or there may be an individual line for each teller. If there are many lines, customers may be served strictly on a first come, first served basis, or some customers may change lines if one is moving faster. The number of tellers may be fixed or variable. These structural assumptions should be verified by actual observation during appropriate time periods plus discussions with managers and tellers regarding bank policies and actual implementation of these policies.

Data assumptions should be based on the collection of reliable data and correct statistical analysis of the data. (Example 9.1 discussed similar issues for a model of a laundromat.) For example, in the bank study previously mentioned, data were collected on:

1. Interarrival times of customers during several 2-hour periods of peak loading ("rush-hour" traffic)
2. Interarrival times during a slack period
3. Service times for commercial accounts
4. Service times for personal accounts

The reliability of the data was verified by consultation with bank managers, who identified typical rush hours and typical slack times. When combining two or more data sets collected at different times, data reliability can be further enhanced by objective statistical tests for homogeneity of data. (Do two data sets $\{X_i\}$ and $\{Y_i\}$ on service times for personal accounts, collected at two different times, come from the same parent population? If so, the two sets can be combined.) Additional tests may be required to test for correlation in the data. As soon as the analyst is assured of dealing with a random sample (i.e., correlation is not present), the statistical analysis can begin.

The procedures for analyzing input data from a random sample were discussed in detail in Chapter 9. Recall that the analysis consisted of three steps:

1. Identifying the appropriate probability distribution
2. Estimating the parameters of the hypothesized distribution
3. Validating the assumed statistical model by a goodness-of-fit test such as the chi-square or Kolmogorov–Smirnov test

The use of goodness-of-fit tests is an important part of the validation of the model assumptions.

10.3.3. Validating Input–Output Transformations

The ultimate test of a model, and in fact the only objective test of the model as a whole, is the model's ability to predict the future behavior of the real system when the model input data match the real inputs and when a policy implemented in the model is implemented at some point in the system. Furthermore, if the level of some input variables (e.g., the arrival rate of customers to a service facility) were to increase or decrease, the model should accurately predict what would happen in the real system under similar circumstances. In other words, the structure of the model should be accurate enough for the model to make good predictions, not just for one input data set, but for the range of input data sets which are of interest.

In this phase of the validation process, the model is viewed as an input–output transformation—that is, the model accepts values of the input parameters and transforms these inputs into output measures of performance. It is this correspondence that is being validated.

Instead of validating the model input–output transformations by predicting the future, the modeler may use past historical data which have been reserved for validation purposes only; that is, if one data set has been used to develop and calibrate the model, it is recommended that a separate data set be used as the final validation test. Thus, accurate "prediction of the past" may replace prediction of the future for the purpose of validating the model.

A model is usually developed with primary interest in a specific set of system responses to be measured under some range of input conditions. For example, in a queueing system, the responses may be server utilization and customer delay, and the range of input conditions (or input variables) may include two or three servers at some station and a choice of scheduling rules. In a production system, the response may be throughput (i.e., production per hour) and the input conditions may be a choice of several machines that run at different speeds, with each machine having its own breakdown and maintenance characteristics. In any case, the modeler should use the main responses of interest as the criteria for validating a model. If the model is used later for a purpose different from its original purpose, the model should be revalidated in terms of the new responses of interest and under the possibly new input conditions.

A necessary condition for the validation of input–output transformations is that some version of the system under study exists, so that system data under at least one set of input conditions can be collected to compare to model predictions. If the system is in the planning stages and no system operating data can be collected, complete input–output validation is not possible. Other types of validation should be conducted, to the extent possible. In some cases, subsystems of the planned system may exist and a partial input–output validation can be conducted.

Presumably, the model will be used to compare alternative system designs, or to investigate system behavior under a range of new input conditions. Assume for now that some version of the system is operating, and that the model of the existing system has been validated. What, then, can be said about the validity of the model of a nonexistent proposed system, or the model of the existing system under new input conditions?

First, the responses of the two models under similar input conditions will be used as the criteria for comparison of the existing system to the proposed system. Validation increases the modeler's confidence that the model of the existing system is accurate. Second, in many cases, the proposed system is a modification of the existing system, and the modeler hopes that confidence in the model of the existing system can be transferred to the model of the new system. This transfer of confidence usually can be justified if the new model is a relatively minor modification of the old model in terms of changes to computer code (it may be a major change for the actual system). Changes in computer code ranging from relatively minor to relatively major include:

1. (Minor) changes of single numerical parameters, such as the speed of a machine, the arrival rate of customers (with no change in distributional form of interarrival times), or the number of servers in a parallel service center
2. (Minor) changes of the form of a statistical distribution, such as the distribution of a service time or a time to failure of a machine
3. (Major) changes in the logical structure of a subsystem, such as a change in queue discipline for a waiting-line model, or a change in the scheduling rule for a job shop model
4. (Major) changes involving a different design for the new system, such as a computerized inventory control system replacing an older non-computerized system, or an automatic computerized storage and retrieval system replacing a warehouse system in which workers pick items manually

If the change to the computer code is minor, such as in items 1 or 2, these changes can be carefully verified and output from the new model accepted with considerable confidence. If a sufficiently similar subsystem exists elsewhere, it may be possible to validate the submodel that represents the subsystem, and then to integrate this submodel with other validated submodels to build a complete model. In this way, partial validation of the substantial model changes in items 3 and 4 may be possible. Unfortunately, there is no way to completely validate the input–output transformations of a model of a nonexisting system. In any case, within time and budget constraints the modeler should use as many validation techniques as possible, including input–output validation of subsystem models if operating data can be collected on such subsystems.

Example 10.2 will illustrate some of the techniques that are possible for input–output validation, and will discuss the concepts of an input variable, uncontrollable variable, decision variable, output or response variable, and input–output transformation in more detail.

EXAMPLE 10.2 (THE FIFTH NATIONAL BANK OF JASPAR)

The Fifth National Bank of Jaspar, as shown in Figure 10.4, is planning to expand its drive-in service at the corner of Main Street. Currently, there is one drive-in window serviced by one teller. Only one or two transactions are allowed at the drive-in window, so it was assumed that each service time was a random sample from some underlying population. Service times $\{S_i, i = 1, 2, \ldots, 90\}$ and interarrival times $\{A_i, i = 1, 2, \ldots, 90\}$ were collected for the 90 customers who arrived between 11:00 A.M. and 1:00 P.M. on a Friday. This time slot was selected for data collection after consultation with management and the teller because it was felt to be representative of a typical rush hour.

Data analysis (as outlined in Chapter 9) led to the conclusion that the arrival

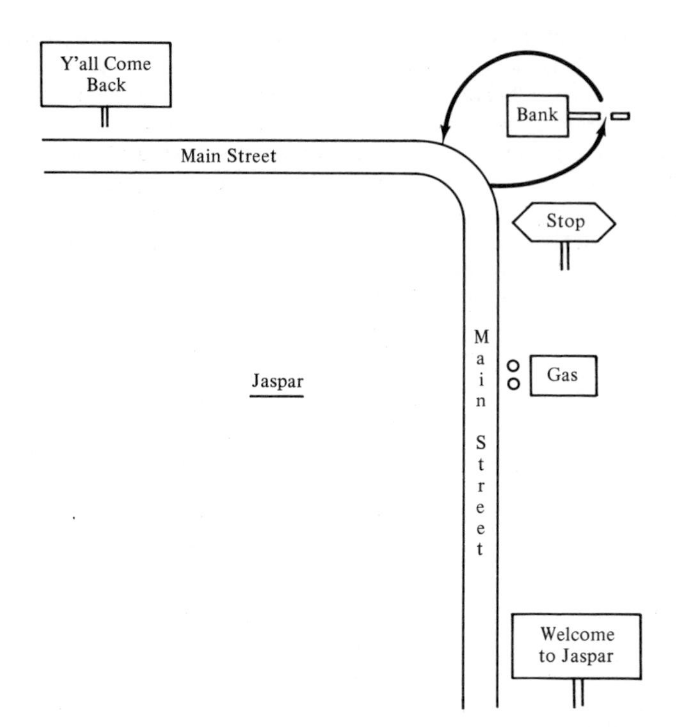

Figure 10.4. Drive-in window at the Fifth National Bank.

process could be modeled as a Poisson process with an arrival rate of 45 customers per hour; and that service times were approximately normally distributed with mean 1.1 minutes and standard deviation 0.2 minute. Thus, the model has two input variables:

1. Interarrival times, exponentially distributed (i.e., a Poisson arrival process) at rate $\lambda = 45$ per hour
2. Service times, assumed to be $N(1.1, (0.2)^2)$

Each input variable has a level: the rate ($\lambda = 45$ per hour) for the interarrival times, and the mean 1.1 minutes and standard deviation 0.2 minute for the service times. The interarrival times are examples of uncontrollable variables (i.e., uncontrollable by management in the real system). The service times are also uncontrollable variables, although the level of the service times may be partially controllable. If the mean service time could be decreased to 0.9 minute by installing a computer terminal, the level of the service time variable becomes a decision variable or controllable parameter. Setting all decision variables at some level constitutes a policy. For example, the current bank policy is one teller ($D_1 = 1$), mean service time $D_2 = 1.1$ minutes, and one line for waiting cars ($D_3 = 1$). ($D_1, D_2, \ldots$ are used to denote decision variables.) Decision variables are under management's control; the uncontrollable variables, such as arrival rate, and actual arrival times, are not under management's control. The arrival rate may change from time to time, but such change is due to external factors not under management control.

A model of current bank operations was developed and verified in close consultation with bank management and employees. Model assumptions were validated, as discussed in Section 10.3.2. The resulting model is now viewed as a "black box" which takes all input variable specifications and transforms them into a set of output or response variables. The output variables consist of all statistics of interest generated by the simulation about the model's behavior. For example, management is interested in the teller's utilization at the drive-in window (percent of time the teller is busy at the window), average delay in minutes of a customer from arrival to beginning of service, and the maximum length of the line during the rush hour. These input and output variables are shown in Figure 10.5, and are listed in Table 10.1 together with some

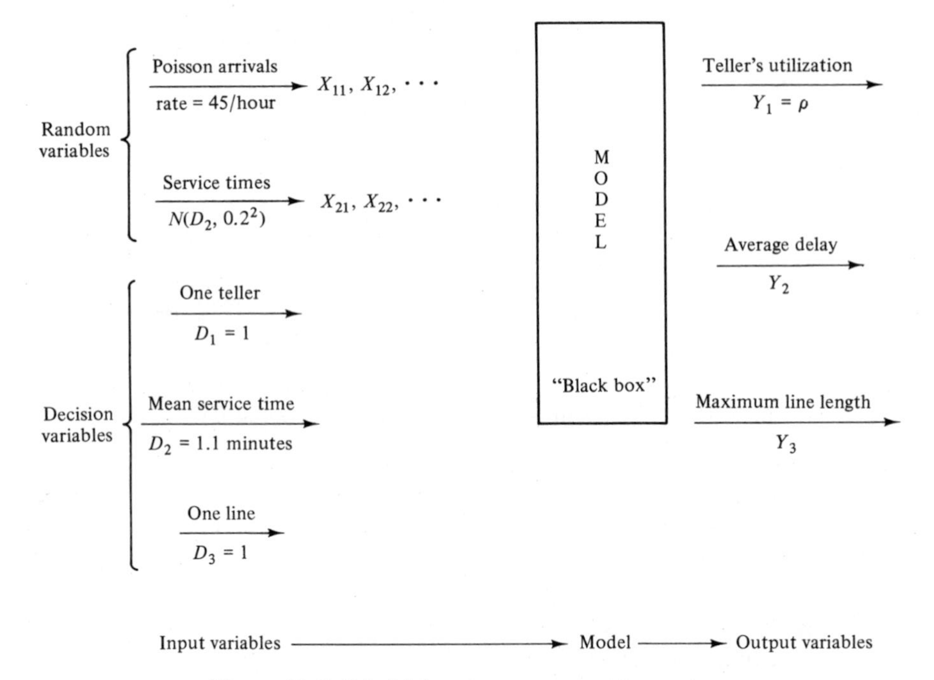

Figure 10.5. Model input–output transformation.

additional output variables. The uncontrollable input variables are denoted by X, the decision variables by D, and the output variables by Y. From the "black box" point of view, the model takes the inputs X and D and produces the outputs Y, namely

$$(X, D) \xrightarrow{f} Y$$

or

$$f(X, D) = Y$$

Here f denotes the transformation that is due to the structure of the model. For the Fifth National Bank study, the exponentially distributed interarrival time generated in the model (by the methods of Chapter 8) between customer $n - 1$ and customer n is denoted by X_{1n}. (Do not confuse X_{1n} with A_n; the latter was an observation made on

Table 10.1. INPUT AND OUTPUT VARIABLES FOR MODEL OF CURRENT BANK OPERATIONS

Input Variables	*Model Output Variables, Y*
D = decision variables X = other variables	Variables of primary interest to management (Y_1, Y_2, Y_3)
Poisson arrivals at rate = 45/hour $X_{11}, X_{12}, \ldots$	Y_1 = teller's utilization Y_2 = average delay Y_3 = maximum line length
Service times, $N(D_2, 0.2^2)$ $X_{21}, X_{22}, \ldots$	Other output variables of secondary interest
D_1 = 1 (one teller) D_2 = 1.1 minutes (mean service time) D_3 = 1 (one line)	Y_4 = observed arrival rate Y_5 = average service time Y_6 = sample standard deviation of service times Y_7 = average length of line

the real system.) The normally distributed service time generated in the model for customer n is denoted by X_{2n}. The set of decision variables, or policy, is $D = (D_1, D_2, D_3) = (1, 1.1, 1)$ for current operations. The output, or response, variables are denoted by $Y = (Y_1, Y_2, \ldots, Y_7)$ and are defined in Table 10.1.

For validation of the input–output transformations of the bank model to be possible, real system data must be available, comparable to at least some of the model output Y of Table 10.1. The system responses should have been collected during the same time period (from 11:00 A.M. to 1:00 P.M. on the same Friday) in which the input data $\{A_i, S_i\}$ were collected. This is important because if system response data were collected on a slower day (say, an arrival rate of 40 per hour), the system responses such as teller utilization (Z_1), average delay (Z_2), and maximum line length (Z_3) would be expected to be lower than the same variables during a time slot when the arrival rate was 45 per hour, as observed. Suppose that the delay of successive customers was measured on the same Friday between 11:00 A.M. and 1:00 P.M., and that the average delay was found to be $Z_2 = 4.3$ minutes.

When the model is run using generated random variates X_{1n} and X_{2n}, it is expected that observed values of average delay, Y_2, should be close to $Z_2 = 4.3$ minutes. However, the generated input values (X_{1n} and X_{2n}) cannot be expected to replicate exactly the actual input values (A_n and S_n) of the real system, but they are expected to replicate the statistical pattern of the actual inputs. Hence, simulation generated values of Y_2 are expected to be consistent with the observed system variable $Z_2 = 4.3$ minutes. Now consider how the modeler might test this consistency.

The modeler makes a small number of statistically independent replications of the model. Statistical independence is guaranteed by using nonoverlapping sets of random numbers produced by the random number generator, or by choosing seeds for each replication independently (from a random number table). The results of six independent replications, each of 2 hours duration, are given in Table 10.2. Observed arrival rate Y_4 and sample average service time Y_5 for each replication of the model are also noted, to be compared to the specified values of 45/hour and 1.1 minutes, respectively. The validation test consists of comparing the system response, namely average delay Z_2 =

Table 10.2. RESULTS OF SIX REPLICATIONS OF THE FIRST BANK MODEL

Replication	Y_4 *(Arrivals/Hour)*	Y_5 *(Minutes)*	Y_2 = *Average Delay (Minutes)*
1	51	1.07	2.79
2	40	1.12	1.12
3	45.5	1.06	2.24
4	50.5	1.10	3.45
5	53	1.09	3.13
6	49	1.07	2.38
Sample mean			2.51
Standard deviation			0.82

4.3 minutes, to the model responses, Y_2. Formally, a statistical test of the null hypothesis

$$H_0: \quad E(Y_2) = 4.3 \text{ minutes}$$

versus (10.1)

$$H_1: \quad E(Y_2) \neq 4.3 \text{ minutes}$$

is conducted. If H_0 is not rejected, then on the basis of this test there is no reason to consider the model invalid. If H_0 is rejected, the current version of the model is rejected and the modeler is forced to seek ways to improve the model, as illustrated by Figure 10.3. As formulated here, the appropriate statistical test is the t test, which is conducted in the following manner:

Step 1. Choose a level of significance α and a sample size n. For the bank model, choose

$$\alpha = 0.05, \qquad n = 6$$

Step 2. Compute the sample mean $\bar{Y}_2$ and the sample standard deviation S over the n replications by equations (9.1) and (9.2):

$$\bar{Y}_2 = \frac{1}{n} \sum_{i=1}^{n} Y_{2i} = 2.51 \text{ minutes}$$

and

$$S = \left[\frac{\sum_{i=1}^{n} (Y_{2i} - \bar{Y}_2)^2}{n - 1} \right]^{1/2} = 0.82 \text{ minute}$$

where Y_{2i}, $i = 1, \ldots, 6$, are as shown in Table 10.2.

Step 3. Get the critical value of t from Table A.4. For a two-sided test such as that in equation (10.1), use $t_{\alpha/2,n-1}$; for a one-sided test, use $t_{\alpha,n-1}$ or $-t_{\alpha,n-1}$ as appropriate ($n - 1$ is the degrees of freedom). From Table A.4, $t_{0.025,5} = 2.571$ for a two-sided test.

Step 4. Compute the test statistic

$$t_0 = \frac{\bar{Y}_2 - \mu_0}{S/\sqrt{n}} \tag{10.2}$$

where μ_0 is the specified value in the null hypothesis, H_0. Here $\mu_0 = 4.3$ minutes, so that

$$t_0 = \frac{2.51 - 4.3}{0.82/\sqrt{6}} = -5.34$$

Step 5. For the two-sided test, if $|t_0| > t_{\alpha/2,n-1}$, reject H_0. Otherwise, do not reject H_0. [For the one-sided test with $H_1: E(Y_2) > \mu_0$, reject H_0 if $t > t_{\alpha,n-1}$; with $H_1: E(Y_2) < \mu_0$, reject H_0 if $t < -t_{\alpha,n-1}$.]

Since $|t| = 5.34 > t_{0.025,5} = 2.571$, reject H_0 and conclude that the model is inadequate in its prediction of average customer delay.

Recall that when testing hypotheses, rejection of the null hypothesis H_0 is a strong conclusion, because

$$P(H_0 \text{ rejected} \mid H_0 \text{ is true}) = \alpha \tag{10.3}$$

and the level of significance α is chosen small, say $\alpha = 0.05$, as was done here. Equation (10.3) says that the probability of making the error of rejecting H_0 when H_0 is in fact true is low ($\alpha = 0.05$); that is, the probability is small of declaring the model invalid when it is valid (with respect to the variable being tested).

The assumptions justifying a t test are that the observations (Y_{2i}) are normally and independently distributed. Are these assumptions met in the present case?

1. The ith observation Y_{2i} is the average delay of all drive-in customers who began service during the ith simulation run of 2 hours, and thus by a central limit theorem effect, it is reasonable to assume that each observation Y_{2i} is approximately normally distributed, provided that the number of customers it is based on is not too small.
2. The observations Y_{2i}, $i = 1, \ldots, 6$, are statistically independent by design, that is, by choice of the random number seeds independently for each replication, or by use of nonoverlapping streams.
3. The t statistic computed by equation (10.2) is a robust statistic; that is, it is approximately distributed as the t distribution with $n - 1$ degrees of freedom, even when $Y_{21}, Y_{22}, \ldots$ are not exactly normally distributed, and thus the critical values in Table A.4 can reliably be used.

Now that the model of the Fifth National Bank of Jaspar has been found lacking, what should the modeler do? Upon further investigation, the modeler realized that the model contained two unstated assumptions:

1. When a car arrived to find the window immediately available, the teller began service immediately.
2. There is no delay between one service ending and the next beginning, when a car is waiting.

Assumption 2 was found to be approximately correct because a service time was considered to begin when the teller actually began service but was not considered to have ended until the car had exited the drive-in window and the next car, if any, had begun service, or the teller saw that the line was empty. On the other hand, assumption 1 was found to be incorrect because the teller had other duties—mainly serving walk-in customers if no cars were present—and tellers always finished with a previous customer before beginning service on a car. It was found that walk-in customers were always present during rush hour; that the transactions were mostly commercial in nature, taking a considerably longer time than the time required to service drive-up customers; and that when an arriving car found no other cars at the window, it had to wait until the teller finished with the present walk-in customer. To correct this model inadequacy, the structure of the model was changed to include the additional demand on the teller's time and data were collected on service times of walk-in customers. Analysis of these data found that they were approximately exponentially distributed with a mean of 3 minutes.

The revised model was run yielding the results in Table 10.3. A test of the null hypothesis $H_0: E(Y_2) = 4.3$ minutes [as in equation (10.1)] was again conducted, according to steps 1 to 5 previously outlined.

Table 10.3. RESULTS OF SIX REPLICATIONS OF THE REVISED BANK MODEL

Replication	Y_4 *(Arrivals/Hour)*	Y_5 *(Minutes)*	Y_2 = *Average Delay (Minutes)*
1	51	1.07	5.37
2	40	1.11	1.98
3	45.5	1.06	5.29
4	50.5	1.09	3.82
5	53	1.08	6.74
6	49	1.08	5.49
Sample mean			4.78
Standard deviation			1.66

Step 1. Choose $\alpha = 0.05$ and $n = 6$ (sample size).

Step 2. Compute $\bar{Y}_2 = 4.78$ minutes, $S = 1.66$ minutes.

Step 3. From Table A.4, the critical value is $t_{0.25,5} = 2.571$.

Step 4. Compute the test statistic $t_0 = (\bar{Y}_2 - \mu_0)/(S/\sqrt{n}) = 0.710$.

Step 5. Since $|t| < t_{0.025,5} = 2.571$, do not reject H_0, and thus tentatively accept the model as valid.

Failure to reject H_0 must be considered as a weak conclusion unless the power of the test has been estimated and found to be high (close to 1). That is, it can only be concluded that the data at hand ($Y_{21}, \ldots, Y_{26}$) were not sufficient to reject the hypothesis $H_0: \mu_0 = 4.3$ minutes. In other words, this test detects no inconsistency between the sample data ($Y_{21}, \ldots, Y_{26}$) and the specified mean μ_0.

The power of a test is the probability of detecting a departure from $H_0: \mu = \mu_0$ when in fact such a departure exists. In the validation context, the power of the test is the probability of detecting an invalid model. The power may also be expressed as 1 minus the probability of a Type II, or β, error, where $\beta = P(\text{Type II error}) = P(\text{failing to reject } H_0 \mid H_1 \text{ is true})$ is the probability of accepting the model as valid when it is not valid.

To consider failure to reject H_0 as a strong conclusion, the modeler would want β to be small. Now, β depends on the sample size n and on the true difference between $E(Y_2)$ and $\mu_0 = 4.3$ minutes, that is, on

$$\delta = \frac{|E(Y_2) - \mu_0|}{\sigma}$$

where σ, the population standard deviation of an individual Y_{2i}, is estimated by S. Tables A.9 and A.10 are typical operating characteristic (OC) curves, which are graphs of the probability of a Type II error $\beta(\delta)$ versus δ for given sample size n. Table A.9 is for a two-sided t test while Table A.10 is for a one-sided t test. Suppose that the modeler would like to reject H_0 (model validity) with probability at least 0.90 if the true mean delay of the model, $E(Y_2)$, differed from the average delay in the system, $\mu_0 = 4.3$ minutes, by 1 minute. Then δ is estimated by

$$\hat{\delta} = \frac{|E(Y_2) - \mu_0|}{S} = \frac{1}{1.66} = 0.60$$

For the two-sided test with $\alpha = 0.05$, use of Table A.9 results in

$$\beta(\hat{\delta}) = \beta(0.6) = 0.75 \qquad \text{for } n = 6$$

To guarantee that $\beta(\hat{\delta}) \leq 0.10$, as was desired by the modeler, Table A.9 reveals that a sample size of approximately $n = 30$ independent replications would be required. That is, for a sample size $n = 6$ and assuming that the population standard deviation is 1.66, the probability of accepting H_0 (model validity), when in fact the model is invalid ($|E(Y_2) - \mu_0| = 1$ minute), is $\beta = 0.75$, which is quite high. If a 1-minute difference is critical, and if the modeler wants to control the risk of declaring the model valid when model predictions are as much as 1 minute off, a sample size of $n = 30$ replications is required to achieve a power of 0.9. If this sample size is too high, either a higher β risk (lower power), or a larger difference δ, must be considered.

In general, it is always best to control the Type II error, or β error, by specifying a critical difference δ and choosing a sample size by making use of an appropriate OC curve. (Computation of power and use of OC curves for a wide range of tests is discussed in Hines and Montgomery [1980].) In summary, in the context of model validation, the Type I error is the rejection of a valid model, and is easily controlled by specifying a small level of significance α (say $\alpha = 0.2$, 0.1, 0.05, or 0.01). The Type II error is the acceptance of a model as valid when it is invalid. For a fixed sample size n, increasing α will decrease

β, the probability of a Type II error. Once α is set, and the critical difference to be detected is selected, the only way to decrease β is to increase the sample size. A Type II error is the more serious of the two types of errors, and thus it is important to design the simulation experiments to control the risk of accepting an invalid model. The two types of error are summarized in Table 10.4, which compares statistical terminology to modeling terminology.

Table 10.4. TYPES OF ERROR IN MODEL VALIDATION

Statistical Terminology	*Modeling Terminology*	*Associated Risk*
Type I: rejecting H_0 when H_0 is true	Rejecting a valid model	α
Type II: failure to reject H_0 when H_1 is true	Failure to reject an invalid model	β

Note that validation is not to be viewed as an either/or proposition, but rather should be viewed in the context of calibrating a model, as conceptually exhibited in Figure 10.3. If the current version of the bank model produces estimates of average delay (Y_2) that are not close enough to real system behavior ($\mu_0 = 4.3$ minutes), the source of the discrepancy is sought, and the model is revised in light of this new knowledge. This iterative scheme is repeated until model accuracy is judged adequate.

10.3.4. INPUT–OUTPUT VALIDATION: USING HISTORICAL INPUT DATA

When using artificially generated data as input data, as was done to test the validity of the bank models in Section 10.3.3, the modeler expects the model to produce event patterns that are compatible with, but not identical to, the event patterns that occurred in the real system during the period of data collection. Thus, in the bank model, artificial input data $\{X_{1n}, X_{2n}, n = 1, 2, \ldots\}$ for interarrival and service times were generated and replicates of the output data Y_2 were compared to what was observed in the real system by means of the hypothesis test stated in equation (10.1). An alternative to generating input data is to use the actual historical record, $\{A_n, S_n, n = 1, 2, \ldots\}$, to drive the simulation model and then to compare model output to system data.

To implement this technique for the bank model, the data $A_1, A_2, \ldots$ and $S_1, S_2, \ldots$ would have to be entered into the model into arrays, or stored on a file to be read as the need arose. Just after customer n arrived at time $t_n = \sum_{i=1}^{n} A_i$, customer $n + 1$ would be scheduled on the future event list to arrive at future time $t_n + A_{n+1}$ (without any random numbers being generated). If customer n were to begin service at time t'_n, a service completion would be scheduled to occur at time $t'_n + S_n$. This event scheduling without random number generation could be implemented quite easily in FORTRAN, GASP,

SLAM, or SIMSCRIPT by using arrays to store the data, and in GPSS by using floating-point "savevalues" in conjunction with an incrementing counter. (Alternatively, GPSS and SIMSCRIPT have the capability to read from a computer file so-called "external events," and this capability could be implemented in FORTRAN quite easily.)

When using this technique, the modeler hopes that the simulation will duplicate as closely as possible the important events that occurred in the real system. In the model of the Fifth National Bank of Jaspar, the arrival times and service durations will exactly duplicate what happened in the real system on that Friday between 11:00 A.M. and 1:00 P.M. If the model is sufficiently accurate, then the delays of customers, lengths of lines, utilizations of servers, and departure times of customers predicted by the model will be close to what actually happened in the real system. It is, of course, the model builder's and model user's judgment that determines the level of accuracy required.

To conduct a validation test using historical input data, it is important that all the input data (A_n, S_n, . . .) and all the system response data, such as average delay (Z_2), be collected during the same time period. Otherwise, the comparison of model responses to system responses, such as the comparison of average delay in the model (Y_2) to that in the system (Z_2), could be misleading. The responses (Y_2 and Z_2) depend on the inputs (A_n and S_n) as well as on the structure of the system, or model. Implementation of this technique could be difficult for a large system because of the need for simultaneous data collection of all input variables and those response variables of primary interest. In some systems, electronic counters and devices have been used to ease the data collection task by automatically recording certain types of data. The following example was based on two simulation models reported in Carson et al. [1981a, b], in which simultaneous data collection and the subsequent validation were both completed successfully.

EXAMPLE 10.3 (THE CANDY FACTORY)

The production line at the Sweet Lil' Things Candy Factory in Decatur consists of three machines which make, package, and box their famous candy. One machine (the candy maker) makes and wraps individual pieces of candy and sends them by conveyor to the packer. The second machine (the packer) packs the individual pieces into a box. A third machine (the box maker) forms the boxes and supplies them by conveyor to the packer. The system is illustrated in Figure 10.6.

Each machine is subject to random breakdowns due to jams and other causes. These breakdowns cause the conveyor to begin to empty or fill. The conveyors between the two makers and the packer are used as a temporary storage buffer for in-process inventory. In addition to the randomly occurring breakdowns, if the candy conveyor empties, a packer runtime is interrupted and the packer remains idle until more candy is produced. If the box conveyor empties because of a long random breakdown of the box machine, an operator manually places racks of boxes onto the packing machine. If the conveyor fills, the corresponding maker becomes idle. The purpose of the model is to investigate the frequency of these operator interventions which require manual

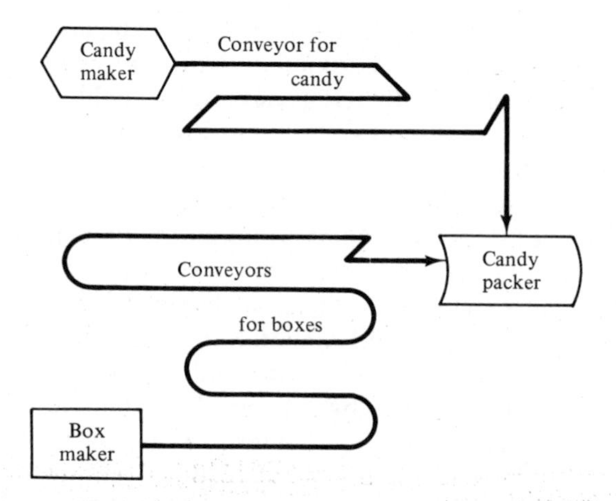

Figure 10.6. Production line at the candy factory.

loading of racks of boxes, as a function of various combinations of individual machines and lengths of conveyor. Different machines have different production speeds and breakdown characteristics, and longer conveyors can hold more in-process inventory. The goal is to hold operator interventions to an acceptable level while maximizing production. As machine stoppages (due to a full or empty conveyor) cause increased damage to the product, this is also a factor in production.

A simulation model of the Candy Factory was developed and a validation effort using historical inputs was conducted. Engineers in the Candy Factory set aside a 4-hour time slot from 7:00 A.M. to 11:00 A.M. to collect data on an existing production line. For each machine, say machine i, time to failure and random downtime data

$$T_{i1}, D_{i1}, T_{i2}, D_{i2}, \ldots$$

were collected. For machine i ($i = 1, 2, 3$), T_{ij} is the jth runtime (or time to failure), and D_{ij} is the successive random downtime. A runtime, T_{ij}, can be interrupted due to a full or empty conveyor (as appropriate) but resumes when conditions are right. Initial system conditions at 7:00 A.M. were recorded so that they could be duplicated in the model as initial conditions at time 0. Additionally, system responses of primary interest—the production level (Z_1) and the number (Z_2) and time of occurrence (Z_3) of operator interventions—were recorded for comparison with model predictions.

The system input data, T_{ij} and D_{ij}, were fed into the model and used as runtimes and random downtimes. The structure of the model determined the occurrence of shutdowns due to a full or empty conveyor, and the occurrence of operator interventions. Model response variables (Y_i, $i = 1, 2, 3$) were collected for comparison to the corresponding system response variables (Z_i, $i = 1, 2, 3$).

The closeness of model predictions to system performance aided the engineering staff considerably in convincing management of the validity of the model. These results are shown in Table 10.5. A simple display such as Table 10.5 can be quite effective in convincing skeptical engineers and managers of a model's validity—perhaps more effective than the most sophisticated statistical methods!

Table 10.5. VALIDATION OF THE CANDY FACTORY MODEL

Response, i	*System, Z_i*	*Model, Y_i*
1. Production level	897,208	883,150
2. Number of operator interventions	3	3
3. Time of occurrence	7: 22, 8: 41, 10: 10	7: 24, 8: 42, 10: 14

With only one set of historical input and output data, only one set of simulated output data can be obtained, and thus no simple statistical tests are possible based on summary measures. But if K historical input data sets are collected, and K observations $Z_{i1}, Z_{i2}, \ldots, Z_{iK}$ of some system response variable, Z_i, are collected, such that the output measure Z_{ij} corresponds to the jth input set, an objective statistical test becomes possible. For example, Z_{ij} could be the average delay of all customers who were served during the time the jth input data set was collected. With the K input data sets in hand, the modeler now runs the model K times, once for each input set, and observes the simulated results $W_{i1}, W_{i2}, \ldots, W_{iK}$ corresponding to Z_{ij}, $j = 1, \ldots, K$. Continuing the same example, W_{ij} would be the average delay predicted by the model when using the jth input set. The available data for comparison appears as in Table 10.6. If the K input data sets are fairly homogeneous, it is reasonable to assume that the K observed differences $d_j = Z_{ij} - W_{ij}$, $j = 1, \ldots, K$, are identically

Table 10.6. COMPARISON OF SYSTEM AND MODEL OUTPUT MEASURES WHEN USING IDENTICAL HISTORICAL INPUTS

Input Data Set	*System Output, Z_{ij}*	*Model Output, W_{ij}*	*Observed Difference, d_j*	*Squared Deviation from Mean, $(d_j - \bar{d})^2$*
1	Z_{i1}	W_{i1}	$d_1 = Z_{i1} - W_{i1}$	$(d_1 - \bar{d})^2$
2	Z_{i2}	W_{i2}	$d_2 = Z_{i2} - W_{i2}$	$(d_2 - \bar{d})^2$
3	Z_{i3}	W_{i3}	$d_3 = Z_{i3} - W_{i3}$	$(d_3 - \bar{d})^2$
.	.	.	.	.
.	.	.	.	.
.	.	.	.	.
K	Z_{iK}	W_{iK}	$d_K = Z_{iK} - W_{iK}$	$(d_K - \bar{d})^2$
			$\bar{d} = \frac{1}{K} \sum_{j=1}^{K} d_j$	$S_d^2 = \frac{1}{K-1} \sum_{j=1}^{K} (d_j - \bar{d})^2$

distributed. Furthermore, if the collection of the K sets of input data was separated in time, say on different days, it is reasonable to assume that the K differences $d_1, \ldots, d_K$ are statistically independent, and hence the differences $d_1, \ldots, d_K$ constitute a random sample. In many cases, each Z_i and W_i is a sample average over customers, so that (by the central limit theorem) the differences $d_j = Z_{ij} - W_{ij}$ are approximately normally distributed with some mean

μ_d and variance σ_d^2. The appropriate statistical test is then a t test of the null hypothesis of no mean difference:

$$H_0: \quad \mu_d = 0$$

versus the alternative of significant difference:

$$H_1: \quad \mu_d \neq 0$$

The proper test is a paired t test (Z_{i1} is paired with W_{i1} since each was produced by the first input data set, and so on). First, compute the sample mean difference $\bar{d}$, and the sample variance S_d^2, by the formulas given in Table 10.6. Then compute the t statistic by

$$t_0 = \frac{\bar{d} - \mu_d}{S_d/\sqrt{K}} \tag{10.4}$$

(with $\mu_d = 0$), and get the critical value $t_{\alpha/2, K-1}$ from Table A.4, where α is the prespecified significance level and $K - 1$ is the number of degrees of freedom. If $|t_0| > t_{\alpha/2, K-1}$, reject the hypothesis H_0 of no mean difference and conclude that the model is inadequate. If $|t_0| \leq t_{\alpha/2, K-1}$, do not reject H_0 and hence conclude that this test provides no evidence of model inadequacy.

EXAMPLE 10.4 (THE CANDY FACTORY, CONTINUED)

Engineers at the Sweet Lil' Things Candy Factory decided to expand the initial validation effort reported in Example 10.3. Electronic devices were installed which could automatically monitor one of the production lines, and the validation effort of Example 10.3 was repeated with $K = 5$ sets of input data. The system and the model were compared on the basis of production level. The results are shown in Table 10.7.

Table 10.7. VALIDATION OF THE CANDY FACTORY MODEL (Continued)

Input Data Set, j	*System Production,* Z_{1j}	*Model Production,* W_{1j}	*Observed Difference,* d_j	*Squared Deviation from Mean,* $(d_j - \bar{d})^2$
1	897,208	883,150	14,058	7.594×10^7
2	629,126	630,550	−1,424	4.580×10^7
3	735,229	741,420	−6,191	1.330×10^8
4	797,263	788,230	9,033	1.362×10^7
5	825,430	814,190	11,240	3.4772×10^7
			$\bar{d} = 5{,}343.2$	$S_d^2 = 7.580 \times 10^7$

A paired t test was conducted to test $H_0: \mu_d = 0$, or equivalently, $H_0: E(Z_1) = E(W_1)$, where Z_1 is system production level and W_1 is the production level predicted by the simulated model. Let the level of significance be $\alpha = 0.05$. Using the results in Table

10.7, the test statistic, as given by equation (10.4), is

$$t_0 = \frac{\bar{d}}{S_d/\sqrt{K}} = \frac{5343.2}{8705.85/\sqrt{5}} = 1.37$$

From Table A.4, the critical value is $t_{\alpha/2,K-1} = t_{0.025,4} = 2.78$. Since $|t_0| = 1.37 < t_{0.025,4} = 2.78$, the null hypothesis cannot be rejected on the basis of this test; that is, no inconsistency is detected between system response and model predictions in terms of mean production level. If H_0 had been rejected, the modeler would have searched for the cause of the discrepancy and revised the model, in the spirit of Figure 10.3.

10.3.5. Input–Output Validation: Using a Turing Test

In addition to statistical tests, or when no statistical test is readily applicable, persons knowledgeable about system behavior can be used to compare model output to system output. For example, suppose that five reports of system performance over five different days are prepared, and simulation output data are used to produce five "fake" reports. The 10 reports should all be in exactly the same format and should contain information of the type that managers and engineers have previously seen on the system. The 10 reports are randomly shuffled and given to the engineer, say, who is asked to decide which reports are fake and which are real. If the engineer identifies a substantial number of the fake reports, the model builder questions the engineer and uses the information gained to improve the model. If the engineer cannot distinguish between fake and real reports with any consistency, the modeler will conclude that this test provides no evidence of model inadequacy. For further discussion and an application to a real simulation, the reader is referred to Schruben [1980]. This type of validation test is commonly called a Turing test. Its use as model development proceeds can be a valuable tool in detecting model inadequacies, and eventually in increasing model credibility as the model is improved and refined.

10.4 Summary

Validation of simulation models is of great importance. Decisions are made on the basis of simulation results; thus, the accuracy of these results should be subject to question and investigation.

Quite often simulations appear realistic on the surface because simulation models, unlike analytic models, can incorporate any level of detail about the real system. To avoid being "fooled" by this apparent realism, it is best to compare system data to model data, and to make the comparison using a wide variety of techniques, including an objective statistical test, if at all possible.

As discussed by Van Horn [1969, 1971], some of the possible validation techniques, in order of increasing cost to value ratios, include:

1. Develop models with high face validity by consulting persons knowledgeable about system behavior on both model structure, model input, and model output. Use any existing knowledge in the form of previous research and studies, observation, and experience.
2. Conduct simple statistical tests of input data for homogeneity, randomness, and goodness of fit to assumed distributional forms.
3. Conduct a Turing test. Have knowledgeable people (engineers, managers) compare model output to system output and attempt to detect the difference.
4. Compare model output to system output by means of statistical tests.
5. After model development, collect new system data and repeat techniques 2 to 4.
6. Build the new system (or redesign the old one) based on simulation results, collect data on the new system, and use this data to validate the model. (Not recommended if this is the only technique used.)
7. Do little or no validation. Implement simulation results without validating. (Not recommended.)

It is usually too difficult, too expensive, or too time consuming to use all possible validation techniques for every model that is developed. It is an important part of the model builder's task to choose those validation techniques most appropriate, both to assure model accuracy and to assure model credibility.

REFERENCES

BALCI, O., AND R. G. SARGENT [1980], "Bibliography on Validation of Simulation Models," *Newsletter—TIMS College on Simulation and Gaming*, Vol. 4, No. 2, pp. 11–15.

BALCI, O., AND R. G. SARGENT [1981], "A Methodology for Cost–Risk Analysis in the Statistical Validation of Simulation Models," *Communications of the Association for Computing Machinery*, Vol. 24, 190–97.

CARSON, J. S., N. WILSON, D. CARROLL, AND C. H. WYSOWSKI [1981a], "A Discrete Simulation Model of a Cigarette Fabrication Process," *Proceedings of the Twelfth Modeling and Simulation Conference*, University of Pittsburgh.

CARSON, J. S., N. WILSON, D. CARROLL, AND C. H. WYSOWSKI [1981b], "Simulation of a Filter Rod Manufacturing Process," *Proceedings of the 1981 Winter Simulation Conference*, Atlanta, Ga.

GAFARIAN, A. V., AND J. E. WALSH [1970], "Methods for Statistical Validation of a Simulation Model for Freeway Traffic near an On-Ramp," *Transportation Research*, Vol. 4, pp. 379–84.

GASS, S. I. [1977], "Evaluation of Complex Models," *Computers and Operations Research*, Vol. 4, pp. 25–37.

HINES, W. W., AND D. C. MONTGOMERY [1980], *Probability and Statistics in Engineering and Management Science*, 2nd ed., Wiley, New York.

LAW, A. M. [1979a], "Validation of Simulation Models, I: An Overview and Survey of Real-World Practice," Technical Report 78–14, Department of Industrial Engineering, University of Wisconsin, Madison.

LAW, A. M. [1979b], "Validation of Simulation Models, II: Comparison of Real-World and Simulation Output Data," Technical Report 78–15, Department of Industrial Engineering, University of Wisconsin, Madison.

NAYLOR, T. H., AND J. M. FINGER [1967], "Verification of Computer Simulation Models," *Management Science*, Vol. 14, pp. 92–101.

ÖREN, T. I. [1981], "Concepts and Criteria to Assess Acceptability of Simulation Studies: A Frame of Reference," *Communications of the Association for Computing Machinery*, Vol. 24, pp. 180–89.

SARGENT, R. G. [1980], "Verification and Validation of Simulation Models," Working Paper 80–013, Department of Industrial Engineering and Operations Research, Syracuse University, Syracuse, N.Y.; also in *Progress in Modelling and Simulation*, ed. by F. E. Cellier, Academic Press, London (1981).

SCHRUBEN, L. W. [1980], "Establishing the Credibility of Simulations," *Simulation*, Vol. 34, pp. 101–5.

SHANNON, R. E. [1975], *Systems Simulation: The Art and Science*. Prentice-Hall, Englewood Cliffs, N.J.

SHECHTER, M., AND R. C. LUCAS [1980], "Validating a Large Scale Simulation Model of Wilderness Recreational Travel," *Interfaces*, Vol. 10, pp. 11–18.

VAN HORN, R. L. [1969], "Validation," in *The Design of Computer Simulation Experiments*, ed. by T. H. Naylor, Duke University Press, Durham, N.C.

VAN HORN, R. L. [1971], "Validation of Simulation Results," *Management Science*, Vol. 17, pp. 247–58.

EXERCISES

1. A simulation model of a job shop was developed to investigate different scheduling rules. To validate the model, the currently used scheduling rule was incorporated into the model and the resulting output compared to observed system behavior. By searching one year's worth of computerized records it was estimated that the average number of jobs in the shop was 22.3 on a given day. Seven independent replications of the model were run, each of 30 days duration, with the following results for average number of jobs in the shop:

18.6 19.9 21.7 20.2 22.1 20.8 22.1

(a) Develop and conduct a statistical test to determine if model output is consistent with system behavior. Use a level of significance of $\alpha = 0.05$.
(b) What is the power of this test if a difference of two jobs is viewed as critical?

What sample size is needed to guarantee a power of 0.8 or higher? (Use $\alpha = 0.05$.)

2. System data for the job shop of Exercise 1 revealed that the average time spent by a job in the shop was approximately 4 working days. The model made the following predictions on seven independent replications, for average time spent in the shop:

3.87 4.06 3.93 3.84 4.01 3.81 3.90

(a) Is model output consistent with system behavior? Conduct a statistical test using a level of significance $\alpha = 0.01$.

(b) If it is important to detect a difference of 0.5 day, what sample size is needed to have a power of 0.90? Interpret your results in terms of model validity or invalidity. (Use $\alpha = 0.01$.)

3. For the job shop of Exercise 1, four sets of input data were collected over four different 10-day periods, together with the average number of jobs in the shop (Z_i) for each period. The input data were used to drive the simulation model for four runs of 10 days each, and model predictions of average number of jobs in the shop (Y_i) were collected, with these results:

i	1	2	3	4
Z_i	16.1	19.2	27.6	21.3
Y_i	18.2	19.7	26.1	22.9

(a) Conduct a statistical test to check the consistency of system output and model output. Use a level of significance of $\alpha = 0.05$.

(b) If a difference of two jobs is viewed as important to detect, what sample size is required to guarantee a probability of at least 0.80 of detecting this difference if it indeed exists? (Use $\alpha = 0.05$.)

4. Obtain at least two of the papers or reports listed in the References, including the one by Van Horn, dealing with validation and verification. Write a short essay comparing and contrasting the various philosophies and approaches to validation.

5. Find several examples of actual simulations reported in the literature in which the authors discuss validation of their model. Is enough detail given to judge the adequacy of the validation effort? If so, compare the reported validation to the criteria set forth in this chapter. Did the authors use any validation technique not discussed in this chapter? (Several potential sources of articles on simulation applications include the two journals *Interfaces* and *Simulation*, and the following proceedings: Winter Simulation Conference Proceedings; Record of Proceedings of the Annual Simulation Symposium (Tampa); various proceedings published by the Institute of Industrial Engineers (IIE) for their fall and spring (annual) conferences; and the Proceedings of the Modeling and Simulation Conference (University of Pittsburgh).

6. Compare and contrast the various simulation languages in their capability to aid the modeler in the often arduous task of debugging and verification (articles discussing the nature of simulation languages may be found in the journals and proceedings listed in Exercise 5 above).

7. (a) Compare validation in simulation to the validation of theories in the physical sciences.

(b) Compare the issues involved and the techniques available for validation of models of physical systems versus models of social systems.

(c) Contrast the difficulties, and compare the techniques, in validating a model of a manually operated warehouse versus a model of an automated storage and retrieval system.

(d) Repeat (c) for a model of a production system involving considerable manual labor and human decision making, versus a model of the same production system after it has been automated.

OUTPUT ANALYSIS FOR A SINGLE MODEL

Output analysis refers to the analysis of data generated by a simulation. Its purpose is to predict the performance of a system, or to compare the performance of two or more alternative system designs. This chapter deals with the analysis of a single model, and Chapter 12 deals with the comparison of two or more models. The need for statistical output analysis is based on the observation that the output data from a simulation exhibits random variability when random number generators are used to produce the values of the input variables; that is, two different streams of random numbers will produce two sets of outputs which (probably) will differ. If the performance of the system is measured by a parameter θ, the result of a set of simulation experiments will be an estimate $\hat{\theta}$ of θ. The accuracy of the estimator $\hat{\theta}$ can be measured by the variance (or standard error) of $\hat{\theta}$. The purpose of the statistical analysis is to estimate this variance, or to determine the number of observations required to achieve a desired accuracy.

A typical output variable, say Y, the total cost of an inventory system per week, should be considered as a random variable with an unknown distribution, and the result of a single run with a 1-week run length provides just one sample observation from the population of all possible observations on Y. By increasing the run length, the sample size can be increased, say to n observations, $Y_1, \ldots, Y_n$, based on a run length of n weeks. However, these observations do not constitute a random sample, since they are not statistically independent. In this case, the ending inventory on hand at the end of a week is the beginning inventory

11

on hand for the next week, and thus the value of Y_i has some influence on the value of Y_{i+1}. The sequence of random variables $Y_1, Y_2, \ldots, Y_n$ is said to be autocorrelated (i.e., correlated with itself). This autocorrelation, which implies a lack of statistical independence, means that classical methods of statistics which assume independence are not directly applicable to the analysis of simulation output data. The methods must be properly modified and the simulation experiments properly designed for valid inferences to be made.

In addition to the autocorrelation present in most simulation output data, the specification of the initial conditions of the system at time 0 may pose a problem for the simulator, and indeed may influence the output data. For example, the inventory on hand and/or the number of backorders at time 0 would most likely influence the value of Y_1, the total cost for week 1. Because of the autocorrelation, these initial conditions would also influence the costs $(Y_2, \ldots, Y_n)$ for subsequent weeks. The specified initial conditions, if not chosen well, may have an especially deleterious effect when attempting to estimate the steady-state performance of a simulation model. For purposes of statistical analysis, the effect of initial conditions is that the output observations may not be identically distributed, and the initial observations may not be representative of steady-state behavior of the system.

Section 11.1 illustrates by example the inherent variability in a stochastic (i.e., probabilistic) discrete-event simulation, and thus demonstrates the need for a statistical analysis of the output. It also discusses the effect of autocorrelation on variance estimation. Section 11.2 distinguishes between two types of simulation—transient versus steady state—and defines commonly used measures of system performance for each type of simulation. Section 11.3 discusses the statistical estimation of performance measures. Section 11.4 discusses the analysis of transient simulations, and Section 11.5 the analysis of steady-state simulations. The main ideas of the chapter are summarized in Section 11.6.

11.1. Stochastic Nature of Output Data

Consider one run of a simulation model over a period of time $(0, T)$. Since the model is an input–output transformation, as illustrated by Figure 10.5, and since some of the model input variables are random variables, it follows that in general the model output variables are random variables. This *stochastic* (or probabilistic) nature of output variables was observed as early as Chapter 2, Exercise 2.8, where the reader was asked to consider the effect of using a different random number stream on the output of an inventory (newspaper) problem. Again in Tables 10.2 and 10.3, the effect of six different random number streams for the six independent replications of the bank model of Example 10.2 was to produce six independent estimates, Y_{2i}, of average delay, which were combined into one sample mean estimator, $\bar{Y}_2$. Table 10.2 clearly exhibits the stochastic nature of the summary measure Y_2.

Two examples are now given to illustrate the nature of the output data from stochastic simulations, and to give a preliminary discussion of several important properties of these data.

EXAMPLE 11.1 (ABLE AND BAKER, REVISITED)

Consider the Able–Baker carhop problem (Example 2.2) which involved customers arriving according to the distribution of Table 2.11 and being served either by Able, whose service-time distribution is given in Table 2.12, or by Baker, whose service-time distribution is given in Table 2.13. The purpose of the simulation is to estimate Able's utilization, ρ, and the mean time spent in the system per customer, w, over the first 2 hours of the workday. Therefore, each run of the model is for a 2-hour period, with the system being empty and idle at time 0. Four statistically independent runs were made, by using four distinct streams of random numbers to generate the interarrival and service times. Table 11.1 presents the results. The estimated utilization for run r is given by $\hat{\rho}_r$, and the estimated average system time by $\hat{w}_r$ (i.e., $\hat{w}_r$ is the sample average time in system for all customers served during run r). Note that, in this sample, the observed utilization ranges from 0.708 to 0.875, and the observed average system time ranges from 3.74 minutes to 4.53 minutes. The stochastic nature of the output data $\{\hat{\rho}_1, \hat{\rho}_2, \hat{\rho}_3, \hat{\rho}_4\}$ and $\{\hat{w}_1, \hat{w}_2, \hat{w}_3, \hat{w}_4\}$ is demonstrated by the results in Table 11.1.

Table 11.1. RESULTS OF FOUR INDEPENDENT RUNS OF 2-HOUR DURATION OF THE ABLE–BAKER QUEUEING PROBLEM

Run, r	*Able's Utilization*, $\hat{\rho}_r$	*Average System Time*, $\hat{w}_r$ (*Minutes*)
1	0.808	3.74
2	0.875	4.53
3	0.708	3.84
4	0.842	3.98

There are two general questions that may be addressed by a statistical analysis, say of the observed utilizations $\hat{\rho}_r$, $r = 1, \ldots, 4$:

1. Estimation of true utilization $\rho = E(\hat{\rho}_r)$ by a confidence interval
2. Testing whether utilization ρ meets some standard, for example, testing whether $\rho \geq 0.85$

These questions are addressed in Section 11.4 for terminating simulations, such as Example 11.1. Classical methods of statistics may be used because $\hat{\rho}_1, \hat{\rho}_2, \hat{\rho}_3$, and $\hat{\rho}_4$ constitute a random sample; that is, they are independent and identically distributed. In addition, since $\rho = E(\hat{\rho}_r)$ is the parameter being estimated, it follows that each $\hat{\rho}_r$ is an unbiased estimate of true mean utilization ρ. The analysis of Example 11.1 is considered in Example 11.10, Section 11.4. A survey of statistical methods applicable to terminating simulations is given by Law [1980]. Additional material may be found in Fishman [1978], Law [1977], and Law and Kelton [1982b].

The next example illustrates the effects of correlation and initial conditions on the estimation of long-run mean measures of performance of a system.

EXAMPLE 11.2

Consider a single-server queue with Poisson arrivals at an average rate of one every 10 minutes ($\lambda = 0.1$ per minute), and service times which are normally distributed with a mean of 9.5 minutes and a standard deviation of 1.75 minutes. (The range of a service time is restricted to ± 5 standard deviations, which excludes the possibility of a negative service time, but which covers well over 99.999% of the normal distribution.) This is an $M/G/1$ queue, which was described and analyzed in Section 5.5.1. By equation (5.13), the long-run server utilization is $\rho = \lambda E(S) = (0.1)(9.5) = 0.95$. The system is simulated for a total of 5000 minutes for the purpose of estimating long-run mean queue length, L_Q, defined by equation (5.6b).

The time interval [0, 5000) is divided into five equal subintervals and an estimator of L_Q is computed over each interval by

$$Y_j = \frac{1}{1000} \int_{(j-1)1000}^{j(1000)} L_Q(t)\, dt, \qquad j = 1, \ldots, 5 \tag{11.1}$$

where $L_Q(t)$ is the (simulated) number of customers in the waiting line at time t; thus, Y_1 is the time-weighted-average number of customers in the queue from time 0 to time 1000, Y_2 is the same average over [1000, 2000), and so on. Equation (11.1) is a special case of equation (5.6b). The observations $\{Y_1, Y_2, Y_3, Y_4, Y_5\}$ provide an example of "batching" raw simulation data, in this case $\{L_Q(t), 0 \le t \le 5000\}$, and the Y_j are called *batch means*. The use of batch means in analyzing output data is discussed in Section 11.5.4.

The simulation results of three statistically independent replications are shown in Table 11.2. (Each replication, or run, uses a distinct stream of random numbers.) For replication 1, Y_{1j} is the batch mean for batch j (the jth interval), as defined by equation (11.1); similarly, Y_{2j} and Y_{3j} are defined for batch j for replications 2 and 3, respectively. Table 11.2 also gives the sample mean over each replication, $\bar{Y}_{r\cdot}$, for replications $r = 1, 2, 3$. That is,

$$\bar{Y}_{r\cdot} = \frac{1}{5} \sum_{j=1}^{5} Y_{rj}, \qquad r = 1, 2, 3 \tag{11.2}$$

It can be shown that each $\bar{Y}_{r\cdot}$ is a time average over [0, 5000) for replication r, as given by equation (5.6b). (*Notation:* The dot, as in the subscript $r\cdot$, indicates summation over the second subscript, and the bar, as in $\bar{Y}_{r\cdot}$, indicates an average.)

Table 11.2 exhibits again the inherent variability in stochastic simulations. Consider the variability within a replication, say within replication 3, in which the average

Table 11.2. BATCHED AVERAGE QUEUE LENGTH FOR THREE INDEPENDENT REPLICATIONS

Batching Interval (Minutes)	*Batch, j*	*Replication* 1, Y_{1j}	2, Y_{2j}	3, Y_{3j}
[0, 1000)	1	3.61	2.91	7.67
[1000, 2000)	2	3.21	9.00	19.53
[2000, 3000)	3	2.18	16.15	20.36
[3000, 4000)	4	6.92	24.53	8.11
[4000, 5000)	5	2.82	25.19	12.62
[0, 5000)		$\bar{Y}_{1\cdot} = 3.75$	$\bar{Y}_{2\cdot} = 15.56$	$\bar{Y}_{3\cdot} = 13.66$

queue length over the batching intervals varies from a low of $Y_{31} = 7.67$ customers during the first 1000 minutes, to a high of $Y_{33} = 20.36$ customers during the third subinterval of 1000 minutes. Table 11.2 also shows the variability between replications. (Compare Y_{15} to Y_{25} to Y_{35}, the average queue lengths over the intervals 4000 to 5000 minutes.)

Suppose for the moment that a simulator makes only one replication of this model and gets the result $\bar{Y}_{1\cdot} = 3.75$ customers as an estimate of mean queue length, L_Q. How accurate is the estimate? This question is usually answered by attempting to estimate the standard error of the point estimate, $\bar{Y}_{1\cdot}$. The simulator may think that $Y_{11}, Y_{12}, \ldots, Y_{15}$ could be regarded as a random sample; however, the sequence is not independent, and in fact it is autocorrelated. If $Y_{11}, \ldots, Y_{15}$ were mistakenly assumed to be independent observations, and their positive autocorrelation ignored, the usual classical methods of statistics might severely underestimate the standard error of $\bar{Y}_{1\cdot}$, possibly resulting in the simulator thinking that a high degree of accuracy had been achieved. On the other hand, $\bar{Y}_{1\cdot}$, $\bar{Y}_{2\cdot}$, and $\bar{Y}_{3\cdot}$ can be regarded as independent observations.

Intuitively, Y_{11} and Y_{12} are correlated because in replication 1 the queue length at the end of the time interval [0, 1000) is the queue length at the beginning of the interval [1000, 2000). Similarly, for any two adjacent batches within a given replication, it can be seen that they are correlated. If the system is congested at the end of one interval, it will be congested for awhile at the beginning of the next time interval. Similarly, periods of low congestion tend to follow each other. Within a replication, say for $Y_{r1}, Y_{r2}, Y_{r3}, \ldots$, high values of a batch mean tend to be followed by high values, and low values by low. This tendency of adjacency of like values is known as positive autocorrelation. The effect of ignoring autocorrelation when it is present is discussed in more detail in Section 11.3.

Now suppose that the purpose of the $M/G/1$ queueing simulation of Example 11.2 is to estimate "steady-state" mean queue length, that is, mean queue length under "typical operating conditions." However, each of the three replications was begun in the empty and idle state. The empty and idle initial state means that within a given replication there will be a higher than "typical" probability of the system being uncongested for times close to 0. The practical effect is that an estimator of L_Q, say $\bar{Y}_{r\cdot}$ for replication r, will be biased low [i.e., $E(\bar{Y}_{r\cdot}) < L_Q$]. The extent of the bias decreases as the run length increases, but for short run length simulations with atypical initial conditions, this initialization bias can lead to misleading results. The problem of initialization bias is discussed further in Section 11.5.1.

Example 11.3 (Fifth National Bank of Jaspar, Revisited)

Reconsider the first run (replication number 1 in Table 10.2) of the bank model. The output variable $Y_{21} = 2.79$ minutes is the average delay of all customers served during the course of the simulation. Thus, Y_{21} is a sample mean of customer delays, say $D_1, D_2, \ldots, D_m$; that is,

$$Y_{21} = \frac{1}{m} \sum_{i=1}^{m} D_i$$

where D_i is the actual delay of customer i, and m is the number of customers served during the 2 hours of the simulation. Chapter 10 did not work directly with the sequence $\{D_i, i = 1, 2, \ldots\}$, but rather with the summary sequence $Y_{21}, Y_{22}, \ldots, Y_{26}$ of average delay on each replication. The reason for avoiding direct statistical analysis of the "raw" output $\{D_i, i = 1, 2, \ldots\}$ is that this sequence is, in general, a nonstationary autocorrelated stochastic process. In contrast, the sequence $Y_{21}, Y_{22}, \ldots$ consists of independent and identically distributed random variables, and thus classical methods of statistical analysis can be applied for estimating standard errors, constructing confidence intervals, and conducting hypothesis tests.

To say that $\{D_i, i = 1, 2, \ldots\}$ is *nonstationary* means that $D_1, D_2, \ldots$ are not identically distributed, and in particular, that $E(D_i) \neq E(D_{i+1})$. This nonstationarity may be caused by the nature of the simulation, or more likely, by arbitrary initialization conditions. For example, if at time 0 (11: 00 A.M.) in the bank model, the teller is idle, the first few customers will have a higher than usual tendency of having small delays. In addition, if customer i has a long delay (D_i is large), the system will probably remain congested until customer $(i + 1)$ arrives, and thus the delay of customer $(i + 1)$, namely D_{i+1}, will also be large. Similarly, if D_i is small, D_{i+1} will tend to be relatively small. As in Example 11.2, this tendency of like values to be followed by like values causes the sequence $D_1, D_2, \ldots$ to be positively autocorrelated. The practical effect of the autocorrelation and of arbitrary initial conditions is similar to the effects in Example 11.2.

11.2. Types of Simulations with Respect to Output Analysis

When analyzing simulation output data, a distinction is made between terminating or transient simulations and steady-state simulations. A *terminating* simulation is one that runs for some duration of time T_E, where E is a specified event (or set of events) which stops the simulation. Such a simulated system "opens" at time 0 under well specified *initial conditions* and "closes" at the *stopping time* T_E. The Able–Baker carhop problem (Example 11.1) is a terminating simulation with $T_E = 2$ hours and initial conditions the empty and idle state.

EXAMPLE 11.4

The Shady Grove Bank opens at 8: 30 A.M. (time 0) with no customers present and 8 of the 11 tellers working (initial conditions) and closes at 4: 30 P.M. (time $T_E =$ 480 minutes). Here, the event E is merely the fact that the bank has been open for 480 minutes. The simulator is interested in modeling the interaction between customers and tellers over the entire day, including the effect of startup, and of closing down at the end of the day.

EXAMPLE 11.5

Consider the Shady Grove Bank of Example 11.4 during the period from 11: 30 A.M. (time 0) to 1: 30 P.M., when it is especially busy. Simulation run length is $T_E =$ 120 minutes. The initial conditions at time 0 (11: 30 A.M.) could be specified in essentially two ways: (1) the real system could be observed at 11: 30 on a number of different days and a distribution of number of customers in system (at 11: 30 A.M.) could be estimated; then these data could be used to load the simulation model with customers at time 0; or (2) the model could be simulated from 8: 30 A.M. to 11: 30 A.M. without collecting output statistics, and the ending conditions at 11: 30 A.M. used as initial conditions for the 11: 30 A.M. to 1: 30 P.M. simulation.

EXAMPLE 11.6

A communications system consists of several components plus several backup components. It is represented schematically in Figure 11.1. Consider the system over a period of time, T_E, until the system fails. The stopping event E is defined by $E =$

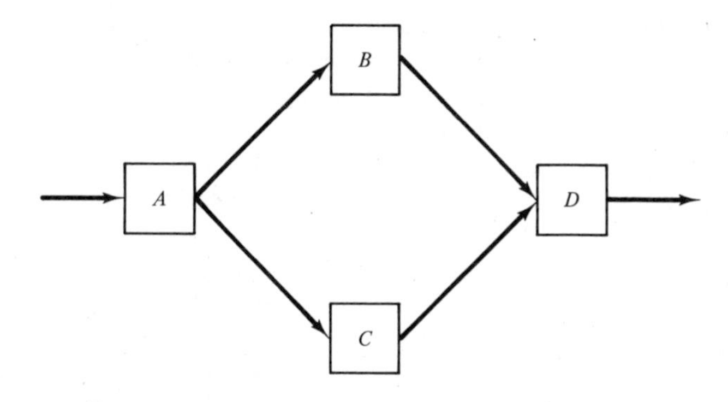

Figure 11.1. Example of a communications system.

{A fails, or D fails, or (B and C both fail)}. Initial conditions are that all components are new at time 0.

Note that in the bank model of Example 11.4, the stopping time $T_E = 480$ minutes is known, but in Example 11.6, the stopping time T_E is generally unpredictable in advance, and, in fact, T_E is probably the response variable of interest, as it represents the total time until the system breaks down. One goal of the simulation might be to estimate $E(T_E)$, the mean time to system failure.

EXAMPLE 11.7

A widget manufacturing process runs continuously from Monday mornings until Saturday mornings. The first shift of each workweek is used to load inventory buffers and chemical tanks with the components and catalysts needed to make the final product (28 varieties of widgets). These components and catalysts are made continually throughout the week, except for the last shift Friday night, which is used for cleanup and maintenance. Thus, most inventory buffers are near empty at the end of the week. During the first shift on Monday, a buffer stock is built up to cover the eventuality of breakdown in some part of the process. It is desired to simulate this system during the first shift (time 0 to time $T_E = 8$ hours) to study various scheduling policies for loading inventory buffers. This is an example of a *transient* (or nonstationary) simulation, as the variables of interest are the in-process inventory levels, which are increasing from zero or near zero (at time 0) to full or near full (at time 8 hours).

When simulating a terminating system, the initial conditions of the system at time 0 must be specified, and the stopping time T_E, or alternatively, the stopping event E, must be well defined. Although it is certainly true that the Shady Grove Bank in Example 11.4 will open again the next day, the simulator has chosen to consider it a terminating system because the object of interest is one day's operation including startup and close down. On the other hand, if the simulator were interested in some other aspect of the bank's operations, such as the flow of money, the system might be considered as a nonterminating one. Similar comments apply to the communications system of Example 11.6. If the failed component were replaced and the system continued to operate, and, most important, if the simulator were interested in studying its long-run behavior, it might be considered as a nonterminating system. In Example 11.6, however, interest is in its short-run behavior, from time 0 until the first system failure at time T_E. Therefore, whether a simulation is considered to be terminating or not depends on both the objectives of the simulation study and the nature of the system.

A *nonterminating system* is a system that runs continuously, or at least over a very long period of time. Examples include assembly lines which shut down infrequently, continuous production systems of many different types, telephone systems and other communications systems, hospital emergency rooms, police dispatching and patrolling operations, fire departments, and continuously operating computer systems.

A simulation of a nonterminating system, however, starts at simulation time 0 under initial conditions defined by the analyst and runs for some analyst specified period of time T_E. Significant problems arise concerning the specification of these initial and stopping conditions, problems that have not yet been completely solved by simulation researchers. Usually, the analyst wants to study steady-state, or long-run, properties of the system, that is, properties which are not influenced by the initial conditions of the model at time 0. A *steady-state* simulation is a simulation whose objective is to study long-run, or steady-state, behavior of a nonterminating system.

EXAMPLE 11.8

Consider the widget manufacturing process beginning with the second shift when the complete production process is under way. It is desired to estimate long-run production levels and production efficiencies. For the relatively long period of 13 shifts, this may be considered as a steady-state simulation. To obtain sufficiently accurate estimates of production efficiency and other response variables, the analyst could decide to simulate for any length of time, T_E (even longer than 13 shifts). That is, T_E is not determined by the nature of the problem (as it was in terminating simulations); rather, it is set by the analyst as one parameter in the design of the simulation experiment.

EXAMPLE 11.9

HAL Inc., a large computer service bureau, has many customers worldwide. Thus, its large computer system with many mainframes and peripherals runs continuously 24 hours per day. Due to an increasing work load, HAL is considering additional CPUs, disk drives, and tape drives in various configurations. HAL systems staff develops a simulation model of the existing system with the current work load, and then explores several possibilities for expanding capacity. HAL is interested in steady-state throughput and utilization of each machine. This is an example of a steady-state simulation. The stopping time, T_E, is not determined by the nature of the problem, but rather by the simulator, either arbitrarily, or with a certain statistical accuracy in mind.

11.3. Measures of Performance and Their Estimation

Consider the estimation of a parameter, θ, (or ϕ) of a simulated system. It is desired to have a point estimate of θ (or ϕ) and an interval estimate of θ (or ϕ). The length of the interval estimate is a measure of the accuracy of the point estimate. The simulation output data is of the form $\{Y_1, Y_2, Y_3, \ldots, Y_n\}$ for estimating θ, or $\{Y(t), 0 \leq t \leq T\}$ for estimating ϕ. For example, Y_i might be the delay of customer i, or the total demand in week i; $Y(t)$ may be the queue length at time t, or the number of backlogged orders at time t. The parameter θ is an ordinary mean; ϕ will be referred to as a time-weighted mean.

11.3.1. Point Estimation

The point estimator of θ based on the data $\{Y_1, \ldots, Y_n\}$ is defined by

$$\hat{\theta} = \frac{1}{n} \sum_{i=1}^{n} Y_i \tag{11.3}$$

where $\hat{\theta}$ is a sample mean based on a sample of size n. The point estimator $\hat{\theta}$ is said to be unbiased for θ if its expected value is θ, that is, if

$$E(\hat{\theta}) = \theta \tag{11.4a}$$

In general,

$$E(\hat{\theta}) = \theta + b \tag{11.4b}$$

where $b = E(\hat{\theta}) - \theta$ is called the bias in the point estimator $\hat{\theta}$. It is desirable to have estimators that are unbiased ($b = 0$), or if this is not possible, as small a bias b as possible relative to the magnitude of θ. Examples of estimators of the form of equation (11.3) include $\hat{w}$ and $\hat{w}_Q$ of equations (5.7) and (5.9), in which case Y_i is the time spent in the (sub)system by customer i.

The point estimator of ϕ based on the data $\{Y(t), 0 \leq t \leq T\}$, where T is the simulation run length, is defined by

$$\hat{\phi} = \frac{1}{T} \int_0^T Y(t)\, dt \tag{11.5}$$

and is called a time average of $Y(t)$ over $[0, T]$. In general,

$$E(\hat{\phi}) = \phi + b \tag{11.6}$$

If $b = 0$, then $\hat{\phi}$ is said to be unbiased for ϕ. Examples of time averages include $\hat{L}$ and $\hat{L}_Q$ of equations (5.6a) and (5.6b), and Y_j of equation (11.1).

Generally, θ and ϕ are regarded as mean measures of performance of the system being simulated. Other measures usually can be put into this common framework. For example, consider estimation of the proportion of days on which sales are lost due to an out-of-stock situation. In the simulation, let

$$Y_i = \begin{cases} 1, & \text{if out of stock on day } i \\ 0, & \text{otherwise} \end{cases}$$

With n equal to the total number of days, $\hat{\theta}$ defined by equation (11.3) is a point estimator of θ, the proportion of out-of-stock days. For a second example, consider estimation of the proportion of time queue length is greater than k_0 customers (k_0 is a specified number). If $L_Q(t)$ represents simulated queue length at time t, then in the simulation define

$$Y(t) = \begin{cases} 1, & \text{if } L_Q(t) > k_0 \\ 0, & \text{otherwise} \end{cases}$$

Then $\hat{\phi}$ defined by equation (11.5) is a point estimator of ϕ, the proportion of time that the queue length is greater than k_0 customers. Thus, it can be seen that estimation of proportions is a special case of the estimation of means.

11.3.2. Interval Estimation

Valid interval estimation requires a method of estimating the variance of the point estimator, $\hat{\theta}$ (or $\hat{\phi}$), in a relatively unbiased fashion. Let $\sigma^2(\hat{\theta}) = \text{var}(\hat{\theta})$ represent the true variance of a point estimator $\hat{\theta}$, and let $\hat{\sigma}^2(\hat{\theta})$ represent an estimator of $\sigma^2(\hat{\theta})$ based on the data $\{Y_1, \ldots, Y_n\}$. Suppose that

$$E[\hat{\sigma}^2(\hat{\theta})] = B\sigma^2(\hat{\theta}) \tag{11.7}$$

where B is called the bias in the variance estimator. It is desirable to have

$$B = 1$$

in which case $\hat{\sigma}^2(\hat{\theta})$ is said to be an unbiased estimator of the variance, $\sigma^2(\hat{\theta})$, of the point estimator. If $\hat{\sigma}^2(\hat{\theta})$ is approximately unbiased ($B \approx 1.0$), then under fairly general conditions, the statistic

$$t = \frac{\hat{\theta} - \theta}{\hat{\sigma}(\hat{\theta})}$$

is approximately t distributed with some number, say f, of degrees of freedom, which is denoted by $t \sim t_f$. Therefore, an approximate $100(1 - \alpha)\%$ confidence interval for θ is given by

$$\hat{\theta} - t_{\alpha/2,f}\hat{\sigma}(\hat{\theta}) \leq \theta \leq \hat{\theta} + t_{\alpha/2,f}\hat{\sigma}(\hat{\theta}) \tag{11.8}$$

where $t_{\alpha,f}$ is the $100(1-\alpha)$ percentage point of a t distribution with f degrees of freedom; that is, $t_{\alpha,f}$ is defined by $P(t \geq t_{\alpha,f}) = \alpha$. A confidence interval of the form of inequality (11.8) will be approximately correct provided that the point estimator is an average of the form of equation (11.3) or (11.5), and is relatively unbiased ($b \approx 0$), and the variance estimator $\hat{\sigma}^2(\hat{\theta})$ is approximately unbiased for $\sigma^2(\hat{\theta})$ (i.e., $B \approx 1.0$).

One of the major problems in simulation output analysis is obtaining approximately unbiased estimates of $\sigma^2(\hat{\theta})$, the variance of the point estimator. There are basically two cases:

Case 1. If $\{Y_1, Y_2, \ldots, Y_n\}$ are statistically independent observations, classical statistical methods can be applied. This case arises when Y_i is an output measure from replication i, and all replications are made independent by use of nonoverlapping random number streams and independently chosen initial conditions. The Able–Baker carhop simulation of Example 11.1 provides one illustration of case 1. In this case, compute $\hat{\theta}$ by equation (11.3) and then compute the sample variance, S^2, by

$$S^2 = \sum_{i=1}^{n} \frac{(Y_i - \hat{\theta})^2}{n-1} \tag{11.9}$$

When the Y_i are independent and identically distributed, the sample variance, S^2, is an unbiased estimator of the population variance $\sigma^2 = \text{var}(Y_i)$ (constant for all $i = 1, \ldots, n$). Since the variance of $\hat{\theta}$ is given by

$$\sigma^2(\hat{\theta}) = \frac{\sigma^2}{n} \tag{11.10}$$

an unbiased estimator of $\sigma^2(\hat{\theta})$, with $f = n - 1$ degrees of freedom, is provided by

$$\hat{\sigma}^2(\hat{\theta}) = \frac{S^2}{n} \tag{11.11}$$

Thus, the confidence interval given by inequality (11.8) is approximately valid provided the point estimate $\hat{\theta}$ is unbiased. The quantity $\hat{\sigma}(\hat{\theta}) = S/\sqrt{n}$, sometimes denoted by s.e.$(\hat{\theta})$, is called the standard error of the point estimator $\hat{\theta}$; the standard error is a measure of the accuracy of a point estimator.

Case 2. If $\{Y_1, \ldots, Y_n\}$ are not statistically independent, S^2/n given by equation (11.11), is a biased estimator of the true variance, $\sigma^2(\hat{\theta})$, of the point estimator. This is almost always the case when $\{Y_1, \ldots, Y_n\}$ is a sequence of output observations from within a single replication. In this situation, $Y_1, Y_2, \ldots$ is an autocorrelated sequence, sometimes called a time series. Example 11.2 (the $M/G/1$ queue) provides an illustration of case 2.

To quantify the effect of autocorrelation on variance estimation, suppose that the time series $Y_1, Y_2, \ldots$ is covariance stationary with mean θ. Intuitively, stationarity is a statistical property which implies that Y_{i+k} is probabilistically related to Y_{i+1} in the same manner as Y_k is related to Y_1. If the model being simulated is in "steady state," the output series would probably be stationary (at least approximately). On the other hand, if a transient situation (such as the production line startup in Example 11.7) were being simulated, the output $Y_1, Y_2, \ldots$ would be nonstationary. Any data that have an upward or downward trend are nonstationary.

For a covariance stationary time series, $Y_1, Y_2, \ldots,$ define the lag k covariance by

$$\gamma_k = \text{cov}\,(Y_1, Y_{1+k}) = \text{cov}\,(Y_i, Y_{i+k}) \tag{11.12a}$$

which, by definition of covariance stationarity, is independent of i. For $k = 0$, γ_k becomes the population variance; that is,

$$\gamma_0 = \text{cov}\,(Y_i, Y_i) = \text{var}\,(Y_i) \tag{11.12b}$$

The lag k autocorrelation is defined by

$$\rho_k = \frac{\gamma_k}{\gamma_0} \tag{11.13}$$

It gives the correlation between any two observations at lag k (i.e., k observations apart). It can be shown that

$$-1 \leq \rho_k \leq 1, \qquad k = 1, 2, \ldots$$

When $\rho_k > 0$ for all k (or most k), the series is said to be positively autocorrelated. The output data of most queueing simulations are positively autocorrelated, as are the successive customer delays $D_1, D_2, \ldots$ in the bank model discussed in Example 11.3. On the other hand, if $\rho_k < 0$, the series $Y_1, Y_2, \ldots$ will exhibit the characteristics of negative autocorrelation. The output of certain inventory simulations sometimes exhibits negative autocorrelation. Two hypothetical stationary series are shown in Figure 11.2, which is a plot of Y_i versus i. Figure 11.2(c) shows an example of a nonstationary time series with an upward trend.

It can be shown for a covariance stationary time series Y_i with sample mean $\hat{\theta}$ as defined by equation (11.3) that the variance of $\hat{\theta}$ is given by

$$\sigma^2(\hat{\theta}) = \text{var}\,(\hat{\theta}) = \text{var}\left(\frac{\sum_1^n Y_i}{n}\right) = \frac{\gamma_0}{n}\left[1 + 2\sum_{k=1}^{n}\left(1 - \frac{k}{n}\right)\rho_k\right] \tag{11.14}$$

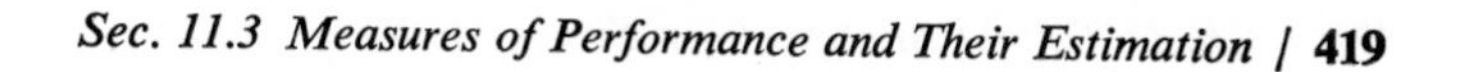

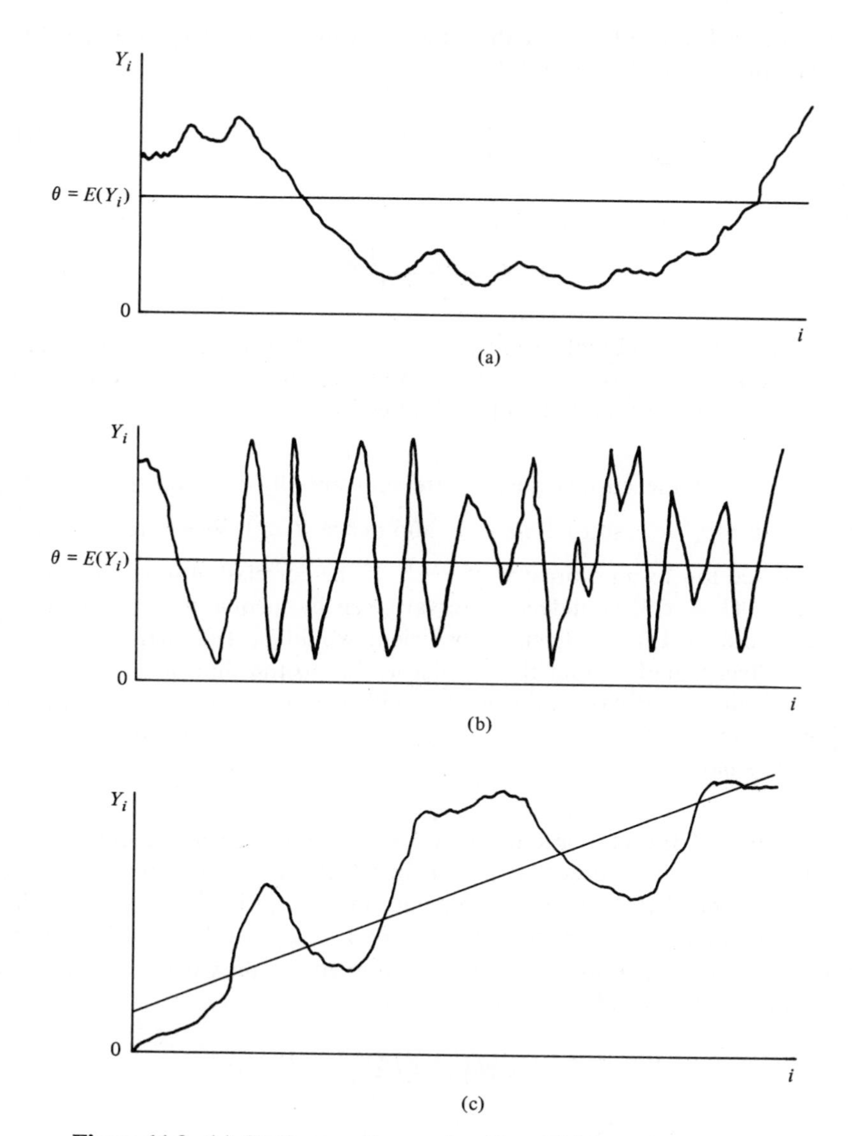

Figure 11.2. (a) Stationary time series Y_i exhibiting positive autocorrelation; (b) stationary time series Y_i exhibiting negative autocorrelation; (c) nonstationary time series with an upward trend.

where ρ_k is the lag k autocorrelation given by equation (11.13), and γ_0 is the population variance defined in equation (11.12b). If the Y_i were independent observations, then $\rho_k = 0$ for $k = 1, 2, 3, \ldots$ and equation (11.14) reduces to the familiar equation (11.10). (Note that $\gamma_0 = \sigma^2$.) Using equation (11.14), it can be shown [Law, 1977] that the expected value of the variance estimator, S^2/n, of equation (11.11) is given by

$$E\left(\frac{S^2}{n}\right) = B\sigma^2(\hat{\theta}) \tag{11.15}$$

where

$$B = \frac{n/c - 1}{n - 1} \tag{11.16}$$

and c is the quantity in brackets in equation (11.14). The effect of the correlation on the estimator S^2/n is evident by an examination of equations (11.14) and (11.16). There are essentially two possibilities:

Case 2a. If the correlations ρ_k are substantially positive, then $c = 1 + 2\sum_{k=1}^{n}(1 - k/n)\rho_k > 1$, so that $n/c < n$, and hence $B < 1$. When this is the case, S^2/n is biased low as an estimator of $\sigma^2(\hat{\theta})$. If the correlation were ignored, the nominal $100(1 - \alpha)\%$ confidence interval given by inequality (11.8) would be too short and its true confidence coefficient would be less than $1 - \alpha$. The practical effect would be that the simulator, due to the shortness of the confidence interval, would have unjustified confidence in the apparent accuracy of the point estimator. If the correlations ρ_k are large, B could be as small as 0.5 or 0.25, or smaller.

Case 2b. If the correlations ρ_k are substantially negative, so that $c < 1$ (c is always ≥ 0), it follows that $B > 1$, in which case S^2/n is biased high for $\sigma^2(\hat{\theta})$. The nominal $100(1 - \alpha)\%$ confidence interval of inequality (11.8) would have true confidence coefficient greater than $1 - \alpha$. In other words, the true accuracy of the point estimator $\hat{\theta}$ would be greater than what is indicated by its variance estimator, S^2/n; that is, in this case

$$\sigma^2(\hat{\theta}) < E\left(\frac{S^2}{n}\right)$$

This error is less serious than the first one, because when negative correlations are present, the estimator $\hat{\theta}$ will tend to be more accurate than is indicated by the variance estimator S^2/n.

Two methods for eliminating or reducing the deleterious effects of autocorrelation when estimating a mean are given in Section 11.5, on analysis of steady-state simulations. Unfortunately, the major simulation languages either use or facilitate the use of S^2 as an estimator of $\sigma^2 = \gamma_0$, the variance of the population. In a GPSS simulation, if the data Y_i are placed into a GPSS Table, GPSS automatically computes S^2 given by Equation (11.9) and prints $S = \sqrt{S^2}$

in the Table output. In SIMSCRIPT, S^2 will automatically be computed if the simulator requests it in a TALLY statement in the preamble. GASP has routines that will compute S^2, and SLAM automatically computes it for certain estimators. If used uncritically in a simulation with positively autocorrelated output data, the downward bias in S^2/n and the resulting shortness of a confidence interval for θ may convey the impression of much greater accuracy than actually is the case. When such positive autocorrelation is present in the output data, the true variance of the point estimator, $\hat{\theta}$, can be many times greater than is indicated by S^2/n.

11.4. Output Analysis for Terminating Simulations

Consider a terminating simulation that runs over a simulated time interval $[0, T_E]$ and results in observations $Y_1, \ldots, Y_n$. The sample size, n, may be a fixed number, or it may be a random variable (say, the number of observations that occur during time T_E). The goal of the simulation is to estimate

$$\theta = E\left(\sum_{i=1}^{n} \frac{Y_i}{n}\right)$$

The method used is called the method of independent replications. The simulation is repeated a total of R times, each run using a different random number stream and independently chosen initial conditions (which includes the case that all runs have identical initial conditions). Let Y_{ri} be the ith observation on replication r, for $i = 1, \ldots, n_r$, and $r = 1, \ldots, R$. For fixed r, $Y_{r1}, Y_{r2}, \ldots$ is an autocorrelated sequence, but for different replications $r \neq s$, Y_{ri} and Y_{sj} are statistically independent. Define a sample mean $\hat{\theta}_r$, for each replication r by equation (11.3), which becomes

$$\hat{\theta}_r = \sum_{i=1}^{n_r} \frac{Y_{ri}}{n_r}, \qquad r = 1, \ldots, R \tag{11.17}$$

The R sample means $\hat{\theta}_1, \ldots, \hat{\theta}_R$ are statistically independent and identically distributed, and are unbiased estimators of θ. Thus, classical methods of confidence interval estimation and hypothesis testing can be applied.

11.4.1. Confidence Interval Estimation for a Fixed Number of Replications

Suppose that R independent replications are made, resulting in the independent estimates defined by equation (11.17). Compute the overall point estimate, $\hat{\theta}$, by

$$\hat{\theta} = \frac{1}{R} \sum_{r=1}^{R} \hat{\theta}_r \tag{11.18}$$

and estimate the variance of $\hat{\theta}$ by equation (11.11), which becomes in this context (with $n = R$),

$$\hat{\sigma}^2(\hat{\theta}) = \frac{1}{(R-1)R}\sum_{r=1}^{R}(\hat{\theta}_r - \hat{\theta})^2 \tag{11.19}$$

This situation is case 1 of Section 11.3.2. A $100(1-\alpha)\%$ confidence interval is given by inequality (11.8) with the degrees of freedom $f = R - 1$. The quantity $\hat{\sigma}(\hat{\theta}) = \sqrt{\hat{\sigma}^2(\hat{\theta})}$ is called the standard error of the point estimate $\hat{\theta}$. Its magnitude is an indication of the accuracy of the point estimator $\hat{\theta}$ of θ. As R increases, the standard error $\hat{\sigma}(\hat{\theta})$ tends to become smaller and approach zero.

When the output data are of the form $\{Y_r(t), 0 \leq t \leq T_E\}$, $r = 1, \ldots, R$, for R independent replications, then analogously to equations (11.17), (11.18), and (11.19), compute

$$\hat{\phi}_r = \frac{1}{T_E}\int_0^{T_E} Y_r(t)\,dt, \qquad r = 1, \ldots, R, \tag{11.20}$$

$$\hat{\phi} = \frac{1}{R}\sum_{r=1}^{R}\hat{\phi}_r \tag{11.21}$$

and

$$\hat{\sigma}^2(\hat{\phi}) = \frac{1}{(R-1)R}\sum_{r=1}^{R}(\hat{\phi}_r - \hat{\phi})^2 \tag{11.22}$$

The confidence interval is analogous to that of inequality (11.8). The time averages $\hat{\phi}_1, \ldots, \hat{\phi}_R$ are statistically independent; $\hat{\phi}$ is an unbiased point estimator of ϕ, where by definition

$$\phi = E\left(\frac{1}{T_E}\int_0^{T_E} Y_r(t)\,dt\right)$$

[ϕ is independent of r since all $Y_r(t)$ are identically distributed]; and $\hat{\sigma}^2(\hat{\phi})$ is an unbiased estimator of $\sigma^2(\hat{\phi}) = \text{var}(\hat{\phi})$. Therefore, the confidence interval of inequality (11.8) is valid, again with degrees of freedom $f = R - 1$.

EXAMPLE 11.10 (THE ABLE-BAKER CARHOP PROBLEM, CONTINUED)

Consider Example 11.1, the Able–Baker carhop problem, with the data for $R = 4$ replications given in Table 11.1. The four utilization estimates, $\hat{\rho}_r$, are time averages of the form of $\hat{\phi}_r$, of equation (11.20). The simulation would produce output data of the form

$$Y_r(t) = \begin{cases} 1, & \text{if Able is busy at time } t \\ 0, & \text{otherwise} \end{cases}$$

and $\hat{\rho}_r = \hat{\phi}_r$ would be computed by equation (11.20) with $T_E = 2$ hours. [Some simulation languages, such as GPSS and SLAM, compute $\hat{\rho}_1, \hat{\rho}_2, \ldots$ automatically for any single server (such as Able), and the simulator need not be concerned with $Y_r(t)$, which is maintained automatically. Other languages, such as SIMSCRIPT and GASP, require that the simulator maintain a variable to indicate when Able's status changes from busy (1) to idle (0), and vice versa. Using this variable, SIMSCRIPT and GASP will automatically compute the time average $\hat{\rho}_r$, if so requested.]

The four average system times, $\hat{w}_1, \ldots, \hat{w}_4$, are of the form of $\hat{\theta}_r$ of equation (11.17), where Y_{ri} is the actual time spent in system by customer i on replication r. (Again, GPSS will automatically compute $\hat{w}_r$, if a queue entity is used, and SIMSCRIPT and GASP will compute $\hat{w}_r$, if requested, provided that the simulator computes Y_{ri} each time a customer leaves the system.)

First, suppose that the analyst desires a 95% confidence interval for Able's true utilization, ρ. Using equation (11.21), compute an overall point estimator

$$\hat{\rho} = \frac{0.808 + 0.875 + 0.708 + 0.842}{4} = 0.808$$

Using equation (11.22), compute the estimated variance of $\hat{\rho}$ by

$$\hat{\sigma}^2(\hat{\rho}) = \frac{(0.808 - 0.808)^2 + \cdots + (0.842 - 0.808)^2}{3(4)} = (0.036)^2$$

Thus, the standard error of $\hat{\rho} = 0.808$ is estimated by s.e.$(\hat{\rho}) = \hat{\sigma}(\hat{\rho}) = 0.036$. Obtain $t_{0.025,3} = 3.18$ from Table A.4, and compute the 95% confidence interval by

$$\hat{\rho} \pm t_{0.025,3}\hat{\sigma}(\hat{\rho})$$
$$0.808 \pm (3.18)(0.036)$$

or with 95% confidence,

$$0.694 \leq \rho \leq 0.922$$

In a similar fashion, compute a 95% confidence interval for mean time in system w:

$$\hat{w} = \frac{3.74 + 4.53 + 3.84 + 3.98}{4} = 4.02 \text{ minutes}$$

$$\hat{\sigma}^2(\hat{w}) = \frac{(3.74 - 4.02)^2 + \cdots + (3.98 - 4.02)^2}{3(4)} = (0.176)^2$$

$$\hat{w} - t_{0.025,3}\hat{\sigma}(\hat{w}) \leq w \leq \hat{w} + t_{0.025,3}\hat{\sigma}(\hat{w})$$

$$3.46 = 4.02 - (3.18)(0.176) \leq w \leq 4.02 + (3.18)(0.176) = 4.58$$

Thus, the 95% confidence interval for w is $3.46 \leq w \leq 4.58$.

EXAMPLE 11.11

To show the effect of increasing run length T_E and/or increasing the number of replications R, the Able–Baker carhop problem of Example 11.10 was simulated for all combinations of 2, 4, and 8 hours, and 4, 8, and 16 replications. The output data for utilization are shown in Table 11.3 for all 16 independent replications. Within each replication, the 8-hour run is a continuation of the 4-hour run, which is a continuation of the 2-hour run. Table 11.4 exhibits the point estimates and their standard errors:

Table 11.3. OUTPUT DATA: ABLE'S OBSERVED UTILIZATION BY REPLICATION AND RUN LENGTH

Replication,	*Run Length, T_E*		
r	*2 Hours*	*4 Hours*	*8 Hours*
1	0.808	0.796	0.785
2	0.875	0.825	0.833
3	0.708	0.787	0.806
4	0.842	0.837	0.833
5	0.742	0.825	0.808
6	0.767	0.775	0.800
7	0.792	0.787	0.794
8	0.950	0.867	0.827
9	0.833	0.821	0.815
10	0.717	0.750	0.821
11	0.817	0.808	0.798
12	0.842	0.746	0.817
13	0.850	0.846	0.854
14	0.850	0.846	0.848
15	0.767	0.783	0.796
16	0.817	0.804	0.813

Table 11.4. ABLE'S ESTIMATED UTILIZATION WITH STANDARD ERROR

$$\hat{\rho} \pm \text{s.e.}(\hat{\rho})$$

FOR VARIOUS RUN LENGTHS AND NUMBER OF REPLICATIONS

Number of Replications,	*Run Length, T_E*		
R	*2 Hours*	*4 Hours*	*8 Hours*
4	0.808 ± 0.036	0.811 ± 0.011	0.814 ± 0.011
8	0.811 ± 0.027	0.812 ± 0.011	0.810 ± 0.006
16	0.811 ± 0.015	0.806 ± 0.009	0.816 ± 0.005

$\hat{\rho} \pm \text{s.e.}(\hat{\rho})$, where $\text{s.e.}(\hat{\rho}) = \hat{\sigma}(\hat{\rho})$ is an estimate of the standard deviation. Theoretically, the standard deviation of $\hat{\rho}$ is given by

$$\sigma(\hat{\rho}) = \sqrt{\text{var}(\hat{\rho})} = \frac{\sigma}{\sqrt{R}}$$

where $\sigma^2 = \text{var}(\hat{\rho}_r)$ is the variance of the population of which $\hat{\rho}_1, \ldots, \hat{\rho}_R$ is a random sample. Since $\text{s.e.}(\hat{\rho})$ is an estimate of $\sqrt{\text{var}(\hat{\rho})} = \sigma(\hat{\rho})$, $\text{s.e.}(\hat{\rho})$ should decrease proportionally to $1/\sqrt{R}$. In other words, the standard error for 8 replications should be approximately $1/\sqrt{2} = 0.71$ as large as that for 4 replications; and the standard error for 16 replications should be approximately $1/\sqrt{4} = 0.5$ as large as that for 4 replications. [The relationship between sample size, R, and standard deviation, $\sigma(\hat{\rho})$, will hold only approximately for $\text{s.e.}(\hat{\rho})$, because a standard error is an estimator and thus it contains a certain amount of random error.] Examination of Table 11.4 shows that this relationship holds approximately for the utilization data. From the standpoint of accuracy, if the simulator desires to reduce s.e.($\hat{\rho}$) by one-half, it is necessary to quadruple the sample size (say from 4 to 16, or 8 to 32).

Increasing the run length T_E will also decrease the true standard deviation, $\sigma(\hat{\rho})$, of the point estimator. As T_E increases in Table 11.4, the estimated standard error, s.e.($\hat{\rho}$), usually decreases, any exception (such as from 4 hours to 8 hours with four replications) being due to the random variation in the data. In a terminating simulation, this option is not available to the simulator, because, by definition, T_E is determined and fixed by the nature of the system being simulated. Nevertheless, the data are presented for illustration purposes. Note that for equal total run length over all replications (RT_E), such as 8 replications of 4 hours each (= 32 hours), and 4 replications of 8 hours each (= 32 hours), that the point estimators have approximately equal accuracy.

11.4.2. Hypothesis Testing

Hypothesis testing was discussed extensively in Section 10.3.3 in the context of validating a model. The discussion here gives two examples of its use in output analysis, but assumes that the reader is familiar with the basic ideas of hypothesis testing as outlined in Example 10.2.

Example 11.12

In the Able–Baker carhop problem of Example 11.1, suppose that management wants Able's utilization to exceed 0.90 (or, equivalently, idle time to be less than 10%). The simulator decides to test the hypothesis

$$H_0: \ \rho \geq 0.90$$

versus the alternative hypothesis

$$H_1: \ \rho < 0.90$$

The simulator specifies that declaring Able's utilization to be less than 0.90, when in fact it is 0.90 or better, should have probability no more than 0.05. In other words, the significance level of the test is

$$\alpha = 0.05 = P(\text{Type I error}) = P(\text{reject } H_0 \,|\, H_0 \text{ true})$$

The simulator also decides to make $R = 4$ independent replications. Application of the five steps in hypothesis testing, outlined after equation (10.1), gives the following:

Step 1. The level of significance is $\alpha = 0.05$ and the sample size is $R = 4$.

Step 2. The sample mean is $\hat{\rho} = 0.808$ and the sample standard deviation is $S = 0.07222$.

Step 3. The critical value is $-t_{0.05,3} = -2.35$ (for this one-sided test).

Step 4. Based on the data in Table 11.4 for $R = 4$ and $T_E = 2$ hours, the test statistic is computed as

$$t_0 = \frac{\hat{\rho} - \rho_0}{S/\sqrt{R}} = \frac{0.808 - 0.90}{0.072/\sqrt{4}} = -2.56$$

Step 5. Since $t_0 < -t_{0.05,3}$, H_0 is rejected. The simulator concludes that with the current design of the system, there is strong evidence in support of the hypothesis that Able's utilization is less than 0.90.

EXAMPLE 11.13

After considerable thought, the carhop management decided that Able's utilization could not be as high as 0.90. They told the simulator that a utilization of 0.82 or higher would be acceptable, but that if the utilization were as low as 0.77, they would like to know it. In designing simulation experiments to meet management's requests, the simulator decided to test the hypothesis

$$H_0: \quad \rho \geq 0.82$$

versus the alternative hypothesis

$$H_1: \quad \rho < 0.82$$

at a level of significance $\alpha = 0.05$. In addition, the simulator required that the sample size be sufficiently large so that if $\rho = 0.77$ were true, the probability of detecting it would be 0.90; that is, the power of the test for the alternative $\rho = 0.77$ is

$$1 - \beta = P(\text{rejecting } H_0 \mid \rho = 0.77) = 0.90$$

To estimate the necessary total sample size, the initial sample of size $R_0 = 4$ is used to estimate

$$\delta = \frac{|0.82 - 0.77|}{\sigma}$$

by

$$\hat{\delta} = \frac{|0.82 - 0.77|}{S} = \frac{0.05}{0.072} = 0.694$$

where S was computed from the four replications used for Example 11.12. Using Table A.10, the OC curve for one-sided t tests, with $\beta = 0.1$ at $\hat{\delta} = 0.694$, yields an approximate sample size $R = 20$. Thus, approximately $20 - 4 = 16$ additional replications are needed to guarantee $\beta = 0.1$ at $\rho = 0.77$ and $\alpha = 0.05$.

11.4.3. CONFIDENCE INTERVALS WITH SPECIFIED ACCURACY

By inequality (11.8), the half-length (h.l.) of a $100(1 - \alpha)\%$ confidence interval for a mean θ, based on the t distribution, is given by

$$\text{h.l.} = t_{\alpha/2,R-1}\hat{\sigma}(\hat{\theta})$$

where $\hat{\sigma}(\hat{\theta}) = S/\sqrt{R}$, S is the sample standard deviation, and R is the number of replications. Suppose that an accuracy criterion ϵ is specified; it is desired to estimate θ by $\hat{\theta}$ with accuracy ϵ with high probability, say at least $1 - \alpha$. Thus it is desired that a sufficiently large sample size, R, be taken to satisfy

$$P(|\hat{\theta} - \theta| < \epsilon) \geq 1 - \alpha$$

When the sample size, R, is fixed, as in Section 11.4.1, no guarantee can be given of the resulting accuracy. But if the sample size can be increased, an accuracy criteria can be specified.

Assume that an initial sample of size R_0 has been observed; that is, the simulator initially makes R_0 independent replications. In practice, R_0 is 2 or larger, but at least 4 or 5 is recommended. The R_0 replications will be used to obtain an initial estimate, S_0^2, of the population variance, σ^2. To meet the half-length criteria, a sample size R must be chosen such that $R \geq R_0$ and

$$\text{h.l.} = \frac{t_{\alpha/2,R-1} S_0}{\sqrt{R}} \leq \epsilon \tag{11.23a}$$

Solving for R in inequality (11.23a), it is seen that R is the smallest integer satisfying $R \geq R_0$ and

$$R \geq \left(\frac{t_{\alpha/2,R-1} S_0}{\epsilon}\right)^2 \tag{11.23b}$$

Since $t_{\alpha/2,R-1} \geq z_{\alpha/2}$, an initial estimate for R is given by

$$R \geq \left(\frac{z_{\alpha/2} S_0}{\epsilon}\right)^2 \tag{11.24}$$

Since $t_{\alpha/2,R-1} \approx z_{\alpha/2}$ for large R (say $R \geq 50$), the second inequality for R is adequate when R is large. After determining the final sample size, R, collect $R - R_0$ additional observations (i.e., make $R - R_0$ additional replications) and form the $100(1 - \alpha)\%$ confidence interval for θ by

$$\hat{\theta} - \frac{t_{\alpha/2,R-1} S}{\sqrt{R}} \leq \theta \leq \hat{\theta} + \frac{t_{\alpha/2,R-1} S}{\sqrt{R}} \tag{11.25}$$

where $\hat{\theta}$ and S are computed based on all R replications, $\hat{\theta}$ by equation (11.18) and $S^2/R = \hat{\sigma}^2(\hat{\theta})$ by equation (11.19). The half-length of the confidence interval given by inequality (11.25) should be approximately ϵ, or smaller; however, with the additional $R - R_0$ observations, the variance estimator S^2 may differ somewhat from the initial estimate S_0^2, possibly causing the half-length to be greater than desired. If the confidence interval (11.25) is too large, the procedure may be repeated, using inequality (11.23b) to determine an even larger sample size.

EXAMPLE 11.14

Suppose that it is desired to estimate Able's utilization in Example 11.1 to within ± 0.04 with probability 0.95. An initial sample of size $R_0 = 4$ is taken, with the results given in Table 11.1 (also, Table 11.3 with $T_E = 2$ hours). An initial estimate of the population variance is $S_0^2 = R_0\hat{\sigma}^2(\hat{\rho}) = 4(0.036)^2 = 0.00518$. (See Example 11.10 or Table 11.4 for the relevant data.) The accuracy criterion is $\epsilon = 0.04$ and the confidence coefficient is $1 - \alpha = 0.95$. From inequality (11.24), the final sample size must be at least as large as

$$\frac{z_{0.025}^2 S_0^2}{\epsilon^2} = \frac{(1.96)^2(0.00518)}{(0.04)^2} = 12.44$$

Next, inequality (11.23b) can be used to test possible candidates (13, 14, . . .) for final sample size, as follows:

R	13	14	15
$t_{0.025,R-1}$	2.18	2.16	2.14
$\frac{t_{0.025,R-1}^2 S_0^2}{\epsilon^2}$	15.39	15.10	14.83

Thus, $R = 15$ is the smallest integer satisfying inequality (11.23b), so $R - R_0 = 15 - 4 = 11$ additional replications are needed. Assuming that 12 additional replications (instead of 11) were made, with the output data $\hat{\rho}_r$ as given in Table 11.3, the half-length for a 95% confidence interval would be given by $t_{0.025,15}\text{s.e.}(\hat{\rho}) = 2.13(0.015) = 0.032$, which is well within the accuracy criterion $\epsilon = 0.04$. The resulting 95% confidence interval for true utilization ρ is $0.811 - 0.032 = 0.779 \leq \rho \leq 0.843 = 0.811 + 0.032$. Note the considerable improvement in accuracy (as measured by confidence interval half-length) compared to Example 11.10, which used a fixed sample size ($R = 4$).

EXAMPLE 11.15

Suppose that management desires to know mean system time, w, in Example 11.1 to within 0.5 minute with probability 0.90. An initial sample of size $R_0 = 4$ is shown in Table 11.1. In Example 11.10, an initial estimate based on this sample yielded the point estimate, $\hat{w}_0$, and sample variance S_0^2 given by

$$\hat{w}_0 = 4.02, \qquad S_0^2 = 4\hat{\sigma}^2(\hat{w}) = 4(0.176)^2 = 0.124$$

Here, $\epsilon = 0.5$ minute and $\alpha = 0.10$. A first estimate of the sample size R needed to meet accuracy criteria ϵ is given by inequality (11.24) as $z_{0.05}^2 S_0^2/\epsilon^2 = (1.645)^2(0.124)/(0.5)^2 = 1.34$, or $R \geq 2$. Using inequality (11.23b) together with $R \geq R_0 = 4$, to find R involves these computations:

R	4
$t_{\alpha/2,R-1}$	2.35
$\frac{t_{0.025,R-1}^2 S_0^2}{\epsilon^2}$	2.74

Therefore, since the initial sample size was $R_0 = 4$, no additional observations are needed to meet the accuracy criteria $\epsilon = 0.5$ minute. The actual half-length of the 90% confidence interval is $t_{0.05,3}S_0/\sqrt{R_0} = 2.35\sqrt{0.124}/\sqrt{4} = 0.41$ minute.

11.5. Output Analysis for Steady-State Simulations

Consider a single run of a simulation model whose purpose is to estimate a *steady-state*, or *long-run*, characteristic of the system. Suppose that the single run produces observations $Y_1, Y_2, \ldots$, which, generally, are samples of an autocorrelated time series. The steady-state (or long-run) measure of performance, θ, to be estimated is defined by

$$\theta = \lim_{n\to\infty} \frac{1}{n} \sum_{i=1}^{n} Y_i \tag{11.26}$$

with probability 1, where the value of θ is independent of the initial conditions. (The phrase "with probability 1" means that essentially all simulations of the model, using different random numbers, will produce series Y_i, $i = 1, 2, \ldots$ whose sample average converges to θ.) For example, if Y_i was the total time job i spent in a job shop, then θ would be the long-run average time a job spends in the shop; the value of θ is independent of shop conditions at time 0.

Of course, the simulator must decide to stop the simulation after some number of observations, say n, have been collected; or the simulator may decide to simulate for some length of time T_E, in which case n is determined (although n may vary from run to run). The sample size n (or T_E) is a *design* choice; it is not inherently determined by the nature of the problem. The simulator will choose simulation run length (n or T_E) with several considerations in mind:

1. The bias in the point estimator due to artificial or arbitrary initial conditions. The bias can be severe if run length is too short, but it generally decreases as run length increases.
2. The desired accuracy of the point estimator, as measured by an estimate of point estimator variability.
3. Budget constraints on computer resources.

The next subsection discusses initialization bias and the following subsections outline two methods of estimating point estimator variability. When discussing one replication (or run), the notation

$$Y_1, Y_2, Y_3, \ldots$$

will be used; if several replications have been made, the output data for replication r will be denoted by

$$Y_{r1}, Y_{r2}, Y_{r3}, \ldots \tag{11.27}$$

11.5.1. Initialization Bias in Steady-state Simulations

There are several ways of reducing the point estimator bias which is caused by using artificial and unrealistic initial conditions in a steady-state simulation.

The first method is to collect data on the system, if it exists, and to use these data to specify more typical initial conditions. This method requires a sometimes large data collection effort. Moreover, if the system being modeled does not exist—for example, if it is a variant of an existing system—this method is impossible to implement. Nevertheless, despite these difficulties, it is recommended that simulators use any available data on existing systems, as this would appear better than assuming the system to be "empty and idle" at time 0.

The second method, possibly used in conjunction with the first, is to divide each simulation run into two phases: first, an initialization phase from time 0 to time T_0, followed by a data collection phase from time T_0 to the stopping time $T_0 + T_E$. That is, the simulation begins at time 0 under specified initial conditions, I_0, and runs for a specified period of time T_0. Data collection on the response variables of interest does not begin until time T_0, and continues until time $T_0 + T_E$. The choice of T_0 is quite important, as the system state at time T_0, denoted by I, should be more representative of steady-state behavior than the original initial conditions, I_0, at time 0. In addition, the length T_E of the data collection phase should be long enough to guarantee sufficiently precise estimates of steady-state behavior. Note that the system state, I, at time T_0 is a random variable, and to say that the system has reached an approximate steady state is to say that the probability distribution of the system state at time T_0 is sufficiently close to the steady-state probability distribution as to make the bias in point estimates of response variables negligible. Figure 11.3 illustrates the two phases of a steady-state simulation. The effect of starting a simulation run of a queueing model in the empty and idle state, as well as several useful plots to aid the simulator in choosing an appropriate value of T_0, are given in the following example.

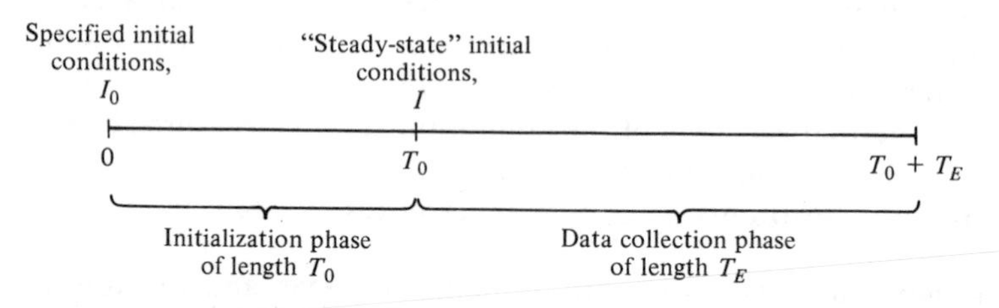

Figure 11.3. Initialization and data collection phases of a steady-state simulation run.

Example 11.16

Consider the $M/G/1$ queue discussed in Example 11.2. A total of 10 independent replications were made ($R = 10$), each replication beginning in the empty and idle state, and total simulation run length on each replication was $T_0 + T_E = 15{,}000$

minutes. The response variable was queue length, $L_Q(t, r)$, at time t, where the second argument, r, denotes the replication ($r = 1, \ldots, 10$). The raw output data were batched, as in Example 11.2, equation (11.1), in batching intervals of 1000 minutes to produce the following batch means:

$$Y_{rj} = \frac{1}{1000} \int_{(j-1)1000}^{j(1000)} L_Q(t, r)\, dt \tag{11.28}$$

for replication $r = 1, \ldots, 10$ and for batch $j = 1, 2, \ldots, 15$. The estimator in equation (11.28) is a time-weighted-average queue length over the interval $[(j-1)1000, j(1000)]$, similar to that in equation (5.6b). The 15 batch means for the 10 replications are given in Table 11.5.

For each batch j, define a sample mean over all replications by

$$\bar{Y}_{\cdot j} = \frac{1}{R} \sum_{r=1}^{R} Y_{rj} \tag{11.29}$$

where R is the number of replications ($R = 10$ here). The averaged batch means $\bar{Y}_{\cdot j}$, $j = 1, \ldots, 15$ are displayed in the third column of Table 11.6. Note that $\bar{Y}_{\cdot 1} = 4.03$ and $\bar{Y}_{\cdot 2} = 5.45$, estimates of mean queue length over the time periods $[0, 1000)$ and $[1000, 2000]$, respectively, are lower in magnitude than all other values of the averaged batch means $\bar{Y}_{\cdot j}$ ($j = 3, \ldots, 15$). The simulator may suspect that this is due to the downward bias in these estimators, which in turn is due to the queue being empty and idle at time 0. This downward bias is further illustrated in the plots that follow.

Figure 11.4 is a plot of the averaged batch means, $\bar{Y}_{\cdot j}$, versus $1000j$, for $j = 1, 2, \ldots, 15$. The actual values, $\bar{Y}_{\cdot j}$, are the discrete set of points in circles, which have been connected by straight lines. Figure 11.4 illustrates the downward bias of the initial observations. As time becomes larger, the effect of the initial conditions on later observations lessens.

Table 11.6 also gives the cumulative average sample mean after no deletion, one deletion, and two deletions. That is, for the average batch means $\bar{Y}_{\cdot j}$, when deleting d observations out of a total of n observations, compute

$$\bar{Y}_{\cdot\cdot}(n, d) = \frac{1}{n-d} \sum_{j=d+1}^{n} \bar{Y}_{\cdot j} \tag{11.30}$$

The results in Table 11.6 for the $M/G/1$ simulation are for $d = 0, 1$, and 2, and $n = d + 1, \ldots, 15$. These cumulative averages with deletion, namely $\bar{Y}_{\cdot\cdot}(n, d)$, are plotted for comparison purposes in Figure 11.5.

From Figures 11.4 and 11.5, it becomes apparent that (1) downward bias is present, and (2) this initialization bias in the point estimator can be reduced by deletion of one or more observations. For the 15 batched and averaged observations, it appears that the first two observations have considerably more bias than any of the remaining ones. The effect of deleting first one and then two batch means is also illustrated in Table 11.6 and Figure 11.5. As expected, the estimators increase in value as more data are deleted; that is, $\bar{Y}_{\cdot\cdot}(15, 2) = 8.43$ and $\bar{Y}_{\cdot\cdot}(15, 1) = 8.21$ are larger than $\bar{Y}_{\cdot\cdot}(15, 0) = 7.94$. It also appears from Figure 11.5 that $\bar{Y}_{\cdot\cdot}(n, d)$ is increasing for $n = 5, 6, \ldots, 11$ (and all $d = 0, 1, 2$), and thus there may be additional significant initialization bias. It seems, however, that deletion of the first two batches removes most of the bias.

Table 11.5. INDIVIDUAL BATCH MEANS (Y_{rj}) FOR $M/G/1$ SIMULATION WITH EMPTY AND IDLE INITIAL STATE

	Batch														
Replication	*1*	*2*	*3*	*4*	*5*	*6*	*7*	*8*	*9*	*10*	*11*	*12*	*13*	*14*	*15*
1	3.61	3.21	2.18	6.92	2.82	1.59	3.55	5.60	3.04	2.57	1.41	3.07	4.03	2.70	2.71
2	2.91	9.00	16.15	24.53	25.19	21.63	24.47	8.45	8.53	14.84	23.65	27.58	24.19	8.58	4.06
3	7.67	19.53	20.36	8.11	12.62	22.15	14.10	9.87	23.96	24.50	14.56	6.08	4.83	16.04	23.41
4	6.62	1.75	12.87	8.77	1.25	1.16	1.92	6.29	4.74	17.43	18.24	18.59	4.62	2.76	1.57
5	2.18	1.32	2.14	2.18	2.59	1.20	4.11	6.21	7.31	1.58	2.16	3.08	2.32	2.21	3.32
6	0.93	3.54	4.80	0.72	2.95	5.56	1.96	2.07	2.74	3.45	14.24	13.39	7.87	0.94	3.19
7	1.12	2.59	5.05	1.16	2.72	5.12	5.03	4.14	4.98	15.81	9.29	2.14	8.72	29.80	28.94
8	1.54	5.94	5.33	2.91	2.69	1.91	3.27	3.61	10.35	9.66	4.13	6.14	7.90	2.61	7.95
9	8.93	4.78	0.74	2.56	9.43	18.63	8.14	1.49	4.51	1.69	12.62	11.28	3.32	3.42	3.35
10	4.78	2.84	10.39	5.87	1.01	2.59	16.77	27.25	26.81	20.96	7.26	2.32	5.04	8.50	9.11

Table 11.6. SUMMARY OF DATA FOR $M/G/1$ SIMULATION: BATCH MEANS AND CUMULATIVE MEANS, AVERAGED OVER 10 REPLICATIONS

Run Length, T	*Batch,* j	*Averaged Batch Mean,* $\bar{Y}_{\cdot j}$	*Cumulative Average (No Deletion),* $\bar{Y}_{\cdot\cdot}(j, 0)$	*Cumulative Average (Delete 1),* $\bar{Y}_{\cdot\cdot}(j, 1)$	*Cumulative Average (Delete 2),* $\bar{Y}_{\cdot\cdot}(j, 2)$
1,000	1	4.03	4.03	—	—
2,000	2	5.45	4.74	5.45	—
3,000	3	8.00	5.83	6.72	8.00
4,000	4	6.37	5.96	6.61	7.18
5,000	5	6.33	6.04	6.54	6.90
6,000	6	8.15	6.39	6.86	7.21
7,000	7	8.33	6.67	7.11	7.44
8,000	8	7.50	6.77	7.16	7.45
9,000	9	9.70	7.10	7.48	7.77
10,000	10	11.25	7.51	7.90	8.20
11,000	11	10.76	7.81	8.18	8.49
12,000	12	9.37	7.94	8.29	8.58
13,000	13	7.28	7.89	8.21	8.46
14,000	14	7.76	7.88	8.17	8.40
15,000	15	8.76	7.94	8.21	8.43

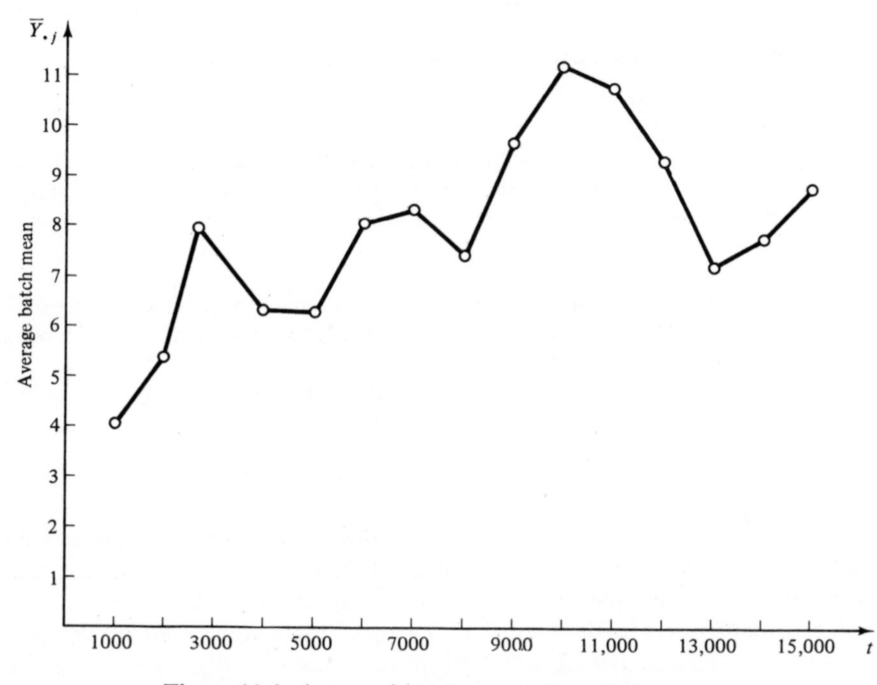

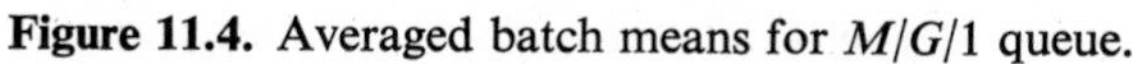
Figure 11.4. Averaged batch means for $M/G/1$ queue.

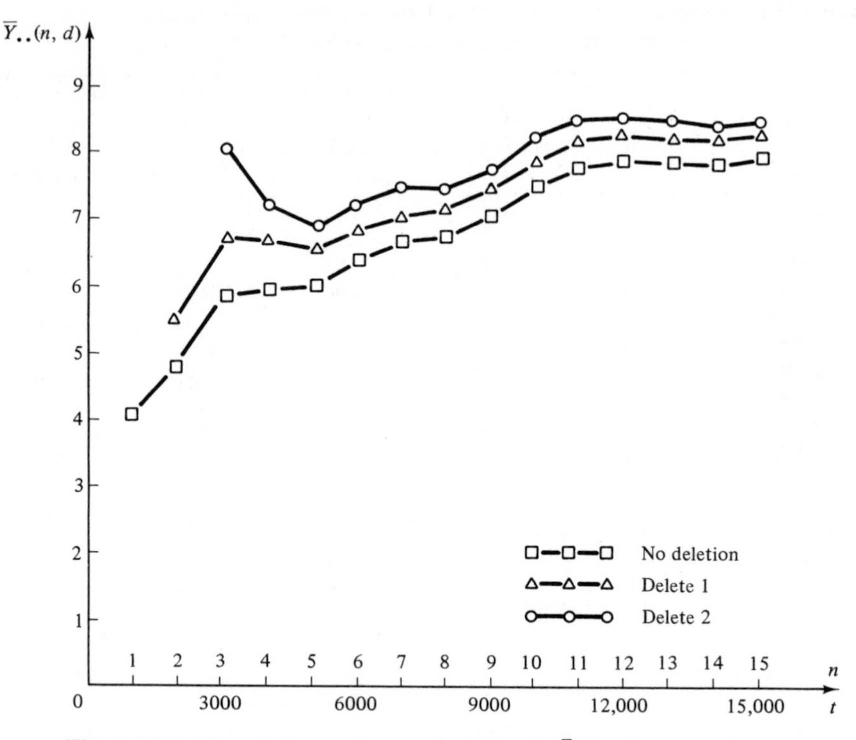

Figure 11.5. Cumulative average queue length $\bar{Y}_{\cdot\cdot}(n, d)$ versus time $1000n$.

Unfortunately, there is no widely accepted, objective, and proven technique to determine how many data to delete to reduce initialization bias to a negligible level. Plots can, at times, be misleading, but they are still recommended. Two points should be kept in mind:

1. Cumulative averages, such as in Figure 11.5, become less variable as more data are averaged. Therefore, it is expected that the left side of the curve will always be less smooth than the right side.
2. Simulation data, especially from queueing models, usually exhibits positive autocorrelation. The more correlation present, the longer it takes for $\bar{Y}_{\cdot j}$ to approach steady state. The positive correlation between successive observations (i.e., batch means) $\bar{Y}_{\cdot 1}, \bar{Y}_{\cdot 2}, \ldots$ can be seen in Figure 11.4.

A survey paper by Gafarian et al. [1978] describes several procedures that have been proposed to handle the problem of initialization bias in point estimators of steady-state parameters. After testing the procedures on sample problems, they conclude that none of the procedures work well in practice. More recently, Kelton [1980] and Kelton and Law [1980] have developed a procedure based on independent replications, deletion of data, and a time-series regression technique. In addition, Schruben [1980] has developed several statistical tests of the hypothesis of no bias. At present, these new procedures have not been thoroughly evaluated and compared, and no existing procedure can be regarded as a complete solution to the problem. On the other hand, the recently developed regenerative method for output analysis eliminates the problem of initialization bias (and also the problem of autocorrelation); however, the method can be applied only to a restricted class of models [Crane and Lemoine, 1977].

11.5.2. Replication Method for Steady-State Simulations

If initialization bias in the point estimator has been reduced to a negligible level, the method of independent replications can be used to estimate point estimator variability and to construct a confidence interval. If, however, significant bias remains in the point estimator and a large number of replications are used to reduce point estimator variability, the resulting confidence interval can be misleading. This happens because bias is not affected by the number of replications (R); it is affected only by deleting more data (i.e., increasing T_0) and/or extending the length of each run (i.e., increasing T_E). Thus, increasing the number of replications (R) may produce shorter confidence intervals around the "wrong point" ($\theta + b$), not around θ.

If the simulator decides to delete d observations of the total of n observations, the point estimator of θ is $\bar{Y}..(n, d)$, defined by equation (11.30). The basic raw output data, $\{Y_{rj}, r = 1, \ldots, R; j = 1, \ldots, n\}$, are exhibited in Table 11.7. Each Y_{rj} is derived in one of the following ways:

Table 11.7. RAW OUTPUT DATA FROM A STEADY-STATE SIMULATION

	Observation						Replication Averages
Replication	*1*	$\cdots$	*d*	*d + 1*	$\cdots$	*n*	
1	$Y_{1,1}$	$\cdots$	$Y_{1,d}$	$Y_{1,d+1}$	$\cdots$	$Y_{1,n}$	$\bar{Y}_{1}.(n, d)$
2	$Y_{2,1}$	$\cdots$	$Y_{2,d}$	$Y_{2,d+1}$	$\cdots$	$Y_{2,n}$	$\bar{Y}_{2}.(n, d)$
$\vdots$	$\vdots$		$\vdots$	$\vdots$		$\vdots$	$\vdots$
R	$Y_{R,1}$	$\cdots$	$Y_{R,d}$	$Y_{R,d+1}$	$\cdots$	$Y_{R,n}$	$\bar{Y}_{R}.(n, d)$
	$\bar{Y}_{\cdot 1}$	$\cdots$	$\bar{Y}_{\cdot d}$	$\bar{Y}_{\cdot d+1}$	$\cdots$	$\bar{Y}_{\cdot n}$	$\bar{Y}..(n, d)$

Case 1. Y_{rj} is an individual observation from replication r; for example, Y_{rj} could be the delay of customer j in a queue, or the response time to job j in a job shop.

Case 2. Y_{rj} is a batch mean from replication r of some number of individual observations. (Batch means are discussed further in Section 11.5.4.)

Case 3. Y_{rj} is a batch mean of a continuous-time process over interval j; for example, as in Example 11.16, equation (11.28) defines Y_{rj} as the time-average (batch mean) number in queue over the interval $[1000(j-1), 1000j)$.

In general, in case 1, the number d of deleted observations and the total number of observations n may vary from one replication to the next, in which case replace d by d_r and n by n_r. For simplicity, assume that d and n are constant over replications. In cases 2 and 3, d and n will be constant.

When using the replication method, each replication is regarded as a single sample for the purpose of estimating θ. For replication r, define

$$\bar{Y}_{r\cdot}(n, d) = \frac{1}{n-d} \sum_{j=d+1}^{n} Y_{rj} \tag{11.31}$$

as the sample mean of all (nondeleted) observations in replication r. Since all replications use different random number streams and all are initialized at time 0 by the same set of initial conditions (I_0), the replication averages

$$\bar{Y}_{1\cdot}(n, d), \ldots, \bar{Y}_{R\cdot}(n, d)$$

are independent and identically distributed random variables; that is, they constitute a random sample from some underlying population having unknown mean

$$\theta_{n,d} = E[\bar{Y}_{r\cdot}(n, d)] \tag{11.32}$$

The overall point estimator, given in equation (11.30), is also given by

$$\bar{Y}_{\cdot\cdot}(n, d) = \frac{1}{R} \sum_{r=1}^{R} \bar{Y}_{r\cdot}(n, d) \tag{11.33}$$

as can be seen from Table 11.7 or from using equation (11.29). Thus, it follows that

$$E[\bar{Y}_{\cdot\cdot}(n, d)] = \theta_{n,d}$$

also. If d and n are chosen sufficiently large, so that $\theta_{n,d} \approx \theta$, then $\bar{Y}_{\cdot\cdot}(n, d)$ is an approximately unbiased estimator of θ. The bias in $\bar{Y}_{\cdot\cdot}(n, d)$ is $b = \theta_{n,d} - \theta$, as defined by equation (11.4b).

For convenience, when the value of n and d are understood, abbreviate $\bar{Y}_{r\cdot}(n, d)$ and $\bar{Y}_{\cdot\cdot}(n, d)$ by $\bar{Y}_{r\cdot}$ and $\bar{Y}_{\cdot\cdot}$, respectively. To estimate the standard error of $\bar{Y}_{\cdot\cdot}$, first compute the sample variance

$$S^2 = \frac{1}{R-1} \sum_{r=1}^{R} (\bar{Y}_{r\cdot} - \bar{Y}_{\cdot\cdot})^2 = \frac{1}{R-1}\left(\sum_{r=1}^{R} \bar{Y}_{r\cdot}^2 - R\bar{Y}_{\cdot\cdot}^2\right) \tag{11.34}$$

The standard error of $\bar{Y}..$ is given by

$$\text{s.e.}(\bar{Y}..) = \frac{S}{\sqrt{R}} \tag{11.35}$$

A $100(1-\alpha)\%$ confidence interval for θ, based on the t distribution, is given by

$$\bar{Y}.. - \frac{t_{\alpha/2,R-1}S}{\sqrt{R}} \leq \theta \leq \bar{Y}.. + \frac{t_{\alpha/2,R-1}S}{\sqrt{R}} \tag{11.36}$$

where $t_{\alpha/2,R-1}$ is the $100(1-\alpha/2)$ percentage point of a t distribution with $R-1$ degrees of freedom. This confidence interval is valid only if the bias, b, of $\bar{Y}..$ is approximately zero.

EXAMPLE 11.17

Consider again the $M/G/1$ queueing simulation of Examples 11.2 and 11.16. Suppose that the simulator decides to make $R = 10$ replications, each of length $T_E = 13{,}000$ minutes, each starting at time 0 in the empty and idle state, and each initialized for $T_0 = 2000$ minutes before data collection begins. The raw output data consist of the batch means defined by equation (11.28). The first two batches are deleted ($d = 2$). The purpose of the simulation is to estimate, by a 95% confidence interval, the long-run time average queue length, denoted by L_Q (or θ).

Table 11.8. DATA SUMMARY FOR $M/G/1$ SIMULATION BY REPLICATION

Replication, r	*Sample Mean for Replication r* (No Deletion) $\bar{Y}_r.(15, 0)$	(Delete 1) $\bar{Y}_r.(15, 1)$	(Delete 2) $\bar{Y}_r.(15, 2)$
1	3.27	3.24	3.25
2	16.25	17.20	17.83
3	15.19	15.72	15.43
4	7.24	7.28	7.71
5	2.93	2.98	3.11
6	4.56	4.82	4.91
7	8.44	8.96	9.45
8	5.06	5.32	5.27
9	6.33	6.14	6.24
10	10.10	10.48	11.07
$\bar{Y}..(15, d)$	7.94	8.21	8.43
$\sum_{r=1}^{R} \bar{Y}_{r.}^2$	826.20	894.68	938.34
S^2	21.75	24.52	25.30
S	4.66	4.95	5.03
$S/\sqrt{10} = \text{s.e.}(\bar{Y}..)$	1.47	1.57	1.59

The replication averages $\bar{Y}_{r.}(15, 2)$, $r = 1, 2, \ldots, 10$, are shown in Table 11.8 in the rightmost column. The point estimator is computed by equation (11.33) as

$$\bar{Y}_{..}(15, 2) = 8.43$$

Its standard error is given by equation (11.35) as

$$\text{s.e.}(\bar{Y}_{..}(15, 2)) = 1.59$$

and using $\alpha = 0.05$ and $t_{0.025,9} = 2.26$, the 95% confidence interval for long-run mean queue length is given by inequality (11.36) as

$$8.43 - 2.26(1.59) \leq L_Q \leq 8.43 + 2.26(1.59)$$

or

$$4.84 \leq L_Q \leq 12.02$$

The simulator may conclude with a high degree of confidence that the long-run mean queue length is between 4.84 and 12.02 customers. The confidence interval computed here as given by inequality (11.36) should be used with caution, because a key assumption behind its validity is that enough data have been deleted to remove any significant bias due to initial conditions—that is, d and n are sufficiently large so that the bias $b = \theta_{n,d} - \theta$ is negligible.

EXAMPLE 11.18

Suppose that in Example 11.17, the simulator had decided to delete one batch ($d = 1$), or no batches ($d = 0$). The quantities needed to compute 95% confidence intervals are shown in Table 11.8. The resulting 95% confidence intervals are computed by inequality (11.36) as follows:

$$(d = 1) \quad 4.66 = 8.21 - 2.26(1.57) \leq L_Q \leq 8.21 + 2.26(1.57) = 11.76$$
$$(d = 0) \quad 4.62 = 7.94 - 2.26(1.47) \leq L_Q \leq 7.94 + 2.26(1.47) = 11.26$$

Note that, for a fixed total sample size, n, two things happen as fewer data are deleted:

1. The confidence interval shifts downward, reflecting the greater downward bias in $\bar{Y}_{..}(15, d)$ as d decreases.
2. The standard error of $\bar{Y}_{..}(n, d)$, namely $S/\sqrt{R}$, decreases as d decreases.

In this example, $\bar{Y}_{..}(n, d)$ is based on a run length of $T_E = 1000(n - d) = 15{,}000 - 1000d$ minutes. Thus, as d decreases, T_E increases, and in effect, the sample mean $\bar{Y}_{..}$ is based on a larger "sample size" (i.e., larger run length). In general, the larger the sample size, the smaller the standard error of the point estimator. This larger sample size can be due to a longer run length (T_E) per replication, or to more replications (R).

Therefore, there is a trade-off between reducing bias and increasing the variance of a point estimator, when the total sample size (R and $T_0 + T_E$) is fixed. The more deletion (i.e., the larger T_0 is and the smaller T_E is, keeping $T_0 + T_E$ fixed), the less bias but greater variance there is in the point estimator.

11.5.3. Accuracy Versus Sample Size in Steady-State Simulations

Suppose it is desired to estimate a long-run performance measure, θ, within an accuracy ϵ with confidence $100(1 - \alpha)\%$. In a steady-state simulation, a specified accuracy may be achieved either by increasing the number of replications (R) or by increasing the run length (T_E). The first solution, by controlling R, is carried out as given in Section 11.4.3 for terminating simulations.

Example 11.19

Consider the data in Table 11.8 for the $M/G/1$ queueing simulation as an initial sample of size $R_0 = 10$. Assuming that $d = 2$, the initial estimate of variance is $S_0^2 = 25.30$. Suppose that it is desired to estimate long-run mean queue length, L_Q, within $\epsilon = 2$ customers with 90% confidence. The final sample size needed must satisfy inequality (11.23b). Using $\alpha = 0.10$ in inequality (11.24) yields an initial estimate:

$$R \geq \left(\frac{z_{0.05}S_0}{\epsilon}\right)^2 = \frac{1.645^2(25.30)}{2^2} = 17.1$$

Thus, at least 18 replications will be needed. Proceeding as in Example 11.14, next try $R = 18$, $R = 19$, . . . as follows:

R	18	19
$t_{0.05,R-1}$	1.74	1.73
$\left(\frac{t_{0.05,R-1}S_0}{\epsilon}\right)^2$	19.15	18.93

Since $R = 19 \geq \{t_{0.05,18}S_0/\epsilon\}^2 = 18.93$ is the smallest integer R satisfying inequality (11.23b), a total sample size of $R = 19$ replications are needed to estimate L_Q within ± 2 customers. Therefore, $R - R_0 = 19 - 10 = 9$ additional replications are needed to achieve the specified accuracy.

An alternative to increasing R is to increase total run length $T_0 + T_E$ within each replication. If the calculations in Section 11.4.3, as illustrated in Example 11.19, indicate that $R - R_0$ additional replications are needed beyond the initial number, R_0, then an alternative is to increase run length $(T_0 + T_E)$ in the same proportion (R/R_0) to a new run length $(R/R_0)(T_0 + T_E)$. Thus, additional data will be deleted, from time 0 to time $(R/R_0)T_0$, and more data will be used to compute the point estimates, as illustrated by Figure 11.6. An advantage of increasing total run length per replication and deleting a fixed proportion $[T_0/(T_0 + T_E)]$ of the total run length is that any residual bias in the point estimator should be further reduced by the additional deletion of data at the beginning of the run. A possible disadvantage of the method is that in order to continue the simulation of all R replications [from time $T_0 + T_E$ to time $(R/R_0)(T_0 + T_E)$],

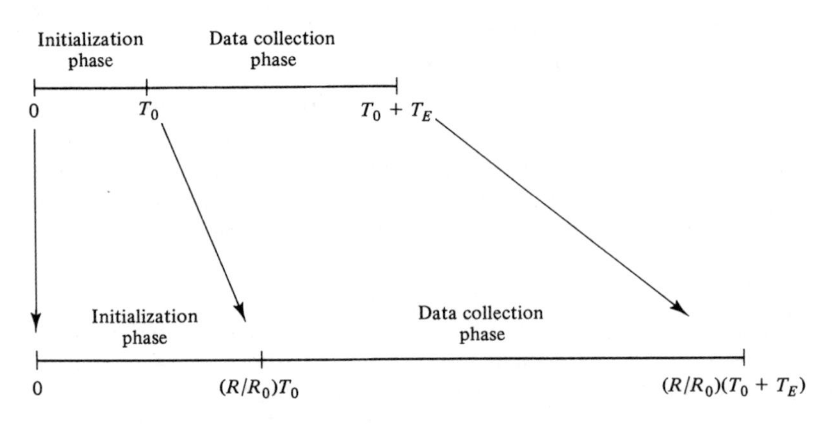

Figure 11.6. Increasing runlength to achieve specified accuracy.

it is necessary to have saved the state of the model at time $T_0 + T_E$ and to be able to restart the model and run it for the additional required time. Otherwise, the simulations would have to be rerun from time 0, which could be excessively expensive for a complex model. Most simulation languages have a restart capability to facilitate intermittent stopping and restarting of a model.

EXAMPLE 11.20

In Example 11.19, suppose that run length was to be increased to achieve the desired accuracy of $\epsilon = 2$ customers. Since $R/R_0 = 19/10 = 1.9$, the run length should be almost doubled to $(R/R_0)(T_0 + T_E) = 1.9(15{,}000) = 28{,}500$ minutes. The data collected from time 0 to time $(R/R_0)T_0 = 1.9(2000) = 3800$ minutes would be deleted, and the data from time 3800 to time 28,500 used to compute new point estimates and confidence intervals.

11.5.4. BATCH MEANS FOR INTERVAL ESTIMATION IN STEADY-STATE SIMULATIONS

One disadvantage of the replication method is that data must be deleted on every replication, and in one sense, deleted data are wasted data, or at least lost information. The method of batch means is based on one long run (versus numerous shorter ones) in which data need be deleted only once. The raw output data are placed in a few (say, 5 to 10) large batches, and the analyst works with these few batch means as if they were independent. As discussed in Sections 11.1 and 11.3, these batch means are not independent; however, if the batch size is sufficiently large, successive batches will be approximately independent, and the bias B in the variance estimator, given by equation (11.16), will be approximately 1. Unfortunately, there is no widely accepted and relatively simple method for choosing an acceptable batch size. Law [1977] found that for a fixed total sample size, it was best to use a very small number of batches of longest possible length. He recommended using five batches. The batch means approach to confidence interval estimation is illustrated in the next example.

EXAMPLE 11.21

Reconsider the $M/G/1$ simulation of Example 11.2. Suppose that the simulator desires to estimate mean queue length, L_Q, by a 95% confidence interval. To illustrate the method of batch means, assume that one run of the model has been made, over the time interval [0, 5000], with the results as shown under replication 1 of Table 11.2. The five observations, namely the five batch means $Y_{11} = 3.61$, $Y_{12} = 3.21$, $Y_{13} = 2.18$, $Y_{14} = 6.92$, and $Y_{15} = 2.82$ will be assumed to be statistically independent (although batch means are not independent, as discussed in Example 11.2). The batch size is 1000 minutes; that is, Y_{1j} is the sample mean over the jth subinterval of length 1000. With this assumption, the computations proceed in the usual way. The point estimate is

$$\bar{Y} = \frac{1}{5} \sum_{j=1}^{5} Y_{1j} = 3.75$$

the sum of squares is $\sum_{j=1}^{5} Y_{1j}^2 = 83.93$, and the sample variance is

$$S^2 = \frac{\sum_{j=1}^{5} Y_{1j}^2 - 5\bar{Y}^2}{4} = \frac{83.93 - 5(3.75)^2}{4} = 3.4$$

and the standard error of $\bar{Y}$ is

$$\text{s.e.}(\bar{Y}) = \frac{S}{\sqrt{5}} = 0.825$$

Thus, a 95% confidence interval is given by

$$\bar{Y} - t_{0.025,4}\text{s.e.}(\bar{Y}) \leq L_Q \leq \bar{Y} + t_{0.025,4}\text{s.e.}(\bar{Y})$$
$$1.46 = 3.75 - 2.78(0.825) \leq L_Q \leq 3.75 + 2.78(0.825) = 6.04$$

Therefore, if the assumptions of an approximately unbiased point estimator ($b \approx 0$) and approximately independent batch means are reasonable, the simulator can assert with 95% confidence that true mean queue length, L_Q, is between 1.46 and 6.04 customers. If these results are not sufficiently accurate for practical use, run length T_E would have to be increased to achieve greater accuracy.

The confidence interval computed in Example 11.21 can be criticized on several grounds. A check was not made on the assumption of no bias ($b \approx 0$) in the point estimator, or the assumption of independence between the batches. In Example 11.16, the bias due to the empty and idle initial condition was investigated for this $M/G/1$ queue, but in that instance many more data were assumed to be available (10 replications, each over a 15,000-minute run length versus one replication over a 5000-minute run length). It was concluded in Example 11.16 that at least two batches should be deleted at the beginning of each replication as a precaution against excessive bias in the point estimator. With the limited data in Example 11.21, it would be difficult to determine the presence, or lack, of bias. In addition, it appears to be a difficult problem to check the independence assumptions. Some work on this problem has been reported by Law and Carson [1979]. A summary of the available techniques for computing confidence intervals in steady-state simulations is given by Law and Kelton [1979, 1982a].

11.6. Summary

This chapter emphasizes the idea that a stochastic discrete-event simulation is a statistical experiment. Therefore, before sound conclusions can be drawn on the basis of the simulation generated output data, a proper statistical analysis is required. The purpose of the simulation experiment is assumed to be to make estimates of one of the performance measures of the system under study. The purpose of the statistical analysis is to acquire some assurance that these estimates are sufficiently accurate for the proposed use of the model.

A distinction was made between terminating simulations and steady-state simulations. It was seen that steady-state simulation output data are more difficult to analyze, because the simulator must address the problem of initial conditions and the choice of run length. Some suggestions were given regarding these problems, but unfortunately no simple, complete, and satisfactory solution exists. Nevertheless, simulators should be aware of the potential problems, and of the possible solutions—namely, deletion of data and increasing the run length. More advanced statistical techniques (not discussed in this text) are given in Crane and Lemoine [1977], Fishman [1978], Law and Carson [1979], and Law and Kelton [1979, 1982a, b].

The statistical accuracy of point estimators can be measured by a standard error estimate, or a confidence interval. The method of independent replications was emphasized. With this method, the simulator generates statistically independent observations, and thus standard statistical methods can be employed. For steady-state simulations, the method of batch means was also briefly discussed.

The main point is that simulation output data contain some amount of random variability, and without some assessment of its magnitude, the point estimates cannot be used with any degree of reliability.

REFERENCES

Crane, M. A., and A. J. Lemoine [1977], *An Introduction to the Regenerative Method for Simulation Analysis*, Lecture Notes in Control and Information Sciences, Vol. 4, Springer-Verlag, New York.

Fishman, G. S. [1978], *Principles of Discrete Event Simulation*, Wiley, New York.

GAFARIAN, A. V., C. J. ANCKER, JR., AND T. MORISAKU [1978], "Evaluation of Commonly Used Rules for Detecting "Steady State" in Computer Simulation," *Naval Research Logistics Quarterly*, Vol. 25, pp. 511–29.

KELTON, W. D. [1980], "The Startup Problem in Discrete-Event Simulation," Technical Report 80–1, Department of Industrial Engineering, University of Wisconsin, Madison.

KELTON, W. D., AND A. M. LAW [1980], "A New Approach for Dealing with the Start-up Problem in Discrete Event Simulation," Technical Report 80–2, Department of Industrial Engineering, University of Wisconsin, Madison.

LAW, A. M. [1977], "Confidence Intervals in Discrete Event Simulation: A Comparison of Replication and Batch Means," *Naval Research Logistics Quarterly*, Vol. 24, pp. 667–78.

LAW, A. M. [1980], "Statistical Analysis of the Output Data from Terminating Simulations," *Naval Research Logistics Quarterly*, Vol. 27, pp. 131–43.

LAW, A. M., AND J. S. CARSON [1979], "A Sequential Procedure for Determining the Length of a Steady-State Simulation," *Operations Research*, Vol. 27, pp. 1011–25.

LAW, A. M., AND W. D. KELTON [1979], "Confidence Intervals for Steady-State Simulations, I: A Survey of Fixed Sample Size Procedures," Technical Report 78–5, Department of Industrial Engineering, University of Wisconsin, Madison.

LAW, A. M., AND W. D. KELTON [1982a] "Confidence Intervals for Steady-State Simulations, II: A Survey of Sequential Procedures," *Management Science*, Vol. 28, No. 5, pp. 550–62.

LAW, A. M., AND W. D. KELTON [1982b], *Simulation Modeling and Analysis*, McGraw-Hill, New York.

SCHRUBEN, L. [1980], "Detecting Initialization Bias in Simulation Output," *Operations Research*, Vol. 30, pp. 569–90.

EXERCISES

1. (a) Consider Example 2.1. Under what circumstances would it be appropriate to use a terminating simulation, versus a steady-state simulation, to analyze this system?
(b) Repeat part (a) for Example 2.2 (the Able–Baker carhop problem).
(c) Repeat part (a) for Example 2.3.
(d) Repeat part (a) for Example 2.4.
(e) Repeat part (a) for Example 2.5.
(f) Repeat part (a) for Example 2.6.
(g) Repeat part (a) for Example 3.4 (the dump truck problem).

2. Suppose that in Example 11.16 the simulator decided to investigate the bias using batch means over a batching interval of 2000 minutes. By definition, a batch mean for the interval $[(j-1)2000, j(2000)]$ is defined by

$$Y_j = \frac{1}{2000} \int_{(j-1)2000}^{j(2000)} L_Q(t)\, dt$$

(a) Show algebraically that such a batch mean can be obtained from two adjacent batch means over the two halves of the interval.

(b) Compute the seven averaged batch means for the intervals [0, 2000), [2000, 4000), . . . for the $M/G/1$ simulation. Use the data ($\bar{Y}_{.j}$) in Table 11.6 (ignoring $Y_{.15} = 8.76$).

(c) Draw plots of the type in Figures 11.4 and 11.5. Does it still appear that deletion of the data over [0, 2000) (the first "new" batch mean) is sufficient to remove most of the point estimator bias?

3. Suppose in Example 11.16 that the simulator could only afford to run 5 independent replications (instead of 10). Use the batch means in Table 11.5 for replications 1 to 5 to compute a 95% confidence interval for mean queue length L_Q. Investigate deletion of initial data. Compare the results using 5 replications to those using 10 replications.

4. In Example 11.15, suppose that management desired 95% confidence (instead of 90%) in the estimate of mean system time w, and the accuracy desired was $\epsilon = 0.4$ minute.

(a) Using the same initial sample of size $R_0 = 4$ (given in Table 11.1), determine the required total sample size.

(b) Assume that the additional $R - R_0$ replications were made. Use the results in the first column of Table 11.3 ($r = 1, 2, \ldots, R_0, \ldots, R$; $T_E = 2$ hours) to compute a 95% confidence interval for w. Compare the half-length to the criteria $\epsilon = 0.4$ minute.

5. Simulate the dump truck problem in Example 3.4. At first make the run length $T_E = 40$ hours. Make four independent replications. Compute a 90% confidence interval for mean cycle time, where a cycle time for a given truck is the time between its successive arrivals to the loader. Investigate the effect of different initial conditions (all trucks initially at the loader queue, versus all at the scale, versus all traveling, versus the trucks distributed throughout the system in some manner).

6. Consider an (M, L) inventory system, in which the procurement quantity, Q, is defined by

$$Q = \begin{cases} M - I & \text{if } I < L \\ 0 & \text{if } I \geq L \end{cases}$$

where I is the level of inventory on hand plus on order at the end of a month, M is the maximum inventory level, and L is the reorder point. Since M and L are under management control, the pair (M, L) is called the inventory policy. Under certain conditions, the analytical solution of such a model is possible, but the computational effort may be prohibitive. Use simulation to investigate an (M, L) inventory system with the following properties. The inventory status is checked at the end of each month. Backordering is allowed at a cost of \$4 per item short per month. When an order arrives, it will first be used to relieve the backorder. The lead time is given by a uniform distribution on the interval (0.25, 1.25) months. Let the beginning inventory level stand at 50 units, with no orders outstanding. Let the holding cost be \$1 per unit in inventory per month. Assume that the inventory position is reviewed each month. If an order is placed, its cost is \$60 + \$5Q, where \$60 is the ordering cost and \$5 is the cost of each item. The time between demands is exponentially distributed with a mean of 1/15 month. The sizes of the demands follow the distribution:

Demand	*Probability*
1	1/2
2	1/4
3	1/8
4	1/8

(a) Make four independent replications, each of run length 100 months preceded by a 12-month initialization period, for the $(M, L) = (50, 30)$ policy. Estimate long-run mean monthly cost with a 90% confidence interval.

(b) Using the results of part (a), estimate the total number of replications needed to estimate mean monthly cost within \$5.

7. Reconsider Exercise 6, except that if the inventory level at a monthly review is zero or negative, a rush order for Q units is placed. The cost for a rush order is $\$120 + \$12Q$, where \$120 is the ordering cost and \$12 is the cost of each item. The lead time for a rush order is given by a uniform distribution on the interval (0.10, 0.25) months.

(a) Make four independent replications for the $(M, L) = (50, 30)$ policy, and estimate long-run mean monthly cost with a 90% confidence interval.

(b) Using the results of part (a), estimate the total number of replications needed to estimate mean monthly cost within \$5.

8. Suppose that the items in Exercise 6 are perishable, with a selling price given by the following data:

On the Shelf (Months)	*Selling Price*
0–1	\$10
1–2	5
>2	0

Thus, any item that has been on the shelf greater than 2 months cannot be sold. The age is determined at the time the demand occurs. If an item is outdated, it is discarded and the next item is brought forward. Simulate the system for 100 months.

(a) Make four independent replications for the $(M, L) = (50, 30)$ policy, and estimate long-run mean monthly cost with a 90% confidence interval.

(b) Using the results of part (a), estimate the total number of replications needed to estimate mean monthly cost within \$5.

At first, assume that all the items in the beginning inventory are fresh. Is this a good assumption? What effect does this "all-fresh" assumption have on the estimates of long-run mean monthly cost? What can be done to improve these estimates? Carry out a complete analysis.

9. Consider the following inventory system:

(a) Whenever the inventory level falls to or below 10 units, an order is placed. Only one order can be outstanding at a time.

(b) The size of each order is Q (where the "optimal" Q is to be determined in Exercise 5 of Chapter 12). Maintaining an inventory costs \$0.50 per day per item in inventory. Placing an order results in a fixed cost of \$10.00.
(c) Lead time is distributed in accordance with a discrete uniform distribution between zero and 5 days.
(d) If a demand occurs during a period when the inventory level is zero, the sale is lost at a cost of \$2.00 per unit.
(e) The number of customers each day is given by the following distribution:

Number of Customers per Day	*Probability*
0	0.23
1	0.41
2	0.22
3	0.14

(f) The demand on the part of each customer is Poisson distributed with a mean of 3 units.
(g) For simplicity, assume that all demands occur at 12 noon and that all orders are placed immediately thereafter. Assume further that orders are received at 5:00 P.M., or after the demand that occurred on that day.

Consider the policy having $Q = 20$. Make five independent replications, each of length 100 days, and compute a 90% confidence interval for long-run mean daily cost. Investigate the effect of initial inventory level and/or existence of an outstanding order on the estimate of mean daily cost. Begin with an initial inventory of $Q + 10$, and no outstanding orders.

10. A store selling Mother's Day cards must decide 6 months in advance on the number of cards to stock. Reordering is not allowed. Cards cost \$0.25 and sell for \$0.60. Any cards not sold by Mother's Day go on sale for \$0.20 for 2 weeks. However, sales of the remaining cards is probabilistic in nature according to the following distribution:

32% of the time, all cards remaining get sold.
40% of the time, 80% of all cards remaining are sold.
28% of the time, 60% of all cards remaining are sold.

Any cards left after 2 weeks are sold for \$0.10. The card shop owner is not sure how many cards can be sold, but thinks it is somewhere (i.e., uniformly distributed) between 200 and 400. Suppose that the card shop owner decides to order 300 cards. Estimate the expected total profit with an error of at most \$5.00. (*Hint:* Make three or four initial replications. Use these data to estimate the total sample size needed. Each replication consists of one Mother's Day.)

11. A very large mining operation has decided to control the inventory of high-pressure piping by a periodic review, order up to M policy, where M is a target level. The annual demand for this piping is normally distributed with a mean of 600 and a variance of 800. This demand occurs fairly uniformly over the year. The lead time for resupply is Erlang distributed of order $k = 2$ with a mean of 2 months.

The cost of each unit is \$400. The inventory carrying charge, as a proportion of item cost on an annual basis, is expected to fluctuate normally about a mean of 0.25 (simple interest) with a standard deviation of 0.01. The cost of making a review and placing an order is \$200, and the cost of a backorder is estimated to be \$100 per unit backordered. Suppose that the inventory level is reviewed every 2 months. Under simplified assumptions, a model for this situation was analyzed in Example 6.9. Use the target level, M, determined in Example 6.9.

(a) Make five independent replications, each of run length 100 months, to estimate long-run mean monthly cost by means of a 90% confidence interval. How much does it differ from that given by the model in Example 6.9?

(b) Investigate the effects of initial conditions. Determine an appropriate number of monthly observations to delete to reduce initialization bias to a negligible level.

12. Consider some number, say N, of $M/M/1$ queues in series. The $M/M/1$ queue, described in Section 5.5, has Poisson arrivals at some rate λ customers per hour, exponentially distributed service times with mean $1/\mu$, and a single server. (Recall that Poisson arrivals means that interarrival times are exponentially distributed.) By $M/M/1$ queues in series, it is meant that upon completion of service at a given server, a customer joins a waiting line for the next server. The system can be shown as follows:

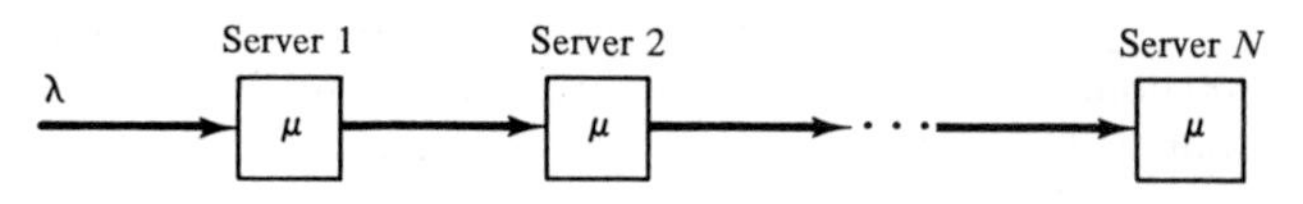

All service times are exponentially distributed with mean $1/\mu$, and the capacity of each waiting line is assumed to be unlimited. Assume that $\lambda = 8$ customers per hour, and $1/\mu = 0.1$ hour. The measure of performance is response time, which is defined to be the total time a customer is in the system.

(a) By making appropriate simulation runs, compare the initialization bias for $N = 1$ (i.e., one $M/M/1$ queue) to $N = 2$ (i.e., two $M/M/1$ queues in series). Start each system with all servers idle and no customers present. The purpose of the simulation is to estimate mean response time.

(b) Investigate the initialization bias as a function of N, for $N = 1, 2, 3, 4$, and 5.

(c) Draw some general conclusions concerning initialization bias for "large" queueing systems when at time 0 the system is assumed to be empty and idle.

13. Jobs enter a job shop in random fashion according to a Poisson process at overall rate 2 every 8-hour day. The jobs are of four types. They flow from work station to work station in a fixed order depending on type, as shown below. The proportions of each type are also shown.

Type	*Flow through Stations*	*Proportion*
1	1 2 3 4	0.4
2	1 3 4	0.3
3	2 4 3	0.2
4	1 4	0.1

Processing times per job at each station depend on type, but all times are (approximately) normally distributed with mean and s.d. (in hours) as follows:

	Station			
Type	1	2	3	4
1	(20, 3)	(30, 5)	(75, 4)	(20, 3)
2	(18, 2)	—	(60, 5)	(10, 1)
3	—	(20, 2)	(50, 8)	(10, 1)
4	(30, 5)	—	—	(15, 2)

Station i will have c_i workers ($i = 1, 2, 3, 4$). Each job occupies one worker at a station for the duration of a processing time. All jobs are processed on a first in, first out basis, and all queues for waiting jobs are assumed to have unlimited capacity. Simulate the system for 800 hours, preceded by a 200-hour initialization period. Assume that $c_1 = 8$, $c_2 = 8$, $c_3 = 20$, $c_4 = 7$. Based on $R = 5$ replications, compute a 97.5% confidence interval for average worker utilization at each of the four stations. Also compute a 95% confidence interval for mean total response time for each job type, where a total response time is the total time that a job spends in the shop.

14. Change Exercise 13 to give priority at each station to the jobs by type. Type 1 jobs have priority over type 2, type 2 over type 3, and type 3 over type 4. Use a run length of 800 hours, an initialization period of 200 hours, and $R = 5$ replications. Compute four 97.5% confidence intervals for mean total response time by type. Also run the model without priorities and compute the same confidence intervals. Discuss the trade-offs when using first in, first out versus a priority system.

15. Consider a single server queue with Poisson arrivals at rate $\lambda = 10.82$ per minute, and normally distributed service times with a mean of 5.1 seconds and a variance of 0.98^2 seconds^2. It is desired to estimate the mean time in the system for a customer who upon arrival finds i other customers in the system; i.e., to estimate

$$w_i = E(W \mid N = i) \qquad \text{for } i = 0, 1, 2, \ldots$$

where W is a typical system time, and N is the number of customers found by an arrival. For example, w_0 is the mean system time for those customers who find the system empty, w_1 is the mean system time for those customers who find one other customer present upon arrival, and so on. The estimate $\hat{w}_i$ of w_i will be a sample mean of system times taken over all arrivals who find i in the system. Plot $\hat{w}_i$ vs i. Hypothesize and attempt to verify a relation between w_i and i.

(a) Simulate for a 10 hour period with empty and idle initial conditions.

(b) Simulate for a 10 hour period after an initialization of one hour. Are there observable differences in the results of (a) and (b)?

(c) Repeat parts (a) and (b) with service times exponentially distributed with mean 5.1 seconds.

(d) Repeat parts (a) and (b) with deterministic service times equal to 5.1 seconds.

(e) Determine the number of replications needed to estimate $w_0, w_1, \ldots, w_6$ with a standard error for each of at most 3 seconds. Repeat parts (a–d) using this number of replications.

16. At Smalltown U. there is one terminal for student use located across campus from the computer center. At 2:00 A.M. one night six students arrive at the terminal to complete an assignment. A student uses the terminal for 10 ± 8 minutes, then leaves to go to the computer center to pick up a listing. There is a 25% chance that the run will be OK and the student will go to sleep. If it is not OK, the student returns to the terminal and waits until it becomes free. The roundtrip from terminal to computer center and back takes 30 ± 5 minutes. The computer becomes inaccessible at 5:00 A.M. Estimate the probability, p, that at least five of the six students will finish their assignment in the 3 hour period. First, make $R = 10$ replications and compute a 95% confidence interval for p. Next determine the number of replications needed to estimate p within $\pm .02$ and make these replications. Recompute the 95% confidence interval for p.

17. Four workers are evenly spaced along a conveyor belt. Items needing processing arrive according to a Poisson process at rate 2 per minute. Processing time is exponentially distributed with a mean of 1.6 minutes. If a worker becomes idle, then he or she takes the first item to come by on the conveyor. If a worker is busy when an item comes by, that item moves down the conveyor to the next worker, taking 20 seconds between two successive workers. When a worker finishes processing an item, the item leaves the system. If an item passes by the last worker, it is recirculated on a loop conveyor and will return to the first worker after 5 minutes.

Management is interested in having a balanced workload; that is, management would like worker utilizations to be equal. Let ρ_i be the long-run utilization of worker i, and let ρ be the average utilization of all workers. Thus, $\rho = (\rho_1 + \rho_2 + \rho_3 + \rho_4)/4$. Using queueing theory, ρ can be estimated by $\rho = \lambda/c\mu$ where $\lambda = 2$ arrivals per minute, $c = 4$ servers and $1/\mu = 1.6$ minutes is the mean service time. Thus, $\rho = \lambda/c\mu = (2/4)1.6 = 0.8$, so on the average, a worker will be busy 80% of the time.

(a) Make 5 independent replications, each of run length 40 hours preceded by a one hour initialization period. Compute 95% confidence intervals for ρ_1 and ρ_4. Draw conclusions concerning workload balance.

(b) Based on the same 5 replications, test the hypothesis $H_0\colon \rho_1 = 0.8$ at a level of significance $\alpha = 0.05$. If a difference of $\pm .05$ is important to detect, determine the probability that such a deviation is detected. In addition, if it is desired to detect such a deviation with probability at least 0.9, determine the sample size needed to do so.

(c) Repeat (b) for $H_0\colon \rho_4 = 0.8$.

(d) Based on the results from (a–c), draw conclusions for management about the balancing of workloads.

COMPARISON AND EVALUATION OF ALTERNATIVE SYSTEM DESIGNS

One of the most important uses of simulation is in the comparison of alternative system designs. Chapter 11 dealt with the accurate estimation of a measure of performance of one system. This chapter discusses a few of the many statistical methods that can be used to compare two or more system designs on the basis of some performance measure. Since the observations of the response variable contain random variation, statistical analysis is needed to determine whether any observed differences are due to differences in design, or merely to the random fluctuation inherent in the models.

The comparison of two system designs is computationally easier than the simultaneous comparison of more than two system designs. Section 12.1 discusses the case of two system designs, using two possible statistical techniques: independent sampling and correlated sampling. Correlated sampling is also known as the common random number technique; simply put, the same random numbers are used to simulate both alternative system designs. If implemented correctly, correlated sampling usually reduces the variance of the estimated difference of the performance measures and thus can provide, for a given sample size, more precise estimates of the mean difference than can independent sampling. Section 12.2 extends the statistical techniques of Section 12.1 to the comparison of more than two system designs, using the Bonferroni approach to confidence interval estimation. For comparison and evaluation of a larger number of alternative system designs, a class of linear statistical models, known as experimental design models, is discussed in Section 12.3. These experimental

12

design models can be especially useful when a large number of system designs are under consideration. They also allow the experimenter to judge the effect of various input parameters on the performance measure.

12.1. Comparison of Two System Designs

Suppose that a simulator desires to compare two possible configurations of a system. In a queueing system, perhaps two possible queue disciplines, or two possible sets of servers, are to be compared. In an inventory system, perhaps two possible ordering policies will be compared. A job shop may have many possible scheduling rules; a production system may have in-process inventory buffers of various capacities. Many other examples of alternative system designs can be provided.

The method of replication will be used to analyze the output data. The mean performance measure for system i will be denoted by $\theta_i (i = 1, 2)$. If it is a steady-state simulation, it will be assumed that deletion of data, or other appropriate techniques, have been used to assure that the point estimators are approximately unbiased estimators of the mean performance measures, θ_i. The goal of the simulation experiments is to obtain point and interval estimates of the difference in mean performance, namely $\theta_1 - \theta_2$. Three methods of computing a confidence interval for $\theta_1 - \theta_2$ will be discussed. But first an example and a general framework will be given.

EXAMPLE 12.1

A vehicle safety inspection station performs three jobs: (1) brake check, (2) headlight check, and (3) steering check. The present system has three stalls in parallel; that is, a vehicle enters a stall, where one attendant makes all three inspections. The present system is illustrated in Figure 12.1(a). Based on data from the existing system, it has been assumed that arrivals occur completely at random (i.e., according to a Poisson process) at an average rate of 9.5 per hour, and that the time for a brake check, a headlight check, and a steering check are normally distributed with means 6.5, 6, and 5.5 minutes, respectively, all having standard deviations of approximately 0.5 minute. There is no limit on the queue of waiting vehicles.

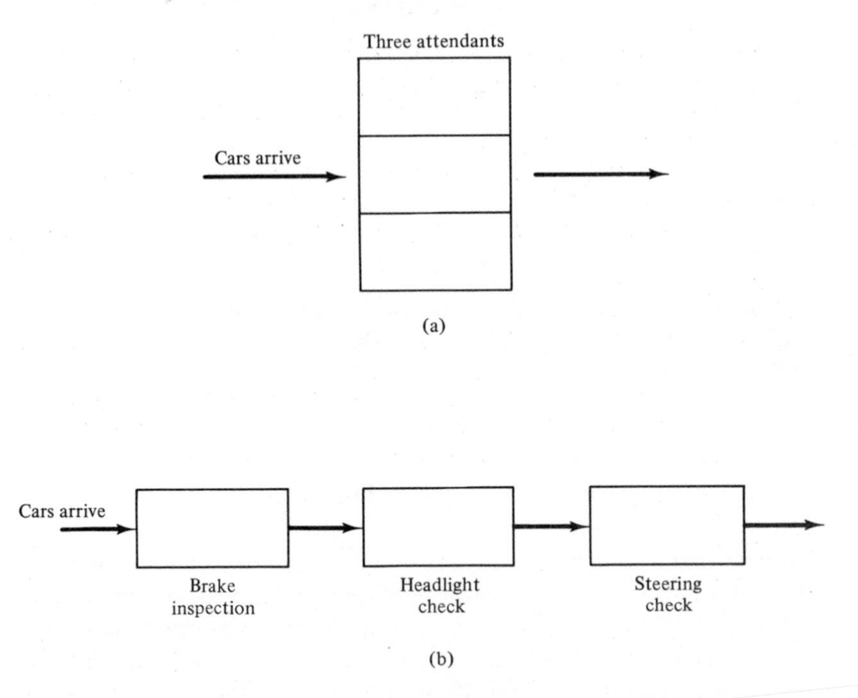

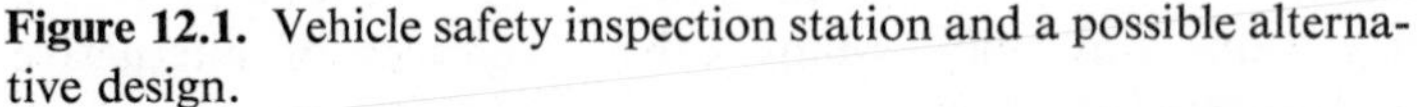

Figure 12.1. Vehicle safety inspection station and a possible alternative design.

An alternative system design is shown in Figure 12.1(b). Each attendant will specialize in a single task, and each vehicle will pass through three work stations in series. No space is allowed for vehicles between the brake and headlight check, or between the headlight and steering check. Therefore, a vehicle in the brake or headlight check must move to the next attendant, while a vehicle in the steering check must exit before the next vehicle can move ahead. Due to the increased specialization of the inspectors, it is anticipated that mean inspection times for each type of check will decrease by 10% to 5.85, 5.4, and 4.95 minutes, respectively, for the brake, headlight, and steering inspections. The Safety Inspection Council has decided to compare the two systems on the basis of mean response time per vehicle, where a response time is defined as the total time from a vehicle arrival until its departure from the system.

When comparing two systems, such as those in Example 12.1, the simulator must decide on a run length $T_E^{(i)}$ for each model ($i = 1, 2$), and a number of replications R_i to be made of each model. From replication r of system i, the simulator obtains an estimate Y_{ri} of the mean performance measure, θ_i. In Example 12.1, Y_{ri} would be the average response time observed during replication r for system i ($r = 1, \ldots, R_i$; $i = 1, 2$). The data, together with the two summary measures, the sample means $\bar{Y}_{\cdot i}$ and the sample variances S_i^2, are exhibited in Table 12.1. Assuming that the estimators Y_{ri} are (at least approximately) unbiased, it follows that

$$\theta_1 = E(Y_{r1}), \quad r = 1, \ldots, R_1; \qquad \theta_2 = E(Y_{r2}), \quad r = 1, \ldots, R_2$$

In Example 12.1, since the Safety Inspection Council is interested in a comparison of the two system designs, the simulator decides to compute a confidence interval for $\theta_1 - \theta_2$, the difference between the two mean performance measures. The confidence interval is used to answer two questions: (1) How large is the mean difference, and how accurate is the estimator of mean difference? (2) Is there a significant difference between the two systems? This second question will lead to one of three possible conclusions:

1. If the confidence interval (c.i.) for $\theta_1 - \theta_2$ is totally to the left of zero, as shown in Figure 12.2(a), there is strong evidence for the hypothesis that $\theta_1 - \theta_2 < 0$, or equivalently $\theta_1 < \theta_2$.

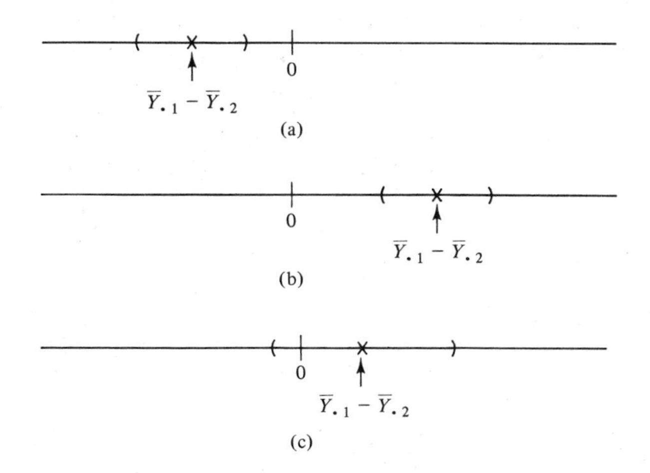

Figure 12.2. Three possible confidence intervals when comparing two systems.

In Example 12.1, $\theta_1 < \theta_2$ means that the mean response time for system 1 (the original system) is smaller than for system 2 (the alternative system).

2. If the c.i. for $\theta_1 - \theta_2$ is totally to the right of zero, as shown in Figure 12.2(b), there is strong evidence that $\theta_1 - \theta_2 > 0$, or equivalently, $\theta_1 > \theta_2$.

In Example 12.1, $\theta_1 > \theta_2$ can be interpreted as system 2 being better than system 1, in the sense that system 2 has smaller mean response time.

3. If the c.i. for $\theta_1 - \theta_2$ contains zero, then, based on the data at hand, there is no strong statistical evidence that one system design is better than the other.

Table 12.1. SIMULATION OUTPUT DATA AND SUMMARY MEASURES WHEN COMPARING TWO SYSTEMS

System	Replication				Sample Mean	Sample Variance
	1	*2*	...	R_i		
1	Y_{11}	Y_{21}	...	$Y_{R_1 1}$	$\bar{Y}_{\cdot 1}$	S_1^2
2	Y_{12}	Y_{22}	...	$Y_{R_2 2}$	$\bar{Y}_{\cdot 2}$	S_2^2

Some statistics textbooks say that the weak conclusion that $\theta_1 = \theta_2$ can be drawn, but such statements can be misleading. A "weak" conclusion is often no conclusion at all. Most likely, if enough additional data were collected (i.e., R_i increased), the c.i. would possibly shift, and definitely shrink in length, until conclusion 1 or 2 would be drawn. In addition to one of these three conclusions, the confidence interval provides a measure of the accuracy of the estimator of $\theta_1 - \theta_2$.

A two-sided $100(1 - \alpha)\%$ c.i. for $\theta_1 - \theta_2$ will always be of the form

$$(\bar{Y}_{\cdot 1} - \bar{Y}_{\cdot 2}) \pm t_{\alpha/2,\nu}\text{s.e.}(\bar{Y}_{\cdot 1} - \bar{Y}_{\cdot 2}) \tag{12.1}$$

where $\bar{Y}_{\cdot i}$ is the sample mean performance measure for system i over all replications:

$$\bar{Y}_{\cdot i} = \frac{1}{R_i}\sum_{r=1}^{R_i} Y_{ri} \tag{12.2}$$

where ν is the degrees of freedom associated with the variance estimator, $t_{\alpha/2,\nu}$ is the $100(1 - \alpha/2)$ percentage point of a t distribution with ν degrees of freedom, and s.e.($\cdot$) represents the standard error of the specified point estimator. To obtain the standard error and the degrees of freedom, the analyst uses one of three statistical techniques. All three techniques assume that the basic data, Y_{ri}, of Table 12.1 are approximately normally distributed. This assumption is reasonable provided that each Y_{ri} is itself a sample mean of observations from replication r (which is indeed the situation in Example 12.1). [Note the similarity between inequality (12.1) and inequality (11.8).]

By design of the simulation experiment, Y_{r1} ($r = 1, \ldots, R_1$) are identically and independently distributed (i.i.d.) with mean θ_1 and variance σ_1^2 (say). Similarly, Y_{r2} ($r = 1, \ldots, R_2$) are i.i.d. with mean θ_2 and variance σ_2^2 (say). The three techniques for computing the confidence interval in inequality (12.1), which are based on three different sets of assumptions, are discussed in the following subsections.

12.1.1. Independent Sampling with Equal Variances

Independent sampling means that different and independent random number streams will be used to simulate the two systems. This implies that all the observations of simulated system 1, namely $\{Y_{r1}, r = 1, \ldots, R_1\}$ are statistically independent of all the observations of simulated system 2, namely $\{Y_{r2}, r = 1, \ldots, R_2\}$. By equation (12.2) and the independence of the replications, the variance of the sample mean, $\bar{Y}_{\cdot i}$, is given by

$$\text{var}(\bar{Y}_{\cdot i}) = \frac{\text{var}(Y_{ri})}{R_i} = \frac{\sigma_i^2}{R_i}, \qquad i = 1, 2$$

When using independent sampling, $\bar{Y}_{\cdot 1}$ and $\bar{Y}_{\cdot 2}$ are statistically independent; hence,

$$\begin{aligned}\text{var}(\bar{Y}_{\cdot 1} - \bar{Y}_{\cdot 2}) &= \text{var}(\bar{Y}_{\cdot 1}) + \text{var}(\bar{Y}_{\cdot 2}) \\ &= \frac{\sigma_1^2}{R_1} + \frac{\sigma_2^2}{R_2}\end{aligned} \tag{12.3}$$

In some cases it is reasonable to assume that the two variances are equal (but unknown in value); that is, $\sigma_1^2 = \sigma_2^2$. The data can be used to test the hypothesis of equal variances (by an F test); if rejected, the method of Section 12.1.2 must be used. In a steady-state simulation, the variance σ_i^2 decreases as the run length $T_E^{(i)}$ increases; therefore, it may be possible to adjust the two run lengths, $T_E^{(1)}$ and $T_E^{(2)}$, to achieve at least approximate equality of σ_1^2 and σ_2^2.

If it is reasonable to assume that $\sigma_1^2 = \sigma_2^2$ (approximately), a two-sample t confidence interval approach can be used. The point estimate of the mean performance difference is

$$\hat{\theta}_1 - \hat{\theta}_2 = \bar{Y}_{\cdot 1} - \bar{Y}_{\cdot 2} \tag{12.4}$$

with $\bar{Y}_{\cdot i}$ given by equation (12.2). Next compute the sample variance for sample i by

$$\begin{aligned}S_i^2 &= \frac{1}{R_i - 1}\sum_{r=1}^{R_i}(Y_{ri} - \bar{Y}_{\cdot i})^2 \\ &= (R_i - 1)^{-1}\left(\sum_{r=1}^{R_i} Y_{ri}^2 - R_i\bar{Y}_{\cdot i}^2\right)\end{aligned} \tag{12.5}$$

Note that S_i^2 is an unbiased estimator of the variance σ_i^2. Since by assumption $\sigma_1^2 = \sigma_2^2 = \sigma^2$ (say), a pooled estimate of σ^2 is obtained by

$$S_p^2 = \frac{(R_1 - 1)S_1^2 + (R_2 - 1)S_2^2}{R_1 + R_2 - 2}$$

which has $\nu = R_1 + R_2 - 2$ degrees of freedom. The c.i. for $\theta_1 - \theta_2$ is then given by expression (12.1) with the standard error computed by

$$\text{s.e.}(\bar{Y}_{\cdot 1} - \bar{Y}_{\cdot 2}) = S_p \sqrt{\frac{1}{R_1} + \frac{1}{R_2}} \tag{12.6}$$

[This standard error is an estimate of the standard deviation of the point estimate, which, by equation (12.3), is given by $\sigma\sqrt{1/R_1 + 1/R_2}$.]

In some cases, the simulator may have $R_1 = R_2$, in which case it is safe to use the c.i. in expression (12.1) with the standard error taken from equation (12.6) even if the variances (σ_1^2 and σ_2^2) are not equal. However, if the variances are unequal and the sample sizes differ, it has been shown that use of the two-sample t c.i. may yield invalid confidence intervals whose true probability of containing $\theta_1 - \theta_2$ is much less than $1 - \alpha$. Thus, if there is no evidence that $\sigma_1^2 = \sigma_2^2$, or if $R_1 \neq R_2$, the approximate procedure in the next subsection is recommended.

12.1.2. Independent Sampling with Unequal Variances

If $\sigma_1^2 \neq \sigma_2^2$, or if the assumption of equal variances cannot safely be made, an approximate $100(1 - \alpha)\%$ c.i. for $\theta_1 - \theta_2$ can be computed as follows. The point estimate and sample variances are computed by equations (12.4) and (12.5). The standard error of the point estimate is given by

$$\text{s.e.}(\bar{Y}_{\cdot 1} - \bar{Y}_{\cdot 2}) = \sqrt{\frac{S_1^2}{R_1} + \frac{S_2^2}{R_2}} \tag{12.7a}$$

with degrees of freedom, ν, approximated by the expression

$$\nu = \frac{(S_1^2/R_1 + S_2^2/R_2)^2}{[(S_1^2/R_1)^2/(R_1 + 1)] + [(S_2^2/R_2)^2/(R_2 + 1)]} - 2 \tag{12.7b}$$

The confidence interval is then given by expression (12.1) using the standard error of equation (12.7a).

12.1.3. Correlated Sampling, or Common Random Numbers

Correlated sampling means that, for each replication, the same random numbers are used to simulate both systems. Thus R_1 and R_2 must be equal, say

$R_1 = R_2 = R$. Thus, for each replication r, the two estimates, Y_{r1} and Y_{r2}, are no longer independent but rather are correlated. Since independent streams of random numbers are used on any two different replications, the pairs (Y_{r1}, Y_{r2}) are mutually independent. (For example, in Table 12.1, the observation Y_{11} is correlated with Y_{12}, but Y_{11} is independent of all other observations.) The purpose of using correlated sampling is to induce a positive correlation between Y_{r1} and Y_{r2} (for each r), and thus to achieve a variance reduction in the point estimator of mean difference, $\bar{Y}_{\cdot 1} - \bar{Y}_{\cdot 2}$. In general, this variance is given by

$$\begin{aligned} \text{var}(\bar{Y}_{\cdot 1} - \bar{Y}_{\cdot 2}) &= \text{var}(\bar{Y}_{\cdot 1}) + \text{var}(\bar{Y}_{\cdot 2}) - 2\,\text{cov}(\bar{Y}_{\cdot 1}, \bar{Y}_{\cdot 2}) \\ &= \frac{\sigma_1^2}{R} + \frac{\sigma_2^2}{R} - \frac{2\rho_{12}\sigma_1\sigma_2}{R} \end{aligned} \tag{12.8}$$

where ρ_{12} is the correlation between Y_{r1} and Y_{r2}. [By definition, $\rho_{12} = \text{cov}(Y_{r1}, Y_{r2})/\sigma_1\sigma_2$, which does not depend on r.]

Now compare the variance of $\bar{Y}_{\cdot 1} - \bar{Y}_{\cdot 2}$ when using correlated sampling [equation (12.8), call it V_{CORR}] to the variance when using independent sampling with equal sample sizes [equation (12.3) with $R_1 = R_2 = R$, call it V_{IND}]. It is seen that

$$V_{\text{CORR}} = V_{\text{IND}} - \frac{2\rho_{12}\sigma_1\sigma_2}{R} \tag{12.9}$$

If correlated sampling works as intended, the correlation ρ_{12} will be positive; hence, the second term on the right side of equation (12.9) will be positive and therefore

$$V_{\text{CORR}} < V_{\text{IND}}$$

That is, the variance of the point estimator will be smaller when using correlated sampling than when using independent sampling. A smaller variance (for the same sample size R) implies that the estimator based on correlated sampling is more accurate.

To compute a $100(1-\alpha)\%$ c.i. with correlated data, first compute the differences

$$D_r = Y_{r1} - Y_{r2} \tag{12.10}$$

which, by the definition of correlated sampling, are i.i.d.; then compute the sample mean difference by

$$\bar{D} = \frac{1}{R}\sum_{r=1}^{R} D_r \tag{12.11}$$

(Thus, it is seen that $\bar{D} = \bar{Y}_{\cdot 1} - \bar{Y}_{\cdot 2}$.) The sample variance of the differences

$\{D_r\}$ is computed by

$$S_D^2 = \frac{1}{R-1}\sum_{r=1}^{R}(D_r - \bar{D})^2$$
$$= (R-1)^{-1}\left(\sum_{r=1}^{R} D_r^2 - R\bar{D}^2\right) \tag{12.12}$$

which has degrees of freedom $\nu = R - 1$. The $100(1-\alpha)\%$ c.i. for $\theta_1 - \theta_2$ is given by expression (12.1) with standard error of $\bar{Y}_{\cdot 1} - \bar{Y}_{\cdot 2} = \bar{D}$ estimated by

$$\text{s.e.}(\bar{D}) = \text{s.e.}(\bar{Y}_{\cdot 1} - \bar{Y}_{\cdot 2}) = \frac{S_D}{\sqrt{R}} \tag{12.13}$$

Since $S_D/\sqrt{R}$ of equation (12.13) is an estimate of $\sqrt{V_{\text{CORR}}}$, and expression (12.6) or (12.7a) is an estimate of $\sqrt{V_{\text{IND}}}$, it can be seen that, when working properly (i.e. when $\rho_{12} > 0$), correlated sampling will generally produce a c.i. which is shorter for a given sample size than the c.i. produced by independent sampling.

For any problem, there are always many ways of implementing common random numbers. It is never enough to simply use the same seed on the random number generator(s). Each random number used in one model for some purpose must be used for the same purpose in the second model; that is, the use of the random numbers must be synchronized. For example, if the ith random number is used to generate a service time at work station 2 for the jth arrival in model 1, the ith random number should be used for the very same purpose in model 2. For queueing systems or service facilities, synchronization of the common random numbers guarantees that both systems face identical work loads—both systems face arrivals at the same instants of time, and these arrivals demand equal amounts of service. (The actual service times of a given arrival in the two models may not be equal, but may be proportional, if the server in one model is faster than the server in the other model.) For an inventory system when comparing two different ordering policies, synchronization guarantees that the two systems face identical demand for a given product. For production or reliability systems, synchronization guarantees that downtimes for a given machine will occur at exactly the same times, and will have identical durations, in the two models. On the other hand, if some aspect of one of the systems is totally different from the other system, synchronization may be inappropriate, or even impossible to achieve. In summary, those aspects of the two system designs which are sufficiently similar should be simulated with common random numbers in such a way that the two models "behave" similarly; but those aspects that are totally different should be simulated with independent random numbers.

Implementation of common random numbers is model dependent, but certain guidelines can be given that will make correlated sampling more likely to yield a positive correlation. The purpose of the guidelines is to ensure that synchronization occurs:

1. Dedicate a random number stream to a specific purpose, and use as many different streams as needed. (Different random number generators, or widely spaced seeds on the same generator, can be used to get two different, nonoverlapping streams.) In addition, assign independently chosen seeds to each stream at the beginning of each run. It is not sufficient to assign seeds at the beginning of the first run and then let the random number generator merely continue for the second and subsequent runs. If conducted in this manner, the first replication will be synchronized but subsequent replications may not be.
2. For systems (or subsystems) with external arrivals: As each entity enters the system, the next interarrival time is generated, and then immediately all random variables (such as service times, order sizes, etc.) needed by the arriving entity and identical in both models are generated in a fixed order and stored as attributes of the entity, to be used later as needed. Apply guideline 1; that is, dedicate one random number stream to these external arrivals and all their attributes.
3. For systems having an entity performing given activities in a cyclic or repeating fashion, assign a random number stream to this entity. (Example: a machine that cycles between two states: up–down–up–down– $\cdots$. Use a dedicated random number stream to generate the runtimes and downtimes.)
4. If synchronization is not possible, or it is inappropriate, for some part of the two models, use independent streams of random numbers for this subset of random variates.

Unfortunately, there is no guarantee that correlated sampling will always induce a positive correlation between comparable runs of the two models. It is known that if, for each input random variate X, the estimators Y_{r1} and Y_{r2} are increasing functions of the random variate X (or both are decreasing functions of X), then ρ_{12} will be positive. The intuitive idea is that both models (i.e., both Y_{r1} and Y_{r2}) respond in the same direction to each input random variate, and this results in positive correlation. This increasing or decreasing nature of the response variables (called monotonicity) with respect to the input random variables is known to hold for certain queueing systems (such as the $G/G/c$ queues), when the response variable is customer delay, so some evidence exists that common random numbers is a worthwhile technique for queueing simulations. (For simple queues, customer delay is an increasing function of service times and a decreasing function of interarrival times.) Wright and Ramsay [1979] reported, however, a negative correlation for certain inventory simulations. In summary, the guidelines should be followed and some reasonable notion that the response variable of interest is a monotonic function of the random input variables should be evident.

EXAMPLE 12.1 (CONTINUED)

The two inspection systems shown in Figure 12.1 will be compared using both independent sampling and correlated sampling, in order to illustrate the greater accuracy of correlated sampling when it works.

Each vehicle arriving to be inspected has four input random variates associated with it:

$$
\begin{aligned}
A_n &= \text{interarrival time between vehicles } n \text{ and } n+1 \\
S_n^{(1)} &= \text{brake inspection time for vehicle } n \text{ in model 1} \\
S_n^{(2)} &= \text{headlight inspection time for vehicle } n \text{ in model 1} \\
S_n^{(3)} &= \text{steering inspection time for vehicle } n \text{ in model 1}
\end{aligned}
$$

For model 2 (of the proposed system), mean service times are decreased by 10%. When using independent sampling, different values of service (and interarrival) times would be generated for model 2. But when using correlated sampling, the random number generator must be used in such a way that exactly the same values are generated for $A_1, A_2, A_3, \ldots$; for service times, $S_n^{(i)}$ $(i = 1, 2, 3)$ could be generated for model 1 and $S_n^{(i)} - 0.1E(S_n^{(i)})$ used in model 2. Alternatively, since normal random variates are usually generated by first generating a standard normal variate and then using equation (8.30), the service times for a brake inspection could be generated by

$$E(S_n^{(1)}) + \sigma Z_n^{(1)} \tag{12.14}$$

where $Z_n^{(1)}$ is a standard normal variate, $\sigma = 0.5$ minute, $E(S_n^{(1)}) = 6.5$ minutes for model 1 and $E(S_n^{(1)}) = 5.85$ minutes for model 2; the other two inspection times would be generated in a similar fashion. To implement (synchronized) common random numbers, the simulator would generate identical $Z_n^{(i)}$ sequences $(i = 1, 2, 3;\ n = 1, 2, \ldots)$ in both models and then use the appropriate version of Equation (12.14) to generate the inspection times. For the synchronized runs, the service times for a vehicle were generated at the instant of arrival (by guideline 2) and stored as an attribute of the vehicle, to be used as needed. Runs were also made with nonsynchronized common random numbers, in which case one random number stream was used as needed.

Table 12.2 gives the average response time for each of $R = 10$ replications, each of run length $T_E = 16$ hours. It was assumed that two cars were present at time 0, waiting to be inspected. Model 2 was run using independent random numbers (column 2I) and common random numbers without synchronization (column 2C*) and with synchronization (column 2C). The purpose of the simulation is to estimate mean difference in response times for the two systems.

For the two independent runs (1 and 2I), it was assumed that the variances were not necessarily equal, so the method of Section 12.1.2 was applied. Sample variances and the standard error were computed by equations (12.5) and (12.7a), yielding

$$S_1^2 = 118.9, \qquad S_{2I}^2 = 244.3$$

and

$$\text{s.e.}(\bar{Y}_{\cdot 1} - \bar{Y}_{\cdot 2I}) = \sqrt{\frac{118.9}{10} + \frac{244.3}{10}} = 6.03$$

Table 12.2. COMPARISON OF SYSTEM DESIGNS FOR THE VEHICLE SAFETY INSPECTION SYSTEM

	Average Response Time for Model				*Observed Differences*	
Replication	*1*	*2I*	*2C**	*2C*	$D_{1,2C^*}$	$D_{1,2C}$
1	29.59	51.62	56.47	29.55	−26.88	0.04
2	23.49	51.91	33.34	24.26	−9.85	−0.77
3	25.68	45.27	35.82	26.03	−10.14	−0.35
4	41.09	30.85	34.29	42.64	6.80	−1.55
5	33.84	56.15	39.07	32.45	−5.23	1.39
6	39.57	28.82	32.07	37.91	7.50	1.66
7	37.04	41.30	51.64	36.48	−14.60	0.56
8	40.20	73.06	41.41	41.24	−1.21	−1.04
9	61.82	23.00	48.29	60.59	13.53	1.23
10	44.00	28.44	22.44	41.49	21.56	2.51
Sample mean	37.63	43.04			−1.85	0.37
Sample variance	118.90	244.33			208.94	1.74
Standard error	6.03				4.57	0.42

with degrees of freedom, ν, equal to 17, as given by equation (12.7b). The point estimate is $\bar{Y}_{\cdot 1} - \bar{Y}_{\cdot 2I} = -5.4$ minutes, and a 95% c.i. [expression (12.1)] is given by

$$-5.4 \pm 2.11(6.03)$$

or

$$-18.1 \leq \theta_1 - \theta_2 \leq 7.3 \qquad (12.15)$$

The 95% confidence interval in inequality (12.15) contains zero, which indicates that there is no strong evidence that the observed difference of −5.4 minutes is due to anything other than random variation in the output data. Thus, if the simulator had decided to use independent sampling, no strong conclusion is possible. In addition, the estimate of $\theta_1 - \theta_2$ is quite inaccurate.

For the two sets of correlated runs (1 and 2C*, and 1 and 2C), the observations are paired and analyzed as given in equations (12.10) through (12.13). The point estimate when not synchronizing the random numbers is given by equation (12.11) as

$$\bar{D} = -1.9 \text{ minutes}$$

the sample variance by $S_D^2 = 208.9$ (with $\nu = 9$ degrees of freedom), and the standard error by s.e.$(\bar{D}) = 4.6$. Thus, a 95% c.i. for the true mean difference in response times, as given by expression (12.1), is

$$-1.9 \pm 2.26(4.6)$$

or

$$-12.3 < \theta_1 - \theta_2 < 8.5 \qquad (12.16)$$

Again no strong conclusion is possible since the confidence interval contains zero. Note, however, that the estimate of $\theta_1 - \theta_2$ is slightly more accurate than that in inequality (12.15).

When complete synchronization of the random numbers was used in run 2C, the point estimate of the mean difference in response times was

$$\bar{D} = 0.4 \text{ minute}$$

the sample variance was $S_{\bar{D}}^2 = 1.7$ (with $\nu = 9$ degrees of freedom), and the standard error was s.e.$(\bar{D}) = 0.4$. A 95% c.i. for the true mean difference is given by

$$0.40 \pm 2.26(0.40)$$

or

$$-0.50 < \theta_1 - \theta_2 < 1.30 \tag{12.17}$$

The confidence interval in inequality (12.17) again contains zero, but it is considerably shorter than the previous two intervals. This greater accuracy in the estimation of $\theta_1 - \theta_2$ is due to the use of synchronized common random numbers. The short length of the interval in inequality (12.17) suggests that the true difference, $\theta_1 - \theta_2$, is close to zero.

As seen by comparing the confidence intervals in inequalities (12.15), (12.16), and (12.17), the width of the confidence interval is reduced by 18% when using non-synchronized common random numbers, and by 93% when using common random numbers with full synchronization. Comparing the estimated variance of $\bar{D}$ using synchronized common random numbers to the variance of $\bar{Y}_{\cdot 1} - \bar{Y}_{\cdot 2}$ using independent sampling shows a variance reduction of 99.5%, which means that to achieve accuracy comparable to that achieved by correlated sampling, a total of approximately $R = 2090$ independent replications would have had to have been made.

The next few examples show how common random numbers can be implemented in other contexts.

EXAMPLE 12.2 (THE DUMP TRUCK PROBLEM, REVISITED)

Consider Example 3.3 (the dump truck problem), shown in Figure 3.7. Each of the trucks repeatedly goes through three activities—loading, weighing, and traveling. Assume that there are eight trucks, and at time 0 all eight are at the loaders. Weighing time per truck on the single scale is uniformly distributed between 1 and 9 minutes; and travel time per truck is exponentially distributed with mean 85 minutes. An unlimited queue is allowed before the loader(s) and before the scale. All trucks can be traveling at the same time. Management desires to compare one fast loader to the two slower loaders currently being used. Each of the slow loaders can fill a truck in 1 to 27 minutes, uniformly distributed. The new fast loader can fill a truck in 1 to 19 minutes, uniformly distributed. The basis for comparison is mean system response time, where a response time is defined as the duration of time from a truck arrival at the loader queue to that truck's departure from the scale.

To implement synchronized common random numbers, a separate and distinct random number stream was assigned to each of the eight trucks. At the beginning of each replication (i.e., at time 0), a new and independently chosen set of eight seeds was specified, one seed for each random number stream. Thus, weighing times and travel times for each truck were identical in both models, and the loading time for a given truck's ith visit to the fast loader was proportional to the loading time in the original system (with two slow loaders). Implementation of common random numbers without synchronization (e.g., using one random number stream to generate all loading, weighing, and travel times as needed) would likely lead to a given random number being used to generate a loading time in model 1 but a travel time in model 2, or vice versa, and from that point on the use of a random number would most likely be different in the two models.

Six replications of each model were run, each of run length $T_E = 40$ hours. The results are shown in Table 12.3. Both independent sampling and correlated sampling were used, to illustrate the advantage of correlated sampling. The first column (labeled model 1) contains the observed average system response time for the existing system

Table 12.3. COMPARISON OF SYSTEM DESIGNS FOR THE DUMP TRUCK PROBLEM

	Average System Response Time for Model			
Replication	*1* (*2 Loaders*)	*2I* (*1 Loader*)	*2C* (*1 Loader*)	*Differences,* $D_{1,2C}$
1	21.38	29.01	24.30	−2.92
2	24.06	24.70	27.13	−3.07
3	21.39	26.85	23.04	−1.65
4	21.90	24.49	23.15	−1.25
5	23.55	27.18	26.75	−3.20
6	22.36	26.91	25.62	−3.26
Sample mean	22.44	26.52		−2.56
Sample variance	1.28	2.86		0.767
Sample standard deviation	1.13	1.69		0.876

with two loaders. The columns labeled 2I and 2C are for the alternative design having one loader; the independent sampling results are in 2I, and the correlated sampling results are in the column labeled 2C. The rightmost column, labeled $D_{1,2C}$, lists the observed differences between the runs of model 1 and model 2C.

For independent sampling assuming unequal variances, the following summary statistics were computed using equations (12.2), (12.5), (12.7a), (12.7b), and (12.1) and the data (in columns 1 and 2I) in Table 12.3:

Point estimate: $\bar{Y}_{.1} - \bar{Y}_{.2I} = 22.44 - 26.52 = -4.08$ minutes
Sample variances: $S_1^2 = 1.28,\ S_{2I}^2 = 2.86$
Standard error: $\text{s.e.}(\bar{Y}_{.1} - \bar{Y}_{.2}) = (S_1^2/R_1 + S_{2I}^2/R_2)^{1/2} = 0.831$
Degrees of freedom: $\nu = 10.22 \approx 10$
95% c.i. for $\theta_1 - \theta_2$: $-4.08 \pm 2.23(0.831)$ or -4.08 ± 1.851
$$-5.93 \leq \theta_1 - \theta_2 \leq -2.23$$

For correlated sampling, implemented by the use of synchronized common random numbers, the following summary statistics were computed using equations (12.11),

(12.12), (12.13) and (12.1), plus the data (in columns 1 and 2C) in Table 12.3:

Point estimate: $\bar{D} = -2.56$ minutes
Sample variance: $S_D^2 = 0.767$
Standard error: $\text{s.e.}(\bar{D}) = S_D/\sqrt{R} = 0.876/\sqrt{6} = 0.358$
Degrees of freedom: $\nu = R - 1 = 5$
95% c.i. for $\theta_1 - \theta_2$: $-2.56 \pm 2.57(0.358)$ or -2.56 ± 0.919
$-3.48 \leq \theta_1 - \theta_2 \leq -1.641$

By comparing the c.i. widths, it is seen that the use of correlated sampling with synchronization reduced c.i. width by 50%. Equivalently, if equal accuracy were desired, independent sampling would require approximately four times as many observations as would correlated sampling, or approximately 24 replications of each model instead of six.

To illustrate how correlated sampling can fail when not implemented correctly, consider the dump truck model again. There were eight trucks, and each was assigned its own random number stream. For each of the six replications, eight seeds were randomly chosen, one seed for each random number stream. Therefore, a total of 48 (6 times 8) seeds were specified for the correct implementation of common random numbers. When the authors first developed and ran this example, eight seeds were specified at the beginning of the first replication only; on the remaining five replications, the random numbers were generated by continuing down the eight original streams. Since comparable replications with one and two loaders required different numbers of random variables, only the first replications of the two models were synchronized. The remaining five were not synchronized. The resulting confidence intervals for $\theta_1 - \theta_2$ when using correlated sampling were approximately the same length, or only slightly shorter, than the confidence intervals when using independent sampling. Therefore, correlated sampling is quite likely to fail in reducing the standard error of the estimated difference unless proper care is taken to guarantee synchronization of the random number streams on all replications.

EXAMPLE 12.3

In Example 2.5, two policies for replacing bearings in a milling machine were compared. In the "pencil and paper" simulation of Tables 2.24 and 2.25, common random numbers were used for bearing lifetimes but independent random numbers were used for repairperson delay times. The bearing-life distribution, assumed discrete in Example 2.5 (Table 2.22), is now more realistically assumed to be continuous on the range 950 to 1950 hours, with the first column of Table 2.22 giving the midpoint of 10 intervals of width 100 hours. The repairperson delay-time distribution of Table 2.23 is also assumed continuous, in the range 2.5 to 17.5 minutes, with interval midpoints as given in the first column. The probabilities of each interval are given in the second column of Tables 2.22 and 2.23.

The two models were run using correlated sampling and, for illustrative purposes, independent sampling, each for $R = 10$ replications with a run length of $T_E = 3$ years. The purpose was to estimate the difference in mean total costs per year, with the cost data given in Example 2.5. The estimated total cost over the 3-year period for the two policies is given in Table 12.4. Policy 1 was to replace each bearing as it failed. Policy

Table 12.4. TOTAL COSTS FOR ALTERNATIVE DESIGNS OF BEARING REPLACEMENT PROBLEM

Replication	*Total Cost over 3 Years for Policy*			*Difference in Total Cost*
r	2	*1I*	*1C*	$D_{1C,2}$
1	6,670	8,505	8,778	2,108
2	6,380	8,764	8,580	2,200
3	6,501	8,978	8,904	2,403
4	6,762	8,960	9,006	2,244
5	6,877	9,440	9,100	2,223
6	6,659	8,764	8,968	2,310
7	6,716	8,787	9,175	2,459
8	7,104	8,977	9,699	2,595
9	6,612	9,145	8,806	2,194
10	6,589	8,680	8,978	2,388
Sample mean	6,687	8,900		2,312
Sample variance	40,178	69,047		21,820

2 was to replace all three bearings whenever one bearing failed. Policy 2 was run first, and then policy 1 was run using independent sampling (column 1I), and using correlated sampling (column 1C). The 95% confidence intervals for mean cost difference are as follows:

Independent sampling: $\$2213 \pm 219$
Correlated sampling: $\$2312 \pm 106$

(The computation of these confidence intervals is left as an exercise for the reader.)

Note that the confidence interval for mean cost difference when using correlated sampling is approximately 50% of the length of the confidence interval based on independent sampling. Therefore, for the same computer costs (i.e., for $R = 10$ replications), correlated sampling produces estimates which are twice as accurate in this example. If correlated sampling were used, the simulator could conclude with 95% confidence that the mean cost difference between the two policies is between \$2206 and \$2418 over a 3-year period.

12.2. Comparison of Several System Designs

Suppose that a simulator desires to compare some number, say K, of alternative system designs. The comparison will be made on the basis of some specified performance measure, θ_i, of system i, for $i = 1, 2, \ldots, K$. Many different staitistical procedures have been developed which can be used to analyze simulation data and draw statistically sound inferences concerning the parameters θ_i. These procedures can be classified as being fixed-sample-size procedures, or sequential sampling (or multistage) procedures. In the first type, a predetermined sample size (i.e., run length and number of replications) is used to draw inferences via hypothesis tests and/or confidence intervals. Examples of fixed-sample-size procedures include the interval estimation of a mean performance measure [as in expression (11.8), Section 11.3], and the interval estimation of the difference in mean performance measures of two systems [as by expression (12.1) in Section 12.1]. Advantages of fixed-sample-size procedures include the known or easily estimated cost in terms of computer time before running the experiments. When computer time is limited, or when conducting a pilot study, a fixed-sample-size procedure may be appropriate. In some cases, clearly inferior system designs may be ruled out at this early stage. A major disadvantage is that no strong conclusion may be possible. For example, the confidence interval may be too wide for practical use, since the width is an indication of the accuracy of the point estimator. A hypothesis test may lead to a failure to reject the null hypothesis, a weak conclusion in general, meaning that there is no strong evidence one way or the other about the truth or falsity of the null hypothesis.

A sequential sampling scheme is one in which more and more data are collected until an estimator with a prespecified accuracy is achieved, or until one of several alternative hypotheses is selected, with the probability of correct selection being larger than a prespecified value. A two-stage (or multistage) procedure is one in which an initial sample is used to estimate how many additional observations are needed to draw conclusions with a specified accuracy. An example of a two-stage procedure for estimating the performance measure of a single system was given in Section 11.4.3.

The proper procedure to use depends on the goal of the simulator. Some possible goals include:

1. Estimation of each parameter, θ_i
2. Comparison of each performance measure, θ_i, to a control, θ_1 (where θ_1 may represent the mean performance of an existing system)
3. All possible comparisons, $\theta_i - \theta_j$, for $i \neq j$
4. Selection of the best θ_i (largest or smallest)

The first three goals will be achieved by the construction of confidence intervals. The number of such confidence intervals is $C = K$, $C = K - 1$, and $C = K(K - 1)/2$, respectively. The fourth goal requires the use of a type of statistical procedure known as a multiple ranking and selection procedure, which is beyond the scope of this text. Procedures to achieve these and other goals are discussed by Kleijnen [1975, Chaps. II and V], who also discusses their relative merit and disadvantages. Law and Kelton [1982] also discuss those selection procedures most relevant to simulation. The next subsection presents a fixed-sample-size procedure which can be used to meet goals 1 to 3 and which is applicable in a wide range of circumstances.

12.2.1. Bonferroni Approach to Multiple Comparisons

Suppose that a total of C confidence intervals are computed, and that the ith interval has confidence coefficient $1 - \alpha_i$. Then the ith confidence interval is a statement that a given interval contains the parameter (or difference of two parameters) being estimated. This statement, call it S_i, may be true or false for a given set of data, but the procedure leading to the interval is designed so that statement S_i will be true with probability $1 - \alpha_i$. When it is desired to make statements about several parameters simultaneously, as in goals 1 to 3, the analyst would like to have high confidence that all statements are true simultaneously. The Bonferroni inequality states that

$$P(\text{all statements } S_i \text{ are true}, i = 1, \ldots, C) \geq 1 - \sum_{j=1}^{C} \alpha_j = 1 - \alpha_E \qquad (12.18)$$

where $\alpha_E = \sum_{j=1}^{C} \alpha_j$ is called the overall error probability. Expression (12.18) can be restated as

$$P(\text{one or more statements } S_i \text{ is false}, i = 1, \ldots, C) \leq \alpha_E$$

or equivalently,

$$P(\text{one or more of the } C \text{ confidence intervals does not contain the parameter being estimated}) \leq \alpha_E$$

Thus, α_E provides an upper bound on the probability of a false conclusion. When conducting an experiment making C comparisons, first select the overall error probability, say $\alpha_E = 0.05$ or 0.10. The individual α_j may be chosen to be equal ($\alpha_j = \alpha_E/C$), or unequal, as desired. The smaller the value of α_j, the wider the jth confidence interval will be. For example, if two 95% c.i.'s ($\alpha_1 = \alpha_2 = 0.05$) are constructed, the overall confidence level will be 90% or greater ($\alpha_E = \alpha_1 + \alpha_2 = 0.10$). If ten 95% c.i.'s are constructed ($\alpha_i = 0.05, i = 1, \ldots, 10$), the resulting overall confidence level could be as low as 50% ($\alpha_E = \sum_{i=1}^{10} \alpha_i = 0.50$), which is far too low for practical use. To guarantee an overall confidence level of 95%, for example, when 10 comparisons are being made, then one solution is to construct ten 99.5% confidence intervals for the parameters (or differences) of interest.

The Bonferroni approach to multiple confidence intervals is based on expression (12.18). A major advantage is that it holds whether the models for the alternative designs are run with independent sampling, or with common random numbers.

The major disadvantage to the Bonferroni approach when making a large number of comparisons is the increased width of each individual interval. For example, for a given set of data and a large sample size, a 99.5% c.i. will be $z_{0.0025}/z_{0.025} = 2.807/1.96 = 1.43$ times longer than a 95% c.i. For small sample sizes, say for a sample size of 5, a 99.5% c.i. will be $t_{0.0025,4}/t_{0.025,4} = 5.598/2.776 = 1.99$ times longer than a 95% c.i. The width of a c.i. is a measure of the accuracy of the estimate. For these reasons, it is recommended that the Bonferroni approach only be used when a small number of comparisons are being made. Ten or so comparisons appears to be the practical upper limit.

Corresponding to goals 1 to 3 above, there are at least three possible ways of using the Bonferroni inequality (12.18) when comparing K alternative system designs:

1. (*Individual c.i.'s*): Construct a $100(1 - \alpha_i)\%$ c.i. for parameter θ_i by expression (11.8), in which case the number of intervals is $C = K$. If independent sampling were used, the K c.i.'s would be mutually independent, and thus the overall confidence level would be $(1 - \alpha_1) \times (1 - \alpha_2) \times \cdots \times (1 - \alpha_C)$, which is larger (but not much larger) than the right side of expression (12.18). It is generally believed that correlated sampling (common random numbers) will make the true overall confidence level larger than the right side of expression (12.18), and usually larger than when using independent sampling. The right side of expression (12.18) can be thought of as giving the worst case (i.e., the lowest possible overall confidence level).
2. (*Comparison to an existing system*): Compare all designs to one specific design, usually to an existing system. That is, construct a $100(1 - \alpha_i)\%$ c.i. for $\theta_i - \theta_1$ ($i = 2, 3, \ldots, K$) using expression (12.1). (System 1 with performance measure θ_1 is assumed to be the existing system). In this case, the number of intervals is $C = K - 1$.
3. (*All possible comparisons*): Compare all designs to each other. That is, for any two system designs $i \neq j$, construct a $100(1 - \alpha_{ij})\%$ c.i. for $\theta_i - \theta_j$. With K designs, the number of confidence intervals computed is $C = \binom{K}{2} = K(K - 1)/2$. The overall confidence coefficient would be bounded below by $1 - \alpha_E = 1 - \sum\sum_{i \neq j} \alpha_{ij}$ [which follows by expression (12.18)].

EXAMPLE 12.4

Reconsider the vehicle inspection station of Example 12.1. Suppose that the construction of enough space to hold one waiting car is being considered. The alternative system designs are:

1. Existing system (parallel stations)
2. No space between stations in series
3. One space between brake and headlight inspection only
4. One space between headlight and steering inspection only

Design 2 was compared to the existing setup in Example 12.1. Designs 2, 3, and 4 are series queues as shown in Figure 12.1(b), the only difference being the number and/or location of a waiting space between two successive inspections. The arrival process and the inspection times are as given in Example 12.1. The basis for comparison will be mean response time, θ_i, for system i, where a response time is the total time it takes for a car to get through the system. Confidence intervals for $\theta_2 - \theta_1$, $\theta_3 - \theta_1$, and $\theta_4 - \theta_1$ will be constructed having an overall confidence level of 95%. The run length T_E has now been set at 40 hours (instead of the 16 hours used in Example 12.1), and the number of replications R of each model is 10. Common random numbers will be used in all models, but this does not affect the overall confidence level, because, as mentioned, the Bonferroni inequality (12.18) holds regardless of the statistical independence or dependence of the data.

Since the overall error probability is $\alpha_E = 0.05$ and $C = 3$ confidence intervals are to be constructed, let $\alpha_i = 0.05/3 = 0.0167$ for $i = 2, 3, 4$. Then use expression (12.1) (with proper modifications) to construct $C = 3$ confidence intervals with $\alpha = \alpha_i = 0.0167$ and degrees of freedom $\nu = 10 - 1 = 9$. The standard error is computed by equation (12.13), since common random numbers are being used. The output data Y_{ri} are displayed in Table 12.5; Y_{ri} is the sample mean response time for replication r

Table 12.5. ANALYSIS OF OUTPUT DATA FOR VEHICLE INSPECTION SYSTEM WHEN USING CORRELATED SAMPLING

Replication,	*Average Response Time for System Design*				*Observed Difference*		
	1,	*2,*	*3,*	*4,*			
r	Y_{r1}	Y_{r2}	Y_{r3}	Y_{r4}	D_{r2}	D_{r3}	D_{r4}
1	63.72	63.06	57.74	62.63	0.66	5.98	1.09
2	32.24	31.78	29.65	31.56	0.46	2.59	0.68
3	40.28	40.32	36.52	39.87	−0.04	3.76	0.41
4	36.94	37.71	35.71	37.35	−0.77	1.23	−0.41
5	36.29	36.79	33.81	36.65	−0.50	2.48	−0.36
6	56.94	57.93	51.54	57.15	−0.99	5.40	−0.21
7	34.10	33.39	31.39	33.30	0.71	2.71	0.80
8	63.36	62.92	57.24	62.21	0.44	6.12	1.15
9	49.29	47.67	42.63	47.46	1.62	6.66	1.83
10	87.20	80.79	67.27	79.60	6.41	19.93	7.60
Sample mean, $\bar{D}_{\cdot i}$					0.80	5.686	1.258
Sample standard deviation, S_{D_i}					2.12	5.338	2.340
Sample variance, $S^2_{D_i}$					4.498	28.498	5.489
Standard error, $S_{D_i}/\sqrt{R}$					0.671	1.688	0.741

on system i ($r = 1, \ldots, 10$; $i = 1, 2, 3, 4$). The differences $D_{ri} = Y_{r1} - Y_{ri}$ are also shown, together with the sample mean differences, $\bar{D}_{\cdot i}$, averaged over all replications as in equation (12.11), the sample variances $S^2_{D_i}$, and standard error. By expression (12.1), the three confidence intervals, with overall confidence coefficient at least $1 - \alpha_E$, are given by

$$\bar{D}_{\cdot i} - t_{\alpha_i/2, R-1}\,\text{s.e.}(\bar{D}_{\cdot i}) \le \theta_1 - \theta_i \le \bar{D}_{\cdot i} + t_{\alpha_i/2, R-1}\,\text{s.e.}(\bar{D}_{\cdot i}), \qquad i = 2, 3, 4$$

The value of $t_{\alpha_i/2,R-1} = t_{0.0083,9} = 2.97$ is obtained from Table A.4 by interpolation. For these data, with 95% confidence, it is estimated that

$$-1.19 \leq \theta_1 - \theta_2 \leq 2.79$$
$$0.67 \leq \theta_1 - \theta_3 \leq 10.71$$
$$-0.94 \leq \theta_1 - \theta_4 \leq 3.46$$

The simulator has high confidence (at least 95%) that all three confidence statements are correct. Note that the c.i. for $\theta_1 - \theta_2$ again contains zero; thus, there is no significant difference between design 1 and design 2, a conclusion which confirms the previous results in Example 12.1. The c.i. for $\theta_1 - \theta_3$ lies completely above zero, which provides strong evidence that $\theta_1 - \theta_3 > 0$, that is, that design 3 is better than design 1 because its mean response time is smaller. Since the c.i. for $\theta_1 - \theta_4$ contains zero, there is no significant difference between designs 1 and 4.

If the simulator now decides that it would be desirable to compare designs 3 and 4, more simulation runs would be necessary because it is not legitimate to decide which confidence intervals to compute after the data have been examined. On the other hand, if the simulator had decided to compute all possible confidence intervals (and had made this decision before collecting the data, Y_{ri}), the number of confidence intervals would have been $C = 6$ and the three c.i.'s above would have been $t_{0.0042,9}/t_{0.0083,9} \approx 3.32/2.97 = 1.12$ times (or 12%) longer. There is always a trade-off between the number of intervals (C) and the width of each interval. The simulator should carefully consider the possible conclusions before running the simulation experiments, and decide on those runs and analyses that will provide the most useful information. In particular, the number of confidence intervals computed should be as small as possible, preferably 10 or less.

For purposes of illustration, 10 replications of each of the four designs were run using independent sampling (i.e., different random numbers for all runs). The results are presented in Table 12.6, together with sample means ($\bar{Y}_{\cdot i}$), sample standard deviations (S_i), and sample variances (S_i^2), plus the observed difference of sample means ($\bar{Y}_{\cdot 1} - \bar{Y}_{\cdot i}$) and the standard error (s.e.) of the observed difference. It is observed that all three confidence intervals for $\theta_1 - \theta_i$ ($i = 2, 3, 4$) will contain zero. Therefore, no strong conclusion is possible based on these data and this sample size. By contrast, a sample size of ten was sufficient, when using correlated sampling, to provide strong evidence that design 3 is superior to design 1.

Note the large increase in standard error of the estimated difference when using independent sampling versus using common random numbers. These standard errors are compared in Table 12.7. In addition, a careful examination of Tables 12.5 and 12.6 illustrates the superiority of correlated sampling. In Table 12.5, in all 10 replications, system design 3 has a smaller average response time than that for system design 1. Comparing replications 1 and 2 in Table 12.5, it can be seen that a random number stream that leads to high congestion and large response times in system design 1, as does the first replication, produces results of similar magnitude across all four system designs. Similarly, when system design 1 exhibits relatively low congestion and low response times, as in the second replication, all system designs produce relatively low average response times. This similarity of results on each replication is due, of course, to the use of common random numbers within replications. By contrast, for independent sampling, Table 12.6 shows no such similarity across system designs. In only 5

Table 12.6. ANALYSIS OF OUTPUT DATA FOR VEHICLE INSPECTION SYSTEM WHEN USING INDEPENDENT SAMPLING

Replication, r	*Average Response Time for System Design* 1, Y_{r1}	2, Y_{r2}	3, Y_{r3}	4, Y_{r4}
1	63.72	59.37	52.00	59.03
2	32.24	50.06	47.04	49.97
3	40.28	60.63	53.21	60.18
4	36.94	46.36	40.88	45.44
5	36.29	68.87	50.85	66.65
6	56.94	66.44	60.42	66.03
7	34.10	27.51	26.70	27.45
8	63.36	47.98	40.12	47.50
9	49.29	29.92	28.59	29.84
10	87.20	47.14	41.62	46.44
Sample mean $\bar{Y}_{\cdot i}$	50.04	50.43	44.14	49.85
S_i	17.70	13.98	10.76	13.64
S_i^2	313.38	195.54	115.74	185.98
$\bar{Y}_{\cdot 1} - \bar{Y}_{\cdot i}$		−0.39	5.89	0.18
s.e.$(\bar{Y}_{\cdot 1} - \bar{Y}_{\cdot i})$		7.13	6.55	7.07

Table 12.7. COMPARISON OF STANDARD ERRORS FOR CORRELATED SAMPLING AND INDEPENDENT SAMPLING FOR THE VEHICLE INSPECTION PROBLEM

Difference in Sample Means	*Standard Error When Using:* *Correlated Sampling*	*Independent Sampling*	*Percentage Increase*
$\bar{Y}_{\cdot 1} - \bar{Y}_{\cdot 2}$	0.67	7.13	1064%
$\bar{Y}_{\cdot 1} - \bar{Y}_{\cdot 3}$	1.69	6.55	388%
$\bar{Y}_{\cdot 1} - \bar{Y}_{\cdot 4}$	0.74	7.07	955%

of the 10 replications is the average response time for system design 3 smaller than that for system design 1, although the average difference in response times across all 10 replications is approximately the same magnitude in each case: 5.69 minutes when using correlated sampling, and 5.89 minutes when using independent sampling. The greater variability of independent sampling is reflected also in the standard errors of the point estimates: ± 1.69 minutes for correlated sampling versus ± 6.55 minutes for independent sampling, an increase of 388%, as seen in Table 12.7. This example illustrates again the advantage of correlated sampling.

As stated previously, correlated sampling does not yield a variance reduction in all simulation models. It is recommended that a pilot study be undertaken and variances estimated to confirm (or possibly deny) the assumption that correlated sampling will reduce the variance (or standard error) of an estimated difference. The reader is referred to the discussion in Section 12.1.3.

Some of the exercises at the end of the chapter provide an opportunity to compare correlated and independent sampling, and to compute simultaneous confidence intervals using the Bonferroni approach.

12.3. Statistical Models for Estimating the Effect of Design Alternatives

This section provides an introduction to the statistical design of simulation experiments. As a complete discussion of relevant statistical designs would require a complete text in itself, the reader is advised to consult other texts for more complete accounts. Some standard texts include Hicks [1973], Box et al. [1978], and Montgomery [1976]. Kleijnen [1975] contains a fairly complete summary of experimental designs relevant to simulation. Many standard experimental designs have little or no importance for simulation experiments because they were developed for physical experiments in which the experimenter lacks complete control over the experiment, or is unable to collect data for certain combinations of factors. These problems can be avoided in simulation experiments, in which the analyst has complete control over experimental conditions. The terminology "statistical design" or "experimental design" should not be confused with the earlier usage of the term "design," which referred to "systems design."

12.3.1. Purposes of the Statistical Design of Experiments

The statistical design of experiments is a set of principles for designing and evaluating experiments so as to maximize the information gained from the experiment. In particular, the analyst may desire to quantify the effect of the decision variables on the response variable, and to determine whether a given decision variable has a significant effect on the response variable. In the statistical literature, the input variables to the system, such as the decision variables, the structural assumptions, and the parameters of the random variables, are called *factors*. Each possible value of a factor is called a *level* of the factor. A combination of factors all at a specified level is called a *treatment*. When a simulation is run with the same treatment but an independent stream of random numbers is used, it is said that an independent replication of the experiment has been made. The purpose of statistical design may now be restated as determining the effect of the various factors on the response variable.

Factors may be classified as *qualitative* or *quantitative*. Examples of qualitative factors include queue discipline, such as FIFO versus priority. A quantitative factor is one that can assume a numerical value, as, for example, the number of parallel servers, the arrival rate, and the ordering policy. In addition, certain factors are under management's control and can be changed at will (but perhaps only at great expense). These factors are collectively called decision variables, or policy variables. Examples include the number of parallel servers and the ordering policy. Other factors, such as the arrival rate for randomly arriving customers or the demand rate, cannot be controlled by management. Nevertheless, it sometimes may be desirable to evaluate the effects of both policy and nonpolicy variables on the response variables. For example, it may be of value to know the effect on total cost (the response variable) for a given ordering policy (a decision variable) if the demand rate (a nonpolicy variable) were to increase. Although a nonpolicy factor such as demand rate cannot be controlled by management in the real-world system, it can be controlled by the analyst in the simulation model. Thus, the model can be used to answer "what if" questions involving both policy and nonpolicy factors.

EXAMPLE 12.5

The hospital emergency room at Dead End, Georgia, currently maintains four physicians on duty, and incoming patients are classified as either NIA (need immediate attention) or CW (can wait). Assume that a doctor sees only one patient at a time and that each patient sees only one of the doctors at a time. Figure 12.3 shows a schematic of the system with patients, represented by circles, waiting in line according to the time of their arrival.

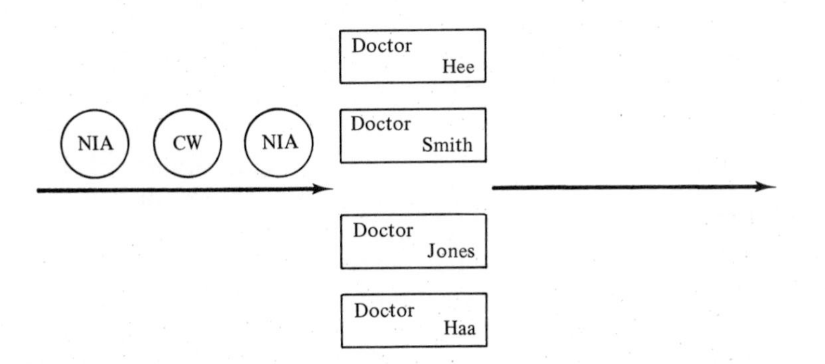

Figure 12.3. Schematic representation of the Dead End Emergency Room.

The response variable of interest is the delay, Y, of type NIA patients, delay meaning time in system before "service" begins. The objective of the model is to predict the effect on Y of two decision variables, D_1 and D_2, where D_1 is the queue discipline, or policy for deciding which patient to see next, and D_2 represents the number of physicians on duty. More explicitly, the alternatives for D_1 are FCFS (first come, first served), or NIA having highest priority (PR), or round robin (RR). (In a round-robin queue discipline, services occur in increments of some parameter δ, say $\delta = 2$ minutes. When a patient gets to the head of the waiting line, the patient receives $\delta = 2$ minutes of service and then joins the back of the line; the patient's remaining service time is reduced by $\delta = 2$ minutes. When a patient's remaining service time is reduced to less than $\delta = 2$ minutes, the patient will see a doctor once more for the remaining service time and then depart. New arrivals join the end of the queue.) The alternatives for D_2 are $D_2 = 4$ or $D_2 = 5$ doctors. Both D_1 and D_2 are examples of input variables, or factors, which are under a policymaker's control. D_1 is a qualitative factor and D_2 a quantitative one. In addition, the model has two other input variables, the arrival process of patients and the demands on the doctor's time made by patients (call it service times), which are usually not controllable by a policymaker. If the interarrival times, X_1, between successive patient arrivals, and the service times, X_2, of patients are generally unpredictable and thus have been modeled by probability distributions, as discussed in Chapters 4, 5, and 9, X_1 and X_2 are called random input variables. The analyst must decide which particular probability distribution is appropriate, and choose its parameters based on available data. If their effect on the response variable is important, the parameters of these probability distributions may be considered as factors (quantitative, nonpolicy factors). One purpose of the model may be to investigate the sensitivity of the response to certain of these nonpolicy factors. For example, what if the estimate of arrival rate used in the model was somewhat low? If the arrival rate is included as a factor, it can be set at two or more values and the sensitivity of the response variable to the arrival rate can be estimated.

The objective of a simulation study such as the Dead End Emergency Room may be (1) to predict the effect of a particular set of decisions D on the response Y, or (2) to select those decisions that have the most important effect on the response Y, or (3) to find the particular set of decisions which yields an optimal response.

When designing the experiments to attain one of these objectives, it is well to keep in mind several other important characteristics of simulation models. Even though certain of the input variables are not under the policymaker's control in terms of the system itself, the particular values of these random variables are under the analyst's control to the extent that the analyst specifies the random number seed and stream to be used. This source of random variation is deliberately introduced into the model to accurately represent system behavior. On the other hand, simulation experiments are not subject to the extraneous variation, such as measurement error, of physical experiments.

Because a simulation is conducted in a completely controlled environment, and in particular the analyst controls the sources of random variation, it is possible to replicate a simulation model under identical conditions, which means that:

1. The values of the decision variables are not changed, and
2. The probability distributions of the random variables are not changed, but
3. The particular random numbers generated are produced from an independent stream of random numbers.

EXAMPLE 12.5 (CONTINUED)

Suppose that the factors are set at these levels:

D_1 = FCFS, D_2 = 4 doctors
X_1 = exponential distribution with mean 5 minutes
X_2 = Erlang distribution of order 2 with mean 18 minutes

In addition, the run length is set at $T_E = 40$ hours. Two computer runs of the model would produce two independent replications under identical conditions provided that the two random number seeds and/or streams were chosen independently. The purpose of such replications is to accentuate the effect on the response Y of the decision variables D and to reduce and isolate the effect of the random variation X. The statistical techniques for estimating the effects of the factors are illustrated in the next two subsections.

12.3.2. Single-Factor Completely Randomized Experimental Designs

First consider a situation where a single factor D may affect the response variable Y. For a queueing system, the single factor might be the queue discipline, which could have three levels, such as FCFS or PRIORITY or round robin. Queue discipline is an example of a qualitative, policy factor. When there is only one factor having some number of levels, say k, the experiment is called a single-factor experiment. The effect of level j of the factor is called τ_j. If the replications of the model at each level and for different levels of the factor are based on independent streams of random numbers, the design is called a *completely randomized design*. The number of replications at each level of the factor do not have to be the same, but are chosen according to cost considerations and the desired power of statistical tests.

The statistical model for the analysis of the single-factor completely randomized experimental design with k treatment levels is

$$Y_{rj} = \mu + \tau_j + \epsilon_{rj} \tag{12.19}$$

($r = 1, 2, \ldots, R_j$; $j = 1, 2, \ldots, k$), where Y_{rj} is observation r of the response variable for level j of the factor, μ is called the overall mean effect, τ_j is the effect due to level j of the factor, ϵ_{rj} is a "random error" in observation r at level j, and R_j is the number of observations made at level j. The model of equation (12.19) essentially views the responses at level j, namely Y_{rj} ($r = 1, \ldots, R_j$), as varying about a mean level $\mu + \tau_j$ by some random amount ϵ_{rj}. In statistics, the terms ϵ_{rj} are called random error, although in this application to simulation data, ϵ_{rj} is merely the random variation of the response Y_{rj} about its mean $\mu + \tau_j$ and should not be thought of as error.

The random deviation terms ϵ_{rj} are assumed to be normally and independently distributed with mean zero and common variance σ^2. The parameters μ and τ_j are assumed to be fixed and to satisfy $\sum_{j=1}^{k} \tau_j = 0$. This model applies when the levels of the factor can be fixed by the analyst, and result in what is called the *fixed-effects* model, which is the model studied here. If the levels of the factor cannot be fixed but instead are chosen at random from some population, the τ_j are assumed to be normally distributed, resulting in the *random-effects* model [Hicks, 1973; Montgomery, 1976; Box et al., 1978].

The initial analysis of a single-factor fixed-effects completely randomized experiment consists of a statistical test of the hypothesis

$$H_0: \quad \tau_j = 0 \qquad (j = 1, 2, \ldots, k) \tag{12.20}$$

that the levels of the factor have no effect on the response. The applicable statistical test is a one-way analysis of variance (ANOVA); the test itself consists of computing an F-statistic and comparing its value to an appropriate critical value. [If there are only $k = 2$ levels of a single factor, the ANOVA is equivalent to the independent t-test in Section 12.1.1. The model in equation (12.19) is designed to handle the case $k > 2$.] If the hypothesis H_0 is not rejected, the analyst may conclude that the mean response is μ for all levels of the factor; that is, the factor has no significant effect on the response variable. If the hypothesis H_0 is rejected, the analyst has reason to believe that the levels of the factor do have some effect upon mean response. After performing an ANOVA and rejecting H_0, the analyst may want to know which means are significantly different, or whether levels 1 and 2 of the factor have significantly different means, for example. Questions about means after ANOVA can be answered by a variety of tests, including the range test, Scheffé's test, and the method of orthogonal contrasts [Hicks, 1973; Montgomery, 1976; Box et al., 1978].

The following example illustrates the analysis of the statistical model of equation (12.19).

Example 12.6

In the emergency room model, Example 12.5, assume that the number of doctors is fixed at 4, and the arrival and service processes are given and fixed. (That is, D_2, X_1, and X_2 are specified.) Then there is only one factor (namely, D_1) with $k = 3$ levels. The simulator decides to make $R_j = 3$ independent replications at each level $j = 1, 2, 3$, for a total of $R = 3(3) = 9$ replications. The statistical model then becomes

$$Y_{rj} = \mu + \tau_j + \epsilon_{rj}, \qquad j = 1, 2, 3, \quad r = 1, 2, 3$$

That is, Y_{11}, Y_{21}, Y_{31} are three independent observations of the response variable Y, the average delay of type NIA patients, when the first ($j = 1$) level of the factor or design alternative, namely, $D_1 =$ FCFS, is in effect; similarly for Y_{12}, Y_{22}, Y_{32}, and for Y_{13}, Y_{23}, Y_{33}.

The one-way ANOVA test using an F statistic of the hypothesis in equation (12.20) can be conducted by using a standard statistical package, such as SAS, SPSS, MINITAB, or BMDP, which are available on many large and small computer systems. For purposes of illustration, all the necessary formulas are given here.

The ANOVA test is based on partitioning the variability of the observed responses, Y_{rj}, into two components, one component due to the level of each factor and one due to the inherent variability of the process being simulated. First, place the observed responses, Y_{rj}, in a layout similar to Table 12.8, and compute the totals $T_{.j}$ and $T_{..}$, and sample means $\bar{Y}_{.j}$ of the jth level, and $\bar{Y}_{..}$,

Table 12.8. DATA LAYOUT FOR ONE-WAY ANOVA

Replication,	*Level j of the Single Factor*						
r	*1*	*2*	...	*j*	...	*k*	
1	Y_{11}	Y_{12}	...	Y_{1j}	...	Y_{1k}	
2	Y_{21}	Y_{22}	...	Y_{2j}	...	Y_{2k}	
.	.	.		.		.	
.	.	.		.		.	
.	.	.		.		.	
R_j	$Y_{R_1 1}$	$Y_{R_2 2}$	...	$Y_{R_j j}$	...	$Y_{R_k k}$	
Totals	$T._1$	$T._2$	...	$T._j$	...	$T._k$	$T..$
Means	$\bar{Y}._1$	$\bar{Y}._2$	...	$\bar{Y}._j$	...	$\bar{Y}._k$	$\bar{Y}..$

the overall sample mean or grand mean. Recall that the dot notation is used to indicate sums over the indicated subscript. For example, $T._j = \sum_{r=1}^{R_j} Y_{rj}$ is the sum of all responses for the jth treatment, and $T.. = \sum_{j=1}^{k} \sum_{r=1}^{R_j} Y_{rj} = \sum_{j=1}^{k} T._j$ is the grand sum of all responses. Letting $R = \sum_{j=1}^{k} R_j$ be the total number of replications, it can be shown that the grand sample mean $\bar{Y}..$ can be computed by any of the following expressions:

$$\bar{Y}.. = \frac{T..}{R} = \sum_{j=1}^{k} \frac{R_j \bar{Y}._j}{R} = \sum_{j=1}^{k} \sum_{r=1}^{R_j} \frac{Y_{rj}}{R}$$

The variation of the response variable about the overall sample mean can be written as

$$Y_{rj} - \bar{Y}.. = (\bar{Y}._j - \bar{Y}..) + (Y_{rj} - \bar{Y}._j) \tag{12.21}$$

which shows the variation to be composed of a component due to deviation of a treatment mean from the grand mean, plus deviation of each response from the sample mean response at its level. Squaring equation (12.21), summing over all r and j, and observing that the cross terms on the right-hand side sum to zero, the final result is

$$\sum_{j=1}^{k} \sum_{r=1}^{R_j} (Y_{rj} - \bar{Y}..)^2 = \sum_{j=1}^{k} R_j(\bar{Y}._j - \bar{Y}..)^2 + \sum_{j=1}^{k} \sum_{r=1}^{R_j} (Y_{rj} - \bar{Y}._j)^2 \tag{12.22}$$

or, more succinctly,

$$SS_{TOTAL} = SS_{TREAT} + SS_E$$

where SS stands for "sum of squares," SS_{TOTAL} is the total sum of squares, SS_{TREAT} is the sum of squares due to the treatments, and SS_E is called the error sum of squares. Note again, however, that the terminology for SS_E may be misleading, as the only "error" involved is the deviation of a single response (Y_{rj}) at level j from the sample mean response ($\bar{Y}._j$) at level j. Nevertheless, this terminology is adopted, as it is standard in statistics.

If the assumption of a common variance is correct, the mean square $MS_E = SS_E/(R - k)$ is an unbiased estimate of the variance σ^2 of the response variable Y; that is, $E(MS_E) = \sigma^2$. If, in addition, the hypothesis in equation (12.20) is true, the treatment mean square $MS_{TREAT} = SS_{TREAT}/(k - 1)$ is also an unbiased estimate of σ^2. In any case, MS_{TREAT} and MS_E are statistically independent. When H_0 is true, SS_{TREAT}/σ^2 and SS_E/σ^2 have chi-square distributions with $(k - 1)$ and $(R - k)$ degrees of freedom, respectively. The test statistic for testing the hypothesis in expression (12.20) is computed by

$$F = \frac{MS_{TREAT}}{MS_E} = \frac{SS_{TREAT}/(k - 1)}{SS_E/(R - k)} \tag{12.23}$$

When H_0 is true, this test statistic has an F distribution with $k - 1$ and $R - k$ degrees of freedom. The ANOVA test of the hypothesis H_0 is to reject H_0 if $F > F_{1-\alpha}$, and fail to reject H_0 if $F < F_{1-\alpha}$, where $1 - \alpha$ is the probability of an F statistic with $k - 1$ and $R - k$ degrees of freedom being below the critical value $F_{1-\alpha}$. (Percentage points of the F distribution with $\alpha = 0.05$ for various degrees of freedom are given in Table A.6.)

The rationale for the test is as follows: If H_0 is true, both mean squares are estimates of σ^2 and hence F would be expected to be close to $E(MS_{TREAT})/E(MS_E) = 1$ for $(1 - \alpha)$ proportion of the time. "Close to 1" is determined by $F < F_{1-\alpha}$. (Note that for $\alpha = 0.05$, all the critical values in Table A.6 are larger than 1.) It can be shown that if H_0 is false, $E(MS_{TREAT}) = \sigma^2 + a$, where a is a positive quantity, in which case F would tend to be considerably larger than 1 and hence H_0 would be rejected with high probability. To summarize, if H_0 is true, the probability of falsely rejecting H_0 is α (called "Type I error"); but if H_0 is false, the probability of rejecting H_0 is greater than α. In addition, if H_0 is indeed false, the probability of rejecting H_0 can be increased by increasing the sample size. When H_0 is rejected, there is strong evidence for a difference in mean response $\mu + \tau_1, \mu + \tau_2, \ldots$ due to the different levels of the factor. The one-way ANOVA is now illustrated for Example 12.6.

EXAMPLE 12.6 (CONTINUED)

As stated earlier, $R = 9$ replications were made, $R_j = 3$ at each of the $k = 3$ levels, of the emergency room model. The results, namely Y_{rj}, the average delay of all NIA patients for replication r and level j, are given in Table 12.9. The ANOVA is shown in Table 12.10. The sums of squares are computed by equation (12.22), or equivalently by these computational formulas:

$$SS_{TOTAL} = \sum_{j=1}^{k} \sum_{r=1}^{R_j} Y_{rj}^2 - R\bar{Y}_{..}^2$$

$$SS_{TREAT} = \sum_{j=1}^{k} R_j \bar{Y}_{.j}^2 - R\bar{Y}_{..}^2$$

$$SS_E = SS_{TOTAL} - SS_{TREAT}$$

The mean squares are computed by

$$MS_{TREAT} = \frac{SS_{TREAT}}{k-1}$$

$$MS_E = \frac{SS_E}{R-k}$$

and the test statistic, F, by equation (12.23). The results of these computations are shown in the ANOVA table of Table 12.10. The critical value for $\alpha = 0.05$ and 2 and

Table 12.9. DATA LAYOUT FOR EMERGENCY ROOM EXAMPLE FOR A SINGLE FACTOR

	Response Variable Y_{rj} at Level j of the Queue Discipline			
Replication, r	$j = 1$ FCFS	$j = 2$ PR	$j = 3$ RR	*Total Mean*
1	21.47	11.11	8.05	
2	34.56	10.04	17.35	
3	23.11	9.17	3.35	
Totals	79.14	30.32	28.75	138.21
Means	26.38	10.11	9.58	15.36

Table 12.10. ANOVA FOR THE EMERGENCY ROOM PROBLEM WITH A SINGLE FACTOR

Source of Variation	*Sum of Squares*	*Degrees of Freedom*	*Mean Squares*	*F*
Treatment	$SS_{TREAT} = 547.22$	$k - 1 = 2$	$MS_{TREAT} = 273.61$	8.00
Error	$SS_E = 205.13$	$R - k = 6$	$MS_E = 34.19$	
Total	$SS_{TOTAL} = 752.35$	$R - 1 = 8$		

6 degrees of freedom is $F_{0.05,2,6} = 5.14$, taken from Table A.6. Since the test statistic, $F = 8.00$, is greater than the critical value, the null hypothesis of no treatment effect is rejected at the $\alpha = 0.05$ level of significance.

Since the test indicates a statistically significant effect due to the factor, the analyst may be interested in estimating $\mu + \tau_j$, the mean of the response variable at level j. The estimates are given by

$$\hat{\mu} = \bar{Y}_{..}$$

$$\hat{\tau}_j = \bar{Y}_{.j} - \bar{Y}_{..}, \qquad j = 1, 2, \ldots, k$$

and thus,

$$\hat{\mu} + \hat{\tau}_j = \bar{Y}_{.j}$$

which is intuitively appealing since the estimate of the mean, $\mu + \tau_j$, at level j is just the sample mean of the observations at level j. In addition, a $100(1 - \alpha)\%$ confidence interval for $\mu + \tau_j$ can be computed by

$$\bar{Y}_{.j} \pm t_{\alpha/2, R-k}\sqrt{\frac{MS_E}{R_j}} \tag{12.24}$$

For Example 12.6, the mean effects are

$$\begin{aligned}\hat{\mu} &= \bar{Y}_{..} = 15.36 \text{ minutes}\\ \hat{\tau}_1 &= 11.02 \text{ minutes}\\ \hat{\tau}_2 &= -5.25 \text{ minutes}\\ \hat{\tau}_3 &= -5.78 \text{ minutes}\end{aligned}$$

and

$$\begin{aligned}\hat{\mu} + \hat{\tau}_1 &= \bar{Y}_{.1} = 26.38 \text{ minutes}\\ \hat{\mu} + \hat{\tau}_2 &= \bar{Y}_{.2} = 10.11 \text{ minutes}\\ \hat{\mu} + \hat{\tau}_3 &= \bar{Y}_{.3} = 9.58 \text{ minutes}\end{aligned}$$

A 95% confidence interval for $\mu + \tau_3$, the mean delay when using the round robin queue discipline, is given by equation (12.24) as

$$9.58 \pm (2.45)\sqrt{\frac{34.19}{3}}$$

or

$$1.31 = 9.58 - 8.27 \leq \mu + \tau_3 \leq 9.58 + 8.27 = 17.85$$

Confidence intervals for the other effects can be computed similarly.

At this point, there may be many additional questions the analyst would like answered, such as: Is there a significant difference between the effect of the round-robin and priority-queue disciplines? That is, the analyst may desire to test the hypothesis

$$H_0: \quad \tau_2 = \tau_3$$

versus

$$H_0: \quad \tau_2 \neq \tau_3$$

These types of questions may be addressed by the method of orthogonal contrasts, Duncan's multiple range test, and other tests, all of which are beyond the scope of this text. The reader is referred to Hines and Montgomery [1980], Hicks [1973], or Montgomery [1976]. These three texts also discuss the effect of departures from assumptions in the analysis of variance. The critical assumptions are that treatment effects are additive, that the random deviations (ϵ_{rj}) are normally distributed with constant variance σ^2, and that all observations are statistically independent. By proper control of the random number streams, this last assumption of independence is assured. If the observations, Y_{rj}, are sample averages over some simulation run length, they should be approximately normally distributed. The critical assumption of constant error variance can be tested by Bartlett's test or other procedures, as discussed in Hines and Montgomery [1980].

12.3.3. Factorial Designs with Two Factors

When more than one factor affects the response variable, a more general model than the single-factor model of Section 12.3.2 is needed. In this subsection, a model with two factors is described.

Consider Example 12.6, in which the queue discipline (Q) and the number of doctors (N) are factors. Suppose that Q has $q = 3$ levels and N has $n = 2$ levels, and that four independent replications of each treatment combination are made. Then there are $q \cdot n = 3 \times 2 = 6$ treatments or factor combinations, requiring $R = 3 \times 2 \times 4 = 24$ independent runs of the simulation. The appropriate statistical model for analyzing the response variable then becomes

$$Y_{ijr} = \mu + Q_i + N_j + QN_{ij} + \epsilon_{ijr} \tag{12.25}$$

where $i = 1, 2$, $j = 1, 2, 3$, and $r = 1, 2, 3, 4$. Here Y_{ijr} is the observation of the response variable, Y, for replication r of level i of the first factor and level j of the second factor. The model in equation (12.25) is also a completely randomized design; that is, it is assumed that all replications both within a treatment and over all treatments are made statistically independent by the correct use of the random number generators. More precisely, the random deviation terms ϵ_{ijr} are assumed to be independently and normally distributed with mean zero and a common variance σ^2.

In equation (12.25), μ is the overall mean effect, Q_i the effect of level i of factor Q, N_j the effect of level j of factor N, and QN_{ij} the interaction effect of level i of factor Q and level j of factor N. In other words, if factor Q is set at level i and factor N at level j, then $\mu + Q_i + N_j + QN_{ij}$ is the mean response and ϵ_{ijr} is the random variation of actual responses about this mean. To interpret the interaction effect, consider Figure 12.4, where the mean delay ($\mu + Q_i + N_j + QN_{ij}$) of type NIA patients for level i of factor Q and level j of factor N is plotted on the vertical axis versus each level of factor Q on the horizontal axis. When there is no interaction present (i.e., $QN_{ij} = 0$) as in Figure 12.4(a), it means that the change in mean delay due to having $D_2 = 4$ versus having $D_2 = 5$ doctors is not affected by the particular queue discipline being used (whether it is D_1 = FCFS, D_2 = PR, or D_3 = RR). In Figure 12.4(a) this is illustrated by the fact that the lines connecting corresponding pairs of points are parallel. Figure 12.4(b) illustrates the case when interaction is present. There is a greater change in mean delay between $D_2 = 4$ and $D_2 = 5$ doctors for D_1 = FCFS than for D_1 = PR. The interaction is illustrated by the two line segments not being parallel. Equivalently, the dotted vertical lines representing the change in mean delay are of unequal lengths. In summary, when no interaction is present, a change in one factor produces the same change in mean response regardless of the level of other factors. When interaction is present, the change in mean response due to changing the level of one factor does depend on the level of other factors. In terms of Example 12.6, the change

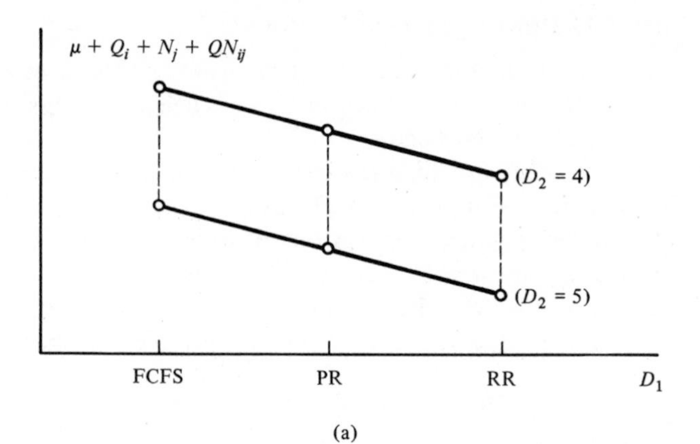

(a)

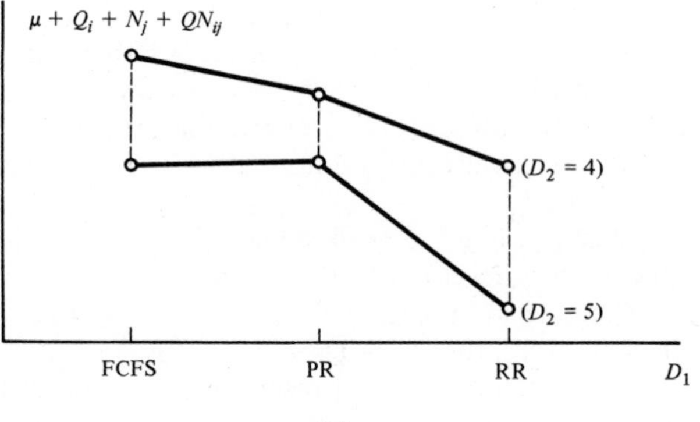

(b)

Figure 12.4. Interaction effect.

in mean delay between level j and level j' of factor N is

$$(\mu + Q_i + N_j + QN_{ij}) - (\mu + Q_i + N_{j'} + QN_{ij'}) = (N_j - N_{j'}) + (QN_{ij} - QN_{ij'})$$

Thus, if interaction is not present, $QN_{ij} = QN_{ij'} = 0$ and the change in mean delay due to a change from level j to level j' of factor N is

$$N_j - N_{j'}$$

which is the same for all levels of factor Q. Figure 12.4 is for purposes of illustration and it is not recommended that a graphical approach be used to detect interaction, because lines can be made to look parallel or nearly so by the choice of scale. An appropriate statistical procedure, namely ANOVA, is necessary for detecting both interaction and significant effects due to the factors.

However, interaction effects cannot be detected unless there is more than one replication for each treatment combination.

To conduct an ANOVA test for the model in equation (12.25), assuming that there are q levels of factor Q and n levels of factor N, and m replications at each treatment level (for a total of $R = qnm$ replications), first compute the following sums of squares:

$$\begin{aligned} SS_{TOTAL} &= \sum_{i=1}^{q} \sum_{j=1}^{n} \sum_{r=1}^{m} (Y_{ijr} - \bar{Y}_{...})^2 \\ SS_Q &= \sum_{i=1}^{q} mn(\bar{Y}_{i..} - \bar{Y}_{...})^2 \\ SS_N &= \sum_{j=1}^{n} mq(\bar{Y}_{.j.} - \bar{Y}_{...})^2 \\ SS_{QN} &= \sum_{i=1}^{q} \sum_{j=1}^{n} m(\bar{Y}_{ij.} - \bar{Y}_{i..} - \bar{Y}_{.j.} + \bar{Y}_{...})^2 \\ SS_E &= SS_{TOTAL} - SS_Q - SS_N - SS_{QN} \end{aligned} \tag{12.26}$$

The same dot notation for sums, and bar over a variable for averaging, is used here that was used for the single-factor model of Section 12.3.2. For example,

$$\bar{Y}_{i..} = \sum_{j=1}^{n} \sum_{r=1}^{m} \frac{Y_{ijr}}{nm}$$

is an estimate of mean response when factor Q is at level i. Since $Y_{...} = \sum_{i=1}^{q} \sum_{j=1}^{n} \sum_{r=1}^{m} Y_{ijr}/qnm$ is the overall sample mean, SS_Q can be interpreted as the variability in response due to factor Q. The other sums of squares can be interpreted similarly.

It can be shown that

$$SS_{TOTAL} = SS_Q + SS_N + SS_{QN} + SS_E \tag{12.27}$$

and thus again the ANOVA procedure partitions the total variability, SS_{TOTAL}, about the mean response into components due to the two factors plus an interaction component plus a random error component. The sums of squares on the right side of equation (12.27) are independently distributed. Thus when each sum of squares is divided by σ^2, the results (SS/σ^2) have a chi-square distribution and appropriate ratios of mean squares have an F distribution. The degrees of freedom (df) and mean squares (MS) are given in Table 12.11. The mean square for error, MS_E, is an unbiased estimate of $\sigma^2 = \text{var}(Y_{ijr})$, and F tests will be made with MS_E in the denominator.

In the model of equation (12.25), there are three hypotheses that can be tested:

$$H_{01}: \quad Q_i = 0 \qquad \text{for all } i \tag{12.28}$$

Table 12.11. ANOVA FOR A TWO-FACTOR EXPERIMENT

Source of Variation	df	SS	MS	F
Factor Q	$q-1$	SS_Q	$MS_Q = \dfrac{SS_Q}{q-1}$	$\dfrac{MS_Q}{MS_E}$
Factor N	$n-1$	SS_N	$MS_N = \dfrac{SS_N}{n-1}$	$\dfrac{MS_N}{MS_E}$
QN interaction	$(q-1)(n-1)$	SS_{QN}	$MS_{QN} = \dfrac{SS_{QN}}{(q-1)(n-1)}$	$\dfrac{MS_{QN}}{MS_E}$
Error	$qn(m-1)$	SS_E	$MS_E = \dfrac{SS_E}{qn(m-1)}$	
Total	$qnm-1$	SS_{TOTAL}		

which says that the queueing discipline has no effect on mean delay;

$$H_{02}: \quad N_j = 0 \qquad \text{for all } j \tag{12.29}$$

which says that the number of doctors has no effect; and

$$H_{03}: \quad QN_{ij} = 0 \qquad \text{for all } i, j \tag{12.30}$$

which says that there is no interaction between the queueing discipline (factor Q) and the number of doctors (factor N).

Table 12.12 gives the results of the $m = 4$ independent replications of Example 12.6 (the Dead End Emergency Room) for each treatment combination,

Table 12.12. AVERAGE DELAY IN MINUTES OF NIA PATIENTS FOR EACH LEVEL OF EACH FACTOR

	Queueing Discipline, Q			
Number of Doctors, N	D_1 = FCFS ($i = 1$)	D_1 = PR ($i = 2$)	D_1 = RR ($i = 3$)	$T_{\cdot j \cdot}$
$D_2 = 4$ ($j = 1$)	21.47	11.11	8.05	
	34.56	10.04	17.35	
	23.11	9.17	3.35	
	22.35	10.27	5.83	176.66
$D_2 = 5$ ($j = 2$)	6.93	6.42	1.00	
	8.62	5.94	2.11	
	9.16	5.49	1.61	
	8.74	4.63	2.44	63.09
$T_{i\cdot\cdot}$	134.94	63.07	41.74	
$T_{\cdot\cdot\cdot}$				239.75

and the sums $T_{i\cdot\cdot}$, $T_{\cdot j\cdot}$, and $T_{\cdot\cdot\cdot}$. Using these sums, compute the sample means as follows:

$$\bar{Y}_{1\cdot\cdot} = \frac{T_{1\cdot\cdot}}{8} = \frac{134.94}{8} = 16.87$$

$$\bar{Y}_{2\cdot\cdot} = \frac{T_{2\cdot\cdot}}{8} = \frac{63.07}{8} = 7.88$$

$$\bar{Y}_{3\cdot\cdot} = \frac{T_{3\cdot\cdot}}{8} = \frac{41.74}{8} = 5.22$$

which are estimates of the mean delay, $\mu + Q_i$, for level i of factor Q ($i = 1, 2, 3$);

$$\bar{Y}_{\cdot 1\cdot} = \frac{T_{\cdot 1\cdot}}{12} = \frac{176.66}{12} = 14.72$$

$$\bar{Y}_{\cdot 2\cdot} = \frac{T_{\cdot 2\cdot}}{12} = \frac{63.09}{12} = 5.26$$

which are estimates of the mean delay, $\mu + N_j$, for level j of factor N ($j = 1, 2$) and finally,

$$\bar{Y}_{\cdots} = \frac{T_{\cdots}}{24} = \frac{239.75}{24} = 9.99$$

which is an estimate of the overall mean, μ.

Now use equations (12.26) to compute the sums of squares SS_{TOTAL}, SS_Q, SS_N, SS_{QN}, and finally by subtraction, SS_E. The results of these computations are shown in the ANOVA Table 12.13.

Table 12.13. ANOVA TABLE FOR EMERGENCY ROOM PROBLEM (3 × 2 FACTORIAL WITH FOUR REPLICATIONS)

Source of Variation	*Sum of Squares*	*Degrees of Freedom*	*Mean Squares*	*F*
Queue discipline	596.32	2	298.16	25.33
Number of doctors	537.45	1	537.45	45.65
QN interaction	175.89	2	87.94	7.47
Error	211.90	18	11.77	
Total	1521.56	23		

To test the hypothesis H_{01} [equation (12.28)] of no effect due to queue discipline, compute

$$F = \frac{MS_Q}{MS_E} = \frac{298.16}{11.77} = 25.33$$

which is significant at the 5% level ($\alpha = .05$), since $F = 25.33 > F_{0.05,2,18} = 3.55$ (the 95% percentage point with 2 and 18 df, taken from Table A.6).

To test the hypothesis H_{02} [equation (12.29)] of no effect due to factor N, compute

$$F = \frac{\text{MS}_N}{\text{MS}_E} = \frac{537.45}{11.77} = 45.65$$

which is also significant at the 5% level, since $F = 45.65 > F_{0.05,1,18} = 4.41$ (the 95% percentage point with 1 and 18 df).

Finally, to test the interaction hypothesis H_{03} [equation (12.30)], compute

$$F = \frac{\text{MS}_{QN}}{\text{MS}_E} = \frac{87.94}{11.77} = 7.47$$

which is also significant at the 5% level, since $F = 7.47 > F_{0.05,2,18} = 3.55$.

These tests indicate that the queue discipline and the number of doctors both have a statistically significant effect on mean delay. In addition, there is significant interaction between the two factors.

12.3.4. Other Experimental Design Models

The two-factor model of equation (12.25) can be generalized to any number of factors. If there are k factors and each factor is restricted to two levels, the model is called a 2^k factorial design, because there will be 2^k treatments, or factor combinations. For example, if $k = 7$, then $2^k = 128$; if two replications are run at each factor combination, a total of $128 \times 2 = 256$ simulation runs would have to be made.

For many complex real-world simulations, such a large number of runs would be too expensive, or simply impossible, given the computer resources available. For this reason, so-called fractional factorial designs have been developed which require a fraction of the number of runs required by a full factorial experiment. Fractional factorial experiments are briefly discussed by Law and Kelton [1982] in the context of simulation, and more thoroughly by the other texts previously referenced [Box et al., 1978; Hicks, 1973]; Montgomery, 1976; Kleijnen, 1975]. Fractional factorial designs are one example of a design useful for screening out unimportant factors at an early stage of experimentation.

12.4. Summary

This chapter provides a basic introduction to the comparative evaluation of alternative system designs based on data collected from simulation runs. It was assumed that a fixed set of alternative system designs had been selected for consideration. Comparisons based on confidence intervals and the use of common random numbers were emphasized. A brief introduction to experimental design models, whose purpose is to evaluate the effect of system alternatives, was also provided. Although beyond the scope of this text, there are many additional topics of potential interest in the realm of statistical analysis techniques relevant to simulation. Some of these topics include:

1. Optimization techniques, for experimentally searching for the "best" system in some sense
2. The regenerative method of estimation, which eliminates the problems of initial conditions and autocorrelation (but which is applicable only to a restricted class of simulation models)
3. Time series methods of estimation, which take autocorrelation into account and thus remove its deleterious effects
4. Ranking and selection procedures, for selecting the "best" system from a given set of systems, or ranking systems from best to worst
5. Variance reduction techniques, which are methods (or "tricks") to improve the statistical efficiency of simulation experiments (common random numbers being an important example)

The reader is referred to Kleijnen [1975] and Law and Kelton [1982] for discussions of these and other topics relevant to simulation.

The most important idea in Chapters 11 and 12 is that simulation output data require a statistical analysis in order to be interpreted correctly. In particular, a statistical analysis can provide a measure of the accuracy of the results produced by a simulation, and can provide techniques for achieving a specified accuracy.

REFERENCES

BOX, G. E. P., W. G. HUNTER, AND J. S. HUNTER [1978], *Statistics for Experimenters: An Introduction to Design, Data Analysis, and Model Building*, Wiley, New York.

HICKS, C. R. [1973], *Fundamental Concepts in the Design of Experiments*, 2nd ed., Holt, New York.

HINES, W. W., AND D. C. MONTGOMERY [1980], *Probability and Statistics in Engineering and Management Science*, 2nd ed., Wiley, New York.

KLEIJNEN, J. P. C. [1975], *Statistical Techniques in Simulation*, Parts I and II, Dekker, New York.

LAW, A. M., AND W. D. KELTON [1982], *Simulation Modeling and Analysis*, McGraw-Hill, New York.

MONTGOMERY, D. C. [1976], *Design and Analysis of Experiments*, Wiley, New York.

WRIGHT, R. D., AND T. E. RAMSAY, JR. [1979], "On the Effectiveness of Common Random Numbers," *Management Science*, Vol. 25, pp. 649–56.

EXERCISES

1. Reconsider the dump truck problem of Example 3.4, which was also analyzed in Example 12.3. As business expands, the company buys new trucks, making the total number of trucks now equal to 16. The company desires to have a sufficient number of loaders and scales so that the average number of trucks waiting at the loader queue plus the average number at the weigh queue is no more than three. Investigate the following combinations of number of loaders and number of scales:

Number of Scales	*Number of Loaders* 2	3	4
1	—	—	—
2	—	—	—

The loaders being considered are the "slow" loaders in Example 12.3. Loading time, weighing time, and travel time for each truck are as previously defined in Example 12.3. Use common random numbers to the greatest extent possible when comparing alternative systems designs. The goal is to find the smallest number of loaders and scales to meet the company's objective of average total queue length no more than three trucks. In your solution, take into account the initialization conditions, run length, and number of replications needed to achieve a reasonable likelihood of valid conclusions.

2. In Exercise 11.6, consider the following alternative (M, L) policies:

M		L: *Low* 30	*High* 40
	Low 50	(50, 30)	(50, 40)
	High 100	(100, 30)	(100, 40)

Investigate the relative costs of these policies using suitable modifications of the simulation model developed in Exercise 11.6. Compare the four system designs on the basis of long-run mean monthly cost. First make four replications of each (M, L) policy, using common random numbers to the greatest extent possible. Each replication should have a 12-month initialization phase followed by a 100-month data collection phase. Compute confidence intervals having an overall confidence level of 90% for mean monthly cost for each policy. Then estimate the additional replications needed to achieve confidence intervals which do not overlap. Draw conclusions as to which is the best policy.

3. Reconsider Exercise 11.7. Compare the four inventory policies studied in Exercise 2, taking the cost of rush orders into account when computing monthly cost.

4. Reconsider Exercise 11.8. Compare the four monthly inventory policies studied in Exercise 2, taking into account the selling price of the perishable items.

5. In Exercise 11.9, investigate the effect of the order quantity on long-run mean daily cost. Since each order arrives on a pallet on a delivery truck, the permissible order quantities, Q, are multiples of 10 (i.e., Q may equal 10, or 20, or 30, . . .). In Exercise 11.9, the policy $Q = 10$ was investigated.

(a) First, investigate the two policies $Q = 10$ and $Q = 50$. Use the run lengths, and so on, suggested in Exercise 11.9. On the basis of these runs, decide whether the optimal Q, say Q^*, is between 10 and 50, or greater than 50. (The cost curve as a function of Q should have what kind of shape?)

(b) Based on the results in part (a), suggest two additional values for Q and simulate the two policies. Draw conclusions. Include an analysis of the strength of your conclusions.

6. In Exercise 11.10, determine the number of cards Q that the card shop owner should purchase to maximize the profit within an error (for total profit) of approximately $5.00 at most. First, make runs for the policy of ordering $Q = 250$ and $Q = 350$ cards. With these results, plus the results of Exercise 11.10 for ordering $Q = 300$ cards, decide on a range of Q worth considering further (e.g., $200 \leq Q \leq 250$, or $250 \leq Q \leq 300$, etc.). Then investigate this restricted range for two additional policies (e.g., $Q = 200$ and $Q = 225$, or $Q = 265$ and $Q = 285$, etc.).

7. In Exercise 11.11, investigate the effect of target level M and review period N on mean monthly cost. Consider two target levels, M, determined by ± 10 from the target level used in Exercise 11.11 (see Example 6.9); and consider review periods N of 1 month and 3 months. Which (N, M) pair is best, based on these simulations?

8. Reconsider Exercises 11.13 and 11.14, which involved the scheduling rules (or queue disciplines) of first in, first out (FIFO) and priority by type (PR) in a job shop. In addition to these two rules, consider a shortest imminent operation (SIO) scheduling rule. For a given station, all jobs of the type with the smallest mean processing time are given highest priority. For example, when using an SIO rule at station 1, jobs are processed in the following order: type 2 first, then type 1, and type 3 last. Two jobs of the same type are processed on a FIFO basis. Develop a simulation experiment to compare the FIFO, PR, and SIO rules on the basis of mean total response time over all jobs.

9. In Exercise 11.13 (the job shop with FIFO rule), determine the minimum number of workers needed at each station to avoid bottlenecks. A bottleneck occurs when average queue lengths at a station steadily increase over time. (Do not confuse increasing average queue length due to an inadequate number of servers with increasing average queue length due to initialization bias. In the former case, average queue length continues to increase indefinitely and server utilization is 1.0. In the latter case, average queue length eventually levels off and server utilization is less than 1.) Report on utilization of workers and total time it takes for a job to get through the job shop, by type and over all types. (*Hint:* If server utilization at a work station is 1.0, and if average queue length tends to increase linearly as simulation run length increases, it is a good possibility that the work station is unstable and therefore is a bottleneck. In this case, at least one additional worker is needed at the work station. Use queueing theory, namely $\lambda / c_i \mu < 1$, to suggest the minimum number of workers needed at station 1. Recall that λ is the arrival rate, $1/\mu$ is the overall mean service time for one job with one worker, and c_i is the number of workers at station i. Attempt to use the same basic condition, $\lambda/c_i\mu < 1$, to determine an initial number of servers at station i for $i = 2, 3, 4$.)

10. (a) Repeat Exercise 9 for the PR scheduling rule (see Exercise 11.14).
(b) Repeat Exercise 9 for the SIO scheduling rule (see Exercise 12.8).
(c) Compare the minimum required number of workers for each scheduling rule: FIFO, versus PR, versus SIO.

11. With the minimum number of workers determined by Exercises 9 and 10 for the job shop of Exercise 11.13, consider adding one worker to the entire shop. This worker can be trained to handle the processing at only one station. At which station should this worker be placed? How does this additional worker affect mean total response time over all jobs? Over type 1 jobs? Investigate the job shop with and without the additional worker for each scheduling rule: FIFO, PR, SIO.

12. In Exercise 11.17, suppose that a buffer of capacity one item is constructed in front of each worker. Design an experiment to determine if this change in system design has a significant impact upon individual worker utilizations (ρ_1, ρ_2, ρ_3 and ρ_4). At the very least, compute confidence intervals for $\rho_1^0 - \rho_1^1$ and $\rho_4^0 - \rho_4^1$, where ρ_i^s is utilization for worker i when the buffer has capacity s.

APPENDIX

TABLES

Table A.1. RANDOM DIGITS

94737	08225	35614	24826	88319	05595	58701	57365	74759
87259	85982	13296	89326	74863	99986	68558	06391	50248
63856	14016	18527	11634	96908	52146	53496	51730	03500
66612	54714	46783	61934	30258	61674	07471	67566	31635
30712	58582	05704	23172	86689	94834	99057	55832	21012
69607	24145	43886	86477	05317	30445	33456	34029	09603
37792	27282	94107	41967	21425	04743	42822	28111	09757
01488	56680	73847	64930	11108	44834	45390	86043	23973
66248	97697	38244	50918	55441	51217	54786	04940	50807
51453	03462	61157	65366	61130	26204	15016	85665	97714
92168	82530	19271	86999	96499	12765	20926	25282	39119
36463	07331	54590	00546	03337	41583	46439	40173	46455
47097	78780	04210	87084	44484	75377	57753	41415	09890
80400	45972	44111	99708	45935	03694	81421	60170	58457
94554	13863	88239	91624	00022	40471	78462	96265	55360
31567	53597	08490	73544	72573	30961	12282	97033	13676
07821	24759	47266	21747	72496	77755	50391	59554	31177
09056	10709	69314	11449	40531	02917	95878	74587	60906
19922	37025	80731	26179	16039	01518	82697	73227	13160
29923	02570	80164	36108	73689	26342	35712	49137	13482
29602	29464	99219	20308	82109	03898	82072	85199	13103
94135	94661	87724	88187	62191	70607	63099	40494	49069
87926	34092	34334	55064	43152	01610	03126	47312	59578
85039	19212	59160	83537	54414	19856	90527	21756	64783
66070	38480	74636	45095	86576	79337	39578	40851	53503
78166	82521	79261	12570	10930	47564	77869	16480	43972
94672	07912	26153	10531	12715	63142	88937	94466	31388
56406	70023	27734	22254	27685	67518	63966	33203	70803
67726	57805	94264	77009	08682	18784	47554	59869	66320
07516	45979	76735	46509	17696	67177	92600	55572	17245
43070	22671	00152	81326	89428	16368	57659	79424	57604
36917	60370	80812	87225	02850	47118	23790	55043	75117
03919	82922	02312	31106	44335	05573	17470	25900	91080
46724	22558	64303	78804	05762	70650	56117	06707	90035
16108	61281	86823	20286	14025	24909	38391	12183	89393
74541	75808	89669	87680	72758	60851	55292	95663	88326
82919	31285	01850	72550	42986	57518	01159	01786	98145
31388	26809	77258	99360	92362	21979	41319	75739	98082
17190	75522	15687	07161	99745	48767	03121	20046	28013
00466	88068	68631	98745	97810	35886	14497	90230	69264

Table A.2. RANDOM NORMAL NUMBERS

0.23	−0.17	0.43	2.18	2.13	0.49	2.72	−0.18	0.42
0.24	−1.17	0.02	0.67	−0.59	−0.13	−0.15	−0.46	1.64
−1.16	−0.17	0.36	−1.26	0.91	0.71	−1.00	−1.09	−0.02
−0.02	−0.19	−0.04	1.92	0.71	−0.90	−0.21	−1.40	−0.38
0.39	0.55	0.13	2.55	−0.33	−0.05	−0.34	−1.95	−0.44
0.64	−0.36	0.98	−0.21	−0.52	−0.02	−0.15	−0.43	0.62
−1.90	0.48	−0.54	0.60	−0.35	−1.29	−0.57	0.23	1.41
−1.04	−0.70	−1.69	1.76	0.47	−0.52	−0.73	0.94	−1.63
−0.78	0.11	−0.91	−1.13	0.07	0.45	−0.94	1.42	0.75
0.68	1.77	−0.82	−1.68	−2.60	1.59	−0.72	−0.80	0.61
−0.02	0.92	1.76	−0.66	0.18	−1.32	1.26	0.61	0.83
−0.47	1.04	0.83	−2.05	1.00	−0.70	1.12	0.82	0.08
−0.40	1.40	1.20	0.00	0.21	−2.13	−0.22	1.79	0.87
−0.75	0.09	−1.50	0.14	−2.99	−0.41	−0.99	−0.70	0.51
−0.66	−1.97	0.15	−1.16	−0.60	0.50	1.36	1.94	0.11
−0.44	−0.09	−0.59	1.37	0.18	1.44	−0.80	2.11	−1.37
1.41	−2.71	−0.67	1.83	0.97	0.06	−0.28	0.04	−0.21
1.21	−0.52	−0.20	−0.88	−0.78	0.84	−1.08	−0.25	0.17
0.07	0.66	−0.51	−0.04	−0.84	0.04	1.60	−0.92	1.14
−0.08	0.79	−0.09	−1.12	−1.13	0.77	0.40	0.69	−0.12
0.53	−0.36	−2.64	0.22	−0.78	1.92	−0.26	1.04	−1.61
−1.56	1.82	−1.03	1.14	−0.12	−0.78	−0.12	1.42	−0.52
0.03	−1.29	−0.33	2.60	−0.64	1.19	−0.13	0.91	0.78
1.49	1.55	−0.79	1.37	0.97	0.17	0.58	1.43	−1.29
−1.19	1.35	0.16	1.06	−0.17	0.32	−0.28	0.68	0.54
−1.19	−1.03	−0.12	1.07	0.87	−1.40	−0.24	−0.81	0.31
0.11	−1.95	−0.44	−0.39	−0.15	−1.20	−1.98	0.32	2.91
−1.86	0.06	0.19	−1.29	0.33	1.51	−0.36	−0.80	−0.99
0.16	0.28	0.60	−0.78	0.67	0.13	−0.47	−0.18	−0.89
1.21	−1.19	−0.60	−1.22	0.07	−1.13	1.45	0.94	0.54
−0.82	0.54	−0.98	−0.13	1.52	0.77	0.95	−0.84	2.40
0.75	−0.80	−0.28	1.77	−0.16	−0.33	2.43	−1.11	1.63
0.42	0.31	1.56	0.56	0.64	−0.78	0.04	1.34	−0.01
−1.50	−1.78	−0.59	0.16	0.36	1.89	−1.19	0.53	−0.97
−0.89	0.08	0.95	−0.73	1.25	−1.04	−0.47	−0.68	−0.87
0.19	0.85	1.68	−0.57	0.37	−0.48	−0.17	2.36	−0.53
0.49	0.32	−2.08	−1.02	2.59	−0.53	0.15	0.11	0.05
−1.44	0.07	−0.22	−0.93	−1.40	0.54	−1.28	−0.15	0.67
−0.21	−0.48	1.21	0.67	−1.10	−0.75	−0.37	0.68	−0.02
−0.65	−0.12	0.94	−0.44	−1.21	−0.06	−1.28	−1.51	1.39
0.24	−0.83	1.55	0.33	−0.59	−1.24	0.70	0.01	0.15
−0.73	1.24	0.40	−0.61	0.68	0.69	0.07	−0.23	−0.66
−1.93	0.75	−0.32	0.95	1.35	1.51	−0.88	0.10	−1.19
0.08	0.16	0.38	−0.96	1.99	−0.20	0.98	0.16	0.26
−0.47	−1.25	0.32	0.51	−1.04	0.97	2.60	−0.08	1.19

Table A.3. CUMULATIVE NORMAL DISTRIBUTION

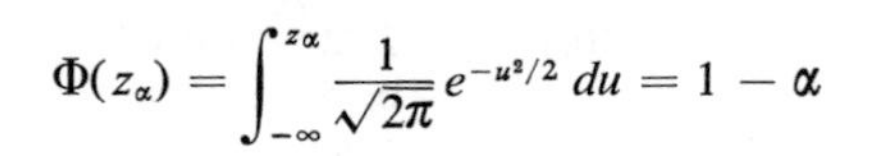

$$\Phi(z_\alpha) = \int_{-\infty}^{z_\alpha} \frac{1}{\sqrt{2\pi}} e^{-u^2/2}\, du = 1 - \alpha$$

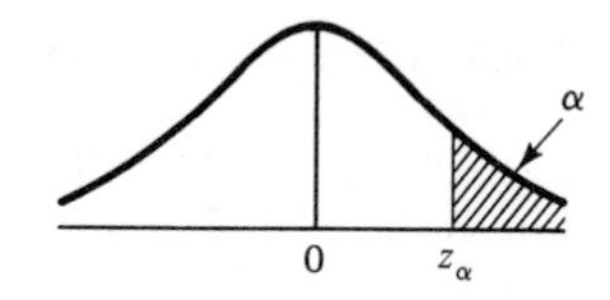

z_α	0.00	0.01	0.02	0.03	0.04	z_α
0.0	0.500 00	0.503 99	0.507 98	0.511 97	0.515 95	**0.0**
0.1	0.539 83	0.543 79	0.547 76	0.551 72	0.555 67	**0.1**
0.2	0.579 26	0.583 17	0.587 06	0.590 95	0.594 83	**0.2**
0.3	0.617 91	0.621 72	0.625 51	0.629 30	0.633 07	**0.3**
0.4	0.655 42	0.659 10	0.662 76	0.666 40	0.670 03	**0.4**
0.5	0.691 46	0.694 97	0.698 47	0.701 94	0.705 40	**0.5**
0.6	0.725 75	0.729 07	0.732 37	0.735 65	0.738 91	**0.6**
0.7	0.758 03	0.761 15	0.764 24	0.767 30	0.770 35	**0.7**
0.8	0.788 14	0.791 03	0.793 89	0.796 73	0.799 54	**0.8**
0.9	0.815 94	0.818 59	0.821 21	0.823 81	0.826 39	**0.9**
1.0	0.841 34	0.843 75	0.846 13	0.848 49	0.850 83	**1.0**
1.1	0.864 33	0.866 50	0.868 64	0.870 76	0.872 85	**1.1**
1.2	0.884 93	0.886 86	0.888 77	0.890 65	0.892 51	**1.2**
1.3	0.903 20	0.904 90	0.906 58	0.908 24	0.909 88	**1.3**
1.4	0.919 24	0.920 73	0.922 19	0.923 64	0.925 06	**1.4**
1.5	0.933 19	0.934 48	0.935 74	0.936 99	0.938 22	**1.5**
1.6	0.945 20	0.946 30	0.947 38	0.948 45	0.949 50	**1.6**
1.7	0.955 43	0.956 37	0.957 28	0.958 18	0.959 07	**1.7**
1.8	0.964 07	0.964 85	0.965 62	0.966 37	0.967 11	**1.8**
1.9	0.971 28	0.971 93	0.972 57	0.973 20	0.973 81	**1.9**
2.0	0.977 25	0.977 78	0.978 31	0.978 82	0.979 32	**2.0**
2.1	0.982 14	0.982 57	0.983 00	0.983 41	0.983 82	**2.1**
2.2	0.986 10	0.986 45	0.986 79	0.987 13	0.987 45	**2.2**
2.3	0.989 28	0.989 56	0.989 83	0.990 10	0.990 36	**2.3**
2.4	0.991 80	0.992 02	0.992 24	0.992 45	0.992 66	**2.4**
2.5	0.993 79	0.993 96	0.994 13	0.994 30	0.994 46	**2.5**
2.6	0.995 34	0.995 47	0.995 60	0.995 73	0.995 85	**2.6**
2.7	0.996 53	0.996 64	0.996 74	0.996 83	0.996 93	**2.7**
2.8	0.997 44	0.997 52	0.997 60	0.997 67	0.997 74	**2.8**
2.9	0.998 13	0.998 19	0.998 25	0.998 31	0.998 36	**2.9**
3.0	0.998 65	0.998 69	0.998 74	0.998 78	0.998 82	**3.0**
3.1	0.999 03	0.999 06	0.999 10	0.999 13	0.999 16	**3.1**
3.2	0.999 31	0.999 34	0.999 36	0.999 38	0.999 40	**3.2**
3.3	0.999 52	0.999 53	0.999 55	0.999 57	0.999 58	**3.3**
3.4	0.999 66	0.999 68	0.999 69	0.999 70	0.999 71	**3.4**
3.5	0.999 77	0.999 78	0.999 78	0.999 79	0.999 80	**3.5**
3.6	0.999 84	0.999 85	0.999 85	0.999 86	0.999 86	**3.6**
3.7	0.999 89	0.999 90	0.999 90	0.999 90	0.999 91	**3.7**
3.8	0.999 93	0.999 93	0.999 93	0.999 94	0.999 94	**3.8**
3.9	0.999 95	0.999 95	0.999 96	0.999 96	0.999 96	**3.9**

Table A.3. Continued

z_α	0.05	0.06	0.07	0.08	0.09	z_α
0.0	0.519 94	0.523 92	0.527 90	0.531 88	0.535 86	**0.0**
0.1	0.559 62	0.563 56	0.567 49	0.571 42	0.575 34	**0.1**
0.2	0.598 71	0.602 57	0.606 42	0.610 26	0.614 09	**0.2**
0.3	0.636 83	0.640 58	0.644 31	0.648 03	0.651 73	**0.3**
0.4	0.673 64	0.677 24	0.680 82	0.684 38	0.687 93	**0.4**
0.5	0.708 84	0.712 26	0.715 66	0.719 04	0.722 40	**0.5**
0.6	0.742 15	0.745 37	0.748 57	0.751 75	0.754 90	**0.6**
0.7	0.773 37	0.776 37	0.779 35	0.782 30	0.785 23	**0.7**
0.8	0.802 34	0.805 10	0.807 85	0.810 57	0.813 27	**0.8**
0.9	0.828 94	0.831 47	0.833 97	0.836 46	0.838 91	**0.9**
1.0	0.853 14	0.855 43	0.857 69	0.859 93	0.862 14	**1.0**
1.1	0.874 93	0.876 97	0.879 00	0.881 00	0.882 97	**1.1**
1.2	0.894 35	0.896 16	0.897 96	0.899 73	0.901 47	**1.2**
1.3	0.911 49	0.913 08	0.914 65	0.916 21	0.917 73	**1.3**
1.4	0.926 47	0.927 85	0.929 22	0.930 56	0.931 89	**1.4**
1.5	0.939 43	0.940 62	0.941 79	0.942 95	0.944 08	**1.5**
1.6	0.950 53	0.951 54	0.952 54	0.953 52	0.954 48	**1.6**
1.7	0.959 94	0.960 80	0.961 64	0.962 46	0.963 27	**1.7**
1.8	0.967 84	0.968 56	0.969 26	0.969 95	0.970 62	**1.8**
1.9	0.974 41	0.975 00	0.975 58	0.976 15	0.976 70	**1.9**
2.0	0.979 82	0.980 30	0.980 77	0.981 24	0.981 69	**2.0**
2.1	0.984 22	0.984 61	0.985 00	0.985 37	0.985 74	**2.1**
2.2	0.987 78	0.988 09	0.988 40	0.988 70	0.988 99	**2.2**
2.3	0.990 61	0.990 86	0.991 11	0.991 34	0.991 58	**2.3**
2.4	0.992 86	0.993 05	0.993 24	0.993 43	0.993 61	**2.4**
2.5	0.994 61	0.994 77	0.994 92	0.995 06	0.995 20	**2.5**
2.6	0.995 98	0.996 09	0.996 21	0.996 32	0.996 43	**2.6**
2.7	0.997 02	0.997 11	0.997 20	0.997 28	0.997 36	**2.7**
2.8	0.997 81	0.997 88	0.997 95	0.998 01	0.998 07	**2.8**
2.9	0.998 41	0.998 46	0.998 51	0.998 56	0.998 61	**2.9**
3.0	0.998 86	0.998 89	0.998 93	0.998 97	0.999 00	**3.0**
3.1	0.999 18	0.999 21	0.999 24	0.999 26	0.999 29	**3.1**
3.2	0.999 42	0.999 44	0.999 46	0.999 48	0.999 50	**3.2**
3.3	0.999 60	0.999 61	0.999 62	0.999 64	0.999 65	**3.3**
3.4	0.999 72	0.999 73	0.999 74	0.999 75	0.999 76	**3.4**
3.5	0.999 81	0.999 81	0.999 82	0.999 83	0.999 83	**3.5**
3.6	0.999 87	0.999 87	0.999 88	0.999 88	0.999 89	**3.6**
3.7	0.999 91	0.999 92	0.999 92	0.999 92	0.999 92	**3.7**
3.8	0.999 94	0.999 94	0.999 95	0.999 95	0.999 95	**3.8**
3.9	0.999 96	0.999 96	0.999 96	0.999 97	0.999 97	**3.9**

Source: W. W. Hines and D. C. Montgomery, *Probability and Statistics in Engineering and Management Science*, Second Ed., © 1980, pp. 592–3. Reprinted by permission of John Wiley & Sons, Inc., New York.

Table A.4. PERCENTAGE POINTS OF THE STUDENTS t DISTRIBUTION WITH v DEGREES OF FREEDOM

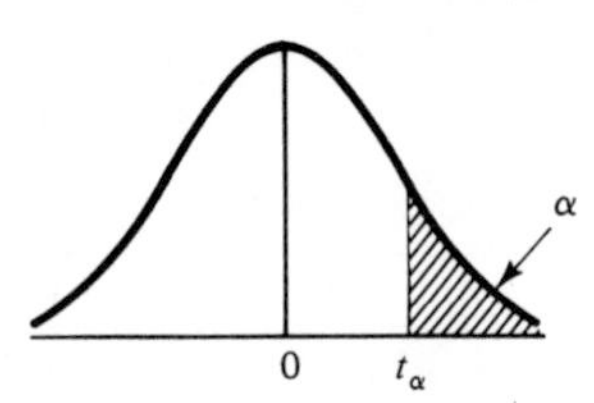

v	$t_{0.005}$	$t_{0.01}$	$t_{0.025}$	$t_{0.05}$	$t_{0.10}$
1	63.66	31.82	12.71	6.31	3.08
2	9.92	6.92	4.30	2.92	1.89
3	5.84	4.54	3.18	2.35	1.64
4	4.60	3.75	2.78	2.13	1.53
5	4.03	3.36	2.57	2.02	1.48
6	3.71	3.14	2.45	1.94	1.44
7	3.50	3.00	2.36	1.90	1.42
8	3.36	2.90	2.31	1.86	1.40
9	3.25	2.82	2.26	1.83	1.38
10	3.17	2.76	2.23	1.81	1.37
11	3.11	2.72	2.20	1.80	1.36
12	3.06	2.68	2.18	1.78	1.36
13	3.01	2.65	2.16	1.77	1.35
14	2.98	2.62	2.14	1.76	1.34
15	2.95	2.60	2.13	1.75	1.34
16	2.92	2.58	2.12	1.75	1.34
17	2.90	2.57	2.11	1.74	1.33
18	2.88	2.55	2.10	1.73	1.33
19	2.86	2.54	2.09	1.73	1.33
20	2.84	2.53	2.09	1.72	1.32
21	2.83	2.52	2.08	1.72	1.32
22	2.82	2.51	2.07	1.72	1.32
23	2.81	2.50	2.07	1.71	1.32
24	2.80	2.49	2.06	1.71	1.32
25	2.79	2.48	2.06	1.71	1.32
26	2.78	2.48	2.06	1.71	1.32
27	2.77	2.47	2.05	1.70	1.31
28	2.76	2.47	2.05	1.70	1.31
29	2.76	2.46	2.04	1.70	1.31
30	2.75	2.46	2.04	1.70	1.31
40	2.70	2.42	2.02	1.68	1.30
60	2.66	2.39	2.00	1.67	1.30
120	2.62	2.36	1.98	1.66	1.29
∞	2.58	2.33	1.96	1.645	1.28

Source: Robert E. Shannon, *Systems Simulation: The Art and Science*, © 1975, p. 375. Reprinted by permission of Prentice-Hall, Inc., Englewood Cliffs, N.J.

Table A.5. PERCENTAGE POINTS OF THE CHI-SQUARE DISTRIBUTION WITH ν DEGREES OF FREEDOM

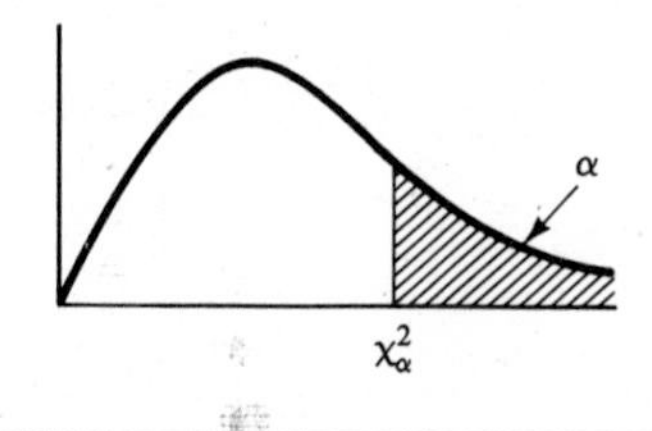

ν	$\chi^2_{0.005}$	$\chi^2_{0.01}$	$\chi^2_{0.025}$	$\chi^2_{0.05}$	$\chi^2_{0.10}$
1	7.88	6.63	5.02	3.84	2.71
2	10.60	9.21	7.38	5.99	4.61
3	12.84	11.34	9.35	7.81	6.25
4	14.96	13.28	11.14	9.49	7.78
5	16.7	15.1	12.8	11.1	9.2
6	18.5	16.8	14.4	12.6	10.6
7	20.3	18.5	16.0	14.1	12.0
8	22.0	20.1	17.5	15.5	13.4
9	23.6	21.7	19.0	16.9	14.7
10	25.2	23.2	20.5	18.3	16.0
11	26.8	24.7	21.9	19.7	17.3
12	28.3	26.2	23.3	21.0	18.5
13	29.8	27.7	24.7	22.4	19.8
14	31.3	29.1	26.1	23.7	21.1
15	32.8	30.6	27.5	25.0	22.3
16	34.3	32.0	28.8	26.3	23.5
17	35.7	33.4	30.2	27.6	24.8
18	37.2	34.8	31.5	28.9	26.0
19	38.6	36.2	32.9	30.1	27.2
20	40.0	37.6	34.2	31.4	28.4
21	41.4	38.9	35.5	32.7	29.6
22	42.8	40.3	36.8	33.9	30.8
23	44.2	41.6	38.1	35.2	32.0
24	45.6	43.0	39.4	36.4	33.2
25	49.6	44.3	40.6	37.7	34.4
26	48.3	45.6	41.9	38.9	35.6
27	49.6	47.0	43.2	40.1	36.7
28	51.0	48.3	44.5	41.3	37.9
29	52.3	49.6	45.7	42.6	39.1
30	53.7	50.9	47.0	43.8	40.3
40	66.8	63.7	59.3	55.8	51.8
50	79.5	76.2	71.4	67.5	63.2
60	92.0	88.4	83.3	79.1	74.4
70	104.2	100.4	95.0	90.5	85.5
80	116.3	112.3	106.6	101.9	96.6
90	128.3	124.1	118.1	113.1	107.6
100	140.2	135.8	129.6	124.3	118.5

Source: Robert E. Shannon, *Systems Simulation: The Art and Science*, © 1975, p. 372. Reprinted by permission of Prentice-Hall, Englewood Cliffs, N.J.

Table A.6. PERCENTAGE POINTS OF THE F DISTRIBUTION WITH $\alpha = 0.05$

ν_1 / ν_2 (Degrees of Freedom for Denominator (ν_2))	Degrees of Freedom for the Numerator (ν_1)																		
	1	2	3	4	5	6	7	8	9	10	12	15	20	24	30	40	60	120	∞
1	161.4	199.5	215.7	224.6	230.2	234.0	236.8	238.9	240.5	241.9	243.9	245.9	248.0	249.1	250.1	251.1	252.2	253.3	254.3
2	18.51	19.00	19.16	19.25	19.30	19.33	19.35	19.37	19.38	19.40	19.41	19.43	19.45	19.45	19.46	19.47	19.48	19.49	19.50
3	10.13	9.55	9.28	9.12	9.01	8.94	8.89	8.85	8.81	8.79	8.74	8.70	8.66	8.64	8.62	8.59	8.57	8.55	8.53
4	7.71	6.94	6.59	6.39	6.26	6.16	6.09	6.04	6.00	5.96	5.91	5.86	5.80	5.77	5.75	5.72	5.69	5.66	5.63
5	6.61	5.79	5.41	5.19	5.05	4.95	4.88	4.82	4.77	4.74	4.68	4.62	4.56	4.53	4.50	4.46	4.43	4.40	4.36
6	5.99	5.14	4.76	4.53	4.39	4.28	4.21	4.15	4.10	4.06	4.00	3.94	3.87	3.84	3.81	3.77	3.74	3.70	3.67
7	5.59	4.74	4.35	4.12	3.97	3.87	3.79	3.73	3.68	3.64	3.57	3.51	3.44	3.41	3.38	3.34	3.30	3.27	3.23
8	5.32	4.46	4.07	3.84	3.69	3.58	3.50	3.44	3.39	3.35	3.28	3.22	3.15	3.12	3.08	3.04	3.01	2.97	2.93
9	5.12	4.26	3.86	3.63	3.48	3.37	3.29	3.23	3.18	3.14	3.07	3.01	2.94	2.90	2.86	2.83	2.79	2.75	2.71
10	4.96	4.10	3.71	3.48	3.33	3.22	3.14	3.07	3.02	2.98	2.91	2.85	2.77	2.74	2.70	2.66	2.62	2.58	2.54
11	4.84	3.98	3.59	3.36	3.20	3.09	3.01	2.95	2.90	2.85	2.79	2.72	2.65	2.61	2.57	2.53	2.49	2.45	2.40
12	4.75	3.89	3.49	3.26	3.11	3.00	2.91	2.85	2.80	2.75	2.69	2.62	2.54	2.51	2.47	2.43	2.38	2.34	2.30
13	4.67	3.81	3.41	3.18	3.03	2.92	2.83	2.77	2.71	2.67	2.60	2.53	2.46	2.42	2.38	2.34	2.30	2.25	2.21
14	4.60	3.74	3.34	3.11	2.96	2.85	2.76	2.70	2.65	2.60	2.53	2.46	2.39	2.35	2.31	2.27	2.22	2.18	2.13
15	4.54	3.68	3.29	3.06	2.90	2.79	2.71	2.64	2.59	2.54	2.48	2.40	2.33	2.29	2.25	2.20	2.16	2.11	2.07
16	4.49	3.63	3.24	3.01	2.85	2.74	2.66	2.59	2.54	2.49	2.42	2.35	2.28	2.24	2.19	2.15	2.11	2.06	2.01
17	4.45	3.59	3.20	2.96	2.81	2.70	2.61	2.55	2.49	2.45	2.38	2.31	2.23	2.19	2.15	2.10	2.06	2.01	1.96
18	4.41	3.55	3.16	2.93	2.77	2.66	2.58	2.51	2.46	2.41	2.34	2.27	2.19	2.15	2.11	2.06	2.02	1.97	1.92
19	4.38	3.52	3.13	2.90	2.74	2.63	2.54	2.48	2.42	2.38	2.31	2.23	2.16	2.11	2.07	2.03	1.98	1.93	1.88
20	4.35	3.49	3.10	2.87	2.71	2.60	2.51	2.45	2.39	2.35	2.28	2.20	2.12	2.08	2.04	1.99	1.95	1.90	1.84
21	4.32	3.47	3.07	2.84	2.68	2.57	2.49	2.42	2.37	2.32	2.25	2.18	2.10	2.05	2.01	1.96	1.92	1.87	1.81
22	4.30	3.44	3.05	2.82	2.66	2.55	2.46	2.40	2.34	2.30	2.23	2.15	2.07	2.03	1.98	1.94	1.89	1.84	1.78
23	4.28	3.42	3.03	2.80	2.64	2.53	2.44	2.37	2.32	2.27	2.20	2.13	2.05	2.01	1.96	1.91	1.86	1.81	1.76
24	4.26	3.40	3.01	2.78	2.62	2.51	2.42	2.36	2.30	2.25	2.18	2.11	2.03	1.98	1.94	1.89	1.84	1.79	1.73
25	4.24	3.39	2.99	2.76	2.60	2.49	2.40	2.34	2.28	2.24	2.16	2.09	2.01	1.96	1.92	1.87	1.82	1.77	1.71
26	4.23	3.37	2.98	2.74	2.59	2.47	2.39	2.32	2.27	2.22	2.15	2.07	1.99	1.95	1.90	1.85	1.80	1.75	1.69
27	4.21	3.35	2.96	2.73	2.57	2.46	2.37	2.31	2.25	2.20	2.13	2.06	1.97	1.93	1.88	1.84	1.79	1.73	1.67
28	4.20	3.34	2.95	2.71	2.56	2.45	2.36	2.29	2.24	2.19	2.12	2.04	1.96	1.91	1.87	1.82	1.77	1.71	1.65
29	4.18	3.33	2.93	2.70	2.55	2.43	2.35	2.28	2.22	2.18	2.10	2.03	1.94	1.90	1.85	1.81	1.75	1.70	1.64
30	4.17	3.32	2.92	2.69	2.53	2.42	2.33	2.27	2.21	2.16	2.09	2.01	1.93	1.89	1.84	1.79	1.74	1.68	1.62
40	4.08	3.23	2.84	2.61	2.45	2.34	2.25	2.18	2.12	2.08	2.00	1.92	1.84	1.79	1.74	1.69	1.64	1.58	1.51
60	4.00	3.15	2.76	2.53	2.37	2.25	2.17	2.10	2.04	1.99	1.92	1.84	1.75	1.70	1.65	1.59	1.53	1.47	1.39
120	3.92	3.07	2.68	2.45	2.29	2.17	2.09	2.02	1.96	1.91	1.83	1.75	1.66	1.61	1.55	1.55	1.43	1.35	1.25
∞	3.84	3.00	2.60	2.37	2.21	2.10	2.01	1.94	1.88	1.83	1.75	1.67	1.57	1.52	1.46	1.39	1.32	1.22	1.00

Source: W. W. Hines and D. C. Montgomery, *Probability and Statistics in Engineering and Management Science*, 2nd ed., © 1980, p. 599. Reprinted by permission of John Wiley & Sons, Inc., New York.

Table A.7. KOLMOGOROV–SMIRNOV CRITICAL VALUES

Degrees of Freedom (N)	$D_{0.10}$	$D_{0.05}$	$D_{0.01}$
1	0.950	0.975	0.995
2	0.776	0.842	0.929
3	0.642	0.708	0.828
4	0.564	0.624	0.733
5	0.510	0.565	0.669
6	0.470	0.521	0.618
7	0.438	0.486	0.577
8	0.411	0.457	0.543
9	0.388	0.432	0.514
10	0.368	0.410	0.490
11	0.352	0.391	0.468
12	0.338	0.375	0.450
13	0.325	0.361	0.433
14	0.314	0.349	0.418
15	0.304	0.338	0.404
16	0.295	0.328	0.392
17	0.286	0.318	0.381
18	0.278	0.309	0.371
19	0.272	0.301	0.363
20	0.264	0.294	0.356
25	0.24	0.27	0.32
30	0.22	0.24	0.29
35	0.21	0.23	0.27
Over 35	$\frac{1.22}{\sqrt{N}}$	$\frac{1.36}{\sqrt{N}}$	$\frac{1.63}{\sqrt{N}}$

Source: F. J. Massey, "The Kolmogorov–Smirnov Test for Goodness of Fit," *The Journal of the American Statistical Association*, Vol. 46, © 1951, p. 70. Adapted with permission of the American Statistical Association.

Table A.8. MAXIMUM LIKELIHOOD ESTIMATES OF THE GAMMA DISTRIBUTION

$1/M$	β	$1/M$	β
0.020	0.0187	5.200	2.755
0.030	0.0275	5.400	2.855
0.040	0.0360	5.600	2.956
0.050	0.0442	5.800	3.056
0.060	0.0523	6.000	3.156
0.070	0.0602	6.200	3.257
0.080	0.0679	6.400	3.357
0.090	0.0756	6.600	3.457
0.100	0.0831	6.800	3.558
0.200	0.1532	7.000	3.658
0.300	0.2178	7.300	3.808
0.400	0.2790	7.600	3.958
0.500	0.3381	7.900	4.109
0.600	0.3955	8.200	4.259
0.700	0.4517	8.500	4.409
0.800	0.5070	8.800	4.560
0.900	0.5615	9.100	4.710
1.000	0.6155	9.400	4.860
1.100	0.6690	9.700	5.010
1.200	0.7220	10.000	5.160
1.300	0.7748	10.300	5.311
1.400	0.8272	10.600	5.461
1.500	0.8794	10.900	5.611
1.600	0.9314	11.200	5.761
1.700	0.9832	11.500	5.911
1.800	1.034	11.800	6.061
1.900	1.086	12.100	6.211
2.000	1.137	12.400	6.362
2.100	1.188	12.700	6.512
2.200	1.240	13.000	6.662
2.300	1.291	13.300	6.812
2.400	1.342	13.600	6.962
2.500	1.393	13.900	7.112
2.600	1.444	14.200	7.262
2.700	1.494	14.500	7.412
2.800	1.545	14.800	7.562
2.900	1.596	15.100	7.712
3.000	1.646	15.400	7.862
3.200	1.748	15.700	8.013
3.400	1.849	16.000	8.163
3.600	1.950	16.300	8.313
3.800	2.051	16.600	8.463
4.000	2.151	16.900	8.613
4.200	2.252	17.200	8.763
4.400	2.353	17.500	8.913
4.600	2.453	17.800	9.063
4.800	2.554	18.100	9.213
5.000	2.654	18.400	9.363
		18.700	9.513
		19.000	9.663
		19.300	9.813
		19.600	9.963
		20.000	10.16

Source: S. C. Choi and R. Wette, "Maximum Likelihood Estimates of the Gamma Distribution and Their Bias," *Technometrics*, Vol. 11, No. 4, Nov. 1969 ©, pp. 688–9. Adapted with permission of the American Statistical Association.

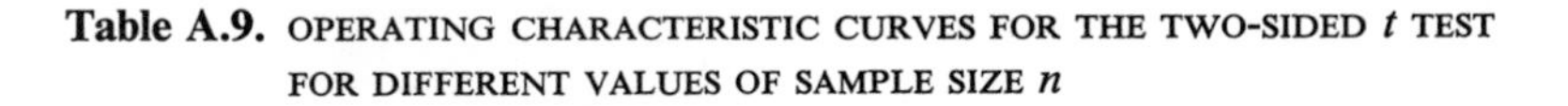

Table A.9. OPERATING CHARACTERISTIC CURVES FOR THE TWO-SIDED t TEST FOR DIFFERENT VALUES OF SAMPLE SIZE n

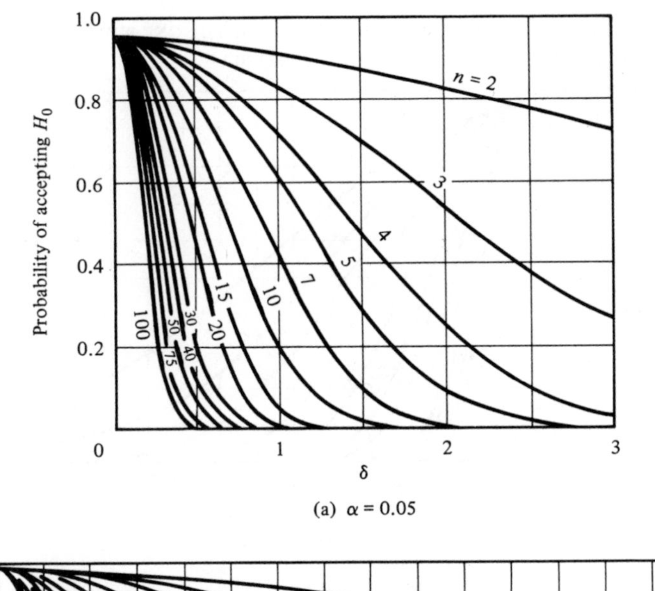

(a) $\alpha = 0.05$

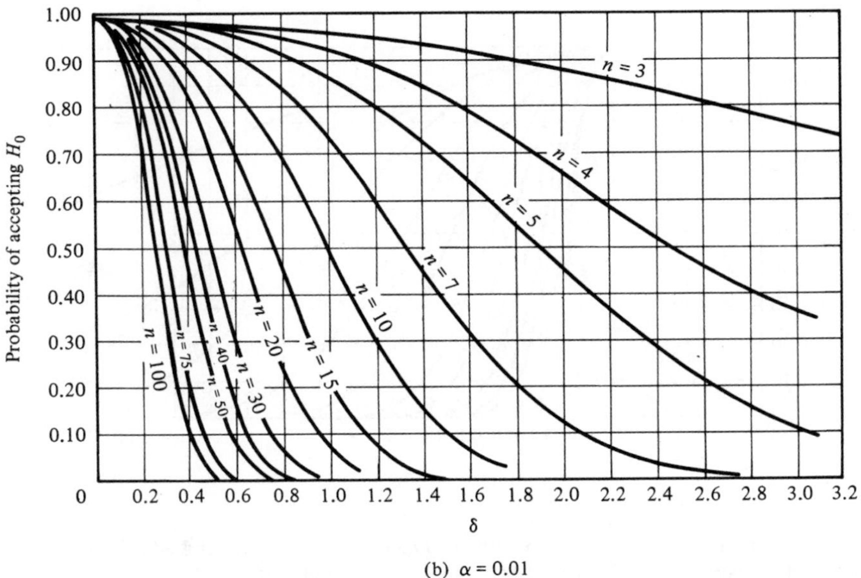

(b) $\alpha = 0.01$

Source: C. L. Ferris, F. E. Grubbs, and C. L. Weaver, "Operating Characteristics for the Common Statistical Tests of Significance," *Annals of Mathematical Statistics*, June 1946. Reproduced with permission of The Institute of Mathematical Statistics.

Table A.10. OPERATING CHARACTERISTIC CURVES FOR THE ONE-SIDED t TEST FOR DIFFERENT VALUES OF SAMPLE SIZE n

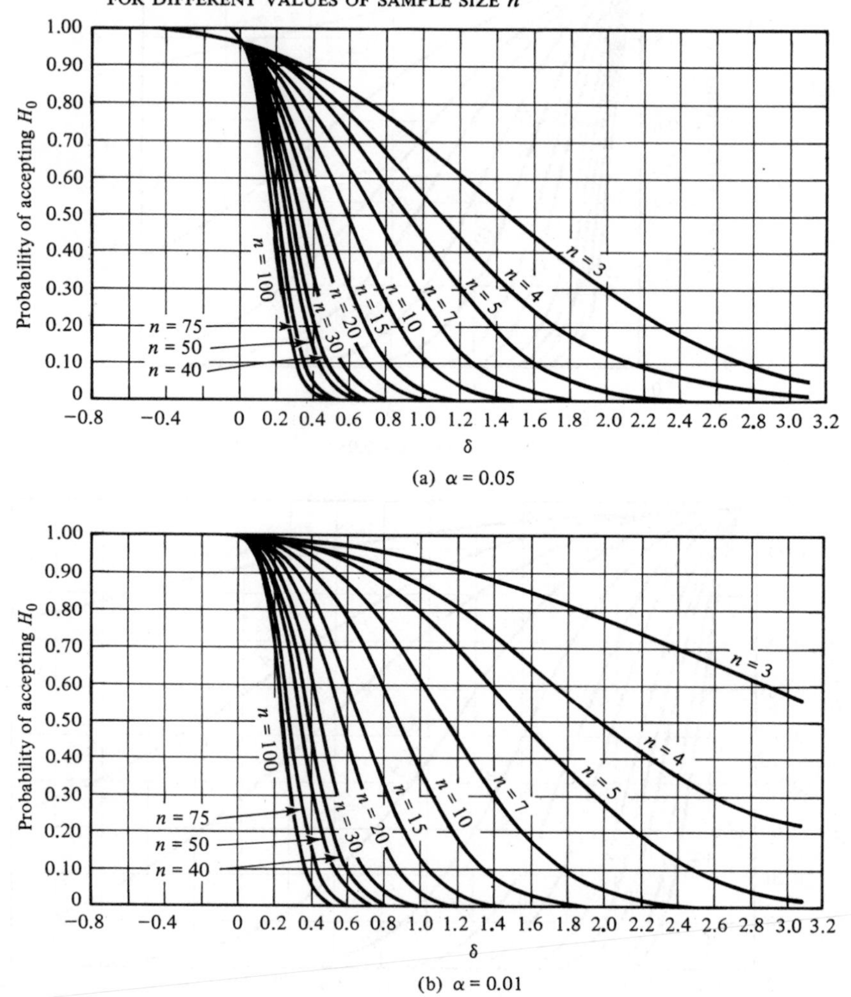

(a) $\alpha = 0.05$

(b) $\alpha = 0.01$

Source: A. H. Bowker and G. J. Lieberman, *Engineering Statistics*, 2nd ed., © 1972, p. 203. Reprinted by permission of Prentice-Hall, Inc., Englewood Cliffs, N.J.

INDEX